$E(X)$ or EX	expected value of random variable X
μ_X	mean of X
$\text{Var}(X)$ or $\sigma^2(X)$	variance of X
σ or σ_X	standard deviation of X
S_n	binomial random variable
$b(x;n,p)$	binomial density function
$B(x;n,p)$	binomial distribution function
$h(x;n,b,w)$	hypergeometric density function
$p(x;\mu)$	Poisson density function
$\mathbf{I}_n$	n-by-n identity matrix
Φ	zero matrix
p_{ij}	transition probability from state E_i to state E_j
$p_j(n)$	absolute probability at time n
$\mathbf{p}(n)$	absolute probability vector
$\mathbf{T}^{(n)}$	n-step transition matrix
$\mathbf{v} = (\pi,\pi,\ldots,\pi)$	limiting distribution for an ergodic Markov chain
$\mathbf{R}^*, \mathbf{B}^*$	optimal strategies
$p(x)$	degree of vertex x
$p_i(x)$	incoming degree of x
$p_o(x)$	outgoing degree of x
$C(i,j)$	capacity of arc (i,j)
$F(i,j)$	flow of arc (i,j)
F_v	value of the flow
$\min[\ ]$	minimum of
$\max[\ ]$	maximum of

FINITE MATHEMATICS

Probability, Programming, Games, and Graphs

CAROLE EISEN

MARTIN EISEN
Temple University

GLENCOE PUBLISHING CO., INC., Encino, California
Collier Macmillan Publishers, London

Glencoe Publishing Co., Inc.
17337 Ventura Boulevard
Encino, California 91316
Collier Macmillan Canada, Ltd.

Library of Congress Catalog Card Number: 76-4049

First Printing 1979

ISBN 0-02-472450-5

Table of Contents

Preface *ix*

CHAPTER 3 Counting Problems, Permutations, and Combinations 70

CHAPTER 4 Dependence and Independence 122

CHAPTER 5 Random Variables 154

CHAPTER 6 Vectors, Matrices, and Linear Equations 210

CHAPTER 7 Markov Chains 260

Preface

This text presents the topics of finite mathematics—probability, linear programming, game theory, and graph theory—and their applications in a manner that makes these topics easily accessible to students without extensive mathematical background. The only prerequisite for using the material in this text is a knowledge of high school algebra.

Many of the techniques presented in finite mathematics courses were originally developed as tools for improving the efficiency of industrial operations. However, it soon became evident that these techniques were applicable to many other fields. For this reason, finite mathematics is no longer studied by business students alone; it is also studied by students of engineering, social science, medical science, and so on. In addition, many students majoring in mathematics now take a course in finite mathematics in order to be better-prepared for the job market.

Nearly all students find finite mathematics interesting when the subject is illustrated by a large number of realistic applications. *Finite Mathematics* provides such diversified applications that students with limited mathematical background find that they can solve problems in genetics, psychology, sociology, education, market research, transportation, statistics, economics, business, medicine, and other areas. (For a list of the applications treated in the text, classified by subject area, see the Answer and Test Booklet.)

All of the material in this text has been class-tested with students at Rutgers University and Temple University. Material from this text has been used successfully in a one-semester finite mathematics course, a two-semester finite mathematics course, a one-semester probability course, and for self-study.

Chapter Independence. Although the text contains more material than can be covered in the usual one-semester finite mathematics course, many of the chapters are independent. This structure allows instructors great flexibility in designing a course that meets the needs of their students. The following table shows the dependence relations among the various chapters. It also indicates those sections that can be regarded as optional.

Chapter	Prerequisites	Optional Sections; Comments
1. Sets	None	Can be omitted by students who have already studied sets.
2. Probability	A knowledge of sets	Sections 2.5–2.7 can be omitted.
3. Counting Problems, Permutations, and Combinations	None	Sections 3.4, 3.6, and 3.9 can be omitted.
4. Dependence and Independence	Sections 2.1–2.4	

Chapter	Prerequisites	Optional Sections; Comments
5. Random Variables	Sections 4.1–4.2	Section 5.1 can be omitted by students who have already studied functions. Sections 5.7 and 5.8 can be omitted.
6. Vectors, Matrices, and Linear Equations	None	Section 6.7 can be omitted.
7. Markov Chains	Sections 4.1–4.2 Sections 6.1–6.3	
8. Linear Programming	None	Section 8.1 can be omitted by students who have already studied graphs of linear equations.
9. Game Theory	Section 6.1; Sections 9.4–9.6 require Sections 5.1–5.4; Section 9.7 requires Chapter 8.	Section 9.7 can be omitted.
10. Graphs and Their Applications	None	Sections 10.4–10.6 can be omitted.
11. Networks and Flows	Section 10.1	Section 11.4 can be omitted.
12. Dynamic Programming	None	This chapter requires greater mathematical maturity than the remainder of the book. Section 12.4, in particular, has been starred as a difficult section.

Specific Features. The style of this text is deliberately simple and readable. Numerous carefully worked-out examples form an integral part of the text, and students are advised to study these examples carefully. Exercises at the end of each section give students an opportunity to practice the principles that they have learned. Many of these exercises use situations that have already been introduced in the worked-out examples. The few difficult or computationally involved exercises are indicated by an asterisk in the margin.

The answers to all odd-numbered exercises are given in the back of the text, so students can check their own answers. Answers to even-numbered exercises are included in the Answer and Test Booklet.

If the instructor feels that students need additional practice in order to understand basic concepts and to master the techniques of a particular section, he or she can assign some of the drill exercises that are provided in the Answer and Test Booklet. In most cases, however, students will have no trouble proceeding directly from the text material itself to the exercises at the end of the section.

Each chapter is followed by a Key Word Review that lists the key words in the order in which they are defined in the chapter. The number of the section in which each key word is defined is listed in parentheses following the word itself.

In the text, each key word appears in boldface type when it is first defined. This way, students wishing to review the chapter will have no difficulty locating the definition of each key word. In addition, a set of Review Exercises appears at the end of each chapter. These allow students to review and integrate their knowledge of the concepts presented in the chapter.

Answer and Test Booklet. An Answer and Test Booklet to accompany this text will be provided by the publisher for each instructor who uses the text. This booklet includes a complete list of the applied problems and examples in the text, a complete set of drill exercises covering each basic concept, and tests based on the text material. The testing program consists of two equivalent tests for each chapter, and two equivalent comprehensive examinations on the entire text. The chapter tests are arranged so that tests from two consecutive chapters easily combine into larger tests suitable for the usual fifty-minute testing period.

Acknowledgments. The authors wish to thank Professor Samuel G. Councilman of California State University at Long Beach for his in-depth review of the original manuscript. Dr. Councilman's many specific, constructive suggestions were of great help to the authors in revising the manuscript. We would also like to thank Tanya Mink, Developmental Editor for Glencoe Publishing Company, who gave us a great deal of helpful advice about the structure and style of the book.

Special thanks are also due to Chris Cagan, who checked the answers to all of the hundreds of exercises and corrected several errors, and to the staff of Glencoe Publishing Company. We would particularly like to thank William Bryden for his encouragement and helpful advice. In addition, the authors would like to express their sincere appreciation to Henrietta Sloane for all that she has done.

This book has been coauthored. The fact that one author's name appears before the other is of no significance.

Carole Eisen

Martin Eisen

FINITE
MATHEMATICS
Probability, Programming,
Games, and Graphs

New York Zoological Society

CHAPTER 1 Sets

INTRODUCTION

The photograph on the facing page shows two groups of animals. Looking at the picture, it is obvious which animals belong to the pride of lions and which belong to the herd of *nyalas* (a type of antelope). The pride of lions and the herd of nyalas are *well-defined* groups of objects—that is, it is absolutely clear which objects belong to each group and which do not. Not all groups of objects are well defined. For example, the group consisting of all of the hungry animals in the picture on the left is not well-defined, because we do not have enough information to decide which animals are hungry and which are not.

In mathematics, the ideas of *set theory* guide the processes of classifying objects into groups, finding precise descriptions for these groups, and describing the relationships among them. These ideas and processes are basic to all of modern mathematics. Because of this, the concepts we develop in this chapter will be used throughout our study of finite mathematics.

1.1 BASIC CONCEPTS

Since this chapter is about sets, we should begin by defining the word *set*. However, in any mathematical system, certain words must remain undefined. For example, in geometry the word *point* is undefined. Since every definition is made in terms of other words, it is impossible to define every word. Therefore, we must start with some words that are not defined, but whose meanings can be made clear by examples. *Set* is one of these words.

Informally, we can describe a **set** as any well-defined collection of objects. The objects are called **elements**, or *members*, of the set. Notice that a set must be a *well-defined* collection of objects. That is, there must be no doubt as to exactly which objects belong to the set. For example, "the first six letters of the English alphabet" are a well-defined collection and hence a set. The elements of this set are a, b, c, d, e, and f.

On the other hand, "all the large cities in the United States" is not a well-defined collection because some people feel that a city of 200,000 is large while others feel that a city of 500,000 is small. Thus, this collection is not a set. However, we can *make* this collection a set by defining a "large city" to be a city with a population of 300,000 or more. With this definition, we can tell whether a city belongs to the set. For example, New York City would be in the set, but Camden, New Jersey, would not.

Sets will usually be denoted by capital letters (e.g., A). Elements of sets will usually be denoted by lowercase letters (e.g., a). If x is an element of the set A, we write

$$x \in A$$

If x is *not* an element of A, we write

$$x \notin A$$

Example 1 illustrates these concepts.

Example 1   Some examples of sets are

  A: The numbers 1, 2, .3, 4, 5, 6, 7
         This set has seven elements; 1 is an element ($1 \in A$), but 10 is not an element
         ($10 \notin A$)

  B: All women who were elected president of the United States before January 1,
         1977

  C: All astronauts who landed on the moon before March 5, 1977

  D: The positive integers $4 \in D$, but $-3 \notin D$

  E: All real numbers

  F: The stockholders of General Electric at noon on February 5, 1977

  G: The first seven positive integers

  H: All people over ten feet tall

Let A be the first set of Example 1. Although it is true that

   strawberry yogurt $\notin A$

this statement is hardly profound. It is more to the point to say that $0 \notin A$; at least 0
resembles the elements of A because it is an integer. When discussing whether an
object is in a set, it is time-consuming and pointless to consider all possible elements.
Therefore, we will want to restrict our attention to those elements needed to solve the
particular problem we are discussing. The set of all these elements is called the
universe.

Definition 1.1   For a particular discussion, the **universe** (or *universal set*) is a set
    to which the elements of all other sets in the discussion must belong. It is
    denoted by Ω (the Greek letter *omega*).

Consider the sets A and G of Example 1. Although they are described differently,
they contain the same elements. Sets such as these are called *equal sets*. Sets that have
no elements, such as H of Example 1, are called *empty sets*.

Definition 1.2   Two sets A and B are **equal sets** if and only if every element of A
    is an element of B, and vice versa.

That is, two sets are equal if and only if they have the same elements.

Definition 1.3   An **empty set** (or *null set*) is a set that contains no elements. An
    empty set is denoted by the symbol $\varnothing$.

The sets B and H of Example 1 are both empty, but are they equal? Suppose
$B \neq H$; then there must be an element of B that does not belong to H, or an element
of H that does not belong to B. Neither of these cases is possible, since both B and H
are empty. It follows that $B = H$. Theorem 1.1 can be proved using a similar
argument.

THEOREM   1.1   All empty sets are equal.

There are three methods that are frequently used to describe sets. The first method is *definition by listing*. If there are only a few elements in the set, the simplest way to describe it is to list the elements and enclose them in a pair of braces, { }. For example, again referring to Example 1,

$$A = \{1,2,3,4,5,6,7\}$$

Although the order of listing is unimportant, one must not repeat an element. Set A may also be written

$$\{7,6,5,4,3,2,1\}, \quad \text{or} \quad \{1,3,5,7,2,4,6\}, \quad \text{or} \quad \{2,4,6,1,3,5,7\}$$

but it may not be written $\{1,1,2,3,4,5,6,7\}$.

If J is the set of letters in the English alphabet, it is possible—but tedious—to list them all. In such cases, braces are again used. However, only enough elements are listed between the braces to clearly define the set. The missing elements are replaced by three dots. For example:

$$J = \{a,b,c,\ldots,z\} \quad \text{and} \quad A = \{1,2,3,\ldots,7\}$$

The set D of Example 1 has no last element and so cannot be described in exactly the preceding manner. However, its elements can be lined up in order with a first element, second element, and so on. Thus, the three dots can be used, not only to indicate that some elements are missing, but also to mean "and so on." Using this notation,

$$D = \{1,2,3,\ldots\}$$

Similarly, the set U of even positive integers can be written

$$U = \{2,4,6,\ldots\}$$

On the other hand, the elements of some sets, such as E of Example 1, cannot be lined up in order and therefore cannot be described by listing. In such cases, we might use the second common method for defining sets, *definition by description*.

We can define a set by describing the distinctive properties of its elements. For example, we can write

$$K = \{x \mid x > 10, x \text{ is an integer}\}$$

The symbol $\mid$ means "such that"; the comma is read "and." The set K, then, is the set of all numbers x such that x is greater than 10 and x is an integer. That is,

$$K = \{11,12,13,\ldots\}$$

Referring to Example 1,

$$D = \{x \mid x \text{ is a positive integer}\}$$

$$A = \{x \mid 0 < x < 8, x \text{ is an integer}\}$$

$$= \{x \mid 0 < x < 8, x \in D\}$$

The third common method for describing sets is to use *Venn diagrams*. In later sections we will study some complicated relationships between sets. To aid our intuition, it is useful to draw diagrams. In a Venn diagram, a set is represented by the interior of a circle or an ellipse (appropriately labeled) as shown in Figure 1.1. If necessary, the elements of the set are represented by dots inside the curve. The universal set Ω is represented by the interior of a rectangle.

The Venn diagram of Figure 1.2 is a pictorial representation of the fact that the sets A and K defined in the preceding discussion have no common element. The diagram of Figure 1.3 shows that every element of set A is an element of set D.

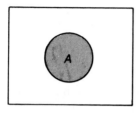
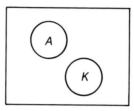
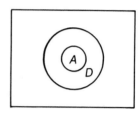

Figure 1.1
Set A is shaded

Figure 1.2
Sets A and K have no common element

Figure 1.3
Every element of A is an element of D

Example 2   The following sets are frequently used in mathematics. Sets D and U have already been defined; they are included here for reference.

$N =$ The positive numbers $= \{x \mid x > 0\}$

$O =$ The nonnegative numbers $= \{x \mid x \geq 0\}$

$P =$ The negative numbers $= \{x \mid x < 0\}$

$Q =$ The integers $= \{\ldots, -3, -2, -1, 0, 1, 2, 3, \ldots\}$

$D =$ The positive integers $= \{x \mid x \in Q,\ x \in N\} = \{1, 2, 3, 4, \ldots\}$

$R =$ The prime numbers $= \{2, 3, 5, 7, 11, 13, 17, 23, \ldots\}$

$S =$ The odd integers $= \{\ldots, -5, -3, -1, 1, 3, 5, \ldots\} = \{2x + 1 \mid x \in Q\}$

$T =$ The even integers $= \{\ldots, -6, -4, -2, 0, 2, 4, 6, \ldots\} = \{2x \mid x \in Q\}$

$U =$ The positive even integers $= \{2, 4, 6, 8, \ldots\} = \{2x \mid x \in D\}$

1.1 EXERCISES

1.   Which of the following collections are sets? If the collection is a set, name two members (if possible). If it is not a set, define enough words in its description to make it a set.

$A_1 =$ All fat people

$A_2 =$ The odd prime numbers

$A_3 =$ All kangaroos who drive cars

$A_4 =$ All beatniks

$A_5 =$ All people who have had at least one heart attack

$A_6 =$ All successful people

$A_7 =$ All students in your class who are on the dean's list

$A_8 =$ All numbers between 0 and 1

2.   Which of the sets in Example 1 and Exercise 1 are empty?

3.   Let $n(X)$ be the number of elements in the set X and let $Z = \{\emptyset\}$.
         (a) Find (*i*) $n(\emptyset)$, (*ii*) $n(Z)$.
         (b) Does $\emptyset = \{\emptyset\}$?

4.   Which of the following statements are true? (Refer to Example 2.)
         (a) $10 \in N$ **(b)** $10 \in P$ **(c)** $0 \notin N$ **(d)** If $x \subseteq D$, then $x \in N$.
         (e) If $x \in Q$, then $x \notin P$.

5.   **(a)** Which of the following sets are equal?
              $V = \{1,2,3\}, \qquad W = \{2,3,1\}, \qquad X = \{\{1\}, \{2,3\}\}, \qquad Y = \{\{1,2\}, \{3\}\},$
              $Z = \{\{3,2\}, \{1\}\}$
         (b) Answer true or false.

              (*i*) $1 \in V$ (*ii*) $1 \in X$ (*iii*) $1 \notin Y$
              (*iv*) $\{1\} \in X$ (*v*) $\{1\} \in Y$
         (c) Find (*i*) $n(V)$, (*ii*) $n(X)$. (See Exercise 3.)

6.   Give an equivalent definition for the following sets.
         A = The integers between 7 and 20   (Do not include the endpoints.)
         B = All even negative integers
         C = The members of your immediate family including yourself
         D = All final grades given by your school
         E = The letters in the word "establishment"

7.   Describe the set of integers between 4 and 12 (not including 4 and 12) in as
         many ways as you can.

1.2 SUBSETS AND COMPLEMENTS

The sets D and Q of Example 2 are not equal because $0 \in Q$ but $0 \notin D$. However, every element of D is an element of Q. Therefore, we say that D is a *subset* of Q. In this section we will study the properties of subsets. We will also study two set operations, *complement* and *relative complement*.

Definition 1.4   A set A is a **subset** of a set B if and only if every element of A is an element of B. This is written $A \subseteq B$.

    If $A \subseteq B$, we can also express this by saying that A *is contained in B*, or that B *contains A*. If A is not a subset of B, we write $A \nsubseteq B$. If $A \subseteq B$ and $A \neq B$, we write $A \subset B$. The set is then said to be a **proper subset** of B.
    Example 3 and Theorems 1.2 and 1.3 follow from the definitions.

Example 3   The subsets of $\{1,2\}$ are the sets $\emptyset$, $\{1\}$, $\{2\}$, and $\{1,2\}$. The first three of these are proper subsets.

THEOREM 1.2   If A is any set, then
(a) $A \subseteq A$
(b) $\emptyset \subseteq A$

THEOREM 1.3   Let A and B be sets. Then $A = B$ if and only if $A \subseteq B$ and $B \subseteq A$.

Although the preceding theorems follow easily from the definitions, the proofs of most theorems are more complicated. However, we can often decide whether a statement about sets is true by drawing a Venn diagram. The Venn diagram of Figure 1.4, for example, represents the fact that $A \subseteq B$ and $B \subseteq C$. It is pictorially clear from this diagram that $A \subseteq C$. In other words, Figure 1.4 indicates that

$$\text{if} \quad A \subseteq B \quad \text{and} \quad B \subseteq C, \quad \text{then} \quad A \subseteq C \tag{1}$$

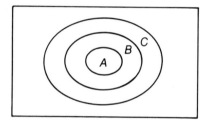

Figure 1.4 $A \subseteq B, B \subseteq C$

Although Figure 1.4 *indicates* that statement (1) is true, it does not *prove* that it is true. A picture is not a proof. *A proof is a logical deduction from the definitions and previously proven theorems.* A proof of statement (1) follows.

THEOREM 1.4   Let A, B, and C be sets. If $A \subseteq B$ and $B \subseteq C$, then $A \subseteq C$.

Proof   We must prove that every element of A is an element of C. Therefore, let x be any element of A. We must show that $x \in C$. Since $x \in A$ and $A \subseteq B$, it follows from Definition 1.4 that $x \in B$. Since $B \subseteq C$, it follows that $x \in C$.

There are five operations that can be applied to sets. Two of these operations, complement and relative complement, will be studied now. The remaining operations will be studied in Sections 1.3 and 1.5.

Suppose Ω is the set of all integers. The set A of even integers and the set B of odd integers are both subsets of Ω. In fact, set B is the set of all integers which are *not* in set A; that is, $B = \{x \,|\, x \notin A\}$. The set B is called the *complement* of A.

Definition 1.5   The **complement** of a set A (with respect to Ω) is the set of all elements of Ω which are *not* in A. The complement of A is symbolized by A' (" A prime ").

Note that $x \in A$ if and only if $x \notin A'$, and that $A' = \{x \,|\, x \notin A\}$. Note also that the complement operation corresponds to the English word "not." For example, let Ω be the set of all people. If A is the set of all people who smoke,

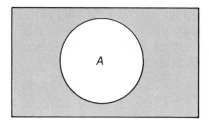

Figure 1.5   The set A' is shaded

then A' is the set of all people who do not smoke. If B is the set of all people who are over six feet tall, then B' is the set of all people who are *not* over six feet tall.

Definition 1.6 is similar to Definition 1.5.

Definition 1.6   The **relative complement** of A in B is the set of all elements of B that are not in A. The relative complement is symbolized $B - A$.

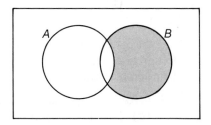

Figure 1.6   The set $B - A$ is shaded

Symbolically, $B - A = \{x \mid x \in B$, and $x \notin A\}$.

Examples 4 and 5 illustrate the definitions of complement and relative complement. Theorem 1.5 lists several important properties of these operations.

Example 4   Let $\Omega = \{1,2,3,4,5\}$, $A = \{1,2\}$, and $B = \{2\}$. Then—

(a) $A' = \{3,4,5\}$ (b) $\Omega - A = \{3,4,5\} = A'$ (c) $A - B = \{1\}$
(d) $B - A = \varnothing$ (e) $(A')' = \{3,4,5\}' = \{1,2\} = A$ (f) $\Omega' = \varnothing$

Example 5   Let Ω be the set of all adults, A be the set of all men, and B be the set of all adults who have diabetes. Then—

(a) set A' is the set of all adults who are *not* men; that is, A' is the set of all women.
(b) set $A - B$ is the set of all men who do not have diabetes.
(c) set $B - A$ is the set of all women who have diabetes.

THEOREM   1.5   Let A be any set. Then—

(a) $(A')' = A$
(b) $\Omega - A = A'$
(c) $\Omega' = \varnothing$
(d) $\varnothing' = \Omega$

The Venn diagram in Figure 1.5 can be used to verify (not prove) part (a) of Theorem 1.5. The shaded area is A'. Then everything in the rectangle which is not shaded is $(A')'$; this is evidently A. Part (b) can be similarly verified. Parts (c) and (d) follow easily from the definitions.

1.2 EXERCISES

1. **(a)** What are the subsets of $\{a,b,c\}$?
   (b) What are the proper subsets?

2. State ten relationships of the type $X \subseteq Y$, $X \subset Y$, or $X = Y$ between pairs of the sets in Example 1.

3. Let $A = \{1,2,3\}$. Which of the following statements are true?
   (a) $\varnothing \in A$ **(b)** $1 \notin A$ **(c)** $\{1\} \subseteq A$
   (d) $\varnothing \subseteq A$ **(e)** $A \subset A$ **(f)** $A \subseteq A$
   (g) $2 \in A$ **(h)** $2 \subseteq A$

4. A research firm is studying the correlation between smoking and cancer. What would be a reasonable universal set for this study? What would be some relevant subsets and their complements?

5. Let $\Omega = \{a,b,c,\ldots,z\}$, $A = \{a,b,c\}$, $B = \{c,d\}$, and $C = \{c,b,a\}$. Find—
   (a) A' **(b)** $\Omega - A$ **(c)** $C - A$ **(d)** $B - A$
   (e) B' **(f)** $A - B$ **(g)** $C - B'$ **(h)** $A - \varnothing$
   (i) $\varnothing - A$ **(j)** $A' - B'$ **(k)** $A' - \varnothing'$

6. Let $\Omega = \{1,2,3,\ldots,10\}$ and $A = \{1,2,3\}$. What are (a) A', (b) Ω', and (c) $\varnothing'$?

7. Let A be any set. Complete the following equations.
   (a) $\Omega - A = \underline{\quad}$ **(b)** $A - A = \underline{\quad}$ **(c)** $A - \varnothing = \underline{\quad}$
   (d) $A - A' = \underline{\quad}$ **(e)** $A - \Omega = \underline{\quad}$ **(f)** $\varnothing' - A = \underline{\quad}$
   (g) $(A' - A)' = \underline{\quad}$

8. Let $\Omega =$ All students at Temple University, $D =$ All Temple University students who are on the dean's list, and $F =$ All Temple University freshmen. State the verbal equivalent of the following sets.
   (a) D' **(b)** $F - D$ **(c)** $D - F$ **(d)** $D' - F'$

9. Use a Venn diagram to verify Theorem 1.5(b).

10. Prove Theorem 1.2 and 1.3.

1.3 UNIONS AND INTERSECTIONS

Complement and relative complement were studied in Section 1.2. Two additional operations, set union and intersection, will be studied in this section.

Definition 1.7   The **union** of two sets A and B is the set of all elements that belong either to A or to B or to both.

The union of A and B is symbolized $A \cup B$, which is read "A union B." Symbolically,

$$A \cup B = \{x \mid x \in A \text{ or } x \in B\}$$

Note that elements belonging to both A and B appear only once in the union.

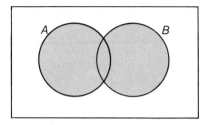

Figure 1.7 The set $A \cup B$ is shaded

Definition 1.8 The **intersection** of sets A and B is the set of all elements that belong to A and B simultaneously.

The intersection of A and B is symbolized by $A \cap B$, or by AB; both are read "A intersection B." Symbolically,

$$A \cap B = \{x \mid x \in A \text{ and } x \in B\}$$

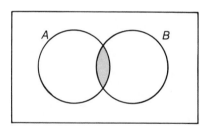

Figure 1.8 The set $A \cap B$ is shaded

Definition 1.9 Two sets A and B are **disjoint** sets if they have no common elements; that is, $A \cap B = \emptyset$.

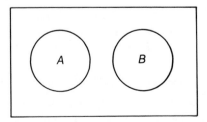

Figure 1.9 Sets A and B are disjoint

Examples 6–8 illustrate these concepts.

Example 6   If $A = \{a,b\}$, $B = \{b,c\}$, and $C = \{c,d\}$, then—

(a) $A \cup A = \{a,b\} = A$ (b) $A \cup B = \{a,b,c\}$ (c) $A \cup C = \{a,b,c,d\}$
(d) $A \cap B = \{b\}$ (e) $(A \cup B) \cap C = \{a,b,c\} \cap \{c,d\} = \{c\}$

A and B are not disjoint, because $A \cap B \neq \varnothing$. However, A and C are disjoint.

Example 7   If $\Omega = \{1,2,3,\ldots\}$, $A = \{1,2,3\}$, and $B = \{2,3,4\}$, then—

(a) $A \cup B = \{1,2,3,4\}$ (b) $B \cup A = \{1,2,3,4\}$
(c) $A \cup B' = \{1,2,3,5,6,7,8,\ldots\}$ (d) $A \cap B' = \{1\} = A - B$
(e) $A \cap \Omega = A$ (f) $A \cup \Omega = \Omega$

Example 8   Let A be any set. Using Venn diagrams, the reader can verify that—

(a) $A \cup A = A$ (b) $A \cap A' = \varnothing$ (c) $A \cup \varnothing = A$
(d) $A \cup \Omega = \Omega$

In the preceding section we learned that the complement operation corresponds to the English word "not." Definitions 1.8 and 1.9 imply that the union and intersection operations correspond to the English words "or" and "and," respectively. These correspondences are summarized in Table 1.1.
Example 9 illustrates these correspondences.

Table 1.1	Operation	Symbol	English equivalent
	Complement	$'$	not
	Union	$\cup$	or
	Intersection	$\cap$	and

Example 9   Suppose one card is picked from a deck of fifty-two cards and the suit is noted. The set of all possible suits is

$$\Omega = \{h,s,d,c\}$$

where h, s, d, and c stand for hearts, spades, diamonds, and clubs, respectively. The various subsets of Ω describe the different types of cards that may be drawn. For example, $H = \{h\}$ represents "the card is a heart," and $S = \{s\}$ represents "the card is a spade."

$H \cup S = \{h,s\}$ represents "the card is a heart *or* a spade."
$H \cap S = \varnothing$ represents "the card is a heart *and* a spade."
$H' = \{s,c,d\}$ represents "the card is *not* a heart."

The sets of Figure 1.10(a) are disjoint. It is clear from this diagram that the number of elements in $A \cup B$ can be found by first finding the number of elements in

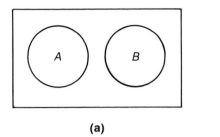

(a)   **(b)**

(a) Sets A and B are disjoint   (b) Sets A and B are not disjoint

Figure 1.10

A, then finding the number of elements in B, and then adding these two numbers. In other words, if we let $n(X)$ be the number of elements in the set X, then

$$n(A \cup B) = n(A) + n(B) \qquad \text{if } A \text{ and } B \text{ are disjoint} \qquad (2)$$

If A and B are disjoint, $A \cup B$ is often written $A + B$. Equation (2) then becomes

$$n(A + B) = n(A) + n(B) \qquad (3)$$

which is easy to remember.

On the other hand, as Figure 1.10(b) indicates, equation (2) does *not* hold if sets A and B are *not* disjoint. It is clear that in this case $n(A) + n(B)$ is greater than $n(A \cup B)$ because the elements of $A \cap B$ are counted twice—once in $n(A)$ and once in $n(B)$. In fact,

$$n(A \cup B) = n(A) + n(B) - n(A \cap B) \qquad (4)$$

Equation (4) applies even if sets A and B are disjoint. If this is the case, then $A \cap B = \varnothing$, and so $n(A \cap B) = 0$. Equation (4) then reduces to equation (2).

The following example shows how to use equations (2) and (4).

Example 10   Let S, C, and K be the sets of all spades, clubs, and kings, respectively, in a deck of cards. Find **(a)** $n(S \cup C)$, **(b)** $n(S \cup K)$.

Solution

(a) Equation (2) can be applied because S and C are disjoint sets. Applying this equation yields

$$n(S \cup C) = n(S) + n(C) = 13 + 13 = 26$$

(b) Equation (2) *cannot* be applied because S and K are *not* disjoint sets. Applying equation (4) yields

$$n(S \cup K) = n(S) + n(K) - n(S \cap K) = 13 + 4 - 1 = 16$$

The following theorem, known as De Morgan's laws, shows how the set operations of union, intersection, and complement are related.

THEOREM 1.6   (*De Morgan's Laws*)

(a) $(A \cup B)' = A' \cap B'$

(b) $(A \cap B)' = A' \cup B'$

Part (a) of this theorem can be verified by using the Venn diagrams in Figure 1.11. In the Venn diagrams, $(A \cup B)'$ and $A' \cap B'$ are represented by identical regions. This gives us good reason to believe that the sets are equal. We repeat that this is a verification, not a mathematical proof.

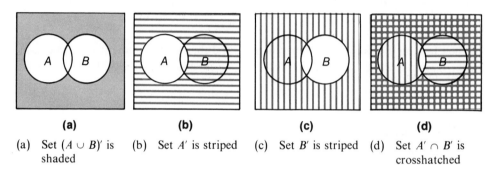

(a)	**(b)**	**(c)**	**(d)**
(a) Set $(A \cup B)'$ is shaded	(b) Set A' is striped	(c) Set B' is striped	(d) Set $A' \cap B'$ is crosshatched

Figure 1.11

Part (b) of Theorem 1.6 can be verified in a similar manner.

Note that when we apply De Morgan's laws, $\cup$ becomes $\cap$, $\cap$ becomes $\cup$, and sets are replaced by their complements.

The following example verifies De Morgan's first law (Theorem 1.6(a)) for a specific case.

Example 11   Let

$$A = \{1,3,5\}, \qquad B = \{1,2,4,6\}, \qquad \text{and} \qquad \Omega = \{1,2,3,4,5,6,7\}$$

Then

$$A \cup B = \{1,2,3,4,5,6\}, \qquad (A \cup B)' = \{7\}$$

$$A' = \{2,4,6,7\}, \qquad B' = \{3,5,7\}, \qquad \text{and} \qquad A' \cap B' = \{7\}$$

It follows that $(A \cup B)' = A' \cap B'$. This verifies De Morgan's first law for these sets.

The preceding definitions and concepts can be easily extended to three sets, four sets, and so on. For example—

(a) The *union of three sets* A, B, and C (denoted $A \cup B \cup C$) is the set of all elements that belong to at least one of A, B, or C. (See Figure 1.12(a).)

(b) The *intersection of three sets* A, B, and C (denoted $A \cap B \cap C$ or ABC) is the set of all elements which belong to A, B, and C simultaneously. (See Figure 1.12(b).)

(c) Three sets A, B, and C are *pairwise disjoint* if $A \cap B = A \cap C = B \cap C = \varnothing$. (See Figure 1.12(c).)

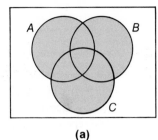 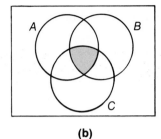 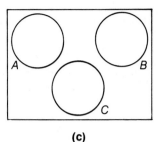

(a)	**(b)**	**(c)**
(a) The set $A \cup B \cup C$ is shaded	(b) The set $A \cap B \cap C$ is shaded	(c) Three pairwise disjoint sets

Figure 1.12

(d) De Morgan's laws for three sets A, B, and C are

$$(A \cup B \cup C)' = A' \cap B' \cap C'; \qquad (A \cap B \cap C)' = A' \cup B' \cup C'$$

The reader can extend these definitions to four sets, five sets, and so on. The following examples illustrate the definitions.

Example 12   If $A = \{a,b\}$, $B = \{b,c,d\}$, $D = \{b,c,d,e\}$, and $\Omega = \{a,b,c, \ldots, z\}$, then—

(a) $A \cup B \cup D = \{a,b,c,d,e\}$ **(b)** $A \cap B \cap D = \{b\}$
(c) $A \cap B' \cap D = \varnothing$ **(d)** $A \cap D \cap B' = \varnothing$
(e) $B' \cap A \cap D = \varnothing$ **(f)** $A' \cup B' \cup D' = \{a,c,d,e, \ldots, z\}$

The sets A, B, and D are not pairwise disjoint, because $A \cap B \neq \varnothing$.

Example 13   Let $\Omega = \{1,2,3, \ldots, 10\}$, $A = \{1,2\}$, $B = \{3,4\}$, and $C = \{5,6\}$. The sets A, B, and C are pairwise disjoint.

Example 14   Let M, P, C, and F be the sets of all students taking math, physics, chemistry, and French, respectively. Then—

(a) The set $M \cup P \cup C \cup F$ is the set of all students who are taking at least one of these subjects.

(b) The set $M \cap P \cap C \cap F$ is the set of all students taking all four of these subjects.

(c) The set $M \cap P' \cap F$ is the set of all students taking math and French but not physics.

The union of five sets A_1, A_2, $\ldots$, A_5 is

$$A_1 \cup A_2 \cup A_3 \cup A_4 \cup A_5$$

Since writing such an expression is tedious, the three-dot method from set notation is frequently used to simplify it to

$$A_1 \cup A_2 \cup \cdots \cup A_5$$

Writing this expression is also time-consuming, so the union is sometimes written

$$\bigcup_{k=1}^{5} A_k \tag{5}$$

The symbol $\bigcup$ in expression (5) indicates that we are taking the *union* of the sets A_k. The particular sets in the union are indicated by the upper and lower indices on the union symbol. These sets are obtained by setting k equal to the lower index (in this case, 1) in A_k and increasing the subscript by 1 until the upper index (in this case, 5) is reached. For example,

$$\bigcup_{k=3}^{5} A_k = A_3 \cup A_4 \cup A_5 \qquad \bigcup_{k=6}^{9} A_k = A_6 \cup A_7 \cup A_8 \cup A_9$$

Note that when an expression like (5) is expanded, the k does not appear. This means that any other letter (e.g., i, j, or n) may be used in place of k. For example, expression (5) can also be written

$$\bigcup_{i=1}^{5} A_i \quad \text{or} \quad \bigcup_{j=1}^{5} A_j$$

A similar notation can be used to write set intersections. For example,

$$\bigcap_{k=3}^{7} A_k = A_3 \cap A_4 \cap A_5 \cap A_6 \cap A_7$$

Example 15 illustrates the use of these notations.

Example 15   Let $A_1 = \{1,2\}$, $A_2 = \{2,3\}$, $A_3 = \{3,4\}$,   and, in   general,   let $A_k = \{k, k+1\}$. Then—

(a) $\displaystyle\bigcup_{k=1}^{5} A_k = A_1 \cup A_2 \cup \cdots \cup A_5 = \{1,2,3,4,5,6\}$

(b) $\displaystyle\bigcap_{k=1}^{5} A_k = \varnothing$

(c) $\displaystyle\bigcup_{j=3}^{8} A_j = \bigcup_{k=3}^{8} A_k = A_3 \cup A_4 \cup \cdots \cup A_8 = \{3,4,\ldots, 8,9\}$

1.3 EXERCISES

1.   Let $\Omega = \{1,2,3,\ldots, 10\}$, $A = \{1,2,4\}$, $B = \{1,2,5,6\}$, and $C = \{4,7\}$. Find—
   (a) $A \cup B$ (b) $A \cap B$ (c) $(A \cup B) \cap C$ (d) $A' \cap B'$
   (e) $A \cup \varnothing$ (f) $B + C$ (g) $n(A)$ (h) $n(B \cup C)$
   (i) $n(A \cup C)$ (j) $A \cup B \cup C$ (k) $A \cap B' \cap C'$

2. Let $\Omega = \{1,2,3,4,\ldots\}$, $A = \{1,2,3,4\}$, $B = \{4,5,6\}$, $C = \{5,6,7\}$, and $D = \{8,9\}$.
   (a) Which three sets are pairwise disjoint?
   (b) Find—

        (i) $A \cap B \cap C$ (ii) $A \cup B \cup D$
        (iii) $A' \cap B' \cap C$ (iv) $A \cup B \cup C \cup D'$

3. Let $\Omega = \{1,2,3,\ldots\}$, $A_1 = \{1,2,3\}$, $A_2 = \{2,3,4\}$, $A_3 = \{3,4,5\}$, and, in general, $A_k = \{k, k+1, k+2\}$. Find—
   (a) A_7 **(b)** $A_1 \cup A_2$ **(c)** $A_1' \cap A_3 \cap A_5$

   (d) $\bigcup_{k=1}^{10} A_k$ **(e)** $\bigcup_{j=3}^{7} A_j$ **(f)** $\bigcap_{j=3}^{7} A_j$

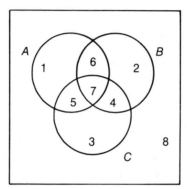

Figure 1.13   Venn diagram for three sets

4. Figure 1.13 is the Venn diagram for three sets. Each of the eight labeled areas is the intersection of three sets; for example, area 1 is $A \cap B' \cap C'$, and area 7 is $A \cap B \cap C$.
   (a) Write the intersections corresponding to the other six areas.
   (b) Use the numerals to identity the areas corresponding to
        (i) $A \cap B'$
        (ii) $B - C$
        (iii) $A \cup (B \cap C')$.

5. Draw the Venn diagrams for the following sets.
   (a) $(A \cap B)'$ **(b)** $A \cap ((B \cup C)')$ **(c)** $(A' \cup B') \cap C'$
   (d) $(A \cap B) - C$ **(e)** $A \cap B \cap C'$ **(f)** $(A \cup B \cup C)'$

6. Use Venn diagrams to verify that $A - B = A \cap B'$.

7. Let A be the set of all intelligent people, B be the set of all people who like math, and C be the set of all college professors.[1] What is the verbal equivalent of—
   (a) $A \cup B$ **(b)** $A \cap B$ **(c)** A' **(d)** $A - B$
   (e) $A \cap B \cap C$ **(f)** $A \cup B' \cup C'$ **(g)** $A' \cap B' \cap C$?

---

1. The reader may justifiably protest that A is not well-defined and hence is not a set. We shall, however, assume that some definition of intelligence has been given (e.g., a person is intelligent if he has an IQ above 100). A similar assumption will be used for all other sets in the remainder of the book.

8. Using the sets of Exercise 7, draw the Venn diagrams that represent—
   (a) all college professors who like math
   (b) all college professors who do not like math
   (c) the statement "All people who like math are intelligent" (Hint: This means that $B \subseteq A$.)
   (d) the statement "All intelligent people like math"
   (e) the statement "If a person is intelligent, he is not a college professor" (Hint: This means that $A \cap C = \varnothing$.)

9. An elementary school teacher is doing a private survey of the effect of television upon the academic performance of her students. Define a relevant universal set Ω and two disjoint subsets.

10. If $A \subseteq B$, what are **(a)** $A \cup B$ and **(b)** $A \cap B$?

11. Simplify the following expressions.
   (a) $A \cup A'$ **(b)** $A \cap A'$ **(c)** $A \cap \Omega$ **(d)** $(A \cap \varnothing) \cup A'$
   (e) $(A - A) \cup A'$ **(f)** $A \cup A$ **(g)** $A \cap A$ **(h)** $(A \cap B)'$

12. The membership of a certain organization includes thirteen men, nine doctors, five women, and six male doctors. Let M, D, and W be the sets of all men, doctors, and women in the organization, respectively. Find—
   (a) $n(M)$ **(b)** $n(M \cap D)$ **(c)** $n(M \cup W)$
   (d) $n(W \cap D)$ **(e)** $n(M \cup D)$

13. A survey is being made of the relationship between heart attacks, obesity, and high blood pressure.
   (a) Define a relevant universal set Ω for this study.
   (b) Define two relevant subsets A and B which are disjoint.
   (c) Define two relevant subsets C and D which are not disjoint.

14. Use Venn diagrams to verify Theorem 1.6(b).

1.4 ARITHMETIC PROPERTIES OF SETS

The operations of set union and intersection have many of the same properties as addition and multiplication of numbers. For example, if a and b are any two numbers, then $a + b = b + a$. This is called the *commutative law* for addition. The reader can easily prove that the corresponding law holds for set union: If A and B are any two sets, then $A \cup B = B \cup A$.

In addition, we know that, if a, b, and c are any three numbers, it does not matter how we group them when adding. That is $(a + b) + c = a + (b + c)$. This is the *associative law* for addition.

Using the Venn diagrams of Figure 1.14, we can easily verify the associative law for set union: If A, B, and C are any three sets, then $(A \cup B) \cup C = A \cup (B \cup C)$. Since the Venn diagrams for $(A \cup B) \cup C$ and $A \cup (B \cup C)$ are identical, we have good reason for believing that the sets $(A \cup B) \cup C$ and $A \cup (B \cup C)$ are equal.

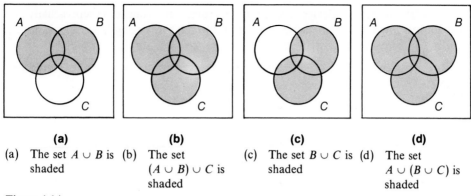

(a) **(b)** **(c)** **(d)**

(a) The set $A \cup B$ is (b) The set (c) The set $B \cup C$ is (d) The set
shaded $(A \cup B) \cup C$ is shaded $A \cup (B \cup C)$ is
shaded shaded

Figure 1.14

The important arithmetic properties of sets are summarized in Table 1.2. The arithmetic laws appear in the second column. The corresponding set laws in the third column are formed by replacing $+$ by $\cup$, $\cdot$ by $\cap$, 0 by $\varnothing$, and 1 by Ω.

Table 1.2

Property	Arithmetic Law	Set-theoretic Law
Commutativity	$a + b = b + a$	$A \cup B = B \cup A$
	$a \cdot b = b \cdot a$	$A \cap B = B \cap A$
Associativity	$(a + b) + c = a + (b + c)$	$(A \cup B) \cup C = A \cup (B \cup C)$
	$(a \cdot b) \cdot c = a \cdot (b \cdot c)$	$(A \cap B) \cap C = A \cap (B \cap C)$
Distributivity	$a \cdot (b + c) = a \cdot b + a \cdot c$	$A \cap (B \cup C) = (A \cap B) \cup (A \cap C)$
Existence of a "zero"	$a + 0 = a$	$A \cup \varnothing = A$
	$a \cdot 0 = 0$	$A \cap \varnothing = \varnothing$
Existence of a "one"	$a \cdot 1 = a$	$A \cap \Omega = A$

There are many other results that are valid in both systems. For example, if a, b, and c are any three numbers, associativity implies that

$$a + (b + c) = (a + b) + c$$

Therefore, we can indicate the addition of three numbers by $a + b + c$ without using parentheses; that is,

$$a + b + c = a + (b + c) = (a + b) + c \tag{6}$$

Using the definitions and the associative law for set union, the reader can easily show that the set law corresponding to equation (6),

$$A \cup B \cup C = A \cup (B \cup C) = (A \cup B) \cup C$$

is also valid. Moreover, similar laws hold for the union of four or more sets, and for the intersection of sets.

Table 1.2 does *not* mean that every law which is true for arithmetic is also true for sets when $+$ is replaced by $\cup$, $\cdot$ by $\cap$, and so on. The analogy between the operations is a useful mnemonic device and nothing more. For example, there is only one distributive law for arithmetic. (See Table 1.2.) However, using Venn diagrams, the reader can show that there are *two* distributive laws for sets:

$$A \cap (B \cup C) = (A \cap B) \cup (A \cap C) \tag{7}$$

and

$$A \cup (B \cap C) = (A \cup B) \cap (A \cup C) \tag{8}$$

The arithmetic law that corresponds to equation (7) is listed in Table 1.2. The arithmetic "law" that corresponds to equation (8) is

$$a + (b \cdot c) = (a + b) \cdot (a + c)$$

This "law" is *not* valid for all numbers a, b, and c, because, for example,

$$1 + (2 \cdot 3) \neq (1 + 2) \cdot (1 + 3)$$

The two distributive laws for sets are stated in Theorem 1.7.

THEOREM 1.7   (Distributive Laws)

(a) $A \cap (B \cup C) = (A \cap B) \cup (A \cap C)$
(b) $A \cup (B \cap C) = (A \cup B) \cap (A \cup C)$

The following examples show how the set-theoretic laws work in specific cases.

Example 16   Let $A = \{1,2,3\}$, $B = \{2,3,4\}$, and $C = \{1,3,4,5,6,7\}$. Then

(a) $A \cup B = B \cup A = \{1,2,3,4\}$
    $A \cap B = B \cap A = \{2,3\}$
(b) $A \cup (B \cup C) = (A \cup B) \cup C = A \cup B \cup C = \{1,2,3,4,5,6,7\}$
    $A \cap (B \cap C) = (A \cap B) \cap C = A \cap B \cap C = \{3\}$
(c) $A \cap (B \cup C) = A \cap \{1,2,3,4,5,6,7\} = \{1,2,3\}$, and
    $(A \cap B) \cup (A \cap C) = \{2,3\} \cup \{1,3\} = \{1,2,3\}$. Therefore,

$$A \cap (B \cup C) = (A \cap B) \cup (A \cap C) = \{1,2,3\}$$

Similarly,

$$A \cup (B \cap C) = (A \cup B) \cap (A \cup C) = \{1,2,3,4\}$$

Example 17   Let A, B, and C be any three sets. The following simplifications result from the distributive laws (Theorem 1.7), De Morgan's laws (Theorem 1.6, page 14), the laws listed in Table 1.2, and the definitions.

(a) $(A \cap A')' = \varnothing' = \Omega$
(b) $(A \cap B) \cup (A \cap B') = A \cap (B \cup B') = A \cap \Omega = A$
(c) $[(A \cap B) \cup B']' = [B' \cup (A \cap B)]' = [(B' \cup A) \cap (B' \cup B)]'$
    $\qquad = [(B' \cup A) \cap \Omega]' = (B' \cup A)' = B \cap A'$
(d) $(A \cup A') = \Omega' = \varnothing$
(e) $(A \cup B \cup C)' \cap (A' \cap B' \cap C') = (A' \cap B' \cap C') \cap (A' \cap B' \cap C')$
    $\qquad = A' \cap B' \cap C'$

1.4 EXERCISES

1.   Let $A = \{a,b,c\}$, $B = \{b,e,f\}$, and $C = \{a,c,f\}$. Find
     (a) $A \cup B$ **(b)** $B \cup A$ **(c)** $A \cap B$ **(d)** $B \cap A$
     (e) $A \cup (B \cup C)$ **(f)** $(A \cup B) \cup C$ **(g)** $A \cup B \cup C$

2.   Let A, B, and C be the sets of Exercise 1. Show that both distributive laws hold for these sets.

3.   Let A, B, and C be any three sets. Simplify the following expressions.
     (a) $(A \cup A') \cap \Omega$ **(b)** $(A \cup \varnothing)'$
     (c) $(A \cap B) \cup (A \cap B')$ **(d)** $(A \cap \varnothing) \cup \Omega$
     (e) $(A \cup B)' \cap (A' \cap B')$ **(f)** $A \cap [(A \cap B) \cup (A \cap C)]$

1.5 CARTESIAN PRODUCTS

In this section, we will study the fifth set operation, Cartesian product. We shall also see that this operation enables us to symbolize many verbal expressions clearly and concisely.

Suppose we are tossing a nickel and a dime. A natural way to record the result of each toss is to write

$$(h,t), \quad (t,h), \quad (t,t), \quad \text{or} \quad (h,h)$$

where (h,t) signifies that the nickel turned up heads and the dime turned up tails; (t,h) signifies that the nickel turned up tails and the dime turned up heads; and so on.

Both (h,t) and (t,h) are examples of **ordered pairs.** "Ordered" signifies that order is important: (h,t) is not the same as (t,h). The entries of an ordered pair are called *components.* For example, the **first component** of the ordered pair (h,t) is h; the **second component** is t.

Definition 1.10   The ordered pairs (a,b) and (c,d) are **equal ordered pairs** if and only if $a = c$ and $b = d$.

For example, the ordered pairs $(2,6)$ and $(2,7)$ are not equal; neither are the ordered pairs $(2,6)$ and $(6,2)$. On the other hand, the ordered pairs $(2,6)$ and $(\frac{4}{2},6.0)$ are equal, because $2 = \frac{4}{2}$ and $6 = 6.0$.

Definition 1.11   The **Cartesian product** of the sets A and B is the set $A \times B$ consisting of all the ordered pairs (a,b) that can be formed by taking the first component a from set A and the second component b from set B.[2]

Symbolically, $A \times B = \{(a,b) \mid a \in A, b \in B\}$.
Example 18 illustrates Definition 1.11.

---

2. The Cartesian product is named after the French mathematician René Descartes (1596–1650).

Example 18 If $A = \{a,b\}$ and $B = \{c,d,e\}$, then

$$A \times B = \{(a,c),\ (a,d),\ (a,e),\ (b,c),\ (b,d),\ (b,e)\}$$
$$B \times A = \{(c,a),\ (d,a),\ (e,a),\ (c,b),\ (d,b),\ (e,b)\}$$
$$A \times A = \{(a,a),\ (a,b),\ (b,a),\ (b,b)\}$$

The above ideas may be generalized to **ordered k-tuples**, that is, ordered arrangements of k objects like $(x_1, x_2, \ldots, x_k)$.

Definition 1.12 Two ordered k-tuples $(a_1, a_2, \ldots, a_k)$ and $(b_1, b_2, \ldots, b_k)$ are *equal* if and only if $a_i = b_i$ for each $i = 1, 2, \ldots, k$.

Definition 1.13 The **Cartesian product** of the sets $A_1, A_2, \ldots, A_k$ is the set $A_1 \times A_2 \times \cdots \times A_k$ consisting of all ordered k-tuples $(a_1, a_2, \ldots, a_k)$ that can be formed by taking the first component a_1 from the set A_1, the second component a_2 from the set A_2, and so on.

Symbolically,

$$A_1 \times A_2 \times \cdots \times A_k$$
$$= \{(a_1, a_2, \ldots, a_k) \,|\, a_1 \in A_1,\ a_2 \in A_2, \ldots, a_k \in A_k\}$$

The Cartesian product $A_1 \times A_2 \times \cdots \times A_k$ can also be written

$$\underset{i=1}{\overset{k}{\mathsf{X}}} A_i$$

The set $A \times A \times \cdots \times A$ (k times) is written A^k.

The following examples illustrate Definition 1.13.

Example 19 If $A = \{a\}$, $B = \{a,b\}$, and $C = \{c\}$, then—

  (a) $A^2 = A \times A = \{(a,a)\}$
  (b) $B^3 = B \times B \times B$
     $= \{(a,a,a),\ (a,a,b),\ (a,b,a),\ (a,b,b),\ (b,a,a),\ (b,a,b),\ (b,b,a),\ (b,b,b)\}$
  (c) $A \times B \times C = \{(a,a,c),\ (a,b,c)\}$
  (d) $C \times B \times A = \{(c,a,a),\ (c,b,a)\}$
  (e) $A \times A \times C = \{(a,a,c)\}$

Example 20 The strike committee of a certain union consists of four members, one from each local. If A, B, C, and D stand for the sets consisting of the members of locals 1, 2, 3, and 4, respectively, then the set of all possible strike committees is the Cartesian product $A \times B \times C \times D$.

Example 21 Students are randomly selected and their sex is noted by writing m or f.

  (a) If one student is selected, the set of possible results is $A = \{m,f\}$.

(b) If three students are selected, the set of all possible results is

$$A^3 = \{(m,m,m), (m,m,f), (m,f,m), (m,f,f), (f,m,m), (f,m,f), (f,f,m), (f,f,f)\}$$

(c) If twenty students are selected, a set of all possible results is A^{20}.

1.5 EXERCISES

1.   If $A = \{1,2,3\}$, $B = \{2,3\}$, and $C = \{4\}$, find—
   (a) $A \times B \times C$ **(b)** $A \times C \times B$ **(c)** $A \times C \times C$
   (d) $C \times C$ **(e)** $C \times \varnothing$ **(f)** C^8

2.   Let $A = \{a,b,c\}$, $B = \{b,c,d\}$, $C = \{a\}$, and $D = \{b,d,e\}$. Find—
   (a) $(A \cap B) \times C$ **(b)** $(B - A) \times D$ **(c)** $(C \cup D) \times C$
   (d) $(B \cap D) \times (B \cap D) \times (B \cap D)$ **(e)** $n(C \times D)$ **(f)** $n(C \times C)$

3.   **(a)** When will $A \times B = B \times A$?
   (b) When will $A \times B = \varnothing$?

4.   Which of the following are equal?
   (a) $\{1,2\}$ and $\{2,1\}$
   (b) $(1,2)$ and $(2,1)$

5.   If a die is tossed once, the set of all possible numbers that might turn up is $A = \{1,2,\ldots,6\}$. Find the set of all possible combinations of numbers that might turn up if the die is tossed (a) 3 times, (b) 5 times, or (c) 100 times.

6.   Two balls are removed one at a time from an urn containing one green ball and one red ball.
   (a) Give the set of all possible color combinations that might result—
      (i)   if the first ball is replaced before the second ball is withdrawn
      (ii) if the first ball is not replaced before the second ball is withdrawn.
   (b) Can the sets of part **(a)** be written as Cartesian products?

Key Word Review

Each key word is followed by the number of the section in which it is defined.

set   1.1
element   1.1
universe   1.1
equal sets   1.1
empty set   1.1
subset   1.2
proper subset   1.2
complement   1.2
relative complement   1.2

union   1.3
intersection   1.3
disjoint sets   1.3
ordered pairs   1.5
first component   1.5
second component   1.5
equal ordered pairs   1.5
Cartesian product   1.5
ordered k-tuples   1.5

CHAPTER 1   **REVIEW EXERCISES**

1.   Define each word in the Key Word Review.

2.   What do the following symbols mean?
   (a) $\varnothing$ **(b)** Ω **(c)** $\in$ **(d)** AB **(e)** $A \cup B$
   (f) $A \cap B$ **(g)** A' **(h)** $A - B$ **(i)** $\subseteq$ **(j)** $\subset$
   (k) $\bigcup_{i=1}^{10} A_i$ **(l)** $n(A)$ **(m)** $A + B$ **(n)** $A \times B$ **(o)** $\bigcap_{i=3}^{13} A_i$

3.   Let $\Omega = \{1,2,3,\ldots,20\}$, $A = \{1,2,3,4\}$, $B = \{2,3,4,5\}$, and $C = \{4\}$. Find—
   (a) $A \cap B$ **(b)** $A \cup B$ **(c)** $A - B$ **(d)** A'
   (e) $A \times C$ **(f)** $A \cap B \cap C$ **(g)** $A \cup B \cup C$ **(h)** C^2
   (i) C^3 **(j)** $n(A' + C)$ **(k)** $n(A \cup B)$ **(l)** $A \times \varnothing$

4.   Using the sets of Exercise 3, find—
   (a) $(A - B) \cup C$ **(b)** $(A \cap B) \times (B \cap C)$ **(c)** $A' \cap B' \cap C'$
   (d) $(A' \cap B') - C'$ **(e)** $(A \cup B \cup C)'$

5.   Let $\Omega = \{1,2,3, \ldots\}$, $A_1 = \{1,2,3\}$, $A_2 = \{2,4,6\}$, $A_3 = \{3,6,9\}$, and, in general, let $A_k = \{k, 2k, 3k\}$. Find—
   (a) A_7 **(b)** $A_1 \cap A_2 \cap A_4$ **(c)** $A_1 \cup A_3 \cup A_5$
   (d) $\bigcup_{k=1}^{9} A_k$ **(e)** $\bigcap_{k=1}^{7} A_k$

6.   Let $A = \{r, s, t, v\}$. Answer true or false:
   (a) $r \in A$ **(b)** $\varnothing \subset A$ **(c)** $y \notin A$
   (d) $\varnothing \subseteq \{r,s\}$ **(e)** $A \not\subset A$ **(f)** $\varnothing \in A$

7.   Draw the Venn diagrams for the following sets.
   (a) $A \cup B$ **(b)** $A \cap B'$ **(c)** $(A \cup B) - C$
   (d) $(A \cap B) - (B \cap C)$

8.   Simplify the following expressions.

(a) $A \cup A'$ (b) $A \cap A'$ (c) $(A - A) \cup A$

(d) $(A \cup A) - B$ (e) $(A \cap \Omega)'$

9.   Simplify the following expressions.

(a) $(A \cup B)' \cap (A \cup C)$ (b) $[(A \cup \Omega) - A'] - A'$

(c) $(A \cup B \cup C)' - C'$

10. Symbolize the set of the first four positive even integers in four different ways.

11. A survey is being made to determine the relationship between salary and sex among employees at a certain university.

(a) Define a relevant universal set Ω for this study.

(b) Define two relevant disjoint subsets A and B.

(c) Define two relevant nondisjoint subsets C and D.

(d) What are the verbal equivalents of (*i*) $C \cup D$, (*ii*) A', and (*iii*) $A' \cap B$?

12. Let $A = \{\varnothing\}$.

(a) Find (*i*) $n(A)$ and (*ii*) $n(\varnothing)$.

(b) Answer true or false:

(*i*) $\varnothing \subseteq A$ (*ii*) $\varnothing \in A$ (*iii*) $\varnothing = A$.

13. A survey was made of employee absenteeism at the Sherman Shirt Factory. Let Ω be the set of all employees, F the set of all female employees, and A the set of all employees who were absent at least ten days last year. State the verbal equivalent of—

(a) F' (b) $F \cup A$ (c) $F \cap A$ (d) $F - A$ (e) $F' \cap A'$

14. Using the sets of Exercise 13, suppose that $n(F') = 50$, $n(F) = 50$, $n(A) = 30$, and $n(A \cap F) = 10$. Find—

(a) $n(F \cup A)$ (b) $n(F \cup F')$ (c) $n(F \cap F')$ (d) $n(A')$

Suggested Readings

1. Breuer, Joseph. *Introduction to the Theory of Sets.* Translated by H. Fehr. Englewood Cliffs, N.J.: Prentice-Hall, 1958.
2. Fraenkel, Abraham A. *Abstract Set Theory.* 3d ed. Amsterdam: North Holland Publishing Co., 1966.
3. Riker, William H. *The Theory of Political Coalitions.* New Haven, Conn.: Yale University Press, 1962.

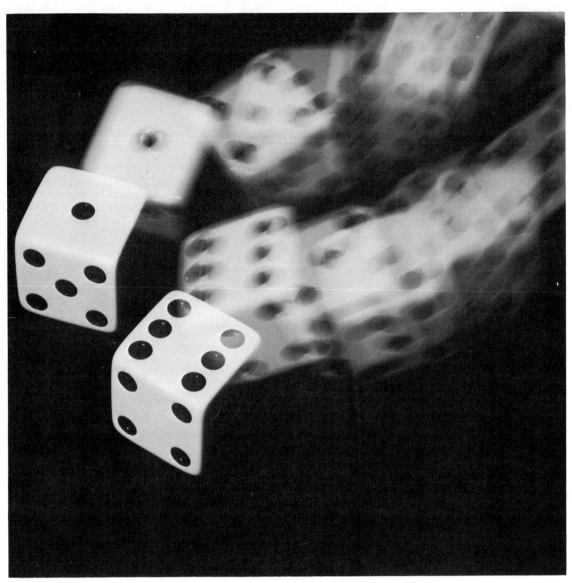

Harold M. Lambert

CHAPTER 2 Probability

INTRODUCTION

What is the probability that the dice on the facing page will land with nine dots face up? What are the odds that they will land with two dots face up? We are all aware—from newspaper reports and television, if not from personal experience—that probability theory is important in gambling and weather prediction. It is also crucial to the study of genetics and to the calculations on which insurance companies base their premiums.

Terms like "probability," "odds," and "chance" are common in everyday speech:

> "The *probability* of rain today is 50 percent."
> "If you smoke, you increase your *chance* of getting lung cancer."
> "What *odds* will you give me that the Oakland
> Raiders will win the Super Bowl?"

Exactly what do these terms mean? This is the question we will answer in this chapter. In the process, we will learn about many new applications for probability theory.

2.1 RANDOM EXPERIMENTS AND SAMPLE SPACES

If you flip a coin, what is the "probability" of getting heads? Almost anyone will tell you that the probability is $\frac{1}{2}$. The reason usually given is that there are two possibilities: heads and tails. Therefore, the total number of ways that the coin may fall is two, while the number of ways that heads may occur is one. This is an application of a frequently used "definition" of probability:

the probability of an event A

$$= \frac{\text{number of favorable cases (ways that } A \text{ can occur)}}{\text{total number of cases}}$$

Although the meaning is intuitively clear, the reader should recognize that this statement is not a mathematical definition. After all, what is an *event* and what is a *favorable case*?

Let us, for the moment, overlook these picky points and simply apply this "definition" to the ordinary occurrence of crossing a dangerous street. Two things can happen: either one crosses safely or one doesn't. Hence the probability of crossing safely is $\frac{1}{2}$. But this is absurd. What went wrong? Note that in this situation the two cases—crossing safely and not crossing safely—are not *equally likely*. Therefore, it appears that the "definition" of probability does not apply when the possible cases are not equally likely.

Suppose that a friend asks you to think of a number. What is the probability that he will guess the number you pick? If we try to apply the "definition," we would have to divide by infinity, because there are infinitely many numbers. Since this is impossible, we cannot apply the "definition" unless the total number of cases is *finite*.

This discussion is beginning to become complicated; we can no longer rely on our intuition. Precise definitions are needed. In order to develop a precise definition of probability, we must understand the concept of a *random experiment*.

In this scientific age of ours, we are accustomed to considering causes and effects. If we see a child playing by an open window on the twentieth floor we are likely to snatch her away from it, because we are aware of what will happen if she falls out. She will undoubtedly fall down; there is no chance that she will fall up, fall sideways, or fly back into the room. Similarly, the results of the following experiments are completely known:

(a) Putting your hand on a very hot stove

(b) Putting an ice cube on a table in a warm room

(c) Punching a hole in an inflated balloon

(d) Raising the temperature of a pot of water past the boiling point

Such experiments are called *deterministic*. What happens, however, when we toss a coin after shaking it thoroughly? Sometimes it lands heads and sometimes tails. Unless we cheat by not shaking the coin properly or by using a two-headed coin, there is no way of determining the result. It may be argued that, if we only knew enough about the initial velocity and position of the coin, the manner in which it was shaken and tossed, and the exact shape and weight of the coin, then the result would be determined. However, in general, this information is not available and so the result is unknown. Experiments such as this are called *nondeterministic*.

In probability theory the result of an experiment is called the **outcome**. Although we do not know the exact outcome of a coin toss, we do know that the coin must fall either heads or tails. Any other possibilities that we might imagine, such as " the coin lands on its side" or "a bird swoops down and steals the coin," are considered so unlikely that they are ignored. If such an unlikely event should occur, the experiment is repeated with another coin and the toss with the unusual outcome is not counted. It follows that a set of all possible outcomes for the coin-tossing experiment is

$$\Omega = \{h,t\}$$

The set Ω is called a *sample space*. Each subset of Ω is called an *event*. Let $H = \{h\}$ be the event "heads is obtained." If a coin is tossed once, we do not know whether or not it will land heads. However, consider what happened when a coin was tossed repeatedly at Temple University. The results are shown in Table 2.1.

It seems clear from the table that, although we cannot predict the outcome of a particular toss, if the coin is tossed a large number of times, the relative frequency of heads is about $\frac{1}{2}$. Hence, if a coin is tossed about 10,000 times, we would expect about 5000 heads. We say that the *empirical probability* of heads is $\frac{1}{2}$. This statement agrees with the intuitive ideas about probability mentioned earlier.

The coin-tossing experiment has the following characteristics:

(1) The experiment can be repeated indefinitely under essentially unchanged conditions.

(2) The experiment is *nondeterministic*; that is, the outcome of each repetition is unknown.

(3) Although we are not able to state what the outcome will be, we are able to describe a set of all possible outcomes.

Table 2.1

Number of Repetitions (n)	Number of Heads (N(H))	Relative Frequency of Heads $\left(\dfrac{N(H)}{n}\right)$
1	0	0.000
20	8	0.400
100	48	0.480
200	101	0.505
300	152	0.507
400	205	0.513
500	254	0.508
800	403	0.504
1400	701	0.501

Any experiment that has these three characteristics is called a **random experiment**.

Suppose R is a random experiment. Property **(3)** implies that we can find a set of all possible outcomes. Usually, many such sets can be found. The one we use will depend upon our interest. For example, suppose we toss a die and receive one of the following payoffs:

(a) One dollar if an even number occurs and nothing otherwise

(b) Four dollars if a number less than 5 occurs and -2 dollars if a number greater than 2 occurs (i.e., we must pay 2 dollars if a number greater than 2 occurs).

(c) Three dollars for each spot that occurs on the die

Appropriate sets of all possible outcomes reflecting the above interests are—

(a) $A = \{e,o\}$, where e and o represent even and odd, respectively

(b) $B = \{f,t\}$, where f and t represent a number less than 5 and a number greater than 2, respectively

(c) $C = \{1,2,3,4,5,6\}$

Each of the preceding sets lists all possible outcomes. However, the outcomes of sets A and C are **mutually exclusive**: in each of these sets no two outcomes can occur at the same time. The outcomes of B, on the other hand, are not mutually exclusive: "a number less than 5" and "a number greater than 2" will occur simultaneously if the die shows a 3 or a 4.

For reasons of mathematical convenience, we will only consider sets of all possible outcomes in which the outcomes are mutually exclusive. Such sets are called *sample spaces*. It follows that A and C are both sample spaces for our die-tossing experiment, but B is not.

Suppose we are interested in payoff **(b)**. We have seen that although set B reflects our interest, it is not a sample space. However, we can also use set C, which is a sample space, to reflect our interest. A number less than 5 can be represented by the subset $\{1,2,3,4\}$ of set C, and a number greater than 2 can be represented by the subset $\{3,4,5,6\}$.

The sample space C can also be used to reflect payoff **(a)**. Even and odd numbers are represented by the subsets $\{2,4,6\}$ and $\{3,5\}$, respectively.

In general, if R is any random experiment, we can always find a set of all possible mutually exclusive outcomes. This set, which is called a sample space of the experiment R, is denoted by Ω.

Definition 2.1   A **sample space** Ω associated with the random experiment E is a set of all possible mutually exclusive outcomes of E.

Example 1   Suppose we stand in front of the student union building at a certain university and note the sex of the first student who appears. This procedure may be repeated indefinitely. In each case, we do not know what the particular outcome will be, but we do know that the student will be either male or female. It follows that this is a random experiment and—if we let m stand for male and f for female—$\Omega = \{m,f\}$.

Example 2   The following are examples of random experiments and their sample spaces:

(a) A coin is tossed twice and the number of heads noted. The sample space is $\Omega = \{0,1,2\}$.

(b) A die is tossed and the number of dots on the uppermost face is recorded; $\Omega = \{1,2,\ldots,6\}$.

(c) A box of 100 screws is picked from the assembly line and the number of defective screws in the box is noted; $\Omega = \{0,1,2,\ldots,100\}$.

(d) One hundred volunteers are selected and tested for diabetes, and the number that have diabetes is recorded; $\Omega = \{0,1,2,\ldots,100\}$.

(e) A ball is withdrawn from an urn containing four red and six white balls and the color of the ball is recorded; $\Omega = \{r,w\}$.

Example 3   A college student is chosen at random, and his or her height (in inches) is recorded. A possible sample space for this experiment is $\Omega = \{t \mid 48 \leq t \leq 96\}$. This set Ω is, of course, a mathematically idealized set and not experimentally realizable. The accuracy of our measurements is naturally limited by the precision of our instruments and by the requirements of the problem. An experimentally realizable sample space for this experiment is $\Omega_1 = \{48,49,\ldots,96\}$. However, a review of the records of the college since its inception may reveal that no one has ever been taller than 80 inches. In this case, the sample space should be $\Omega_2 = \{48,49,50,\ldots,80\}$. If more precision is desired, $\Omega_3 = \{48,48.5,49,49.5,\ldots,80\}$ might be used. Each of these answers is correct.

It is clear from Example 3 that we should not speak of *the* sample space associated with an experiment, but instead speak of *a* sample space. Many different sample spaces may be correct. However, each sample space has the property that it is *exhaustive* and its elements are *mutually exclusive*. That is, no other outcomes are possible (although some of the outcomes listed may be impossible) and no two outcomes can occur at the same time.

Example 4   A coin is tossed until a head appears, and the number of tosses required to obtain the first head is noted.

Solution   It is clear that the head may appear for the first time on the first toss, the second toss, and so on. Hence, 1, 2, 3, and so on are clearly elements of the sample space Ω. However, it is also conceptually possible that a head will never appear, no matter how often we toss the coin. (We shall see in Chapter 5 that the probability of this event is zero.) To be complete, our sample space must contain a symbol for this occurrence. Let us arbitrarily denote it by ∞. Hence, $\Omega = \{\infty, 1, 2, 3, 4, \ldots\}$, where $\{\infty\}$ represents "a head never occurs."

We saw that when a coin is tossed once, we cannot tell whether it will land heads or tails; but if it is tossed frequently, the relative frequency of heads is about $\frac{1}{2}$. Similarly, if any random experiment is performed repeatedly, the individual outcomes seem to occur in a haphazard manner. However, after a large number of repetitions, a definite regularity becomes apparent. This regularity, which will be discussed in Section 2.3, makes a definition of probability possible.

2.1 EXERCISES

Define a sample space for each of the following random experiments.

1. Five cards are chosen from an ordinary fifty-two-card deck, and the number of aces is recorded.

2. A class at Rutgers University is chosen (e.g., Finite Math, Section 1) and the number of women in the class is recorded. (Assume that no class at Rutgers can have more then 200 students.)

3. A fair coin is tossed until the same result appears twice in succession.

4. There are two balls in an urn, one red and one white. A ball is picked and its color is noted.

5. The number of raisins in a 15-ounce box of raisin bran is recorded. (Assume that no box can contain more than 500 raisins.)

6. The numbers 1–366 are written on slips of paper and each slip is put in a capsule. The capsules are thoroughly mixed. Then a capsule is removed from the pile and the number written on the slip it contains is recorded.

7. Two coins are tossed and the number of heads is noted.

8. A hospital in Denver is picked at random and the number of babies born in that hospital on that day is recorded. (Assume that no more than ten babies have ever been born in one day at a Denver hospital.)

9. A car of a particular make and model is selected and test driven on a prescribed route for 100 miles. Its gas mileage (miles per gallon) is recorded. Give three different sample spaces for this experiment. (Assume that the car cannot get less than 5 or more than 40 miles per gallon.)

2.2 EVENTS

In Chapter 1 we learned how to use sets to describe certain verbal expressions. In particular, we learned that the set operations of union, intersection, and complement correspond to the English words "or," "and," and "not." This knowledge will now be applied to random experiments.

In Section 2.1 we saw that a random experiment can be described by its associated sample space Ω. In this section we will see that the subsets of Ω describe statements about the outcomes of the experiment. For this reason, subsets of Ω are called events.

Definition 2.2 Any subset of a sample space Ω is called an **event**.

In common English, an event is just a statement about the outcomes of a random experiment. For example, in the experiment described in Example 1, "the student selected is male" is an event. Subsets containing only one element are called **elementary events**, or *elementary outcomes*. All other subsets except $\varnothing$ are called **compound events**. Example 5 illustrates the correspondence between subsets of Ω and verbal statements about outcomes.

Example 5 A die is tossed and we are asked to note the number of dots on the uppermost face. A sample space for this experiment is $\Omega = \{1,2,3,4,5,6\}$. The following are examples of the correspondence between verbal descriptions of events and subsets of Ω.

Verbal description of event	Corresponding subset of Ω
T: A 2 appears	$\{2\}$
E: An even number appears	$\{2,4,6\}$
S: A 7 appears	$\varnothing$
L: A number larger than 3 appears	$\{4,5,6\}$
O: A number that is even *or* larger than 3 appears	$E \cup L = \{2,4,5,6\}$
A: A number that is even *and* larger than 3 appears	$E \cap L = \{4,6\}$
N: A 2 does *not* appear	$T' = \{1,3,4,5,6\}$
B: A number less than 8 appears	Ω

The event T of Example 5 is an elementary event. (There are five other elementary events.) Events E, L, O, A, N, and Ω are compound events.

Generalizing from Example 5, it is clear that every English statement about the outcomes of the die-tossing experiment is equivalent to a subset of Ω, and vice versa.

An event A is said to *occur* if the result of an experiment corresponds to an element of A. In Example 5 event E will occur if a 2 turns up. Event E will also occur if a 4 or a 6 turns up.

Since the set operations of union, intersection, and complement correspond to the English words "or," "and," and "not," it follows that events have the following properties:

Properties of Events

Let A and B be events.

(a) $A \cup B$ is the event that occurs if and only if A *or* B (or both) occur.

(b) $A \cap B$ is the event that occurs if and only if A *and* B occur.

(c) A' is the event that occurs if and only if A does *not* occur.

These properties may be extended to the union and intersection of three sets, four sets, and so on.

Since every subset of Ω is an event, both $\varnothing$ and Ω are events. The event Ω, which is certain to occur, is called the **sure event**. The event $\varnothing$, which cannot occur is called the **impossible event**. The empty set is the only event which is neither compound nor elementary.

Definition 2.3

(a)   Two events A and B are **mutually exclusive** (or *disjoint*) if and only if they cannot occur together; that is, if and only if $A \cap B = \varnothing$.

(b)   The n events $A_1, A_2, \ldots, A_n$ are **mutually exclusive** (or *pairwise disjoint*) if and only if $A_i \cap A_j = \varnothing$ whenever $i \neq j$.

Example 6 illustrates these concepts.

Example 6   A ball is drawn from an urn containing three red, two black, and three green balls. A sample space for this experiment is $\Omega = \{r,g,b\}$. Consider the following events:

Verbal description of event	*Corresponding subset of* Ω
R: A red or black ball is obtained	$\{r,b\}$
G: A green ball is obtained	$\{g\}$
B: A black or green ball is obtained	$\{b,g\}$
Y: A yellow ball is obtained	$\varnothing$
M: A red, black, or green ball is obtained	Ω

Events R and G are mutually exclusive because $R \cap G = \varnothing$. However, G and B are not mutually exclusive; they can occur together. The event Y, which cannot occur, is the impossible event. The event M, which must occur, is the sure event.

2.2 EXERCISES

1.   Using the sample space of Example 5, write the following events in set notation.
   (a) A number between 1 and 6 appears. (Do not include 1 and 6.)
   (b) A 1 appears.

2. Write the following events as combinations of the eight events of Example 5. (Combine the events of Example 5 using union, intersection, complement, and difference.)

(a) An odd number appears.

(b) A 1 or 3 appears.

(c) A 5 appears.

(d) An odd number or a 2 appears.

(e) An odd number and a 2 appear.

3. All of the members of a certain club are doctors, lawyers, or executives. Suppose a member is randomly chosen.

(a) State a sample space for this experiment.

(b) Give a verbal description of each of the following events.

(i) The impossible event

(ii) The sure event

4. A coin is tossed three times. A sample space for this experiment is

$$\Omega = \{(h,h,h),\ (h,h,t),\ (h,t,h),\ (h,t,t),\ (t,h,h),\ (t,h,t),\ (t,t,h),\ (t,t,t)\}$$

where $\{(h,h,t)\}$ is the event " two heads are followed by a tail."

(a) State two elementary events.

(b) How many elementary events are there?

(c) Give a verbal description of each of the following events.

(i) $\{(h,h,h)\}$

(ii) $\{(h,h,h),\ (h,h,t),\ (h,t,h),\ (h,t,t)\}$

(iii) $\{(h,t,t),\ (t,t,h),\ (t,h,t)\}$

5. Consider the experiment of Exercise 4.

(a) Verbally describe two mutually exclusive events.

(b) Verbally describe two compound events.

(c) Give the subset of Ω that corresponds to each of the following events:

(i) At least one tail occurs.

(ii) A tail occurs on the second toss.

(iii) A head occurs on the first toss and a tail occurs on the second toss.

6. A die is rolled four times.

(a) Define a suitable sample space Ω.

(b) Let A be the event that an even number appears on the first roll, and let B be the event that an even number appears on the third roll. Are A and B mutually exclusive?

(c) Give the verbal equivalent of—

(i) $A \cap B$ (ii) A' (iii) $B \cap A'$ (iv) $A - B$

7. Consider the series electrical circuit shown in Figure 2.1. Suppose that C_1 and C_2 each have two states, operative (1) and inoperative (0). The circuit will be operative if and only if both C_1 and C_2 are operative. A sample space that describes the possible states of the components of the circuit is $\Omega = \{(0,0), (0,1), (1,0), (1,1)\}$, where $\{(0,1)\}$ is the event "C_1 is inoperative and C_2 is operative."

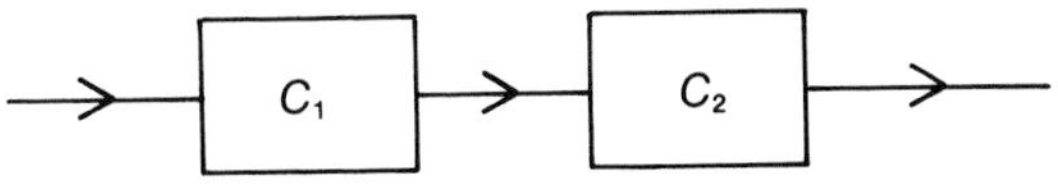

Figure 2.1

Let E_1 be the event "the entire circuit is operative" and let E_2 be the event "either C_1 or C_2 or both are operative."
(a) Are E_1 and E_2 mutually exclusive?
(b) What is the relationship between E_1' and E_2'?
(c) Give verbal descriptions of E_1' and E_2'.

2.3 RELATIVE FREQUENCIES AND EMPIRICAL AND SUBJECTIVE PROBABILITIES

Before we define "the probability of an event A," let us discuss what we mean by this expression intuitively.

Suppose we say that the probability that a three will turn up when a die is tossed is $\frac{1}{6}$. Intuitively, this means that if we toss the die a large number of times, we expect the relative frequency of threes to be about $\frac{1}{6}$. In other words, we expect a three to appear approximately 1 out of every 6 times. For example, if we toss the die 6000 times, we expect about $\frac{1}{6} \times 6000 = 1000$ threes; if we toss the die 12,000 times, we expect about $\frac{1}{6} \times 12,000 = 2000$ threes.

This does not mean that about 2000 threes will occur *every* time we toss the die 12,000 times. We might run into a streak of bad luck and get no threes at all. (This will occur extremely infrequently.) However, if the experiment of tossing the die 12,000 times is repeated a large number of times (say a million), we would get approximately 2000 threes in the majority of these experiments.

In this section we will study the *empirical* and *subjective* definitions of probability. In Section 2.4 a third definition, Laplace's definition of probability, will be discussed. All three of these definitions agree with our intuitive ideas of probability. Since each definition predicts the relative frequency of an event, we will begin by studying relative frequency.

Suppose a random experiment is repeated n times (e.g., a die is tossed 1000 times, or a card is selected 800 times, with replacement, from a bridge deck). Let A be any event and let $N(A)$ be the number of times that A occurred. Then

$$f_n(A) = \frac{N(A)}{n}$$

is called the **relative frequency** of A. For example, if 5 threes appear when a die is tossed 31 times, then

$$f_{31}(T) = \tfrac{5}{31}$$

where $T = \{3\}$.

Theorem 2.1 lists some of the properties of $f_n(A)$.

THEOREM 2.1   The relative frequency $f_n(A)$ has the following properties:

(a) $0 \le f_n(A) \le 1$
(b) $f_n(\Omega) = 1$
(c) $f_n(A + B) = f_n(A) + f_n(B)$ if A and B are disjoint
(d) $f_n(\varnothing) = 0$
(e)   If $A \subseteq B$, then $f_n(A) \le f_n(B)$.
(f) $f_n(A') = 1 - f_n(A)$

Notice that the sets A and B of part (c) are disjoint. We saw in Chapter 1 that in this case we may use $A + B$ to denote $A \cup B$. This notation makes part (c) much easier to remember.

Proof   We will prove parts (b) and (d) and leave the remainder for the problems. Since Ω is certain to occur every time, $N(\Omega) = n$, so

$$f_n(\Omega) = \frac{N(\Omega)}{n} = \frac{n}{n} = 1$$

Similarly, $N(\varnothing) = 0$, so

$$f_n(\varnothing) = \frac{N(\varnothing)}{n} = \frac{0}{n} = 0$$

Consider the coin-tossing experiment of page 29. The relative frequency of heads, $f_n(H)$, is given in the last column of Table 2.1. Notice that, as the number n of repetitions increases, $f_n(H)$ appears to oscillate about the number 0.5. This number is called the *empirical probability of heads* and is denoted $P(H)$.

In general, let R be any random experiment and let A be any event. It has been observed that, as the number of repetitions of R increases, $f_n(A)$ tends to oscillate about a definite number. This number is called the **empirical probability** of A and is denoted $P(A)$.

As stated here, this oscillation of $f_n(A)$ about the number $P(A)$ is not a mathematical conclusion, but an empirical fact. That is, this oscillation has been observed to happen in a great many cases.

It can also be proved that properties (a)–(f) of Theorem 2.1 still hold when $f_n(A)$ is replaced by the empirical probability $P(A)$.

The reader may now wonder how the empirical probability of an event A may be found. In practical situations, a random experiment is repeated a large number of times. *The relative frequency of A, $f_n(A)$, is then used as an approximation of $P(A)$.* This approximation generally improves as the number n of repetitions increases.[1]

The following examples illustrate these concepts.

---

1. In the coin-tossing experiment of Table 2.1, for example, $f_{200}(H) = 0.505$ is a slightly better approximation of $P(H) = 0.5$ than $f_{300}(H) = 0.507$. However, $f_{1400}(H) = 0.501$ is a much better approximation than $f_{20}(H) = 0.480$. It can be proved that the probability that $f_n(H)$ and $P(H)$ differ by any fixed small number approaches 0 as n gets larger and larger. (This fact is called the *weak law of large numbers*.)

Table 2.2

Student Category	Favored (F)	Opposed (O)	No Opinion (A)	Total
Undergraduate (U)	400	150	50	600
Graduate (G)	80	120	50	250
Medicine (M)	20	20	10	50
Law (L)	60	40	0	100
Total	560	330	110	1000

Example 7   One thousand students at Hohum University were polled to determine their attitudes toward a proposed student strike. Table 2.2 shows the results of the poll.

If a student is selected at random, what is the empirical probability that he or she—

(a) favors striking

(b) is a medical student

(c) has no opinion about the strike

(d) is a graduate student opposed to the strike

(e) is either a law student who opposes the strike or a medical student who opposes the strike?

Solution

(a) Out of 1000 students polled, 560 favor a strike. The relative frequency of students who favor a strike (event F) is therefore

$$f_{1000}(F) = \frac{N(F)}{n} = \frac{560}{1000} = 0.56$$

This relative frequency is an *approximation* of the empirical probability $P(F)$. If more students were sampled, the approximation would be better. Based on this sample,

$$P(F) \approx f_{1000}(F) = 0.56$$

The symbol $\approx$ means "approximately."

(b) Similarly,

$$P(M) \approx f_{1000}(M) = \frac{N(M)}{n} = \frac{50}{1000} = 0.05$$

(c)

$$P(A) \approx f_{1000}(A) = \frac{N(A)}{n} = \frac{110}{1000} = 0.11$$

(d) The number of graduate students who are opposed to the strike, $N(G \cap O)$, is found in the intersection of the row labeled G and the column labeled O in Table 2.2; therefore, $N(G \cap O) = 120$. It follows that

$$P(G \cap O) \approx f_{1000}(G \cap O) = \frac{N(G \cap O)}{n} = 0.12$$

(e) The events $M \cap O$ and $L \cap O$ are disjoint. Therefore, we can apply Theorem 2.1(c) to find the probability of their union.

$$(M \cap O) + (L \cap O)$$

Recall that in Chapter 1 we learned that the notation AB could be used to denote the intersection of the sets A and B. Using this notation, $(M \cap O) + (L \cap O)$ simplifies to $MO + LO$. Then, applying Theorem 2.1(c), we have

$$P(MO + LO) \approx f_{1000}(MO + LO)$$

$$= f_{1000}(MO) + f_{1000}(LO) = \frac{20}{1000} + \frac{40}{1000} = 0.06$$

Example 8   A die was rolled 600 times and 99 threes (event T) and 102 fours (event F) occurred. Suppose the die is rolled again. Based on these results, estimate the empirical probability that—

(a) a 3 will appear
(b) a 3 will not appear
(c) a 3 or a 4 will appear
(d) a 7 will appear
(e) a number less than 8 will appear.

Solution
(a)

$$P(T) \approx f_{600}(T) = \frac{N(T)}{600} = \frac{99}{600} = 0.165$$

(b) Using Theorem 2.1(f) and part **(a)**,

$$P(T') \approx f_{600}(T') = 1 - f_{600}(T) = 1 - 0.165 = 0.835$$

(c) Using Theorem 2.1(c),

$$P(T + F) \approx f_{600}(T + F) = f_{600}(T) + f_{600}(F)$$

$$= \frac{102}{600} + \frac{99}{600} = \frac{201}{600} = 0.335$$

(d) The event "a 7 will appear" is the impossible event $\emptyset$. As mentioned before, Theorem 2.1 holds when the relative frequency f_n is replaced by the empirical probability P. It therefore follows from Theorem 2.1(d) that $P(\emptyset) = 0$.

(e) The event "a number less than 8 will appear" is the sure event Ω. It follows from Theorem 2.1(b) that $P(\Omega) = 1$.

Suppose that we are told that 450 heads were obtained when a coin was tossed 1000 times. Using this information, we are able to calculate the empirical probability that a head will occur in the future:

$$P(H) = \frac{450}{1000} = 0.45$$

Now suppose that we are not told the exact number of times that the coin was tossed, nor the exact number of heads that were obtained. Instead, assume that we are told only the *percentage* of heads that occurred (45 percent) when a coin was tossed a large number of times. In this case, we would not know if 450 heads occurred out of 1000 tosses, or if 900 heads occurred out of 2000 tosses, or if 4500 heads occurred out of 10,000 tosses. We would, however, still know that the relative frequency of heads was 45 out of every 100, or 0.45. We would therefore still be able to calculate the probability of obtaining a head.

In general, suppose that an experiment is repeated a large number of times. The empirical probability that a certain event will occur can be calculated if we know the *percentage* of times that the event occurred in the past. For example, if it is found—after keeping records for a long time—that 30 percent of all babies born in a certain hospital have brown hair, then the empirical probability that a baby born in that hospital will have brown hair is 0.30. If 28 percent of all students at a large university change their majors in the freshmen year, then the empirical probability that a student at that university will change his or her major in the freshmen year is 0.28.

These examples show how empirical probabilities can be calculated. However, there are two common misconceptions about empirical probabilities that can lead to incorrect conclusions about experiments and results.

The first of these misconceptions involves lost data. To illustrate the kind of situation in which this misconception can occur, suppose that a scientist has an apparatus that is repeating the same experiment over and over, and that she plans to use the results of these repeated trials to calculate an empirical probability. If the scientist goes out to lunch and leaves the apparatus operating but recording no data, the noon hour's results will be lost. The misconception arises if we assume that the loss of these results will affect the empirical probability. In fact, *the empirical probability will remain the same even if the results of some trials are not counted.*

The second misconception is one frequently held by gamblers. For example, some gamblers keep careful records of the winning numbers on a roulette wheel. They know that the probability that an even number will win is the same as the probability that an odd number will win. Therefore, if they observe that even has occurred many more times than odd, they immediately bet on odd. These people feel that, since the probabilities are the same, an odd number should turn up now in order to make up for its previous infrequent occurrences.

The roulette wheel, however, has no memory and no desire to make amends for past mistakes. Speaking mathematically, *the empirical probability remains the same no matter what has occurred in the past.*

In order to understand this, let us consider the same problem in the context of tossing a fair coin. Suppose that the coin is tossed 20 times with the following results:

$$N(H) = 14 \qquad N(T) = 6$$

where $H = \{h\}$ and $T = \{t\}$. One might feel now compelled to bet on tails, reasoning that $P(T)$ can only approach 0.5 if more tails than heads appear from now on.

This, however, is not true. The value of $P(T)$ will approach 0.5 if, as mathematically expected, an equal number of tails and heads appear in the tosses following the twentieth toss. Suppose, for example, that the number of heads equals the number of tails during the next 80, 280, and 480 tosses. The data that would be obtained in this case are shown in Table 2.3. Notice that although equal numbers of head and tails appeared in the trials following the first 20 trials, the relative frequency of tails still approaches 0.5. Therefore, a run of heads does not mean that tails is more likely to appear on the next trial.

Table 2.3

n	$N(H)$	$N(T)$	$f_n(H)$	$f_n(T)$
20	14	6	0.700	0.300
100	54	46	0.540	0.460
300	154	146	0.513	0.487
500	254	246	0.508	0.492

If one were uncertain about the fairness of the coin, a run of heads *might* indicate that the coin is biased in favor of heads. It would therefore be more logical to bet on heads. If, however, one *knew* that the coin was fair, *$P(T)$ would always be $\frac{1}{2}$, no matter what had occurred previously.*

Table 2.3 illustrates another important fact about empirical probability: *The relative frequency of an event A is only an estimate of the empirical probability of A.* One must observe a large number of trials before one can get close approximation.[2]

There are many situations in which a definition of probability is needed, but the empirical definition cannot be applied. For example, Lloyd's of London is famous for insuring almost anything; but how does one calculate the probability that a famous singer will fail to appear for an important concert? Suppose that in the past he has never failed to appear for an important engagement. Does that mean that the probability is zero, and so no premium should be charged? What is the probability that Elizabeth Taylor's million-dollar diamond will be stolen, or that a controversial political group's convention at a university will result in damage to the school?

Suppose that you are interested in the probability that you will pass a certain math course. You might consider your past experience with math courses and the comments of former students about the teacher. However, given the same information, a friend's estimate of your chances of passing is likely to differ from yours.

In all these cases, *subjective probability* must be applied. A **subjective probability** is just an educated guess as to the relative frequency of an event—a personal evaluation based, hopefully, upon sound judgment and past experience with similar situations.

---

2. Methods for estimating the number of trials can be found in probability and statistics books; e.g., reference 1 in the Suggested Readings at the end of this chapter.

If you are playing poker with a friend and feel, without having kept records, that in the past when he has played with you he has bluffed approximately 10 percent of the time, then the subjective probability that he is bluffing now is 0.1. If you feel that four out of every ten students with your capabilities will pass a certain course taught by a certain teacher, then the subjective probability that you will pass is 0.4.

Since subjective probabilities are estimations of relative frequencies, Theorem 2.1 should hold when the relative frequency $f_n(A)$ is replaced by the subjective probability $P(A)$. Therefore, this theorem can be used to check that the subjective probabilities are assigned properly. (For example, each probability should be a number that is greater than or equal to 0 and less than or equal to 1, and the probability of the impossible event should be 0.)

2.3 EXERCISES

1. **(a)** An investigation of 1000 used typewriters of a certain make revealed that 50 of them had 3s that were out of line. Based on these results, estimate the probability that a randomly chosen used typewriter of this make has the same defect.

    (b) Suppose that 30 percent of all students entering a certain college pass the English proficiency examination. What is the probability that a randomly chosen student fails this exam?

2. In a survey of the religious attitudes of college students, the following data were obtained:

	Male	Female
Very religious	340	200
Moderately religious	400	250
Not at all religious	160	100

(a) Based on these data, estimate the probability that a student picked at random is (*i*) very religious, (*ii*) male, (*iii*) a moderately religious woman.

(b) If 100 students are randomly selected, approximately how many will be (*i*) male, (*ii*) moderately religious, (*iii*) very religious men?

3. Suppose a survey to determine whether money buys happiness results in the following data:

Family salary per year (in thousands of dollars)	Very happy (V)	Moderately happy (M)	Unhappy (U)	Depressed (D)
Less than 10 (L)	20	100	40	40
Between 10 and 20 (B)	50	300	75	75
Between 20 and 30 (T)	40	90	35	35
More than 30 (G)	10	50	30	10

Let V, M, U, and D be the events that a randomly chosen person is very happy, moderately happy, unhappy, or depressed, respectively.

(a) Based on these data, estimate the following:

(i) $P(M)$ (ii) $P(MB)$ (iii) $P(M \cup B)$ (iv) $P(MD)$

(v) $P(M \cup D)$

(b) If 1000 people are randomly selected, calculate approximately how many will be—

(i) unhappy

(ii) very happy

(iii) very happy members of a family whose family income is less than 10,000 dollars per year.

4. Put two red marbles and one white marble in a container and mix thoroughly. Choose one marble (without looking, of course), record its color, and replace the marble. Repeat this experiment thirty times. Let R be the event "a red marble is chosen."

(a) Calculate $f_5(R)$, $f_{10}(R)$, $f_{20}(R)$, and $f_{30}(R)$.

(b) What are the relative frequencies when your results are combined with those of everyone in your class?

(c) Estimate $P(R)$.

5. Prove parts (a), (c), (e), and (f) of Theorem 2.1.

6. According to newspaper reports, Elizabeth Taylor was once charged $5000 to insure her million-dollar diamond ring for one day, while Temple University was charged $3000 to insure its multi million-dollar buildings during a two-day Black Panther convention. Can you explain why the first premium was higher?

7. What is your estimate of the probability that the next president will be a Republican? Compare your answer with that of your classmates.

8. The frequencies with which the letters of any written language occur are always about the same in any large sample. Here is a list of the percent frequencies of the letters in English:

Letters	% Frequency	Letters	% Frequency	Letters	% Frequency
E	13.0	S,H	6.0	W,G,B	1.5
T	9.0	D	4.0	V	1.0
A,O	8.0	L	3.5	K,X,J	0.5
N	7.0	C,U,M	3.0	Q,Z	0.2
I,R	6.5	F,P,Y	2.0		

(a) What are the frequencies of the letters in the following coded sentences?

ADDQ WDLT. ZDBAM WDLQ FCTUBND CU UWD
EXGCL UWDCUDE. MDU ADHO UX TXLBFD.

(b) Can you decode the message? (Each letter in the code stands for a different English letter.)

9. Flip a coin until a run of three heads appears. Then record whether the next toss is a head (H) or tail (T). Repeat this experiment twenty times.
   (a) Calculate $f_{10}(H), f_{15}(H), f_{20}(H), f_{10}(T), f_{15}(T)$, and $f_{20}(T)$.
   (b) Combine your results with those of your classmates.
   (c) Estimate $P(H)$ and $P(T)$.

2.4 LAPLACE'S DEFINITION OF PROBABILITY

Two definitions of probability were given in Section 2.3, a subjective definition and an empirical one. In this section we will study a third definition of probability. Like the other two definitions of probability, this definition predicts the relative frequency of events. It was first precisely stated by Pierre Simon de Laplace (1749–1827).

Definition 2.4 Let Ω be the sample space of a random experiment and let A be an event. If—

(a) $n(\Omega)$ is finite, that is, if Ω has a finite number of elements, and

(b) the outcomes are equally likely,

then the **probability** of A, denoted $P(A)$, is defined by

$$P(A) = \frac{n(A)}{n(\Omega)}$$

Since $n(A)$ is the number of cases favorable to A, while $n(\Omega)$ is the total number of cases possible, Definition 2.4 is often written as follows:

Definition 2.5 Let A be an event. If—

(a) the total number of cases is finite, and

(b) the outcomes are equally likely,

then the **probability** of A, denoted $P(A)$, is defined as follows:

$$P(A) = \frac{\text{number of favorable cases (ways that } A \text{ can occur)}}{\text{total number of cases}}$$

Notice that Definition 2.4 can only be used if $n(\Omega)$ is *finite*; that is, only if we can count the number of elements in Ω. This means that Definition 2.4 can be used if the outcomes are equally likely and Ω has 5 or 20, or 1,000,000 elements, but it cannot be used if Ω is infinite (e.g., if Ω is the set of all integers or the set of all numbers).

Most of the terms used in Definition 2.4 are well-defined. We now know what a random experiment is and what an event is. However, we still do not know what "equally likely" means. Intuitively, we might say that two events are equally likely if their probabilities are the same. However, this definition involves circular reasoning: we need the concept of "equally likely" to define probability, but we also need the concept of probability to define "equally likely." This is obviously unsatisfactory. In this book, we will therefore have to let "equally likely" be an undefined term. The following discussion will clarify what this term means intuitively.

Suppose that we toss a particular coin 1000 times. If the events "heads" and "tails" are equally likely, we expect *approximately* 500 heads and 500 tails. Unfortunately, even if we obtain 500 heads and 500 tails, we cannot be absolutely certain that the events "heads" and "tails" are equally likely for that coin. It may still be true that the coin is slightly biased in favor of one side or the other. Moreover, the fact that we obtain 450 heads and 550 tails does not prove that the coin is biased in favor of tails. Even if the events "heads" and "tails" are equally likely for that coin, we will sometimes get more heads than tails, and at other times more tails than heads.

How, then, can we ever say that heads and tails are equally likely outcomes of tossing a coin? The solution of this problem is to base our thinking, not on any particular coin, but on an ideal coin. We will *assume* (unless otherwise stated) that the coin under discussion is perfectly balanced. Such coins are called *fair*; others are called *biased*.

The concept of an ideal coin need not worry us, because we are all already familiar with idealized situations in mathematics. In geometry, for example, we deal with ideal points. No matter how small the dot representing a point is drawn, it is still larger than the ideal point. Similarly, no silo can be made in the shape of a perfect cylinder, but the formula for the volume of a perfect cylinder will allow us to calculate how much wheat an imperfectly shaped silo will hold. Although probability formulas also depend on ideal concepts, insurance companies have made fortunes by applying them to the real world.

Hence, unless we state that the coin being discussed is biased, we will *assume* that obtaining heads and obtaining tails are equally likely events. Similarly, if a die is tossed, we will assume that it is an ideal die; that is, that all of the six outcomes are equally likely.

In practical situations, however, we certainly cannot assume that a particular coin or die is fair. This assumption must be tested by actually tossing the coin or die a large number of times. As we have already mentioned, no matter what the results are, we can never be absolutely certain that the object is fair or is biased. However, a statistical analysis can ascertain, for example, that the *probability* that the coin is fair is at least 95 percent. One can then decide whether or not this probability is high enough to accept the hypothesis that the coin is fair. This decision will, of course, depend upon the accuracy required in our particular situation.

Although we have agreed to assume (unless otherwise stated) that our coins and our dice are fair, the reader should remember that not all sets of outcomes are equally likely. Suppose a manufacturer packages bolts in thousand-lot cases, and suppose that a case is selected at random and the defective bolts in the case are counted. Then

$$\Omega = [0,1,2,3\ldots, 1000\}$$

is a relevant sample space, but hopefully the elementary events are not equally likely.

The following examples will show how to apply Definition 2.4.

Example 9   A die is tossed once. A sample space in which the elementary events are equally likely is $\Omega = \{1,2,3,4,5,6\}$. Calculate the probability that

(a) a 2 occurs (event T)
(b) an odd number occurs (event O)
(c) a number larger than 3 occurs (event L)
(d) an 8 occurs (event E).

Solution

(a) Since the problem does not state that the die is biased, we will assume—as discussed on page 44—that the die is fair. Therefore, Definition 2.4 can be used because the elementary events are equally likely and $n(\Omega)$ is finite. Applying Definition 2.4 yields

$$P(T) = \frac{n(T)}{n(\Omega)} = \frac{1}{6}$$

because $T = \{2\}$ has 1 element and Ω has 6 elements.

(b) Since $O = \{1,3,5\}$, it follows that

$$P(O) = \frac{n(O)}{n(\Omega)} = \frac{3}{6} = \frac{1}{2}$$

(c) Since $L = \{4,5,6\}$,

$$P(L) = \frac{3}{6} = \frac{1}{2}$$

(d) Since $E = \varnothing$, $n(E) = n(\varnothing) = 0$. Therefore,

$$P(E) = \frac{0}{6} = 0$$

Example 10   A red die and a blue die are tossed.

(a) The sum of the spots on the upturned faces is recorded. A suitable sample space is $\Omega_1 = \{2,3,4,\ldots,12\}$.
(b) The number of spots on the red die and the blue die are noted separately. A suitable sample space is $\Omega_2 = \{(x,y)\,|\,x,y = 1,2,\ldots,6\}$ where, for example, the ordered pair $(2,4)$ means that a 2 occurred on the red die and a 4 on the blue die. The elements of Ω_2 are listed in Table 2.4. Are the elementary events of Ω_1 and Ω_2 equally likely?

Solution

(a) The elementary events of Ω_1 are not equally likely. A 2 can be obtained only if a 1 appears on each die. A 7, however, can be obtained in six ways: a 1 on the red die and a 6 on the blue; a 2 on the red die and a 5 on the blue; and so on. It is intuitively clear, therefore, that a 7 is much more likely to occur than a 2. The reader can check this conclusion empirically by working Exercise 1 at the end of this section.
(b) The elementary events of Ω_2 are equally likely.

Table 2.4

Number on the Red Die	Number on the Blue Die					
	1	2	3	4	5	6
1	(1,1)	(1,2)	(1,3)	(1,4)	(1,5)	(1,6)
2	(2,1)	(2,2)	(2,3)	(2,4)	(2,5)	(2,6)
3	(3,1)	(3,2)	(3,3)	(3,4)	(3,5)	(3,6)
4	(4,1)	(4,2)	(4,3)	(4,4)	(4,5)	(4,6)
5	(5,1)	(5,2)	(5,3)	(5,4)	(5,5)	(5,6)
6	(6,1)	(6,2)	(6,3)	(6,4)	(6,5)	(6,6)

Example 11   Two dice are tossed. Find the probability that the sum of the numbers showing on the dice is—

(a) 2 (event T)   **(b)** 7 (event S)   **(c)** an odd number (event A)

Solution   We can imagine that one of the dice is painted red and the other is painted blue. Then the two sample spaces Ω_1 and Ω_2 of Example 10 can be associated with this experiment. However, if we wish to find the probability of an event by applying Definition 2.4, we cannot use Ω_1 because the elementary events of this sample space are not equally likely. We can, however, use Ω_2.

By simply counting the number of elements listed in Table 2.4, we see that $n(\Omega_2) = 36$.

(a) Since $T = \{(1,1)\}$, $n(T) = 1$, and

$$P(T) = \frac{n(T)}{n(\Omega_2)} = \frac{1}{36}$$

(b) Since $S = \{(1,6), (2,5), (3,4), (4,3), (5,2), (6,1)\}$, $P(S) = \frac{6}{36}$.
(c) Similarly, $P(A) = \frac{18}{36}$.

In Section 2.3, we learned that the information needed to calculate the empirical probability of an event is sometimes given as a percentage. This is also true for the information needed to apply Laplace's definition of probability. Suppose, for example, that we wish to know the probability that a picture taken by Bob is out of focus. If we are told that 30 percent of the pictures that Bob has taken are out of focus, we do not know whether 30 out of 100 pictures are out of focus, whether 60 out of 200 pictures are out of focus, or even how many pictures were taken. We do, however, know that the ratio of favorable cases (out-of-focus pictures) to the total number of cases (total number of pictures) is 30 out of 100, or $\frac{30}{100} = 0.3$. Applying Laplace's definition of probability, it then follows that the probability that a picture taken by Bob is out of focus is 0.3.

In general, Laplace's definition of probability can be applied if we know the *percentage of favorable cases.* For example, if 25 percent of all students at a certain school are boys, then the probability that a randomly chosen student is a boy is 0.25; if 42 percent of all animals entered in a children's pet show are dogs and the winner is randomly chosen, then the probability that the winner is a dog is 0.42.

Theorem 2.2 lists some of the properties of the probability P defined by Laplace. These properties are similar to the properties of the relative frequency f_n listed in Theorem 2.1.

THEOREM 2.2   Let A and B be events. The probability P defined in Definition 2.4 has the following properties:

      (a) $0 \leq P(A) \leq 1$

      (b) $P(\Omega) = 1$

      (c) $P(A + B) = P(A) + P(B)$ if A and B are disjoint

      (d) $P(\varnothing) = 0$

      (e)   If $A \subseteq B$, then $P(A) \leq P(B)$.

      (f) $P(A') = 1 - P(A)$

Proof   We will prove (b) and (e); the proofs of the remaining parts of the theorem will be left for the exercises.

(b)

$$P(\Omega) = \frac{n(\Omega)}{n(\Omega)} = 1$$

(e)   If $A \subseteq B$, then every element of A is an element of B. It follows that $n(A) \leq n(B)$. Since $n(\Omega) > 0$,

$$P(A) = \frac{n(A)}{n(\Omega)} \leq \frac{n(B)}{n(\Omega)} = P(B)$$

The following examples illustrate the usefulness of Theorem 2.2.

Example 12   Two dice are tossed (see Example 11). Find the probability that the sum of the numbers showing on the two dice is—

(a) an even number (event E)

(b) a 2 or a 7 (event A)

(c) at least 3 (event L)

(d) at least 3 or an even number (event B)

Solution

(a) Using the sets defined in Example 11, $E = A'$. It therefore follows from Theorem 2.2 (f) and Example 11 (c) that

$$P(E) = P(A') = 1 - P(A) = 1 - \tfrac{18}{36} = \tfrac{1}{2}$$

(b) Since $A = T + S$, it follows from Theorem 2.2(c) and Example 11 that

$$P(A) = P(T + S) = P(T) + P(S) = \tfrac{1}{36} + \tfrac{6}{36} = \tfrac{7}{36}$$

(c) Since $L = T'$, it follows from Theorem 2.2 (f) and Example 11 that

$$P(L) = P(T') = 1 - P(T) = 1 - \tfrac{1}{36} = \tfrac{35}{36}$$

(d) The events E and L are not disjoint. Therefore, Theorem 2.2(c) cannot be applied. However, since $B = E \cup L = \Omega$, it follows from Theorem 2.2(b) that $P(B) = P(\Omega) = 1$.

Example 13 Most roulette wheels in the United States have compartments numbered 1–36 inclusive and two more that are blank. The blanks are not considered either even or odd. The wheel is turned; the ball spins around and lands on one of the numbers or on one of the blanks. Find the probability that the ball—

(a) lands on a 4 (event F)
(b) does not land on a 4 (event N)
(c) lands on a blank (event B)
(d) lands on a 4 or on a blank (event S)
(e) lands on a 37 (event T)
(f) lands on an even number, an odd number, or a blank (event A).

Solution We assume that the wheel is not rigged; that is, the ball is just as likely to land on any particular compartment as on any other. Hence, the elementary events in the sample space

$$\Omega = \{b_1, b_2, 1, 2, 3, \ldots, 36\}$$

are equally likely, and so Laplace's definition can be used. (The symbols b_1 and b_2 refer to the first and second blanks, respectively.) Applying Definition 2.4 and Theorem 2.2, we obtain the following probabilities.

(a) $P(F) = \frac{1}{38}$
(b) $P(N) = P(F') = 1 - P(F) = 1 - \frac{1}{38} = \frac{37}{38}$
(c) $P(B) = \frac{2}{38}$
(d) $P(S) = P(F + B) = P(F) + P(B) = \frac{1}{38} + \frac{2}{38} = \frac{3}{38}$
(e) $P(T) = P(\varnothing) = 0$
(f) $P(A) = P(\Omega) = 1$

Three different definitions of probability have been studied: Laplace's, empirical, and subjective. In order to apply Laplace's definition, we must assume that the elementary events are equally likely. For example, we must assume that the coin is fair or that the die is unbiased. We repeat that in practical situations this assumption must be checked statistically by actually tossing the coin, rolling the die, or repeating the experiment a large number of times. If the elementary events are equally likely, then the empirical probability and the probability obtained by using Laplace's definition will be the same (see Exercise 1 following this section). Both will predict the relative frequencies of events.

The third definition of probability, subjective probability, is only used when the other two definitions cannot be applied. This is because it is only an educated guess at the relative frequency of an event, and so may vary from person to person. However, it can be an invaluable tool. For example, if scientists someday agree on subjective probabilities for life on the various planets, these probabilities, combined with factors of distance, cost, environment, and so on, could affect the goals and design of our space program.

2.4 EXERCISES

1. Toss two dice thirty-six times and count the number of times a total of i dots appears on both dice ($i = 2,3,\ldots,12$). Combine your results with those of the other members of your class. On the basis of your experience, are these events equally likely? Which is most likely? Which is least likely? How do these results compare with the results predicted by Laplace's definition?

2. **(a)** Which of the sample spaces of Example 2 (page 30) have equally likely elementary events?
    (b) To which sample spaces of Example 2 can Laplace's definition of probability be applied?

3. Can Laplace's definition be applied to the sample spaces of Examples 3 and 4, pages 30–31?

4. Three coins are tossed.
    (a) Find a sample space for this experiment in which the elementary events are *not* equally likely.
    (b) Find a sample space for this experiment in which the elementary events *are* equally likely.
    (c) What is the probability of obtaining—
      (*i*)   exactly one head
      (*ii*) no heads
      (*iii*) no tails
      (*iv*) at least four heads
      (*v*)   two or more tails?
    (d) If the three coins are tossed 2400 times, on approximately how many tosses will exactly one head occur?

5. A card is drawn from a well-shuffled deck. Find the probability of drawing—
    (a) a spade
    (b) a spade or a king
    (c) a spade or a diamond.

6. A doctor once said to his patient: "You have a dangerous disease. In fact, the probability of surviving it is one-tenth. However, you needn't worry. You're very lucky to have come to me. I've just treated nine people with this disease and they all died." Would you have felt comforted? Why?

7. Prove parts (a), (c), and (f) of Theorem 2.2. (See page 47.)

8. Compare Theorems 2.1 and 2.2, pages 36 and 47. In what way is the probability $P(A)$ defined by Laplace the same as the relative frequency $f_n(A)$? How are they different?

9. A coin is tossed three times.
    (a) Give a sample space in which the elementary events are equally likely.

(b) What is the probability that—
    (*i*) the first toss is a head
    (*ii*) the second toss is a head
    (*iii*) at least one toss is a head
    (*iv*) exactly one toss is a head?

10. An urn contains two red, three black, and four green balls. If a ball is randomly chosen, find the probability that it is—
    (a) red   **(b)** black   **(c)** red or black   **(d)** not red

11. A firm has 600 male and 400 female employees. Sixty percent of the men drive American-made cars, 30 percent drive Japanese-made cars, and 10 percent drive European-made cars. Seventy percent of the women drive American-made cars, 25 percent drive Japanese-made cars, and 5 percent drive European-made cars. If an employee is chosen at random, find the probability that the employee—
    (a) drives a European-made car
    (b) is male
    (c) is male and drives a Japanese-made car
    (d) does not drive a Japanese-made car
    (e) is male or drives a European-made car (or both)
    (f) drives a European-made car or a Japanese-made car

2.5 GENETICS

Gregor Mendel (1822–84), one of the founders of the theory of genetics, studied pairs A–B of contrasting properties of peas (e.g., green–yellow, tall–short, etc.). Mendel found that, when a green pea was crossed with a yellow pea, the resulting offspring were green. In general, when peas with property A were crossed with peas of the contrasting property B, all of the hybrid peas had property A, or all had property B. The property that the hybrid peas had was named the *dominant* property; the contrasting property was called the *recessive* property. When the hybrid peas were crossed, their offspring had the properties shown in Table 2.5.

Table 2.5

Contrasting Properties	Dominant		Recessive		Total	Percent Recessive
Seed form	round	5474	wrinkled	1850	7324	25.26
Color of reserve material	yellow	6022	green	2001	8023	24.94
Form of seed coat	inflated	882	wrinkled	299	1181	25.31
Color of seed coat	gray	705	white	224	929	24.11
Color of unripe pods	green	428	yellow	152	580	26.21
Position of flower	axial	651	terminal	207	858	24.13
Length of stem	tall	787	dwarf	277	1064	26.03
		14,949		5010	19,959	
		(74.90%)		(25.10%)		

Adapted From "Mendelism," 1953.

Mendel naturally wondered why, in every case, approximately 25 percent of the hybrid peas had the recessive characteristics. He theorized that there are factors (now called *genes*) that determine each inheritable characteristic. In higher organisms, certain traits called *simple Mendelian characters* are determined by a single pair of genes (one from each parent). This pair is called the *genotype* for that characteristic.

Since the pea pods studied by Mendel were either yellow or green, Mendel theorized that there are two different kinds of genes affecting color, G and y. (The capital G signifies that green is the dominant color.) Three genotypes are therefore possible: GG, yy, and $Gy = yG$.[3] He further theorized that both the pure green (GG) and hybrid green (Gy or yG) plants would *look* green. However, unlike the pure green plant, the hybrid green plant could have yellow offspring. The only plants that would look yellow are those with genotype yy.

Theoretically, if a pure green plant (GG) is crossed with a pure yellow plant (yy), each offspring inherits a G gene from the pure green parent and a y gene from the other parent. Hence, all offspring have genotype Gy and so appear green. This, of course, agrees with Mendel's observations.

The cross of the pure green and pure yellow plants is illustrated in the diagram below. The possible genetic contribution of the female parent appears across the top of the diagram, while that of the male parent appears along the left side. The possible genotypes of their offspring appear in the interior.

	G	G
y	yG	yG
y	yG	yG

When members of the hybrid generation are crossed, the theoretical genotypes of the offspring are given in the interior of the diagram below.

	G	y
G	GG	Gy
y	yG	yy

Mendel theorized not only that the three genotypes should appear, but also that they should appear in the proportions indicated in the last diagram; that is, one-fourth should be pure green (GG), one-half should be hybrid green (Gy), and one-fourth should be pure yellow (yy). Judging from appearances, one-fourth of the plants should be yellow. This agrees with Mendel's experimental results. Succeeding generations were crossed and the observed results again approximated the theoretical ones.

If we assume that the four cases in the preceding experiment (GG, Gy, yG, yy) are all equally likely (and this assumption is justified by experience), we can apply Laplace's definition of probability. Hence, if two hybrid plants are crossed, the probability that a particular offspring is yellow is $\frac{1}{4}$. The probability that it is green is, of course, $\frac{3}{4}$.

---

3. Both genotypes Gy and yG mean that a gene for green was inherited from one parent and a gene for yellow was inherited from the other parent. Since we do not care which parent gave which gene, we say that the two genotypes Gy and yG are equal.

Example 14   Wavy hair in mice is a recessive characteristic; straight hair is dominant. If a wavy-haired mouse (*ww*) is mated with a hybrid straight-haired mouse (*Sw*), find the probability that their first offspring—

(a) has wavy hair (event *A*)
(b) has straight hair (event *B*)
(c) has genotype *SS* (event *C*).

Solution   The possible genotypes of the offspring are found in the interior of the following diagram:

	w	*w*
S	*Sw*	*Sw*
w	*ww*	*ww*

(a) The only mice with wavy hair are those with genotype *ww*. Since there are four equally likely genotypes and two favor wavy hair, $P(A) = \frac{2}{4}$.
(b) Since straight hair is dominant, a mouse will have straight hair if it has either genotype *Sw* or genotype *SS*. Since there are two favorable cases and four possible cases, $P(B) = \frac{2}{4}$.
(c) This genotype is impossible. Hence, $P(C) = P(\varnothing) = 0$.

In some cases, neither characteristic is dominant. If red shorthorn cattle (genotype *RR*) are crossed with white shorthorn cattle (genotype *WW*), the offspring are neither red nor white but roan (genotype *WR*). That is, the offsprings' coats have a red base color lightened by an admixture of white hairs.

Example 15   If a red shorthorn is crossed with a roan, what is the probability that their second offspring is—

(a) red (event *R*)
(b) white (event *W*)
(c) roan (event *N*)?

Solution   The possible genotypes of each offspring are given in the diagram below. It does not matter whether the offspring is the first, the second, or any other.

	R	*R*
R	*RR*	*RR*
W	*WR*	*WR*

Using the above diagram, the probabilities are as follows:

(a) $P(R) = \frac{2}{4}$
(b) $P(W) = 0$
(c) $P(N) = \frac{2}{4}$

Thus far, we have mentioned only the role played by genes in inheritance. The genes, however, are part of a structure called the *chromosome*. Each species has a certain number of chromosome pairs in each nonsex cell. Humans, for example, have twenty-three pairs. Like the genes, one chromosome from each pair is inherited from each parent. All the genes on the same chromosome are said to be *linked*. If an offspring inherits a chromosome from its father then, in general (there are exceptions to this rule), it inherits all the genes on that chromosome. In mammals such as man, sex is determined by a pair of chromosomes called the *sex chromosomes*. Females have two similar chromosomes each denoted X, while males have an X chromosome and a smaller Y chromosome.[4] The genes on the X chromosome and the traits they determine are called *sex-linked*. If X^a and X^b denote two different genes for a sex-linked trait, then three different genotypes are possible for a female: X^aX^a, X^aX^b, and X^bX^b. Since the sex-linked genes are carried by the X chromosome only, the only two genotypes possible for males are X^aY and X^bY. If the a trait is recessive, it will appear in females whose genotype is X^aX^a and in males whose genotype is X^aY. If the b trait is dominant, it will appear in females of genotypes X^aX^b and X^bX^b and in men of genotype X^bY.

Example 16 Hemophilia is a recessive sex-linked disease. Let h and N denote the hemophilic and normal genes, respectively. The following genotypes are possible:

Male		*Female*	
X^hY	hemophilic man	X^hX^h	hemophilic woman
X^NY	normal man	X^hX^N	carrier
		X^NX^N	completely normal woman

A *carrier* is a person who does not have the disease herself but can transmit it to her children; she appears normal. A completely normal woman is one who does not have the disease and is not a carrier.

Example 17 If an apparently normal couple have a hemophilic son, what is the probability of each of the following events?

$A =$ Their next child will be hemophilic

$B =$ Their next son will have hemophilia

$C =$ Their next daughter will be a carrier

$D =$ Their next child is completely normal

Solution Since the father appears normal, his genotype must be X^NY. Since the mother appears normal, her genotype is either X^hX^N or X^NX^N. However, since

---

4. This, like many other items in this section, is a simplification. Exceptions exist to almost every genetic rule. Actually, some men, called *supermales*, have the chromosome pattern XYY. Some people are even $XXXXY$, XXX, $XXXYY$, etc. Chromosome patterns such as these are often associated with severe medical problems such as leukemia, mental retardation, schizophrenia, and congenital heart disease.

her son is hemophilic, her genotype must be $X^h X^N$. The possible genotypes for their next offspring are given in the interior of the diagram below.

	X^h	X^N
X^N	$X^N X^h$	$X^N X^N$
Y	$X^h Y$	$X^N Y$

From the diagram, we see that $P(A) = \frac{1}{4}$ and $P(D) = \frac{2}{4}$. There are only two possibilities for the genotype of their next son, $X^h Y$ and $X^N Y$. Of these two genotypes, one favors hemophilia. It follows that $P(B) = \frac{1}{2}$. Similarly, $P(C) = \frac{1}{2}$.

A couple can use the probability that they will have a normal child, as determined by a genetic counselor, to decide whether or not to adopt rather than have a natural child.

2.5 EXERCISES

1. Huntington's chorea, a nervous disorder, is transmitted by a dominant gene C. Suppose a man has genotype nC where n denotes the normal gene. Find the probability that his next child will have the disease if—
    (a) his wife has genotype nC
    (b) his wife has genotype nn.

2. If a hybrid green plant (genotype Gy) is crossed with a pure yellow plant (genotype yy), what is the probability that a particular seed will produce a yellow plant?

3. Cystic fibrosis results from the presence of two recessive genes. If the parents are apparently normal, but each has a single gene for the disease, what is the probability that their first child will (a) have the disease, (b) be a carrier for the disease?

4. In a certain plant, tall is dominant over short. If two tall plants were crossed and one-third of the offspring were short, what was the genetic make-up of the parents? Why?

5. Albinism in human beings is a simple Mendelian character with nonalbinism being dominant. If two apparently normal parents have an albino child—
    (a) what are the genotypes of the parents
    (b) what is the probability that their next child will be albino?

6. Ignoring factors in the prenatal environment, who determines the sex of the child, the father or the mother?

7. A certain plant may have red (genotype RR), pink (genotype RW), or white (genotype WW) flowers. If two pink plants are crossed, what is the probability that a particular seed will produce (a) a red plant, (b) a white plant, (c) a pink plant? (Assume the seed actually grows and produces flowers.)

8. Certain traits that are not sex-linked are *sex-influenced*—that is, due to the presence of sex hormones in the internal environment, the trait is dominant in one sex but recessive in the other. Pattern baldness, for example, is *dominant in men*, but *recessive in women.*

    (a) If a woman who has pattern baldness marries a man with one gene for pattern baldness, what is the probability that—

    (*i*) their first son will have pattern baldness

    (*ii*) their first daughter will have pattern baldness

    (*iii*) their first daughter will be a carrier of pattern baldness?

    (b) Redo the above probabilities under the assumption that the husband has *two* genes for pattern baldness.

9. A woman has a brother who has pseudohypertrophic muscular dystrophy, a recessive sex-linked disease. She and both of her parents appear normal. What is the probability that she is a carrier? (Hint: See Example 16.)

10. A woman is a carrier of common red–green colorblindness, a recessive sex-linked disease. If she marries a normal man, what is the probability that—

    (a) their first son will be color-blind

    (b) their second daughter will be color-blind?

11. Redo Exercise 10 under the assumption that the woman marries a red–green color-blind man.

12. A woman has Aldrich's syndrome, a recessive sex-linked disease. What are the genotypes of her parents?

13. If a hemophilic man marries a woman who is a carrier, what is the probability that their first child is—

    (a) a boy who is hemophilic

    (b) a girl who is a carrier

    (c) hemophilic

    (d) completely normal (neither hemophilic nor a carrier)?

    (See Example 16.)

2.6 PROBABILITY THEOREMS

In the preceding sections we discussed three definitions of probability: Laplace's definition, empirical probability, and subjective probability. We have seen that each of these definitions of probability has the characteristics listed in the following theorem.

THEOREM 2.3   Let A and B be events. The three definitions of probability have the following properties:

(a) $0 \leq P(A) \leq 1$

(b) $P(\Omega) = 1$

(c) $P(A + B) = P(A) + P(B)$ if A and B are mutually exclusive

(d) $P(\varnothing) = 0$

(e) If $A \subseteq B$, then $P(A) \leq P(B)$.

(f) $P(A') = 1 - P(A)$

In this section we will use these properties to prove additional theorems. Since properties (a)–(f) are true for each definition of probability, the validity of the new theorems will in no way depend on which definition of probability we used.

Theorem 2.4 extends property (c) of probability to three mutually exclusive events.

THEOREM 2.4 Let A, B, and C be any three mutually exclusive events. Then

$$P(A + B + C) = P(A) + P(B) + P(C)$$

Proof Let $D = A + B$. Then C and D are mutually exclusive events. (See Figure 2.2.) It follows from Theorem 2.3(c) that

$$P(A + B + C) = P((A + B) + C) = P(D + C)$$
$$= P(D) + P(C) = P(A + B) + P(C)$$
$$= P(A) + P(B) + P(C)$$

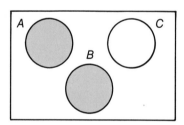

Figure 2.2 Set D is shaded

Theorem 2.4 can be similarly extended to four mutually exclusive events, five mutually exclusive events, and so on. In particular,

$$P(A + B + C + D) = P(A) + P(B) + P(C) + P(D)$$

and

$$P(A + B + C + D + E) = P(A) + P(B) + P(C) + P(D) + P(E)$$

where A, B, C, D, and E are mutually exclusive.

Notice that Theorem 2.3(c) and Theorem 2.4 hold only if the events being considered are mutually exclusive. Theorems 2.5 and 2.6 can be applied to any events.

THEOREM 2.5 Let A and B be any events. Then

$$P(A \cup B) = P(A) + P(B) - P(A \cap B)$$

Proof Notice that

$$A \cup B = (A - B) + (B - A) + (A \cap B)$$
$$A = (A - B) + (A \cap B)$$

and

$$B = (B - A) + (A \cap B)$$

Therefore,

$$P(A \cup B) = P(A - B) + P(B - A) + P(A \cap B) \tag{1}$$
$$P(A) = P(A - B) + P(A \cap B) \tag{2}$$
$$P(B) = P(B - A) + P(A \cap B) \tag{3}$$

Subtracting equations (2) and (3) from equation (1) yields

$$P(A \cup B) - P(A) - P(B) = -P(A \cap B) \tag{4}$$

If $P(A) + P(B)$ is added to both sides of equation (4), the theorem follows.

THEOREM 2.6 If A, B, and C are any events, then

$$P(A \cup B \cup C) = P(A) + P(B) + P(C) - P(A \cap B)$$
$$- P(A \cap C) - P(B \cap C) + P(A \cap B \cap C)$$

The proof of Theorem 2.6 will be left for the exercises.
Let $n(X)$ be the number of elements in a set X. In Chapter 1 we learned that

$$n(A \cup B) = n(A) + n(B) - n(A \cap B)$$
$$n(A + B) = n(A) + n(B) \quad \text{if } A \text{ and } B \text{ are mutually exclusive}$$

Theorems 2.3(c) and 2.5 now tell us that similar laws hold for probabilities; that is,

$$P(A \cup B) = P(A) + P(B) - P(A \cap B)$$
$$P(A + B) = P(A) + P(B) \quad \text{if } A \text{ and } B \text{ are mutually exclusive}$$

Examples 18–23 illustrate how the preceding theorems can be applied.

Example 18 A teacher has found that the following table gives the probability that x students will be absent on any given day.

Number of students absent	0	1	2	3	4	>4
Probability	0.40	0.40	0.10	0.05	0.05	0

For example, the probability that zero students will be absent tomorrow is 0.4; the probability that two students will be absent is 0.1; and so on. Find the probability that—

(a) exactly two students will be absent (event A)
(b) at least two students will be absent (event B)
(c) at most one student will be absent (event C).

Solution   Let A_i be the event "exactly i students are absent." Using the table and Theorem 2.4—

(a) $P(A) = P(A_2) = 0.1$
(b) $P(B) = P(A_2) + P(A_3) + P(A_4) = 0.1 + 0.05 + 0.05 = 0.2$
(c) $P(C) = P(A_0) + P(A_1) = 0.4 + 0.4 = 0.8$

Example 19   What is the probability that the class of Example 18 will have exactly six absences tomorrow?

Solution   Let M be the event "there are more than four absences" and A_6 be the event "there are exactly six absences." Then $A_6 \subseteq M$, and $P(M) = 0$ from the chart in Example 18. Parts **(e)** and **(a)** of Theorem 2.3 imply that

$$P(A_6) \le P(M) = 0 \qquad \text{and} \qquad 0 \le P(A_6)$$

It follows that

$$0 \le P(A_6) \le 0$$

and so $P(A_6) = 0$.

Example 20   Al, Bob, and Carl are the only contestants in a swimming race. If you decide that Al and Bob are evenly matched, but Carl is twice as likely to win as Bob, what is the probability that—
(a) Bob wins
(b) either Al or Carl wins?

Solution

(a) Let A, B, C be the events "Al wins the race," "Bob wins the race," "Carl wins the race," respectively. Since Al and Bob are evenly matched, $P(A) = P(B)$; since Carl is twice as likely to win as Bob, $P(C) = 2[P(B)]$; and since there are no other people in the race, $A + B + C = \Omega$. It therefore follows from Theorem 2.4 and Theorem 2.3 **(b)** that

$$1 = P(\Omega) = P(A + B + C) = P(A) + P(B) + P(C)$$
$$= P(B) + P(B) + 2[P(B)] = 4[P(B)]$$

and so $P(B) = \frac{1}{4}$.

(b) Since $A + C = B'$, it follows from Theorem 2.3 **(f)** that

$$P(A + C) = P(B') = 1 - P(B) = 1 - \tfrac{1}{4} = \tfrac{3}{4}$$

Example 21   Forty percent of the members of a certain club are doctors (D), 30 percent are men (M), 30 percent are women doctors, and every member is either a doctor, a lawyer (L), or a teacher (T). If a member is randomly chosen, what is the probability that the member is—

(a) a woman (W)

(b) a woman or a doctor

(c) a lawyer or a teacher?

Solution   It follows from the given information that $P(D) = 0.4$, $P(M) = 0.3$, $P(W \cap D) = 0.3$, and $\Omega = L + D + T$.

(a) Using Theorem 2.3 (f),

$$P(W) = P(M') = 1 - P(M) = 1 - 0.3 = 0.7$$

(b) Using Theorem 2.5 and part (a),

$$P(W \cup D) = P(W) + P(D) - P(W \cap D) = 0.7 + 0.4 - 0.3 = 0.8$$

(c) Since $\Omega = L + T + D$, it follows that $L + T = D'$. Hence,

$$P(L + T) = P(D') = 1 - P(D) = 1 - 0.4 = 0.6$$

In some problems, such as those in the next two examples, the application of Theorems 2.3–2.6 can be simplified by using Venn diagrams.

Example 22   Thirty percent of the students in a certain school take algebra (A), 40 percent take biology (B), and 10 percent take both. Find (a) $P(A \cup B)$, (b) $P(A')$, (c) $P(A - B)$.

Solution   This problem may be solved by the methods used in the previous examples. It may also be solved by first drawing a Venn diagram like the one shown in Figure 2.3.

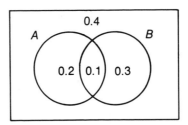

Figure 2.3

It follows from the given information that $P(A) = 0.3$, $P(B) = 0.4$, and $P(A \cap B) = 0.1$. We may illustrate the fact that $P(A \cap B) = 0.1$ by placing 0.1 on the set $A \cap B$ in Figure 2.3. Since $P(A) = 0.3$, and $P(B) = 0.4$, it is now clear from the diagram that

$$P(A \cap B') = 0.3 - 0.1 = 0.2 \quad \text{and} \quad P(B \cap A') = 0.4 - 0.1 = 0.3$$

These probabilities are also indicated on Figure 2.3. Since $P(\Omega) = 1$, it is now clear that

$$P(B' \cap A') = 1 - [0.2 + 0.1 + 0.3] = 0.4$$

Using Figure 2.3, Theorem 2.3(c), and Theorem 2.4, we see that—

(a) $P(A \cup B) = 0.2 + 0.1 + 0.3 = 0.6$
(b) $P(A') = 0.3 + 0.4 = 0.7$
(c) $P(A - B) = 0.2$

Notice that in the preceding example we labeled $A \cap B$ with its corresponding probability before labeling the remaining regions of the Venn diagram. This is the only way to begin solving this problem. We cannot begin, for example, by labeling A with $P(A) = 0.3$, because some portion of this 0.3 is the probability of $A \cap B$, and some is the probability of $A \cap B'$. Unless we first determine what these two portions are, we cannot unambiguously label the regions of the Venn diagram.

Example 23   A survey of a large number of adults living in a certain town resulted in the following statistics: 59 percent are men (M); 20 percent are Republicans (R); 61 percent are over 40 (A); and 14 percent are Republican men. Nineteen percent are men who are not over 40; 11 percent are Republicans who are over 40; and 10 percent are Republican men who are over 40. If an adult resident of the town is randomly chosen, what is the probability that this person is—

(a) a Republican who is not over 40
(b) a woman who is not over 40
(c) a woman who is not a Republican
(d) a person who is either male, or Republican, or over 40?

Solution   Recall that in Chapter 1 we learned that the notation ABC can be used to denote the intersection of three sets A, B, and C, and that AB can be used to denote the intersection of two sets A and B. Using this notation, the results of the survey can be rewritten as follows:

$$P(M) = 0.59, \quad P(R) = 0.20, \quad P(A) = 0.61, \quad P(RM) = 0.14$$
$$P(MA') = 0.19, \quad P(RA) = 0.11, \quad P(RMA) = 0.10$$

Figure 2.4 shows the Venn diagram for the three sets A, M, and R. The fact that $P(MRA) = 0.1$ is illustrated by placing 0.10 on the set MRA. (See Figure 2.4(a).) Since $P(RM) = 0.14$, it is now clear that $P(RMA') = 0.14 - 0.10 = 0.04$. (See Figure 2.4(b)). Proceeding in a similar manner results in the complete Venn diagram of Figure 2.4(c). It is now clear from this diagram that—

(a) $P(RA') = 0.05 + 0.04 = 0.09$
(b) $P(M'A') = 0.05 + 0.15 = 0.20$
(c) $P(M'R') = 0.20 + 0.15 = 0.35$
(d) $P(M \cup R \cup A) = 1 - 0.15 = 0.85$

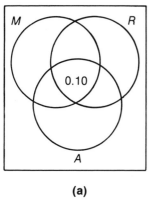
(a)

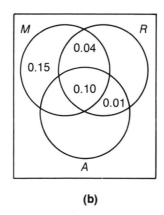
(b)

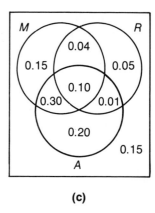
(c)

Figure 2.4

Alternatively, this problem may be solved by using the theorems of this section without using Venn diagrams. For example, since

$$R = RA + RA'$$

it follows that

$$P(R) = P(RA + RA') = P(RA) + P(RA')$$

and so

$$P(RA') = P(R) - P(RA) = 0.20 - 0.11 = 0.09$$

2.6 EXERCISES

1. If 25 percent of the students at a certain University are men, 40 percent of the students are married, and 10 percent of the students are married men, what is the probability that a randomly chosen student is either married or male?

2. In an experiment on heredity, pairs of cream-colored guinea pigs were mated, and the colors of their offspring were noted. Based on their investigation, the scientists listed the following outcomes and their probabilitities:

Color of descendent	Probability
White (W)	$\frac{1}{4}$
Cream (C)	$\frac{1}{2}$
Yellow (Y)	$\frac{1}{4}$

(a) Define a sample space for this experiment.
(b) Are the elementary outcomes equally likely?
(c) Calculate $(i)\,P(W \cup Y)$, $(ii)\,P(W')$, and $(iii)\,P(W \cap Y)$.

3. One thousand raffle tickets, each bearing a different number from 1 to 1000, are put into an urn. A ticket is then randomly selected. What is the probability that the number on the winning ticket is—
(a) even
(b) greater than 90
(c) even, or greater than 90
(d) even, and greater than 90
(e) at most 90
(f) greater than 1000?

4. Suppose $P(A) = 0.65$, $P(B) = 0.35$, $P(C) = 0.69$, $P(AB) = 0.30$, $P(BC) = 0.14$, $P(AC) = 0.40$, and $P(ABC) = 0.10$. Calculate—

(a) $P(A')$ (b) $P(A \cup B)$ (c) $P(A \cup B \cup C)$
(d) $P(\varnothing)$ (e) $P(\Omega)$

5. The probability that a store will sell N television sets during the week is given in the following table.

N	0	1	2	3	4	5
Probability	0.25	0.35	0.20	0.10	0.05	0.05

What is the probability that next week the store will sell—
(a) exactly one television set
(b) at least one television set
(c) at most one television set
(d) more than five television sets?

2.7 ODDS

In gambling one often hears the term "odds." Suppose Lucky Lou offers you odds of $r : s$ in favor of an event E. This means that—

(1) Lou will pay you r dollars if E does not occur
(2) You must pay Lou s dollars if E does occur

Clearly, certain odds will favor Lou, while others will favor you. In this section we will see what odds will make a bet *fair*. (A bet is *fair* if, unlike a bet at a gambling casino or race track, neither you nor your opponent expects to make any profit in the long run.)

Example 24 will lead us to a mathematical definition of odds. After discussing this definition, we will show that a bet offered at the odds specified by the definition is a fair bet.

Example 24   A ball is randomly chosen from an urn, its color is noted, and the ball is then replaced. After many repetitions, the observer decides that $P(R) = \frac{1}{4}$

and $P(G) = \frac{3}{4}$, where R and G are the events "a red ball is chosen," "a green ball is chosen," respectively. In the long run, what is the ratio of green balls to red balls?

Solution   If 4000 balls were picked, we would expect about $4000(\frac{3}{4}) = 3000$ green balls and about $4000(\frac{1}{4}) = 1000$ red balls, a ratio of $3 : 1$ in favor of green balls. If 40,000 balls were picked, we would expect about 30,000 green balls and 10,000 red balls. Again, the ratio is $3 : 1$. In general, if N balls are taken, we would expect about $N \cdot [P(G)]$ green balls and $N \cdot [P(R)]$ red balls. The ratio would therefore be $N[P(G)] : N[P(R)]$. Since

$$\frac{N \cdot [P(G)]}{N \cdot [P(R)]} = \frac{P(G)}{P(R)} = \frac{\frac{3}{4}}{\frac{1}{4}} = \frac{3}{1}$$

the ratio is always $3 : 1$.

It is reasonable to assume that R and G of Example 24 are the only possible outcomes, because $P(R) + P(G) = 1$. The ratio of green balls to red balls is therefore $P(G) : P(G')$. This ratio is called the *odds in favor* of G. Generalizing this idea to any event E, we arrive at the following definition.

Definition 2.6   Let E be an event whose probability is not 1. The **odds in favor** of E are $r : s$ if and only if

$$\frac{r}{s} = \frac{P(E)}{P(E')}$$

Example 25 shows how to use this definition.

Example 25   A die is tossed. What are the odds in favor of—

(a) a 3 (event T)
(b) an even number (event E)
(c) a number larger than 2 (event L)?

Solution

(a) Since $P(T) = \frac{1}{6}$, $P(T') = \frac{5}{6}$. Therefore,

$$\frac{P(T)}{P(T')} = \frac{\frac{1}{6}}{\frac{5}{6}} = \frac{1}{5}$$

It follows that the required odds are $1 : 5$.

(b) Similarly,

$$\frac{P(E)}{P(E')} = \frac{\frac{1}{2}}{\frac{1}{2}} = \frac{1}{1}$$

and so the odds in favor of E are $1 : 1$.

(c) The odds in favor of L are $2 : 1$ because $P(L) = \frac{4}{6}$.

Example 26 shows that the odds offered in a bet will be fair if and only if they are the odds of Definition 2.6.

Example 26   The probability that David hits the bull's-eye when shooting at a target is 0.4. Suppose he offers you odds of $r : s$ that he will hit the bull's-eye. Find values of r and s that will make the bet fair.

Solution   If David shoots 1000 times, you can expect him to hit approximately $(0.4)(1000)$ bull's-eyes. Each time he hits the bull's-eye you will have to pay him s dollars. It follows that your total loss will be about $L = (0.4)(1000)s$. Similarly, your total gain will be about $G = (0.6)(1000)r$.

In general, if David shoots N times, where N is some large number, he will hit approximately $(0.4)N$ bull's-eyes. Your total loss will therefore be approximately $L = (0.4)Ns$. Similarly, your total gain will be about $G = (0.6)Nr$.

If this bet is fair, your total gain G after a large number of shots should approximately equal your total loss L. That is, if the bet is fair,

$$(0.4)Ns = (0.6)Nr$$

Dividing both sides of this equation by $(0.6)Ns$, we see that the bet will be fair if and only if

$$\frac{r}{s} = \frac{0.4}{0.6} = \frac{P(B)}{P(B')}$$

where B is the event "David hits the bull's-eye." For example, the bet will be fair if $r = 4$ and $s = 6$.

Notice that the bet of Example 26 is fair if and only if the odds given are the odds of Definition 2.6. A similar argument will show that *any* bet is fair if and only if it is offered at the odds of Definition 2.6.

Definition 2.6 shows how the odds in favor of an event E can be calculated if $P(E)$ is known. Example 27 shows how $P(E)$ can be calculated if the odds in favor of E are known.

Example 27   Suppose that the odds in favor of A are $2 : 3$. Find $P(A)$.

Solution
Applying Definition 2.6,

$$\frac{P(A)}{P(A')} = \frac{2}{3}$$

Therefore,

$$3P(A) = 2P(A') = 2[1 - P(A)] = 2 - 2P(A)$$

It follows that

$$5P(A) = 2$$

and so

$$P(A) = \tfrac{2}{5}$$

The method used in Example 27 can be used to prove the next theorem.

THEOREM   2.7   If the odds in favor of the event A are $r : s$, then

$$P(A) = \frac{r}{r + s}$$

Example 28 is an application of Theorem 2.7.

Example 28   According to a certain stock-market analyst, the odds are $3 : 1$ in favor of an increase in the Dow Jones average within the next week of 20 percent or more (event D). Find $P(D)$.

Solution   Applying Theorem 2.7 with $r = 3$ and $s = 1$ yields

$$P(D) = \frac{3}{3 + 1} = \frac{3}{4}$$

2.7 EXERCISES

1.   The odds in favor of A are $3 : 5$.
     (a) What is $P(A)$?
     (b) What are the odds in favor of A'?

2.   Suppose that the probability of A was found by using Laplace's definition. Prove that the odds in favor of A are given by the ratio $n(A) : n(A')$.

3.   Joe offers odds of $3 : 1$ that he will pass math, $4 : 1$ that he will pass math or French, and $1 : 4$ that he will pass math and French. Assuming that these odds are fair, what is the probability that Joe will—
     (a) not pass math
     (b) pass math and French
     (c) pass French?

4.   When yellow mice are mated together, two-thirds of the progeny are yellow and one-third are *agouti* (i.e., they have a yellow band of pigment near the tip of the hair). What are the odds in favor of the event "the first offspring of two yellow mice is agouti"?

5.   Suppose that $P(A) = 0.1$. Find the odds in favor of A and those in favor of A'.

Key Word Review

Each key word is followed by the number of the section in which it is defined.

outcome   **2.1**
random experiment   **2.1**
mutually exclusive   **2.1, 2.2**
sample space   **2.1**
event   **2.2**
elementary events   **2.2**
compound events   **2.2**
sure event   **2.2**

impossible event   **2.2**
relative frequency   **2.3**
empirical probability   **2.3**
subjective probability   **2.3**
probability (Laplace's
   definition)   **2.4**
odds in favor   **2.7**

CHAPTER 2   **REVIEW EXERCISES**

1.   Define each word in the Key Word Review.

2.   What do the following symbols mean?

   (a) $f_n(A)$ **(b)** $P(T)$ **(c)** $n(E)$

3.   A die is tossed twice.

   (a) Define a sample space Ω for this experiment in which the elementary events are equally likely.
   (b) Write the following events in set notation:
       A = the first toss is a 3
       B = the total is an even number.
   (c) Write verbal descriptions of $A \cup B$ and A'.
   (d) Name two mutually exclusive events.
   (e) How many elementary events are there?
   (f) State two compound events.

4.   A die is tossed 600 times, yielding the following results:

Number on die	1	2	3	4	5	6
Frequency	99	100	104	98	97	102

   Find the relative frequency of—

   (a) A = A 5 appears   **(b)** B = An even number appears
   (c) A' **(d)** $A \cap B$ **(e)** $A \cup B$

5.   Joe offers $1 : 4$ odds that A will occur and $1 : 5$ odds that B will occur. He knows that A and B cannot both occur. What odds should he give that A or B will occur?

6. Congenital agammaglobulinemia is a recessive sex-linked disease in which the body fails to form antibodies. If an apparently normal couple marry and have a son with this disease, what is the probability that—

   (a) their next son will have the disease

   (b) their next child will be completely normal (not diseased and not a carrier)?

7. Sickle-cell anemia is a recessive but not sex-linked hereditary trait. If two apparently normal people have a child with sickle-cell anemia, what is the probability that—

   (a) their next child will not have the trait

   (b) their next son will be a carrier (i.e., will not have the trait, but be able to transmit it to his children)?

8. In order to test the appeal of a new cereal before marketing it, the cereal company asked a random sample of 1000 people to taste Crunchy Munchies. The following statistics were obtained:

Category	Hated It	Liked It	Loved It
Male child	40	70	100
Female child	50	80	110
Male adult	130	70	50
Female adult	150	100	50

   Based on this sample, estimate the probability that a person chosen at random—

   (a) hates the cereal

   (b) is a child who loves the cereal

   (c) is a female adult who likes the cereal.

9. Seventy percent of the faculty of a certain university are women (W), 20 percent are professors (T), and 5 percent are male professors. Calculate—

   (a) $P(W')$ (b) $P(W' \cap T)$ (c) $P(W' \cup T)$

10. A boy has three red, two black, four green, and five white marbles. If he randomly selects a marble, what are the odds in favor of the following events?

   (a) A red marble is chosen

   (b) A red or white marble is chosen

   (c) A black marble is not chosen.

11. Suppose $P(ABC') = 0.10$, $P(A'B'C') = P(A'BC) = 0.05$, $P(AB) = P(C) = 0.30$, $P(B) = 0.65$, and $P(AC) = 0.21$. Calculate—

(a) $P(B \cup C)$ (b) $P(BC)$ (c) $P(B')$ (d) $P(A - B)$
(e) $P(A \cup B \cup C)$ (f) $P(ABC)$

Suggested Readings

1. Eisen, Martin, and Eisen, Carole. *Probability and Its Applications.* New York: Quantum Press, 1975.
2. Feller, William. *An Introduction to Probability Theory and Its Applications.* Vol. 1. 3d ed. New York: John Wiley & Sons, 1968, Vol. 2, 2d ed. 1971.
3. Parzen, Emanuel. *Modern Probability Theory and Its Applications.* New York: John Wiley & Sons, 1960.

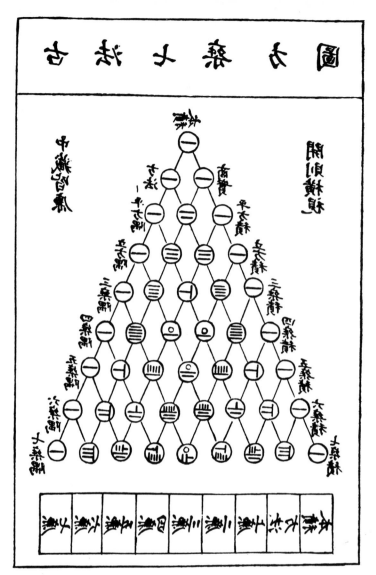

Joseph Needham, Cambridge University Press

CHAPTER 3 Counting Problems, Permutations, and Combinations

INTRODUCTION

The diagram on the facing page first appeared in the front of a fourteenth-century Chinese manuscript. The author of the manuscript called this diagram "The Old Method Chart of the Seven Multiplying Squares," but a modern mathematician would probably call it "Pascal's triangle," after a French mathematician, Blaise Pascal, who developed a similar diagram in the seventeenth century. The only difference between this diagram and Pascal's (see Figure 3.9) is that the Chinese author used Chinese numerals on his diagram, while Pascal used Arabic numerals.

For both the medieval Chinese author and the modern mathematician, the important thing about this diagram is not how it is written, but how it can be used to calculate quickly the numbers of objects in certain sets. Pascal's triangle is only one of the sophisticated methods of counting that we will develop in this chapter. We will find that all of these methods can be used to simplify probability calculations.

3.1 TWO COUNTING PRINCIPLES

Before applying Laplace's definition of probability, the number of elements in certain sets must be counted. In Example 10 of Chapter 2 we calculated the number of ways that two dice may be tossed, $n(\Omega)$, by listing all the elements of the sample space Ω. However, if Ω has a large number of elements, such a complete listing is impractical. Therefore, we must develop other methods for counting the elements of large sets. In this section we will study two theorems that are useful in calculating the number of elements in a set. These theorems are called the *first* and *second counting principles*. We will also see how Venn diagrams can be used to solve counting problems.

Example 1 will lead us to a statement of the first counting principle.

Example 1 Students in a certain university are classified by sex—male (m) and female (f)—and by class—freshmen (1), sophomore (2), junior (3), and senior (4). How many different categories are there?

Solution The various categories are conveniently illustrated by the tree diagram in Figure 3.1.

Since there are 2 branches in the first stage and each of these branches has 4 branches attached to it, the tree has $2 \cdot 4 = 8$ endpoints; each endpoint corresponds to a different category. The 8 categories are listed to the right of the tree diagram.

In general, we can think of performing a selection process in stages. For example, in Example 1 we first select one of the two letters m or f and then select one of the four numbers 1, 2, 3, 4. From Figure 3.1 it is clear that the total number of choices is $2 \cdot 4 = 8$. Theorem 3.1 generalizes this result.

THEOREM 3.1 (First Counting Principle)
Suppose k operations are to be performed sequentially. If the first operation can be performed in r_1 distinct ways, and for

each of these ways, the second operation can be performed in r_2 distinct ways, and so on, up to the kth operation, which can be performed in r_k distinct ways, then the k operations can be performed sequentially in $r_1 r_2 \cdots r_k$ different ways.

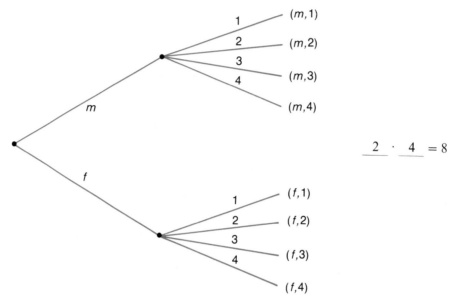

$$\underline{2 \quad \cdot \quad 4} = 8$$

Figure 3.1   Tree diagram for Example 1

The following examples are applications of Theorem 3.1.

Example 2   There are three roads from John's house to the traffic circle and two roads from the traffic circle to the library. How many different ways can John go from the house to the library and back again?

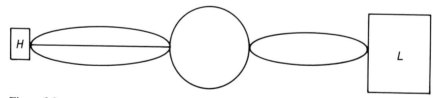

Figure 3.2

Solution   Figure 3.2 illustrates the routes from John's house to the library. For each of the 3 ways that John can get from the house to the traffic circle, there are 2 ways of going from the circle to the library. It follows from Theorem 3.1 that there are $3 \cdot 2 = 6$ ways of going from the house to the library. For each of these 6 ways, there are 6 ways of going from the library to the house. Applying Theorem 3.1 again, there are $6 \cdot 6 = 36$ different routes to the library and back.

Example 3   When a salesperson cannot find the shoe that a customer wants in the correct size, the customer complains that a proper shoe store should stock every style in every color, width, and size. If the store has twenty different styles and each style comes in fourteen sizes, three widths, and five colors, how many different shoes would the store have to stock?

Solution   This problem may also be solved by applying Theorem 3.1. The application of this theorem is simplified by using a two-step approach equivalent to drawing a tree diagram.

   The first step is to represent each stage in the sequence of operations by a blank. In this case, the operations are choosing style, size, width, and color. Therefore, four blanks are needed.

$$\underline{\qquad}\ \ \underline{\quad}\ \ \underline{\quad}\ \ \underline{\quad}$$

   For the second step, instead of drawing branches at each point of a given stage as we would in a tree diagram, we write the number of branches at each stage above the blank corresponding to that stage. First the number of ways that the first choice can be made (i.e., the number of styles) is written above the first blank.

$$\underset{\underline{\qquad}}{20}\ \ \underline{\quad}\ \ \underline{\quad}\ \ \underline{\quad}$$

The blanks for the remaining steps are then filled in the same way.

$$\underset{\underline{\qquad}}{20}\ \ \underset{\underline{\quad}}{14}\ \ \underset{\underline{\quad}}{3}\ \ \underset{\underline{\quad}}{5}$$

   This diagram reminds us that the first operation was performed in 20 distinct ways; for each of these ways, the second operation was performed in 14 distinct ways; for each of these ways, the third operation was performed in 3 distinct ways; and for each of these ways the fourth operation was performed in 5 distinct ways. Applying Theorem 3.1, the store would have to stock $20 \cdot 14 \cdot 3 \cdot 5 = 4200$ different pairs of shoes.

Example 4   Each monogram produced by a manufacturer consists of two initials.

   (a) How many different monograms are there?
   (b) How many monograms consist of two distinct letters?

Solution

   (a) Since each initial can be selected in 26 ways, the number of monograms can be found by using the following diagram:

$$\underset{\underline{\qquad}}{26}\ \ \underset{\underline{\qquad}}{26}$$

   The answer is therefore $26 \cdot 26 = 676$ different monograms.

   (b) The first initial can occur in 26 ways. However, if the letters are to be distinct, the last initial cannot be the same as the first. Therefore, once the first letter is chosen, there are only 25 ways left to choose the second. Applying Theorem 3.1, the number of monograms consisting of distinct letters is $26 \cdot 25 = 650$.

Example 5 shows how the first counting principle can be used to find the number of elements in any Cartesian product.

Example 5   Let $A = \{1, 2, \ldots, 6\}$ and $B = \{h, t\}$. Calculate $n(A \times B)$.

Solution

$$A \times B = \{(a, b) \mid a \in A, b \in B\}$$

We need to find out how many ordered pairs (a, b) there are. Since the first component a is an element of A, it can be any one of 6 numbers. For each of these numbers, the second component b can be any one of 2 letters. Applying Theorem 3.1,

$$n(A \times B) = 6 \times 2 = 12 = n(A) \times n(B)$$

The following theorem can be proved using the same reasoning used to solve Example 5.

THEOREM 3.2   Let $A_1, A_2, \ldots, A_k$ be any k sets. Then

$$n(A_1 \times A_2 \times \cdots \times A_k) = n(A_1) \times n(A_2) \times \cdots \times n(A_k)$$

In particular, if A and B are any two sets,

$$n(A \times B) = n(A) \times n(B)$$

Example 6 is an application of Theorem 3.2.

Example 6   Let $A = \{1, 2, 3\}$, $B = \{4, 5\}$, and $C = \{1, 3, 7, 8\}$. Calculate—

   (a) $n(A \times B)$ **(b)** $n(A \times B \times C)$ **(c)** $n(A^4)$ **(d)** $n(A \times \varnothing)$

Solution   Applying Theorem 3.2—

   (a) $n(A \times B) = n(A) \times n(B) = 3 \times 2 = 6$
   (b) $n(A \times B \times C) = n(A) \times n(B) \times n(C) = 3 \times 2 \times 4 = 24$
   (c) $n(A^4) = n(A \times A \times A \times A) = n(A) \times n(A) \times n(A) \times n(A) = 3^4 = 81$
   (d) $n(A \times \varnothing) = n(A) \times n(\varnothing) = 3 \times 0 = 0$. Alternatively, since $A \times \varnothing = \varnothing$,
   $n(A \times \varnothing) = n(\varnothing) = 0$.

The following example shows how Theorems 3.1 and 3.2 can be used to calculate probabilities.

Example 7   If three dice are thrown, what is the probability of getting—

   (a) three 4s (event S)
   (b) two 4s and a 3 (event T)?

Solution   A relevant sample space in which the elementary events are equally likely is $\Omega = A^3$, where $A = \{1,2,3,\ldots,6\}$. Applying Theorem 3.2 yields $n(\Omega) = n(A \times A \times A) = 6^3 = 216$.

(a) Since $S = \{(4,4,4)\}$, $P(S) = \frac{1}{216}$.

(b) Similarly, since $T = \{(4,4,3), (4,3,4), (3,4,4)\}$, $n(T) = 3$ and $P(T) = \frac{3}{216}$.

Sometimes students are so impressed by Theorem 3.1 that, whenever they have a problem involving two events, they immediately multiply the number of elements in one event by the number of elements in the other event. This is, of course, poor mathematics. The results of Theorem 3.1 can be used *only* if the hypotheses of the theorem apply to the problem. The following example shows that care must be taken when using counting techniques.

Example 8

(a) If each of four cats has three kittens, how many cats will there be?

(b) If a cat fancier has four Siamese cats and three Persian cats, how many cats does he have?

Solution

(a) Applying Theorem 3.1, there are $4 \cdot 3 = 12$ cats.

(b) Let S and P represent the sets of Siamese and Persian cats, respectively. Clearly,

$$n(S + P) = 4 + 3 = 7$$

so there are 7 cats in all.

Notice that the number of elements in the set $S + P$ of Example 8(b) is $n(S) + n(P)$, not $n(S) \cdot n(P)$. In situations like Example 8(b), the following theorem should be used instead of Theorem 3.1.

THEOREM   3.3   (Second Counting Principle)

If $A_1, A_2, \ldots, A_n$ are pairwise disjoint sets, then

$$n(A_1 + A_2 + \cdots + A_n) = n(A_1) + n(A_2) + \cdots + n(A_n)$$

In particular, if A and B are disjoint sets, then

$$n(A + B) = n(A) + n(B)$$

The last equation of Theorem 3.3 was first discussed in Chapter 1. (See equation (3), page 13.)

Notice that

$$n(A \times B) = n(A) \times n(B)$$

and

$$n(A + B) = n(A) + n(B)$$

In some situations, both Theorems 3.1 and 3.3 must be used, as the following examples illustrate.

Example 9   How many different numbers can be formed from the digits 1, 2, 3, and 4 if repetition of digits is not allowed?

Solution   The number may have either 1, 2, 3 or 4 digits. Let A_i be the set of all numbers that have exactly i digits and satisfy the conditions of the problem. Using Theorems 3.1 and 3.3, the total number of different numbers is

$$n(A_1 + A_2 + A_3 + A_4) = n(A_1) + n(A_2) + n(A_3) + n(A_4)$$
$$= 4 + 4 \cdot 3 + 4 \cdot 3 \cdot 2 + 4 \cdot 3 \cdot 2 \cdot 1 = 64$$

Example 10   A biologist is studying families with six or fewer children. He lets bg represent a family with two children where the older is a boy and the younger a girl, while $gggb$ represents a family with four children where the youngest is the only boy. How many different kinds of families is he studying?

Solution   The set of all families F in the study may be partitioned into those with no children (F_0), those with one child (F_1), and so on. It follows that $F = F_0 + F_1 + \cdots + F_6$. Applying the second counting principle,

$$n(F) = n(F_0) + n(F_1) + \cdots + n(F_6)$$

Since the first child in the family may be either a boy or a girl, the second child in the family may be either a boy or a girl, and so on, $n(F_6)$ may be found by using the following diagram and the first counting principle:

$$\underline{\;2\;} \quad \underline{\;2\;} \quad \underline{\;2\;} \quad \underline{\;2\;} \quad \underline{\;2\;} \quad \underline{\;2\;}$$

It follows that $n(F_6) = 2^6$. The values of $n(F_0)$, $n(F_1)$, and so on may be calculated in the same way. Therefore,

$$n(F) = 1 + 2 + 2^2 + 2^3 + 2^4 + 2^5 + 2^6 = 127$$

Just as the filling-the-blanks device developed in Example 3 helps us to apply the first counting principle, Venn diagrams can be used to help us apply the second counting principle. The following examples show how this is done.

Example 11   In a test of some rocketship circuits, ten were found to be defective. Seven had worn-out transistors, five had faulty connections, and four had broken wires. One had both a worn-out transistor and a faulty connection but good wires, one had a faulty connection and a broken wire but good transistors, two had both worn-out transistors and broken wires but good connections, and three only had worn-out transistors. How many had—

(a) worn-out transistors but good connections
(b) all three defects
(c) broken wires only?

Solution   This problem can be solved by using a method similar to the method used in Examples 22 and 23 of Chapter 2, pages 59–60. As in those examples, Venn diagrams will be used. However, instead of labeling events with their probabilities, we will label them with the number of elements they contain. Then, instead of applying Theorem 2.4, we will apply Theorem 3.3.

Let T, C, and W be the sets of circuits with worn-out transistors, faulty connections, and broken wires, respectively. We need to find $n(TC')$, $n(TCW)$, and $n(T'C'W)$.

The Venn diagram for the three sets T, C, and W is shown in Figure 3.3(a).

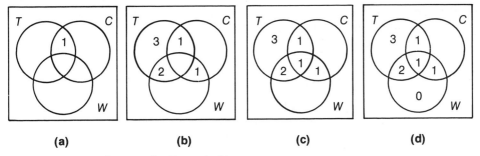

Figure 3.3   Venn diagrams for Example 11

From the information given, we know that $n(\Omega) = 10$, $n(T) = 7$, $n(C) = 5$, $n(W) = 4$, $n(TCW') = 1$, $n(CWT') = 1$, $n(TWC') = 2$ and $n(TW'C') = 3$. Since $n(T) = 7$, we might be tempted to put a 7 in the circle representing T. A quick look at Figure 3.3(a), however, shows that T is really partitioned into four proper subsets: TCW, TCW', $TC'W'$, and TWC'. We therefore cannot tell exactly where to put the 7. The set TCW', however, is not partitioned, and we know that $n(TCW') = 1$. We may therefore put a 1 in the area which designates TCW', as indicated in Figure 3.3(a). Similarly, we can enter the number of elements in the sets CWT', TWC', and $TW'C'$, as shown in Figure 3.3(b).

(a) It is clear from Figure 3.3(b) and Theorem 3.3 that

$$n(TC') = n(TWC') + n(TC'W') = 2 + 3 = 5$$

(b) From Figure 3.3(b) and Theorem 3.3, we see that

$$n(T) = n(TCW) + n(TW'C') + n(TCW') + n(TWC')$$

$$= n(TCW) + 3 + 1 + 2$$

or

$$n(T) = n(TCW) + 6$$

Therefore,

$$n(TCW) = n(T) - 6 = 7 - 6 = 1$$

(See Figure 3.3(c).)

(c) Since $n(W) = 4$, it follows from Figure 3.3(c) and Theorem 3.3 that $n(WT'C') = 0$. (See Figure 3.3(d).)

Example 12   A school has 1000 students. Three hundred of these students are on the dean's list (*D*), 600 are men (*M*), 300 are freshmen (*F*), 50 are female freshmen who are on the dean's list, 80 are freshmen on the dean's list, 100 are men on the dean's list, and 150 are freshmen men. Find the number of students who are—

(a) women who are not freshmen and who are not on the dean's list
(b) not freshmen
(c) female freshmen.

Solution   The Venn diagram in Figure 3.4 can be constructed by using the given information and Theorem 3.3.

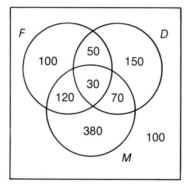

Figure 3.4   Venn diagram for Example 12

Using the Venn diagram, we see that—

(a) $n(M'F'D') = 100$
(b) $n(F') = 150 + 70 + 380 + 100 = 700$
(c) $n(M'F) = 50 + 100 = 150$

3.1 EXERCISES

1.   How many different license plates can be made if each license plate has two letters followed by three numbers?

2.   To win the daily triple at a local race track, one must choose the three winning horses in a single race in the proper order. If there are ten horses in a race, and you know nothing about horses, what is the probability that you will win the triple by guessing?

3.   A young child who cannot read decides to help his mother by putting her four letters into the four addressed envelopes. If the four envelopes are addressed to different people, what is the probability that he will put every letter into the correct envelope? (He knows enough to put exactly one letter into each envelope.)

4. Sheila hates to cook but cannot afford to eat out. She has decided to prepare quick meals using canned soups and vegetables. Her supermarket stocks 15 different canned soups, ten different canned vegetables, five different meats suitable for broiling, and eight flavors of ice cream. How many different menus consisting of soup, vegetable, broiled meat, and ice cream can she prepare?

5. Two children want to bake a cake for Mother's Day. If the supermarket carries ten different cake mixes, eight different frosting mixes, and four different kinds of decoration, how many different kinds of cake complete with frosting and decoration can they make?

6. At an office party, each employee brings a gift, and the gifts are randomly distributed so that each employee receives one. If there are ten employees, what is the probability that—
    (a) Peter will be stuck with the gift Valerie brought
    (b) every employee will get his or her own gift?

7. The *New York Times* reported that a woman in the Bronx had her apartment burglarized so frequently that she kept adding locks to her door. When eight locks were picked, the locksmith advised her to lock some of the locks and leave others open. The burglar then would be kept busy locking the open locks while unlocking the locked ones. In how many ways can the woman lock her door if she doesn't want to either lock all the locks or leave them all open? (P.S. This method worked.)

8. A man has r children, and each of these children has r children, and each of these has r children.
    (a) How many great-grandchildren does he have?
    (b) Assuming that the man and all his descendants are still living (and compatible), how many plates should he set for Christmas dinner? (Everyone except the great-grandchildren is married and brings his or her spouse.)
    (c) How large is the family (*i*) if $r = 6$, (*ii*) if $r = 2$?

9. Little Lisa has a set of four blocks; each bears one of the letters of her name. If she arranges the blocks randomly, what is the probability that she spells her name?

10. Consider the rocketship circuits of Example 11.
    (a) Complete the Venn diagram.
    (b) How many circuits had—

        (*i*) only faulty connections
        (*ii*) either faulty connections or broken wires
        (*iii*) both faulty connections and broken wires?

11. Suppose a study of the relationship between high cholesterol, overweight, and heart disease results in the following data for 56,000 men between the ages of 40 and 60: 7000 men have high cholesterol and heart disease, but are not overweight; 10,000 men have high cholesterol and are not overweight; 29,000 men have high cholesterol; 35,000 men are overweight; 21,000 men

have heart disease; 3000 men are not overweight and do not have high cholesterol, but do have heart disease; and 12,000 men have high cholesterol and heart disease. Of the men surveyed, how many—

(a) are not overweight

(b) are overweight men with heart disease and high cholesterol

(c) are overweight or have heart disease?

12.   Based on the study of Exercise 11, what is the probability that a man between the ages 40 and 60—

(a) is overweight but does not have high cholesterol

(b) does not have heart disease

(c) is not overweight, does not have heart disease, and does not have high cholesterol

(d) has high cholesterol or is overweight?

13.   Suppose that a survey of 1300 people in New York City resulted in the following statistics: 450 people read the *Wall Street Journal*, 650 read the *Daily News*, 450 read the *New York Times*, 100 read all three newspapers, 300 read only the *Daily News*, 200 read the *New York Times* and the *Daily News*, and 250 read the *New York Times* and the *Wall Street Journal*. How many people in the survey—

(a) read none of these newspapers

(b) read only the *New York Times*

(c) read both the *New York Times* and the *Wall Street Journal*, but not the *Daily News*

(d) cannot read?

14.   Based on the statistics of Exercise 13, what is the probability that a randomly chosen New York resident reads—

(a) only the *Daily News* and the *Wall Street Journal*

(b) the *New York Times* or the *Daily News*

(c) at least one of these papers?

15.   If $P(AB) = 0.4$, $P(A'B') = 0.2$ and $P(B) = 0.7$, find—

(a) $P(A - B)$ **(b)** $P(A)$ **(c)** $P(B')$ **(d)** $P(A \cup B)$

3.2 PERMUTATIONS OF DISTINCT OBJECTS

It is often necessary, as in Exercises 3, 6**(b)**, and 9 of Section 3.1, to calculate the number of different orders in which a given number of objects can be arranged. In this section the first counting principle (Theorem 3.1) will be applied to this problem. Example 13 shows how this is done.

Example 13   A child has a set of fifteen different letters. How many three-letter "words" (some may be nonsensical) can she make from them?

Solution   The first letter in the word may be chosen in 15 ways; the second in 14 ways; the third in 13. Applying Theorem 3.1, the total number of words is $15 \cdot 14 \cdot 13 = 2730$.

In Example 13 we needed to calculate the number of ways that 3 out of 15 objects can be arranged when the order of the arrangement is important. Notice that *order* is important; the arrangement EAT is not the same as ATE, TEA, or TAE, even though they all use the same letters. Definition 3.1 generalizes this concept of ordered arrangements.

Definition 3.1   A **permutation** *of n things taken r at a time* is an ordered arrangement of r of the n things. A permutation of n things taken n at a time is called a *permutation* (of n things).

Example 14

(a) The permutations of the letters *aaf* are

aaf, afa, faa

(b) The permutations of the letters *aaef* taken two at a time are

aa,   ae,   af,   ef,   fe,   ea,   fa

(c) The permutations of *abc* are

abc,   bca,   acb,   bac,   cab,   cba

Notice that the n objects need not be distinct. If the objects are distinct, as in Example 14(c), the following theorem enables us to calculate the number of permutations. The proofs of the theorem and its corollary (Theorem 3.5) follow from Theorem 3.1 using the same reasoning we used in Example 13.

THEOREM   3.4   The number of permutations of n distinct objects taken r at a time, designated $_nP_r$, is given by the formula

$$_nP_r = n(n-1)(n-2) \cdots (n-r+1) \qquad r = 1, 2, \ldots, n$$

The following theorem is a corollary of Theorem 3.4.

THEOREM   3.5

$$_nP_n = n(n-1) \cdots 1$$

$_nP_n$ is usually written P_n.

Note that $_nP_r$ has exactly r factors. To find $_7P_2$, for example, two factors must be multiplied, with each factor being one less than the preceding one. In other words, $_7P_2 = 7 \cdot 6 = 42$.

Examples 15–18 are applications of Theorems 3.4 and 3.5.

Example 15

   (a) Example 13 can be solved by applying Theorem 3.4 with $n = 15$ and $r = 3$. This yields $_{15}P_3 = 15 \cdot 14 \cdot 13$ words, which agrees with our previous solution.

   (b) The number of permutations of the letters abc is $P_3 = 3 \cdot 2 \cdot 1 = 6$. (See Example 14(c).)

   (c) Theorem 3.4 cannot be used to find the number of permutations of the letters aaf because these letters are not distinct.

Example 16   In the game of Scrabble, each player has seven letters. A bonus is given if a player makes a word out of all seven letters. John decides that, in order to get the bonus, all he needs to do is to examine every possible permutation of his letters. If all his letters are different, how many permutations must he look at? If it takes him ten seconds to form a new permutation and decide whether or not it is a word, how long will it take him to look at all the possibilities?

Solution   Using Theorem 3.5, the number of permutations is $7 \cdot 6 \cdot 5 \cdot 4 \cdot 3 \cdot 2 \cdot 1 = 5040$. It will take him 50,400 seconds, or 14 hours, to look at all the possibilities.

Example 17   A club with 750 members must choose 3 people to be president, secretary, and treasurer. In how many ways can this be done if no one can hold more than one position?

Solution   The president may be chosen in 750 ways; once this has been done, there are 749 ways of choosing the secretary; there are then 748 ways left to choose the treasurer. It follows from the first counting principle that the answer is $750 \cdot 749 \cdot 748 = {}_{750}P_3$.

   Alternatively, we may let the permutation ABC mean that A is chosen president, B secretary, and C treasurer. The problem then reduces to finding the number of ways of permuting 750 distinct objects 3 at a time. Applying Theorem 3.4, the answer is $_{750}P_3$ as before.

   The following example shows that Theorem 3.4 must be applied with care. If the objects to be arranged are not distinct, Theorem 3.4 cannot be used.

Example 18   In how many ways can the three officers of Example 17 be chosen if a person can hold any number of positions?

Solution   As in Example 17, let the permutation ABC mean that A is chosen as president, B as secretary, and C as treasurer. Theorem 3.4 cannot be applied because the objects to be arranged are not distinct. For example, the permutation AAA can occur, because A can serve in all three positions. This problem can, however, be solved by applying the first counting principle. Since there are 750 choices for president, 750 choices for secretary, and 750 choices for treasurer, the total number of ways of choosing all three officers is $750 \cdot 750 \cdot 750 = (750)^3$.

Whenever we compute $_nP_k$, we have to write down a product of k consecutive integers. It is therefore convenient to use the following special notation when discussing permutations.

Definition 3.2 If n is any nonnegative integer, the notation $n!$ (read **n factorial**) is defined as follows:

$$0! = 1$$

$$n! = n(n-1)(n-2)\cdots 3\cdot 2\cdot 1 \qquad n = 1, 2, 3, \ldots$$

In other words, if $n = 1, 2, 3, \ldots$, $n!$ is the product of all integers from n down to 1. For example,

$1! = 1$	$4! = 4\cdot 3! = 24$
$2! = 2\cdot 1 = 2$	$5! = 5\cdot 4! = 120$
$3! = 3\cdot 2\cdot 1 = 6$	$6! = 6\cdot 5! = 720$

Notice that $(n+1)!$ is just $(n+1)$ times $n!$.

It is clear that the values of $n!$ rapidly become large. Tables of $n!$ are found in most books of mathematical tables. For large values of n, $n!$ can be calculated using a computer. If n is very large, there are approximation formulas for $n!$ (e.g., Stirling's formula).[1]

Example 19 shows how to use Definition 3.2.

Example 19

   (a) In how many ways can the letters in the word SPINDLE be arranged?
   (b) How many arrangements begin with S and end with E?

Solution

   (a) Applying Theorem 3.5, the number of arrangements is $P_7 = 7\cdot 6\cdot 5\cdot 4\cdot 3\cdot 2\cdot 1 = 7!$
   (b) Since the first and last letters are fixed, we need to permute the remaining five distinct letters. Applying Theorem 3.5, the number of arrangements is $P_5 = 5!$.

Care must be taken with circular arrangements, as the following examples illustrate.

Example 20

   (a) In how many ways may six people be seated around a circular table?
   (b) If there are three women and three men and all arrangements are equally likely, what is the probability that every woman sits between two men?

---

1. See reference 1 in the Suggested Readings at the end of this chapter.

Solution

(a) Unlike the problem of seating people in a row, there is no first position. If we fix any one person, say Adam, what is important is not where Adam sits, but who sits on his left, who sits two seats from him on his left, and so on. For example, the arrangements in Figure 3.5 are identical, since in

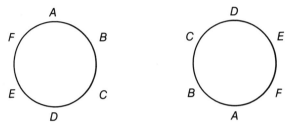

Figure 3.5

both cases Betty (*B*) is sitting to the left of Adam (*A*), Carole (*C*) is sitting to the left of Betty, and so on. Note that the arrangement in Figure 3.6 is different from either of those in Figure 3.5. Since there are 5 choices for the seat to the left of Adam, 4 choices for the next seat, and so on, the total number of ways is $5! = 120$.

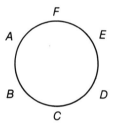

Figure 3.6

(b) The favorable arrangements are illustrated in Figure 3.7. If we fix the circled woman, there are then 3 choices for the man on her left, 2 choices for the next woman, 2 choices for the following man, and so on. The number of favorable ways is therefore $3 \cdot 2 \cdot 2 \cdot 1 \cdot 1 = 12$. Applying

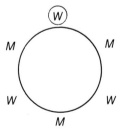

Figure 3.7

Laplace's definition of probability, we have

$$P = \frac{12}{120} = \frac{1}{10}$$

In Example 20**(a)**, we saw that the number of ways that 6 distinct objects can be arranged in a circle is $(6 - 1)! = 5!$. Using a similar method, we can prove that *the number of ways n distinct objects can be arranged in a circle is $(n - 1)!$*. As usual, this theorem must be applied with care. The following example illustrates this point.

Example 21   In how many ways may eight differently colored interlocking beads be put together to form a necklace?

Solution   If we were merely arranging 8 beads in a circle, the answer would be $(8 - 1)! = 7!$. However, since the necklace can be turned over, we cannot distinguish between clockwise and counterclockwise arrangements. For example, if the arrangement of Figure 3.8(a) is turned over, it becomes the arrangement of Figure 3.8(b). Both these arrangements must therefore be considered the same arrangement.

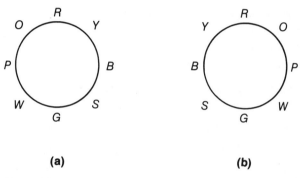

(a)   **(b)**

Figure 3.8

Therefore, although *both* of these arrangements must be counted when forming a circle, only *one* of them can be counted when forming a necklace. There are consequently only half as many ways of forming necklaces as there are of forming circles. Hence, the total number of different necklaces is $\frac{1}{2}(7!)$.

3.2 EXERCISES

1.   Calculate—

   (a) $8!$ **(b)** $\dfrac{7!}{3!}$ **(c)** $_8P_3$ **(d)** P_7 **(e)** $\dfrac{_8P_3}{_4P_2}$ **(f)** $_{1000}P_2$

2.   If the four volumes of a book are placed on a bookshelf in random order, what are the odds in favor of their being in the correct order?

3.   Prove that there are two people in Boston, Massachusetts, with the same first and last initials.

4.   How many different signals can a ship display if the ship has six different flags and a signal consists of four different flags hoisted on a vertical rope?

5. **(a)** How many different batting orders can a nine-man baseball team have?

   (b) If all batting orders are equally likely, what is the probability that the pitcher bats last?

6. In how many ways can eight children form a circle to play ring-around-the-rosy?

7. How many four-digit even integers can be formed using 0, 1, 2, ..., 6 if—
   (a) the digits may be repeated?

   (b) no digit may be repeated?   (Hint: Consider two cases—the one where the last digit is 0 and the one where the last digit is not 0.)
   (Hint: 0123 is a three-digit number.)

8. A child has four alphabet blocks, each bearing one of the letters N, E, S, or T. If he places them in a row at random, what is the probability that he forms an English word?

9. In how many ways can three men and four women be seated—
   (a) on a bench

   (b) around a round table
   if the men insist upon sitting together?

10. In how many ways can fourteen differently colored interlocking beads be put together to form a necklace?

11. A girl has twelve interlocking beads. Three are different shades of red, four are different shades of green, and five are different shades of blue. How many different necklaces can she form if beads of the same color must be grouped together?

12. Prove that—

   (a) $n! = n \cdot (n - 1)!$

   (b) $_nP_r = \dfrac{n!}{(n - r)!}$

   (c) $P_n = n!$

3.3 PERMUTATIONS OF NONDISTINCT OBJECTS

Theorem 3.4 tells us how to permute the letters in the words

CHEMIST,   LAWYER,   and   BAKER

However, it cannot be used to find the number of permutations of the letters in

MATHEMATICIAN

because the letters in this word are not *distinct* (e.g., there are two I's and three A's). The problem of finding the number of permutations of *n* objects when the objects are *not* all distinct will be studied in this section. Example 22 shows how one such problem can be solved. This example will lead us to a theorem for solving such problems in general.

Example 22   How many ways may the letters in the word

$$\text{MATHEMATICIAN} \tag{1}$$

be arranged?

Solution   We cannot simply apply Theorem 3.4 because the letters are not distinct. Instead, let x be the number of ways of arranging the letters, and let us distinguish the various M's, A's, T's, and I's by indexing them:

$$M_1A_1T_1HEM_2A_2T_2I_1CI_2A_3N \tag{2}$$

Fix one of the x ways of permuting (1); for example,

$$\text{AAAMMTTIIENHC} \tag{3}$$

This arrangement corresponds to exactly $3!\,2!\,2!\,2!$ different ways of permuting (2). To see this, note that there are $3!$ ways of permuting the three A's; for each of these, there are $2!$ ways of permuting the two M's; for each of these, there are $2!$ ways of permuting the two T's; and for each of these there are $2!$ ways of permuting the two I's. For example,

$$A_2A_1A_3M_1M_2T_1T_2I_1I_2ENHC \quad\text{and}\quad A_3A_2A_1M_2M_1T_2T_1I_1I_2ENHC$$

are two permutations of (2) that both correspond to (3).

In the same way, for each of the x ways of permuting (1), there are $3!\,2!\,2!\,2!$ ways of permuting (2); so altogether there are

$$3!\,2!\,2!\,2!\,x$$

ways of permuting (2). By applying Theorem 3.5, we also know that the number of ways of permuting (2) is $13!$. It follows that

$$3!\,2!\,2!\,2!\,x = 13!$$

Solving for x yields

$$x = \frac{13!}{3!\,2!\,2!\,2!}$$

The next theorem can be proved using the same reasoning.

THEOREM   3.6   The number of permutations of n things when there are k_1 things of one kind, k_2 things of a second kind, and so on, up to k_r things of an rth kind $(k_1 + k_2 + \cdots + k_r = n)$, is

$$\frac{n!}{k_1!\,k_2!\,\cdots\,k_r!}$$

Examples 23 and 24 are applications of Theorem 3.6.

Example 23   In Example 22 there were 3 things of a first kind (A's), 2 things of a second kind (M's), 2 things of a third kind (T's), 2 things of a fourth kind (I's), and 1 thing of a fifth, sixth, seventh, and eighth kind (E, N, H, and C). Applying Theorem 3.6 with $n = 13$, $k_1 = 3$, $k_2 = k_3 = k_4 = 2$, $k_5 = k_6 = k_7 = k_8 = 1$, and $r = 8$ yields

$$\frac{13!}{3!2!2!2!}$$

as before. (Notice that $13 = 3 + 2 + 2 + 2 + 1 + 1 + 1 + 1$.)

Example 24   If four dice are tossed, what is the probability that the sum of the numbers on the faces will be 22 (event E)?

Solution   The total number of ways of casting 4 dice is 6^4. Event E is the sum of two disjoint events, A and B, where

> A is the event " three 6s and a 4 appear "
>
> B is the event "two 6s and two 5s appear"

Hence $n(E) = n(A + B) = n(A) + n(B)$.

The number of elements in A is equal to the number of ways of permuting three 6s and one 4. For example, the permutation 6 6 6 4 means that a 4 appears on the fourth die and a 6 appears on all the others. Applying Theorem 3.6,

$$n(A) = \frac{4!}{3!1!} = 4$$

The number of elements in B equals the number of permutations of two 6s and two 5s. By Theorem 3.6,

$$n(B) = \frac{4!}{2!2!} = 6$$

It follows that $n(E) = 6 + 4 = 10$, and so

$$P(E) = \frac{10}{6^4}$$

3.3 EXERCISES

1.   If six dice are rolled, what is the probability that—
     (a) every possible number appears exactly once
     (b) the sum of the faces is at least 34
     (c) exactly five 3s appear?

2.   (a) In how many ways can the letters in the word INDIVISIBILITY be arranged?
     (b) How many ways begin with I and end with Y?

3. If five dice are rolled, what is the probability that—
    (a) no two dice show the same face
    (b) two 6s and three 4s are obtained
    (c) no 6s are obtained?

4. Galileo became interested in probability when he was asked by some gamblers to explain why, when three dice were tossed, the sum of the numbers on the faces was ten more often than it was nine. Why is this true?

3.4 PROBABILITY AND SCIENTIFIC EXPERIMENTS

There is a popular saying that statistics can be used to prove anything. For example, newspapers frequently quote the statistical results of scientific experiments to support headlines like "Breakfast Cereals Found Nutritionally Worthless" or "Cranberry Sauce Causes Cancer." No sooner have the readers thrown out their entire supply of breakfast cereal and cranberry sauce, than new articles quoting a second set of experts refute the previous statements. What are the poor readers to believe?

The fact is, scientific experiments must be carefully designed and analyzed in order to ensure valid results. This design and analysis requires a solid knowledge of statistics. Unfortunately, many people feel that anyone with a knowledge of arithmetic and a pencil is a statistician, in the same way that many people feel that anyone who can walk and talk is qualified to be an actor. This is not so; an individual can be a fine scientist and still not have the foggiest notion of proper experimental design.

Consider what you do before you deal the cards in a bridge game: you shuffle them as thoroughly as possible. Is this just a superstitious ritual? Of course not. You are trying to randomize the cards. That is, you are trying to make each hand just as likely as every other hand.

The following example will illustrate how randomization and probability theory apply to the design of scientific experiments.

Example 25   In the 1940s, a paper was published by a biologist which reported a substantial improvement in plants watered with a fluorescein solution. (Fluorescein is an organic dye.) Later the work was discredited because of the complete lack of randomization in the experiment. How can such an experiment be conducted so as to ensure the reliability of the results?

One way to achieve randomization is the following: Line up eighteen flower pots with seedlings and thoroughly shuffle nine red and nine black cards. Place a card in front of each flower pot. If the card is red, place a sticker on the bottom of the pot indicating that the plant is to get the fluorescein treatment. In order to provide equality with respect to heat, light, and watering, alternate the two types of flower pots and change the position of the flower pots regularly. When it is time to see if the treatment has had any effect, have a colleague randomly rearrange the pots, number them from 1 to 18, and remove the labels after recording which pots have received the treatment. Now order the pots according to excellence. Although excellence is an immeasurable quality, at least knowledge of which plants have received treatment will not prejudice your arrangement. Your colleague can now tell you whether the plants you have rated as the best are the ones that received the treatment.

Question 1　　*If the nine best plants received the treatment, can we be certain that the treatment was effective?*

Suppose the treatment has no effect. We will see that it is possible, just by chance, for the nine best plants to be exactly the ones that were treated. Because of the care taken to randomize the experiment and ensure equal light, temperature, and water for each seedling, we may assume that if the treatment has no effect, then all ways of ordering the plants are equally likely. We can represent these various orderings using T to represent treated plants and U to represent untreated plants. Thus,

$$TTTTTTTTTUUUUUUUUU$$

is the ordering in which the nine best plants are precisely the ones that were treated with fluorescein, while

$$TUTUTUTUTUTUTUTUTU$$

is the ordering in which the treated and untreated plants alternate in quality.

To see if it is possible for the first nine plants to be the treated ones if the treatment has no effect, we need to calculate the probability of this event. Since all orderings of the plants are equally likely, we can use Laplace's definition of probability. The probability of the event "the first nine plants are the treated ones" is the same as the probability that, when nine T's and nine U's are arranged at random, the first nine letters are T's.

Applying Theorem 3.6, the total number of ways of arranging 18 things when 9 are of one kind and 9 are of another is

$$\frac{18!}{9!\,9!} = 48{,}620$$

Since only one of these arrangements is favorable to the event, the probability is

$$\frac{1}{48{,}620} = 0.0000206$$

Thus, although the probability is small, it is possible for the nine best plants to be the treated ones, even if the treatment itself has no effect. We therefore cannot be *certain* that this arrangement, which represents the best possible experimental result, was not caused by chance. We can only say that it is *very likely* that the treatment is effective, because events whose probability is small do sometimes occur. (For example, if a million lottery tickets are sold, we know that someone must win, even though the probability that any one individual has the winning ticket is only 0.000001.)

Question 2　　*If we cannot be certain that the treatment is worthwhile, how can we decide whether to accept or reject the treatment?*

Ideally, if the treatment were effective, the nine best plants would have received the treatment. However, plants are not machines and such a deterministic result will, in general, not be obtained. We cannot reasonably expect a treatment (or fertilizer or drug, etc.) to be 100 percent effective. A pattern like

$$TTTTTTTTUUUUUUUUUT$$

might indicate that one of the treated plants had some unidentified defect. Perhaps it would have done even worse without the treatment. A pattern such as

$$TTTTTTUUTTTUUUUUUU$$

might indicate that two of the untreated plants were genetically superior to three of the treated ones.

Fortunately, a method has been devised (by statistician Frank Wilcoxon) for dealing with the large number of possible patterns. To apply this method to the plant experiment, start by numbering the plants from best to worst and then adding up the numbers associated with the treated plants. The following list shows three examples of this association.

Pattern	*Associated Sum*
(1) $T\ T\ T\ T\ T\ T\ T\ T\ T\ U\ U\ U\ U\ U\ U\ U\ U\ U$ $1\ 2\ 3\ 4\ 5\ 6\ 7\ 8\ 9$	$1 + 2 + 3 + \cdots + 9 = 45$
(2) $T\ T\ T\ T\ T\ T\ T\ T\ U\ T\ U\ U\ U\ U\ U\ U\ U$ $1\ 2\ 3\ 4\ 5\ 6\ 7\ 8\ \ \ 10$	$1 + 2 + 3 + \cdots + 8 + 10 = 46$
(3) $T\ T\ T\ T\ T\ T\ T\ U\ U\ T\ T\ U\ U\ U\ U\ U\ U$ $1\ 2\ 3\ 4\ 5\ 6\ 7\ \ \ \ \ 10\ 11$	$1 + 2 + \cdots + 7 + 10 + 11 = 49$

The lowest sum, 45, occurs when all the treated plants are among the first nine. Even in this ideal situation, we cannot be certain that the treatment is effective. As we saw in the discussion following Question 1, if the treatment were completely ineffective, this ideal result would still occur approximately 1 out of every 48,620 times.

For any given number, it is possible to compute the probability that the sum associated with the treated plants is less than or equal to that given number, under the assumption that the treatment is ineffective. If this probability is small, as it is for the sum 45, we know that it is very likely (although not certain) that the treatment is effective. Wilcoxon has compiled a table of these probabilities. In later examples and exercises, we will compute some of the entries in Wilcoxon's table ourselves.

To use Wilcoxon's approach, we must decide upon a **critical number** c before the experiment is begun. If the sum r that is actually obtained is less than or equal to c, we will accept the value of the treatment. If it is greater than c, we will reject it.

Question 3   *How is the critical number c determined?*

The critical number c depends upon how large a risk we are willing to take that we accept as effective a treatment that is actually worthless. We saw in Question 1 that, even if $c = 45$ (the lowest possible critical number), there is a risk of 0.00002 that the accepted treatment is worthless and the result was obtained by chance.

If we let $c = 47$, then we will consider the treatment effective if the associated sum is $r \leq 47$. We will reject the treatment if $r > 47$. As we will see in Exercise 5 following this section, the probability that an ineffective treatment yields the result $r \leq 47$ is

0.00008. By choosing $c = 47$, we are therefore running a risk of 0.00008, or 0.008 percent, that the treatment is ineffective and the result obtained was due to chance. The larger the critical number c, the larger the probability that we accept an ineffective treatment because of random results. This probability is called the **level of significance** of the test.

The risk that one is willing to tolerate depends upon the situation. If one is running a gambling casino, one might be willing to run a risk of 0.01 that the casino will lose money on a particular day. However, if there is a probability of 0.000001 that an atomic pile being used to provide fuel will explode, the risk would be intolerable.

The most common level of significance used by statisticians is 5 percent; 1 percent is often used as a high level of significance.

Example 26   Suppose that the treatment is ineffective. What is the probability that the associated sum is—

(a) 46 or less
(b) 47 or less?

Solution

(a) Let A_i be the event "the sum r is exactly i."

$$P(46 \text{ or less}) = P(A_{45} + A_{46}) = P(A_{45}) + P(A_{46})$$

Since A_{45} and A_{46} can each be obtained in exactly one way—

$TTTTTTTTTUUUUUUUUU$

and

$TTTTTTTTUTUUUUUUUU$, respectively—

and the total number of arrangements is 48,620,

$$P(46 \text{ or less}) = \frac{1}{48,620} + \frac{1}{48,620} = \frac{2}{48,620}$$

(b) A_{47} can be obtained in two ways,

$TTTTTTTTUUTUUUUUUUU$

and

$TTTTTTTUTTUUUUUUUU$

Hence,

$$P(47 \text{ or less}) = P(A_{45}) + P(A_{46}) + P(A_{47})$$

$$= \frac{1 + 1 + 2}{48,620} = \frac{4}{48,620} = 0.00008$$

Question 4   *Why do we consider the probability of events like "the sum is 47 or less" instead of the probability of events like "the sum is exactly 47"?*

Let us first consider a similar question that may be more familiar to the reader.

Suppose that you took a college entrance exam that yields scores between 0 and 800. If you knew that only 0.0001 of the people taking the test received the same grade you did, would that prove that you did well on the exam? Maybe the other 99.99 percent of the people did better!

The testing service usually reports your *percentile* score. From this score, you can calculate the percentage of people whose scores were greater than or the same as yours. For example, if you are told that you are in the 99th percentile, you know that only $100 - 99 = 1$ percent of the people taking the exam did as well as or better than you. This would show that you did well on the exam.

Similarly, if we want to show that an associated sum of 47 indicates that fluorescein treatment is beneficial, it is best to consider the event "the sum is 47 *or less*," because 45 and 46 are even better results than 47. If, on the other hand, we obtain an associated sum of 90 and wanted to show that this indicates that the fluorescein treatment is injurious to plants, we should consider the event "the sum is 90 *or more*."

3.4 EXERCISES

1.   The following table gives the probabilities that various numbers of employees of a certain firm will be absent on any given day.

Number absent	0	1	2	3	4	5	6	7	8	9	10	11	>11
Probability	0.01	0.03	0.06	0.15	0.25	0.30	0.10	0.04	0.03	0.01	0.007	0.008	0.005

Using a 5-percent significance level, which of the following events indicate—
(a) an epidemic or other unusual cause of absence
(b) a payday or other special cause for attendance?

A: Exactly seven employees are absent
B: More than eight employees are absent
C: Less than two employees are absent
D: Exactly ten employees are absent
E: Exactly one employee is absent

2.   What are the sums associated with the following arrangements of treated and untreated plants?
(a) *TTTTTTUTUTUTUUUUUU*
(b) *TTUTTUTTUTTUTUUUUU*
(c) *UUUUUUTTUTUTUTTTTT*

3.   Using Wilcoxon's system, what is the largest sum that can be associated with the outcome of an experiment having nine treated and nine untreated plants?

4.   Give two patterns associated with the sum $r = 48$.

5.   Compile Wilcoxon's table for the numbers 45–50.

6. (a) If twenty-six plants are used in an experiment similar to the one described in this section, what is the probability that chance alone will cause the thirteen treated ones to be judged best?

    (b) What is the probability that the treated plants will be judged best by chance if thirty plants are used? What is the probability if one hundred plants are used?

7. If eighteen plants are used in the experiment, what is the probability that eight or more of the treated plants are among the nine judged best?

8. Most roulette wheels in the United States have compartments numbered 1–36 inclusive and two more numbered zero.

    (a) What is the most probable number?

    (b) What is the probability of this number?

9. Answer the questions of Exercise 8 for a roulette wheel having 10,000 numbers plus two zeros.

10. The roulette wheel at Monte Carlo carries compartments numbered 1–36 and one numbered zero. The zero is not considered to be either even or odd. It was reported that "even" once came up twenty-eight times in succession. What is the probability of this happening? (Assume that there is nothing wrong with the wheel, so that all numbers are equally likely to appear.)

*11. A woman claims that she is able to tell by tasting whether the sugar or the tea was added first to a cup of sweetened tea. Assume that she does not have this ability.

    (a) If she is told that four out of eight cups have been made by adding the tea first, what is the probability that—

    (i) she guesses all correctly

    (ii) she guesses at most two wrong?

    (b) Answer the above questions under the assumption that she is told that six out of twelve cups have had the tea added first.

*12. Answer the questions of Exercise 11 under the assumption that the woman is not told how many cups have had the tea added first. Consider both experiments: the one in which four out of eight cups have been made by adding the tea first, and the one in which six out of twelve cups have been made that way.

3.5 COMBINATIONS

In Section 3.2 *ordered* arrangements were discussed. Sometimes, however, order is not important, as the following example shows.

Example 27 A club contains fifty members. In how many ways can a three-member committee be chosen to plan a Memorial Day picnic?

Solution Let x be the number of ways that the committee may be chosen. The committee will be the same whether Sam is chosen first, Peter second, and

Karen third (*SPK*), or Peter first, Karen second, and Sam third (*PKS*). There-fore, x is not the number of ordered arrangements (*permutations*), but the number of unordered arrangements (*combinations*). The single combination $\{S,K,P\}$ corresponds to the 3! permutations *SKP, SPK, KSP, KPS, PSK,* and *PKS.*

In fact, for each of the x ways that the committee may be selected, there are 3! ways of ordering the chosen members. It follows that the number of ordered arrangements is $3!x$. Since we also know (by Theorem 3.4) that the number of ordered arrangements is $_{50}P_3$, it follows that $_{50}P_3 = 3!x$. Solving for x yields

$$x = \frac{_{50}P_3}{3!} = \frac{50!}{3!\,47!} = \frac{50 \cdot 49 \cdot 48}{3 \cdot 2 \cdot 1} = 19{,}600$$

One must be careful when deciding whether order matters in a particular prob-lem. If the problem is vaguely worded, a good case can be made for either answer. For example, suppose that we want to know how many different games of tennis can be arranged if there are four players, $A_1, A_2, A_3,$ and A_4. If $A_i A_j$ means that player A_i plays player A_j and player A_i serves first (or plays on the sunny court, etc.), then order matters, and the problem is one of permutations. The number of arrangements is $_4P_2 = 12$. If who plays first does not matter, and we are merely interested in which two players are paired, then order does not matter, and the problem is one of combinations. Since each unordered pair $A_i A_j$ corresponds to two ordered pairs, $A_i A_j$ and $A_j A_i$, the number of arrangements is $\frac{12}{2} = 6$.

Definition 3.3   A **combination** *of n things taken r at a time* is a selection of r of the n things without regard to order.

The combination of a and b may be denoted $\{a,b\}$, or simply ab. The set notation emphasizes the fact that order is unimportant. The second notation may be used if it is clear that a combination, not a permutation, is meant.

Example 28

(a) The combinations of the letters a, a, b, b, c taken two at a time are

$aa, \quad ab, \quad ac, \quad bc, \quad bb$

(b) The combinations of the letters a, b, c, d taken two at a time are

$ab, \quad ac, \quad ad, \quad bc, \quad bd, \quad cd$

The objects in a combination need not all be different. However, if they are all different, the following theorem can be applied. The proof is similar to the solution of Example 27.

THEOREM   3.7   The number of combinations of n distinct things taken r at a time is

$$_nC_r = \frac{n!}{r!(n-r)!} \qquad r = 1, 2, \ldots, n$$

In particular,

$$_nC_1 = \frac{n!}{1!(n-1)!} = \frac{n(n-1)!}{(n-1)!} = n$$

because there are n ways of selecting one of the n objects. Similarly,

$$_nC_n = \frac{n!}{n!0!} = 1$$

because there is only one way of selecting n of the n objects.

Note that $_nC_r$ is only defined if $r = 1, 2, \ldots, n$. However, if we let $r = 0$ in Theorem 3.7, we get

$$_nC_0 = \frac{n!}{0!(n-0)!} = \frac{n!}{1 \cdot n!} = 1$$

It is therefore convenient to make the following definition:

$$_nC_0 = 1$$

This is also clear intuitively, because there is only one way zero things can be selected from a collection of n things. Theorem 3.7 now holds for $r = 0, 1, 2, \ldots, n$.

The following corollary of Theorem 3.7 simplifies the computation of $_nC_r$. Its use is illustrated in Example 29.

THEOREM 3.8

$$_nC_r = {_nC_{n-r}} = \frac{_nP_r}{r!} \qquad r = 1, 2, \ldots, n \tag{4}$$

Theorem 3.8 follows from Theorem 3.7 and Exercise 11 at the end of this section.

Example 29

(a) Applying Theorem 3.7,

$$_5C_3 = \frac{5!}{3!2!} = \frac{5 \cdot 4 \cdot 3 \cdot 2 \cdot 1}{(3 \cdot 2 \cdot 1)(2 \cdot 1)} = 10$$

Alternatively, applying Theorem 3.8,

$$_5C_3 = \frac{_5P_3}{3!} = \frac{5 \cdot 4 \cdot 3}{6} = 10$$

as before.

(b) Applying Theorem 3.8 and part (a), $_5C_2 = {_5C_3} = 10$.

(c) Applying Theorem 3.8,

$$_9C_6 = {_9C_3} = \frac{9!}{3!6!} = \frac{9 \cdot 8 \cdot 7 \cdot 6 \cdot 5 \cdot 4 \cdot 3 \cdot 2 \cdot 1}{(3 \cdot 2 \cdot 1)(6 \cdot 5 \cdot 4 \cdot 3 \cdot 2 \cdot 1)} = 84$$

(d) Applying Theorem 3.8,

$$_{10}C_8 = {}_{10}C_2 = \frac{_{10}P_2}{2!} = \frac{10 \cdot 9}{2} = 45$$

(e) Applying Theorem 3.8,

$$_{25}C_{20} = {}_{25}C_5 = \frac{_{25}P_5}{5!} = \frac{25 \cdot 24 \cdot 23 \cdot 22 \cdot 21}{5 \cdot 4 \cdot 3 \cdot 2 \cdot 1} = 53{,}130$$

(f) Applying equation (4), $_{10}C_0 = 1$.

The following examples show how Theorem 3.7 can be used in conjunction with the first counting principle.

Example 30

(a) In how many ways can a committee of 3 students and 3 teachers be chosen from a student body of 500 and a faculty of 40?

(b) If all ways are equally likely, what is the probability that Ricardo and his favorite instructor, Mr. Sims, will both be on the committee?

Solution

(a) The 3 students can be chosen from the 500-member student body in $_{500}C_3$ ways; for each of these ways, the 3 teachers can be chosen in $_{40}C_3$ ways. Applying the first counting principle, the total number of ways is $_{500}C_3 \cdot {}_{40}C_3$.

(b) Since Ricardo must be on the committee, the 2 other students must be chosen from 499 students. This can be done in $_{499}C_2$ ways. Similarly, the number of ways that the teachers can be chosen is $_{39}C_2$. It follows from the first counting principle that the number of favorable ways is $_{499}C_2 \cdot {}_{39}C_2$. Using part **(a)**, the probability is

$$\frac{_{499}C_2 \cdot {}_{39}C_2}{_{500}C_3 \cdot {}_{40}C_3} = \frac{\left(\dfrac{_{499}P_2}{2!}\right)\left(\dfrac{_{39}P_2}{2!}\right)}{\left(\dfrac{_{500}P_3}{3!}\right)\left(\dfrac{_{40}P_3}{3!}\right)}$$

$$= \frac{\left(\dfrac{499 \cdot 498}{2}\right)\left(\dfrac{39 \cdot 38}{2}\right)}{\left(\dfrac{500 \cdot 499 \cdot 498}{3 \cdot 2}\right)\left(\dfrac{40 \cdot 39 \cdot 38}{3 \cdot 2}\right)} = \frac{9}{20{,}000}$$

Example 31   Suppose that Mr. Sims refuses to serve on the committee of Example 30 if Donald is on the committee. In how many ways can the committee be chosen?

Solution   The number of ways that the committee may be chosen is equal to the total number of ways that 3 students and 3 teachers may be chosen (as found in

the solution to Example 30(**a**)) minus the number of choices in which a particular student and a particular teacher are on the committee (as calculated in
the solution to Example 30(**b**)). Hence, the number of ways the committee may
be chosen is

$$_{500}C_3 \cdot {}_{40}C_3 - {}_{499}C_2 \cdot {}_{39}C_2$$

Example 32   In five-card stud poker, an ordinary deck of 52 cards is well-shuffled
and 5 cards are dealt to each of two or more players. The 5 cards are called a
hand and the hands are ranked in the following order (the lower the number,
the better the hand).

(1)   A *straight flush* is a hand in which the cards are of the same suit and in
sequence (e.g., the ace, 2, 3, 4, and 5 of hearts, or the 10, jack, queen, king,
and ace of spades). The last card in the sequence cannot be a 2, 3, or 4.

(2)   *Four of a kind* is a hand having four cards with the same face value (e.g.,
four 2s and a queen).

(3)   *A full house* is a hand having two cards with one face value and three cards
with a second face value (e.g., two queens and three kings).

(4)   In a *flush*, all cards are of the same suit, but they are not in sequence (e.g.,
the 2, 4, 5, queen, and king of hearts).

(5)   A *straight* is a hand in which all the cards are in sequence but are not of the
same suit (e.g., the 2, 3, 4, and 5 of hearts and the 6 of spades).

(6)   In *three of a kind*, there are three cards with the same face value and two
other different cards (e.g., three queens, a 4 and a 6).

(7)   *Two pairs* is a hand having two cards with one face value, two other cards
with a second face value, and a fifth card with a third face value (e.g., two
3s, two 2s, and a 9).

(8)   *One pair* is a hand having two cards with the same face value and three
other cards with different face values (e.g., two 3s, a 2, a 4, and a 9).

(9)   *Nothing* is a hand which is unlike any of the preceding hands.

What is the probability of being dealt each of the preceding hands? Is the order
of ranking justified?

Solution   The total number of ways of getting a hand in poker is just the number
of ways of choosing 5 cards from a set of 52 cards, $_{52}C_5$.

We shall calculate the probabilities for hands (1)–(3); the remaining cases
will be left as an exercise.

(1)   There are 4 ways that the suit may be chosen. Once the suit is chosen, there
are 10 choices for the smallest card in the sequence (ace, 2, 3, ..., 10). Once
the smallest card is selected, the remaining cards are uniquely determined.
By Theorem 3.1, the number of favorable ways is $4 \cdot 10 = 40$, so

$$P(\text{straight flush}) = \frac{40}{_{52}C_5}$$

(2)   Since the face value of the quartet may be chosen in 13 ways, and there are

48 choices for the remaining card, there are $13 \cdot 48 = 624$ ways of obtaining the hand. It follows that

$$P(\text{four of a kind}) = \frac{624}{{}_{52}C_5}$$

Since the probability of getting a straight flush is less than the probability of getting four of a kind, this ranking is justified.

(3) The face value of the triplet may be chosen in 13 ways; the face value of the pair may then be chosen in 12 ways. We must still choose the suits of the cards. Suppose that the cards in the triplet are kings and the cards in the pair are queens. We must choose 3 of the 4 kings in the deck and 2 of the 4 queens. It follows that the number of favorable ways is $13 \cdot 12 \cdot {}_4C_3 \cdot {}_4C_2 = 3744$. Therefore,

$$P(\text{full house}) = \frac{3744}{{}_{52}C_5}$$

This is six times the probability of receiving four of a kind, so the ranking is again justified.

3.5 EXERCISES

1. Calculate—

    (a) ${}_7C_3$ (b) $\dfrac{{}_6C_2}{{}_3C_2}$ (c) ${}_{1999}C_{1998}$ (d) ${}_5C_0$

    (e) ${}_{100}C_{100}$ (f) ${}_0C_0$

2. Calculate—
    (a) ${}_8C_2$ (b) ${}_{19}C_2$ (c) ${}_{12}C_1$ (d) ${}_{20}C_3$ (e) ${}_8C_3$
    (f) ${}_8C_5$ (g) ${}_8P_5$ (h) ${}_{18}C_{18}$

3. A polltaker must question the occupants of ten houses on Michigan Avenue. If there are forty houses on the street, in how many ways may this be done?

4. There are fifteen left shoes and twenty right shoes in a closet. In how many ways can fifteen pairs be formed if
    (a) it does not matter to which right shoe each left shoe is matched
    (b) each left shoe must be paired with the matching right shoe?

5. A business executive would like to take five outfits to wear on a business trip. If she has fourteen different outfits, in how many ways can she choose the ones to take?

6. Complete the solution to Example 32.

7. A company tests its cases of lightbulbs by choosing a sample of 10 bulbs from the case. (The case contains 100 bulbs.) If more than 1 defective bulb is found in the sample of 10, the case is rejected.
    (a) In how many ways can the sample be chosen?

(b) If the case contains 3 defective bulbs, what is the probability that it is rejected?

8. A union has fifty male and forty female members. In how many ways can a ten-member committee be chosen if—
   (a) there are no restrictions on the committee's membership
   (b) the committee must contain exactly four women
   (c) the committee must contain at least four women
   (d) the first person chosen is to be the chairperson?

9. In the game of bridge, thirteen cards are dealt from an ordinary fifty-two-card deck to each of four players, north (N), south (S), east (E), and west (W).
   (a) What is the probability that N's cards are all of the same suit?
   (b) What is the probability that N has no hearts?
   (c) If N has six hearts, what is the probability—as far as N is concerned—that his partner has at least one heart?

10. If there are twenty children performing at a piano recital, what is the probability that the three Eisen children will be among the first ten performers so that the Eisens can leave at the intermission?

11. Prove that—

   (a) $_nC_k = {}_nC_{n-k}$

   (b) $_nC_k = {}_{n-1}C_{k-1} + {}_{n-1}C_k$

   (c) $2^n = {}_nC_0 + {}_nC_1 + \cdots + {}_nC_n$

3.6 PASCAL'S TRIANGLE

Theorems 3.7 and 3.8 of the previous section tell us how to calculate $_nC_k$. However, this method is tedious if there are many values $_nC_k$ to be calculated in one problem. In this section we will learn another method for evaluating $_nC_k$.

Several numerical values of $_nC_k$ appear in Figure 3.9. This array is known as **Pascal's triangle**.

$$
\begin{array}{ccccccc}
{}_0C_0 & & & & & & & & & 1 \\
{}_1C_0 & {}_1C_1 & & & & & & & 1 & 1 \\
{}_2C_0 & {}_2C_1 & {}_2C_2 & & & & & 1 & 2 & 1 \\
{}_3C_0 & {}_3C_1 & {}_3C_2 & {}_3C_3 & & & 1 & 3 & 3 & 1 \\
{}_4C_0 & {}_4C_1 & {}_4C_2 & {}_4C_3 & {}_4C_4 & 1 & 4 & 6 & 4 & 1 \\
{}_5C_0 & {}_5C_1 & {}_5C_2 & {}_5C_3 & {}_5C_4 & {}_5C_5 & 1 & 5 & 10 & 10 & 5 & 1
\end{array}
$$

Figure 3.9   Pascal's triangle

The observant reader may already have figured out the next line of the triangle.

Certainly it is clear that there will be seven numbers in the row and that the first and last numbers will be 1. These and other properties of the triangle are summarized below.

Properties of Pascal's Triangle

(a)   *The triangle is symmetric.* That is, the numbers equidistant from each end in a given row are equal. For example, $_5C_2 = {}_5C_3$ and $_3C_1 = {}_3C_2$. This property follows from Theorem 3.8.

(b)   *The first and last numbers in each row are 1.* This is because $_nC_0 = {}_nC_n = 1$.

(c)   *Each number except the 1s at the ends of each line is the sum of the two numbers appearing to the right and left of it in the line above.* For example,

$$_5C_1 = {}_4C_0 + {}_4C_1 \quad \text{and} \quad {}_4C_3 = {}_3C_2 + {}_3C_3$$

In general,

$$_nC_k = {}_{n-1}C_{k-1} + {}_{n-1}C_k$$

(d)   *The sum of the numbers in the* $(n + 1)$*st row is* 2^n. For example, the sum of the numbers in the first row is $2^0 = 1$; the sum of the numbers in the second row is $2^1 = 2$. In general, the sum of the numbers in the $(n + 1)$st row is

$$_nC_0 + {}_nC_1 + {}_nC_2 + \cdots + {}_nC_n = 2^n$$

(e)   *The nth row has n entries.*

(f)   *The* $(n + 1)$*st row gives the values of* $_nC_0, {}_nC_1, {}_nC_2, \ldots, {}_nC_n$. This is the row that has n as its second entry.

Using properties (b) and (c), one can easily figure out any line of the triangle from the preceding one. The seventh line is

$$1 \quad 1+5 \quad 5+10 \quad 10+10 \quad 10+5 \quad 5+1 \quad 1$$

or

$$1 \quad 6 \quad 15 \quad 20 \quad 15 \quad 6 \quad 1 \tag{5}$$

The usefulness of Pascal's triangle in computing probabilities is illustrated by the following examples.

Example 33   What is the probability that exactly two heads will be obtained when four coins are tossed?

Solution   The total number of outcomes from tossing 4 coins is $2 \times 2 \times 2 \times 2 = 2^4$, because there are 2 choices for the first coin (heads or tails), 2 choices for the second coin (heads or tails), and so on. Two heads will be obtained if exactly 2 of the 4 coins are heads; that is, in $_4C_2$ ways. It follows that

$$P(2 \text{ heads}) = \frac{_4C_2}{2^4}$$

A numerical answer for $P(2$ heads$)$ can be found by using the fifth line of Pascal's triangle. This is the line that has 4 as its second entry. (Notice that 4 coins were tossed.) Applying property (f) of the triangle, this row gives the values of $_4C_0, _4C_1, \ldots, _4C_4$. It follows that the numerator of $P(2$ heads$)$, $_4C_2$, is the third entry of this row, $_4C_2 = 6$. Applying property (d) of the triangle, the denominator of $P(2$ heads$)$, 2^4, is the sum of the entries in this row: $2^4 = 1 + 4 + 6 + 4 + 1 = 16$. The required probability is therefore $P(2$ heads$) = \frac{6}{16}$.

In general, if n coins are tossed, the probability that exactly k heads will be obtained is

$$P(k \text{ heads}) = \frac{_nC_k}{2^n}$$

The numerical value of $P(k$ heads$)$ is found in the $(n + 1)$st row of the triangle. This is the row whose second entry is n. The numerator is the $(k + 1)$st entry of the row; the denominator is the sum of the entries in the row.

Example 34   If six coins are tossed, what is the probability of obtaining—

(a) exactly two heads (A)
(b) four or more heads (B)
(c) at least one head (C)
(d) no heads (D)
(e) exactly four tails (E).

Solution   Since *six* coins are tossed, we must look at the *seventh* line of the triangle. (See equation (5) on page 101.) This is the line having 6 as its second entry. The sum of the entries in this row (64) will be the denominator for each of the probabilities in parts (a)–(e). The entries in the row are the number of ways that 0, 1, 2, 3, 4, 5, 6 heads can be obtained. Hence—

(a) $P(A) = \frac{15}{64}$
(b) $P(B) = P(4 \text{ heads}) + P(5 \text{ heads}) + P(6 \text{ heads}) = (15 + 6 + 1)/64 = \frac{22}{64}$
(c) $P(C) = 1 - P(0 \text{ heads}) = 1 - \frac{1}{64} = \frac{63}{64}$
(d) $P(D) = \frac{1}{64}$
(e) Since obtaining exactly four tails is the same as obtaining exactly two heads,

$$P(E) = P(A) = \frac{15}{64}$$

A sample space associated with the experiment "a coin is tossed n times" in which the elementary events are equally likely is

$$\Omega = \{(x_1, x_2, \ldots, x_n) \mid x_i = h \text{ or } t\}$$

We have seen that Pascal's triangle can be used to find the probability that exactly k *heads* are obtained. A similar argument shows that it can also be used to find the probability that exactly k *tails* are obtained. The value of $P(E)$ in Example 34 (e), for example, can be found from the fifth entry of the seventh row. Since the triangle is symmetric, the fifth and third entries are identical; therefore, $P(E) = \frac{15}{64}$ as before.

If h and t are replaced by two other letters, then the sample space Ω for the coin-tossing experiment can be used to describe other experiments, and Pascal's triangle can be used to compute probabilities. Before applying Pascal's triangle, however, one must be certain that the elementary events are equally likely.

Example 35   In an experiment designed to ascertain whether or not newborn chicks can differentiate shapes, a chick was allowed to peck seven times at 100 pieces of paper—50 circles and 50 triangles. If we assume that the chick cannot differentiate shapes, the probability that it pecks at a triangle is $\frac{50}{100} = \frac{1}{2}$. Therefore, we may use the eighth line of Pascal's triangle to calculate the probability P_k that it pecks at a triangle k of the seven times. Using (5) and property (c) of the triangle, the eighth line is

$$1 \quad 7 \quad 21 \quad 35 \quad 35 \quad 21 \quad 7 \quad 1$$

and so $P_0 = \frac{1}{128}$, $P_1 = \frac{7}{128}$, $P_2 = \frac{21}{128}$, and so on. The probability that the chick pecks at a triangle five or more times out of seven is $\frac{21}{128}$.

In a series of experiments similar to this one (allowing more pecks, of course), it was discovered that chicks have a definite preference for circles. This ability to distinguish shapes probably helps them peck at grain instead of gravel.

3.6 EXERCISES

1.   Use Pascal's triangle to calculate—
     (a) $_6C_2$ **(b)** $_7C_5$ **(c)** $_7C_4$ **(d)** $_8C_2$ **(e)** $_8C_6$ **(f)** $_8C_5$

2.   If seven coins are tossed, what is the probability of obtaining—
     (a) exactly three heads
     (b) four or more heads
     (c) no tails
     (d) at least one head?

3.   Suppose that the chick of Example 35 is allowed to peck nine times at 100 pieces of paper—50 circles and 50 triangles.
     (a) If the chick is unable to differentiate shapes, what is the probability that—
         (*i*)   every peck is at a circle
         (*ii*) six or more pecks are at circles
         (*iii*) exactly six pecks are at triangles
         (*iv*) exactly eight pecks are at triangles?
     (b) Using a 5-percent significance level and a 1-percent high significance level, which of the above results are significant? Which are highly significant?

4.   Suppose that 50 percent of the children in a certain school are boys. If eight students are randomly chosen, what is the probability that—
     (a) exactly five are boys

(b) at least three are boys
(c) at most two are girls
(d) none are boys
(e) more than three are girls?

3.7 COMBINATIONS WITH REPETITIONS

Theorem 3.7, Theorem 3.8, and Pascal's triangle can all be used to find the number of combinations of n distinct objects *if the objects are not repeated*. For example, they can be used to find the number of ways that the letters

$$a, b, c, d$$

can be combined two at a time. There are $_4C_2 = 6$ combinations:

$$ab, \quad ac, \quad ad, \quad bc, \quad bd, \quad cd$$

However, the theorem, its corollary, and Pascal's triangle cannot be used to find the number of ways that the letters a, b, and c may be combined two at a time *if the letters may be repeated*; that is, if combinations such as

$$aa, \quad bb, \quad cc, \quad \text{and} \quad dd$$

are allowed. In this section, we will learn how to find the number of such combinations; that is, the number of combinations of n distinct objects *if repetitions are allowed*. The following example shows one way of doing this.

Example 36   A store sells five different flavors of ice cream. In how many ways can three ice-cream cones be ordered?

Solution   The answer is not simply $_5C_3$, because the flavors may be repeated. (For example, we can order 3 chocolate ice-cream cones.) There are 5 ways of ordering 3 cones all of the same flavor; there are $5 \cdot 4 = 20$ ways of ordering 2 cones of one flavor and 1 of a second flavor; and there are $_5C_3 = 10$ ways of ordering 3 cones of three different flavors. It follows that the total number of ways that the cones may be ordered is $5 + 20 + 10 = 35$.

Although the method used to solve Example 36 worked easily in that case, it would be tedious to apply if there were 10 different flavors of ice cream and 7 ice-cream cones. The next example demonstrates a method that can be easily generalized and that can therefore be applied to more difficult cases.

Example 37 (*Alternate method for solving the problem of Example 36*)   Let five flavors of ice cream—strawberry, chocolate, vanilla, peach, and butter pecan—be designated by the *spaces* between the following four vertical lines.

$$S \quad C \quad V \quad P \quad B$$
$$| \quad | \quad | \quad |$$

A dot that appears to the left of the first vertical line (in the strawberry place) represents a strawberry cone, a dot in the space marked V represents a vanilla cone, and so on. The configuration

$$\cdot \mid \cdot\cdot \mid \quad \mid \quad \mid$$

means that one cone is strawberry and two are chocolate (we have omitted the labels). The configuration

$$\cdot\cdot\cdot \mid \quad \mid \quad \mid \quad \mid$$

means that the cones are all strawberry.

It is clear that every arrangement of four lines and three dots corresponds to an order of ice-cream cones, and vice versa. It follows that the total number of ways of ordering ice-cream cones is the number of ways of arranging four things of one kind (lines) and three things of a second kind (dots). There are

$$\frac{7!}{3!4!} = 35$$

possible arrangements of this type.

In this example we combined $n = 5$ different things $r = 3$ at a time, *allowing repetitions*. The method used to solve the example can be easily altered to prove the following theorem.

THEOREM   3.9   The number of ways of combining n different things r at a time, *allowing repetitions*, is

$$_{n+r-1}C_r$$

For example, applying Theorem 3.9, the number of ways that the letters

$$a, b, c, d$$

can be combined two at a time if repetitions are allowed is $_{4+2-1}C_2 = {}_5C_2 = 10$. The ten ways are

$$ab, \quad ac, \quad ad, \quad bc, \quad bd, \quad cd, \quad aa, \quad bb, \quad cc, \quad dd$$

A second illustration of how to apply Theorem 3.9 is given in the next example.

Example 38   The letters A, B, and O denote three basic blood types found in human beings. If a person has genotype $AO = OA$, for example, he has inherited an A gene from one parent and an O gene from the other.

(a) How many genotypes are there?

(b) How many different matings of genotypes are possible (e.g., $AA \times AO$)?

Solution

(a) The number of genotypes is the number of combinations of 3 distinct things (A, B, and O) taken 2 at a time *allowing repetitions*, because combinations like AA are possible. Applying Theorem 3.9, this number is $_{3+2-1}C_2 = {_4}C_2 = 6$.

(b) The number of different matings is the number of combinations of the 6 genotypes taken 2 at a time with repetitions. This is $_{6+2-1}C_2 = {_7}C_2 = 21$.

Theorems 3.1, 3.4, 3.7, and 3.9 can all be applied to problems of finding the number of ways that r objects can be selected from n objects. In order to decide which of the four theorems should be used for a given problem, readers should ask themselves the following two questions:

(a) Is order important?

(b) Are repetitions allowed?

The following summary may be helpful:

The Number of Ways of Selecting r Objects from n Distinct Objects

(a) Order is important (permutations)

    (*i*) Repetitions are not allowed $_nP_r$ Theorem 3.4 or Theorem 3.1

    (*ii*) Repetitions are allowed n^r Theorem 3.1

(b) Order is not important (combinations)

    (*i*) Repetitions are not allowed $_nC_r$ Theorem 3.7

    (*ii*) Repetitions are allowed $_{n+r-1}C_r$ Theorem 3.9

In addition, the number of permutations of n things when there are k_1 things of one kind, k_2 things of a second kind, and so on, up to k_r things of an rth kind $(k_1 + k_2 + \cdots + k_r = n)$ is

$$\frac{n!}{k_1! k_2! \cdots k_r!}$$

(See Theorem 3.6 on page 87.)

3.7 EXERCISES

1.   A candy manufacturer makes forty different kinds of miniature chocolates and packs them in boxes containing thirty candies each.

    (a) How many different assortments are possible?

    (b) If all assortments are equally likely and a box is randomly chosen, what is the probability that all the chocolates in the box are the same kind?

2.   An ice-cream company sells thirty-four different flavors of ice cream. How many two-scoop ice-cream cones can be made if the order of the scoops is—

    (a) not important

    (b) important?

3.   A Monster Sundae contains three scoops of ice cream, two different syrups, whipped cream, and either nuts or chocolate sprinkles. If the store has twenty-four flavors of ice cream, seven different syrups, and three different kinds of nuts, how many different Monsters can be ordered?

4.   **(a)** A restaurant serves seven different kinds of vegetables. In how many ways can two portions of vegetables be ordered if—
      (*i*) the vegetables must be different
      (*ii*) the vegetables may be the same?
      Assume that the order of the vegetables is unimportant—for example, peas and carrots is the same as carrots and peas.

      (b) Answer the preceding questions if the order of the vegetables is important—for example, if peas and carrots means one customer gets peas while a second customer gets carrots, but carrots and peas means the first customer gets carrots while the second gets peas.

3.8 THE BINOMIAL THEOREM AND SUMMATION NOTATION

By straightforward multiplication, the reader can easily verify that

$$(x + y)^0 = 1$$
$$(x + y)^1 = 1x + 1y$$
$$(x + y)^2 = 1x^2 + 2xy + 1y^2$$
$$(x + y)^3 = 1x^3 + 3x^2y + 3xy^2 + 1y^3$$

If we write the coefficients of the preceding expansions in the form of a triangle, a familiar pattern is obtained.

$$1$$
$$1 \quad 1$$
$$1 \quad 2 \quad 1$$
$$1 \quad 3 \quad 3 \quad 1$$

This is, of course, the first few lines of Pascal's triangle.

Based on the expansions for $(x + y)^0$, $(x + y)^1$, and so on, the reader may suspect that each term in the expansion of $(x + y)^4$ has the form

$$c_k x^{4-k} y^k \qquad k = 0, 1, 2, 3, 4$$

and that the coefficients c_k can be obtained by calculating the next line of the triangle. That is, the reader may suspect that

$$(x + y)^4 = 1x^4 + 4x^3y + 6x^2y^2 + 4xy^3 + 1y^4$$

This can be verified by multiplying $(x + y)^3$ by $(x + y)$.

It now seems clear that there is a mathematical relationship between Pascal's triangle and expansions of powers of binomial expressions of the form $(x + y)^n$. The following example will lead us to a theorem that states this relationship.

Example 39   Expand $(x + y)^5$.

Solution

$$(x + y)^5 = (x + y)(x + y)(x + y)(x + y)(x + y) \tag{6}$$

If the right side of equation (6) is multiplied out, it is clear that we will obtain a sum of products such as

$$xxxyx, \; xxyxx, \; xxxxx$$

and so on. Each product is found by taking an x or y from the first factor of $(x + y)^5$, an x or y from the second factor, and so on. In order to find the expansion, we must therefore find all products of the form

$$z_1 z_2 z_3 z_4 z_5 \qquad \text{where each } z_i \text{ is either } x \text{ or } y \tag{7}$$

One such product is $x^5 = xxxxx$, another is

$$x^4 y = xxxxy = xxxyx = xxyxx = xyxxx = yxxxx$$

Collecting similar terms, the first few terms of the expansion are

$$(x + y)^5 = x^5 + x^4 y + x^4 y + x^4 y + x^4 y + x^4 y + x^3 y^2 + \cdots$$
$$= x^5 + 5x^4 y + \cdots$$

In general, it is clear that every term of the sum will have the form

$$x^{5-k} y^k$$

because, if exactly k of the z_i's in expression (7) are y, the remaining $5 - k$ factors must be x. It follows that

$$(x + y)^5 = x^5 + 5x^4 y + c_2 x^3 y^2 + c_3 x^2 y^3 + c_4 xy^4 + c_5 y^5$$

The coefficient c_k is the number of times that the term $x^{5-k} y^k$ appears when $(x + y)^5$ is expanded. For example, c_2 is the number of times

$$x^3 y^2 = xxxyy = xxyyx = xyxyx = \cdots$$

appears. To find c_2, we must find the number of ways of permuting three x's and two y's. Applying Theorem 3.6, page 87,

$$c_2 = \frac{5!}{3!\,2!} = {}_5 C_2$$

Similarly, $c_k = {}_5 C_k$, and so

$$(x + y)^5 = {}_5 C_0 x^5 + {}_5 C_1 x^4 y + {}_5 C_2 c^3 y^2 + {}_5 C_3 x^2 y^3 + {}_5 C_4 xy^4 + {}_5 C_5 y^5$$

The coefficients $_5C_k$ are found in the sixth line of Pascal's triangle. (See Figure 3.9.) Using the triangle,

$$(x + y)^5 = x^5 + 5x^4y + 10x^3y^2 + 10x^2y^3 + 5xy^4 + y^5$$

The following theorem can be proved using a similar method.

THEOREM 3.10 (Binomial Theorem)   If n is a nonnegative integer,

$$(x + y)^n = {}_nC_0x^n + {}_nC_1x^{n-1}y + {}_nC_2x^{n-2}y^2 + \cdots + {}_nC_ny^n$$

To remember the pattern of Theorem 3.10, observe that—

(a) there are $n + 1$ terms
(b) the powers of x start at n and *decrease* to 0; the powers of y start at 0 and *increase* to n
(c) in each term, the sum of the power of x and the power of y is n
(d) the coefficients come from the $(n + 1)$st row of Pascal's triangle.

The following example shows how to apply Theorem 3.10.

Example 40   Expand $(x + y)^6$.

Solution   The coefficients $_6C_k$ of the expansion appear in the seventh line of Pascal's triangle. (See equation (5), page 101.) Using Theorem 3.10,

$$(x + y)^6 = x^6 + 6x^5y + 15x^4y^2 + 20x^3y^3 + 15x^2y^4 + 6xy^5 + y^6$$

In Examples 39 and 40 it was necessary to write out the sum of a large number of terms. Instead of writing out each term, it is convenient to have a shorthand notation. If $a_1, a_2, \ldots, a_n$ are numbers, the sum

$$a_1 + a_2 + \cdots + a_n$$

is also written

$$\sum_{i=1}^{n} a_i$$

The symbol $\sum$ is the capital Greek letter *sigma* and indicates that the sequence of numbers $a_1, a_2, \ldots, a_n$ is to be *summed*. The range of numbers to be summed is indicated by the limit indices above and below the symbol (in this case, 1 and n). The particular numbers to be summed are obtained by setting the i in a_i equal to the lower limit index (in this case, 1) and increasing the subscript by 1 until the upper limit index (in this case, n) is reached. The limits, of course, can be any integers as long as the lower limit is less than the upper limit. This notation, which is similar to the notation used for the union of mutually disjoint sets, is called **summation notation**.

Example 41 illustrates the summation notation.

Example 41

(a) $\displaystyle\sum_{i=1}^{5} a_i = a_1 + a_2 + a_3 + a_4 + a_5$

(b) $\displaystyle\sum_{i=-1}^{6} a_i = a_{-1} + a_0 + a_1 + \cdots + a_6$

(c) $\displaystyle\sum_{j=s}^{t} b_j = b_s + b_{s+1} + b_{s+2} + \cdots + b_t \qquad (s < t)$

(d) $\displaystyle\sum_{k=1}^{4} 2^k = 2^1 + 2^2 + 2^3 + 2^4 = 30$

(e) $\displaystyle\sum_{k=2}^{10} \frac{1}{k} = \frac{1}{2} + \frac{1}{3} + \frac{1}{4} + \cdots + \frac{1}{10}$

The expansion of

$$\sum_{j=1}^{4} 2^j = 2^1 + 2^2 + 2^3 + 2^4$$

is exactly the same as the expansion of

$$\sum_{k=1}^{4} 2^k$$

When the expansion is written out, neither a j nor a k appears. The j and the k are therefore called *dummy variables*. In general, it does not matter what letter is used to index the numbers that are to be summed.

The following examples show how summation notation can be used when expanding $(x + y)^n$ and calculating probabilities.

Example 42   Using summation notation, the binomial expansion can be written

$$(x + y)^n = \sum_{k=0}^{n} {}_nC_k x^{n-k} y^k \qquad n = 0, 1, 2, 3, \ldots \tag{8}$$

Example 43   Expand $(a - b)^{10}$.

Solution   This is in the form $(x + y)^n$ where $x = a$, $y = -b$, and $n = 10$. Applying equation (8),

$$(a - b)^{10} = [a + (-b)]^{10} = \sum_{k=0}^{10} {}_{10}C_k a^{10-k} (-b)^k$$

$$= \sum_{k=0}^{10} (-1)^k \, {}_{10}C_k a^{10-k} b^k$$

$$= {}_{10}C_0 a^{10} - {}_{10}C_1 a^9 b + {}_{10}C_2 a^8 b^2 - {}_{10}C_3 a^7 b^3$$

$$+ \cdots + {}_{10}C_{10} b^{10}$$

$$= a^{10} - 10a^9 b + 45a^8 b^2 - 120a^7 b^3 + \cdots + b^{10}$$

Example 44 Expand $(a^2 + 3b)^8$.

Solution Applying equation (8) with $x = a^2$, $y = 3b$, and $n = 8$ yields

$$(a^2 + 3b)^8 = \sum_{k=0}^{8} {}_8C_k(a^2)^{8-k}(3b)^k$$

$$= a^{16} + 24a^{14}b + 252a^{12}b^2 + \cdots + 3^8b^8$$

Example 45 If twenty coins are tossed, what is the probability of obtaining more than six heads?

Solution Let A_i be the event "exactly i heads are obtained." Using a method similar to that used in Examples 33 and 34, pages 101–102, the reader can show that

$$P(A_k) = \frac{{}_{20}C_k}{2^{20}}$$

It follows that

$$P(\text{more than 6 heads}) = P(A_7) + P(A_8) + \cdots + P(A_{20})$$

$$= \sum_{k=7}^{20} P(A_k) = \sum_{k=7}^{20} \frac{{}_{20}C_k}{2^{20}}$$

3.8 EXERCISES

1. Expand the following expressions:
    (a) $(a + b)^7$ (b) $(x - y)^7$ (c) $(2x - 3y)^7$ (d) $(x^2 - y)^7$

2. Expand the following expressions:
    (a) $(a^3 - 4b)^3$ (b) $(1 - 2x^2)^4$ (c) $(2x^2 + 3y)^4$

3. Using the binomial theorem, calculate $(1.01)^5$. (Hint: Write 1.01 as $1 + 0.01$.)

4. Using the binomial theorem, calculate $(1.1)^6$ to five decimal places.

5. Write the following sums in $\sum$ notation.
    (a) $1 + 2 + 3 + 4 + \cdots + 50$
    (b) $1 - \frac{1}{2} + \frac{1}{3} - \frac{1}{4} + \cdots + \frac{1}{99}$
    (c) $4 + 9 + 16 + 25 + \cdots + 100$
    (d) $1 + 3 + 5 + 7 + \cdots + 55$

6. Expand the following expressions:
    (a) $\displaystyle\sum_{k=-2}^{10} k^3$ (b) $\displaystyle\sum_{k=4}^{88} (-1)^k 2k$

7.   Calculate—

(a) $\displaystyle\sum_{k=4}^{5} k^2$ (b) $\displaystyle\sum_{k=3}^{6} (-1)^k 2k$ (c) $\displaystyle\sum_{k=1}^{9} \left(\frac{1}{k} - \frac{1}{k+1} \right)$ (d) $\displaystyle\sum_{k=1}^{5} 2^k$

8.   What is the probability of obtaining at least three sixes when ten dice are tossed?

3.9 GEOMETRIC AND ARITHMETIC SERIES

In Section 3.8 we learned how to write the sum of n numbers $a_1, a_2, \ldots, a_n$ compactly, i.e., in the form

$$\sum_{i=1}^{n} a_i$$

We saw that this notation is convenient when stating theorems or the results of problems. However, writing a sum in a compact form does not tell us what the total is. In this section we will study two kinds of sums that occur frequently and learn how to evaluate them.

Example 46 illustrates the two types of sums that we will study in this section.

Example 46

(a) Bill Lewis contributed 5 dollars to a charity in 1971, 7 dollars in 1972, 9 dollars in 1973, and so on. How much money had he contributed by 1976?

(b) Naomi Fukada contributed 1 dollar to her political party in 1971, 2 dollars in 1972, 4 dollars in 1973, and so on. How much money had she contributed by 1976?

Solution   Let L and F be the respective amounts (in dollars) that Bill Lewis and Naomi Fukada contributed by 1976. Then—

(a) $L = 5 + 7 + 9 + 11 + 13 + 15 = 60$ (9)

(b) $F = 1 + 2 + 4 + 8 + 16 + 32 = 63$ (10)

The sums in equations (9) and (10) are examples of *series*. Each of the summands is called a *term* of the series. For example, the terms of (9) are 5, 7, 9, and so on. The series in (9) is called an **arithmetic series** because each term is obtained by adding a fixed number, called the **common difference**, to the preceding term. In equation (9) the common difference is 2. In equation (10), however, there is no common difference; each term is 2 *times* the preceding term. Series like equation (10), in which each term is a fixed multiple of the preceding term, are called **geometric series**. The fixed multiplier is called the **common ratio**. The general forms of arithmetic and geometric series are as follows:

Arithmetic Series $a + (a + d) + (a + 2d) + \cdots + [a + (n-1)d]$ (11)

Geometric Series $a + ar + ar^2 + ar^3 + \cdots + ar^{n-1}$ (12)

where a is the first term, n is the number of terms, d is the common difference, and r is the common ratio.

Notice that the common difference d of a given arithmetic series can always be determined by *subtracting* any two successive terms. The common ratio r of a given geometric series can always be determined by *dividing* any two successive terms. Examples of geometric and arithmetic series follow.

Example 47

(a) The series in equation (9) is an arithmetic series with $a = 5, d = 7 - 5 = 2$, and $n = 6$. Notice that $9 - 7 = 11 - 9 = 13 - 11 = 15 - 13 = 2$, also.

(b) The series in equation (10) is a geometric series with $a = 1, r = \frac{2}{1} = 2$, and $n = 6$. Notice that $\frac{4}{2} = \frac{8}{4} = \cdots = \frac{32}{16} = 2$, also.

Example 48

(a) $2 + 6 + 10 + 14 + 18 + 22 + 26$ is an arithmetic series with $a = 2$, $d = 6 - 2 = 4$, and $n = 7$.

(b) $2 + 8 + 32 + 128$ is a geometric series with $a = 2, r = \frac{8}{2} = 4$, and $n = 4$.

Example 49   Which of the following series are arithmetic, which are geometric, and which are neither arithmetic nor geometric?

(a) $2 + 5 + 8 + 11 + 14 + 17$

(b) $2 + 6 + 8 + 12$

(c) $2 + 6 + 18 + 54 + 162$

Solution

(a) $2 + 5 + 8 + 11 + 14 + 17$ is an arithmetic series because each term is 3 more than the preceding one.

(b) $2 + 6 + 8 + 12$ is neither arithmetic nor geometric. It is not arithmetic because the difference of the first two terms is $6 - 2 = 4$ while the difference of the second and third terms is $8 - 6 = 2$. It is not geometric because the quotient of the first two terms is $\frac{6}{2} = 3$ while the quotient of the second and third terms is $\frac{8}{6}$.

(c) $2 + 6 + 18 + 54 + 162$ is a geometric series because each term is 3 times the previous one.

The calculation of the sum in equation (9) is relatively simple because the arithmetic series has only a few terms. Summing series with a large number of terms, however, is time-consuming and tedious. The following examples will lead us to formulas for finding the sum of any arithmetic series.

Example 50   Let $s = 1 + 3 + 5 + \cdots + 201$. Find s.

Solution   This is an arithmetic series with $a = 1$, $d = 2$, and $n = 101$. Writing the terms in reverse order and adding yields

$$
\begin{array}{rl}
s = & 1 + \quad 3 + \quad 5 + \cdots + 199 + 201 \\
s = & 201 + 199 + 197 + \cdots + \quad 3 + \quad 1 \\
\hline
2s = & 202 + 202 + 202 + \cdots + 202 + 202
\end{array}
$$

Hence

$$2s = 101(202)$$

and so

$$s = \frac{101(202)}{2} = (101)^2$$

Similarly, we can prove that the sum of any arithmetic series is

$$a + (a + d) + (a + 2d) + \cdots + [a + (n - 1)d] = \frac{n}{2}[2a + (n - 1)d] \quad (13)$$

Let l be the last term in an arithmetic series and let s be its sum. Then it follows from series (11) that

$$l = a + (n - 1)d$$

Applying formula (13),

$$s = \frac{n}{2}[2a + (n - 1)d] = \frac{n}{2}[a + a + (n - 1)d] = \frac{n}{2}[a + l]$$

It follows that a second formula for finding the sum of an arithmetic series is

$$a + (a + d) + (a + 2d) + \cdots + l = \frac{n}{2}[a + l] \quad (14)$$

This formula is often easier to remember.

The next example shows how formula (14) can be used to find the sum of the first n integers.

Example 51   Calculate $s = 1 + 2 + 3 + \cdots + n$.

Solution   This is an arithmetic series with $a = d = 1$ and $l = n$. Applying formula (14) yields

$$s = 1 + 2 + 3 + \cdots + n = \frac{n}{2}[1 + n] = \frac{n(n + 1)}{2} \quad (15)$$

Alternatively, the same answer can be obtained by applying formula (13).

Example 52 will lead us to a formula for finding the sum of any geometric series.

Example 52   Calculate $s = 1 + 2 + 4 + \cdots + 2^{100}$.

Solution   This is a geometric series with $a = 1$, $r = 2$, and $n = 100$. If we multiply each term by the common ratio 2, we get

$$s = 1 + 2 + 4 + 8 + \cdots + 2^{100} \tag{16}$$

$$2s = \quad 2 + 4 + 8 + \cdots + 2^{100} + 2^{101} \tag{17}$$

Subtracting equation (16) from equation (17) yields $s = 2^{101} - 1$.

Similarly, we can prove that the sum of any geometric series is

$$a + ar + ar^2 + \cdots + ar^{n-1} = \frac{a(1 - r^n)}{1 - r} \qquad \text{if } r \neq 1 \tag{18}$$

Formulas (13), (14), (15), and (18) are summarized in the following theorem.

THEOREM   3.11

Arithmetic Series

(a) $a + (a + d) + (a + 2d) + \cdots + [a + (n - 1)d] = \dfrac{n}{2}[2a + (n - 1)d]$

(b) $a + (a + d) + (a + 2d) + \cdots + l = \dfrac{n}{2}[a + l]$

(c) $1 + 2 + \cdots + n = \dfrac{n(n + 1)}{2}$

Geometric Series

(d) $a + ar + ar^2 + \cdots + ar^{n-1} = \dfrac{a(1 - r^n)}{1 - r} \qquad \text{if } r \neq 1$

where a is the first term, l is the last term, n is the number of terms, d is the common difference, and r is the common ratio.

The following examples show how to apply Theorem 3.11.

Example 53   Calculate—

   (a) $\displaystyle\sum_{k=1}^{20} k$ **(b)** $\displaystyle\sum_{k=3}^{10} 2k$ **(c)** $\displaystyle\sum_{k=4}^{20} (\tfrac{3}{5})^k$

Solution

   (a) Applying Theorem 3.11(c) with $n = 20$ yields

$$\sum_{k=1}^{20} k = \frac{20(21)}{2} = 210$$

(b) This is an arithmetic series with $a = 6$, $n = 8$, and $d = 2$. Applying Theorem 3.11(a) yields

$$\sum_{k=3}^{10} 2k = \frac{8}{2}[12 + 7(2)] = 104$$

Alternatively, applying Theorem 3.11(b) with $l = 20$ yields

$$\sum_{k=3}^{10} 2k = \frac{8}{2}[6 + 20] = 4(26) = 104$$

as before.

(c) This is a geometric series with $a = (\frac{3}{5})^4$, $r = (\frac{3}{5})$, and $n = 17$. Applying Theorem 3.11(d) yields

$$\sum_{k=4}^{20} (\tfrac{3}{5})^k = \frac{(\frac{3}{5})^4[1 - (\frac{3}{5})^{17}]}{1 - (\frac{3}{5})} = \tfrac{5}{2}(\tfrac{3}{5})^4[1 - (\tfrac{3}{5})^{17}]$$

Example 54   A coin is tossed twenty times. What is the probability that the first head will occur on an even-numbered toss?

Solution   Let A_i be the event "the first head occurs on the ith toss." For the first head to occur on the second toss, a tail must be followed by a head (th); hence

$$P(A_2) = \frac{1}{2^2}$$

The first head will occur on the fourth toss if and only if three tails are followed by a head ($ttth$). Hence

$$P(A_4) = \frac{1}{2^4}$$

Similarly,

$$P(A_n) = \frac{1}{2^n} \qquad (n = 2, 4, 6, \ldots, 20)$$

It follows that the required probability is

$$P = P(A_2 + A_4 + A_6 + \cdots + A_{20}) = P(A_2) + P(A_4) + \cdots + P(A_{20})$$

$$= \frac{1}{2^2} + \frac{1}{2^4} + \frac{1}{2^6} + \cdots + \frac{1}{2^{20}} = \frac{1}{4} + \frac{1}{16} + \frac{1}{64} + \cdots + \frac{1}{4^{10}}$$

This is a geometric series with $r = a = \frac{1}{4}$ and $n = 10$. Applying Theorem 3.11(d) yields

$$P = \frac{\frac{1}{4}[1 - (\frac{1}{4})^{10}]}{1 - \frac{1}{4}} = \tfrac{1}{3}[1 - (\tfrac{1}{4})^{10}]$$

The following theorem is useful in working with sums. It can be easily proved by expanding both sides of the equations in each part.

THEOREM   3.12

(a) $\displaystyle\sum_{k=1}^{n} (a_k + b_k) = \sum_{k=1}^{n} a_k + \sum_{k=1}^{n} b_k$

(b) $\displaystyle\sum_{k=1}^{n} ca_k = c\left(\sum_{k=1}^{n} a_k\right)$

(c) If $a_k = c$ for all k, then $\displaystyle\sum_{k=1}^{n} a_k = nc$.

The following examples are applications of Theorem 3.12.

Example 55   Find—

(a) $\displaystyle\sum_{k=1}^{10} 2$ (b) $\displaystyle\sum_{k=1}^{100} 1$

Solution

(a) This is in the form $a_k = 2$ for all k. Applying Theorem 3.12 (c) with $n = 10$ yields

$$\sum_{k=1}^{10} 2 = 10(2) = 20$$

(b) Similarly, applying Theorem 3.12(c) yields

$$\sum_{k=1}^{100} 1 = 100(1) = 100$$

Example 56   By Theorems 3.11 and 3.12—

(a) $\displaystyle\sum_{k=1}^{10} 3k = 3\left(\sum_{k=1}^{10} k\right) = 3\left(\frac{10 \cdot 11}{2}\right) = 165$

(b) $\displaystyle\sum_{n=1}^{8} (n + 2^n) = \sum_{n=1}^{8} n + \sum_{n=1}^{8} 2^n = \frac{8 \cdot 9}{2} + 2\left(\frac{1 - 2^8}{1 - 2}\right) = 36 + 510 = 546$

(c) $\displaystyle\sum_{n=1}^{100} (2n + 1) = 2\sum_{n=1}^{100} n + \sum_{n=1}^{100} 1 = 2\left(\frac{100 \cdot 101}{2}\right) + 100 = 10{,}200$

See Example 55(**b**).

(d) $\displaystyle\sum_{k=6}^{100} k = \sum_{k=1}^{100} k - \sum_{k=1}^{5} k = \frac{100(101)}{2} - \frac{5 \cdot 6}{2} = 5035$

Alternatively, Theorem 3.11(a) can be applied with $a = 6$, $n = 95$, and $d = 1$; or Theorem 3.11(b) can be applied with $a = 6$, $d = 1$, and $l = 100$.

3.9 EXERCISES

1. Which of the following series are arithmetic? Which are geometric?
   (a) $1 + 3 + 5 + 7 + 8$ **(b)** $2 + 5 + 8 + 11 + 14 + 17$
   (c) $1 + 2 + 4 + 8 + 16$ **(d)** $1 + 2 + 5 + 9 + 12$

2. Let G be a geometric series with five terms. Find G if its first term is 4 and its common ratio is (a) 3, (b) $\frac{1}{2}$.

3. Let A be an arithmetic series with six terms. Find A if its first term is 3 and its common difference is (a) 2, (b) -1.

4. Find the following sums:

   (a) $\sum_{k=1}^{10} k$ **(b)** $\sum_{k=1}^{10} 2^k$ **(c)** $\sum_{k=1}^{50} \frac{1}{4^k}$ **(d)** $\sum_{k=1}^{100} (4^k - k)$

5. Calculate the following sums:

   (a) $\sum_{n=1}^{30} 4n$ **(b)** $\sum_{n=1}^{40} (2n + 3)$ **(c)** $\sum_{r=1}^{15} \frac{4}{2^r}$

   (d) $5 + 9 + 13 + 17 + \cdots + 117$

6. Calculate the following sums:

   (a) $\sum_{k=10}^{100} k$ **(b)** $\sum_{n=3}^{40} (2n - 6)$ **(c)** $\sum_{i=5}^{100} (3i + 4^i)$

7. Which of the series in Exercises 4–6 are (a) arithmetic, (b) geometric?

8. Prove Theorem 3.12.

*9. A coin is tossed twenty times. What is the probability that the first head will appear—
   (a) on or before the twentieth toss
   (b) on an odd-numbered toss
   (c) after twelve or more tosses, but on or before the twentieth toss?

Key Word Review

Each key word is followed by the number of the section in which it is defined.

permutation 3.2
n factorial 3.2
critical number 3.4
level of significance 3.4
combination 3.5
Pascal's triangle 3.6

summation notation 3.8
arithmetic series 3.9
common difference 3.9
geometric series 3.9
common ratio 3.9

CHAPTER 3   **REVIEW EXERCISES**

1. Define each word in the Key Word Review.

2. What do the following symbols mean?

  (a) $_nC_k$ (b) $_nP_k$ (c) P_n (d) $n!$ (e) $\sum\limits_{i=1}^{n} a_i$

3. Calculate—

  (a) $_{20}P_3$ (b) $_{14}C_2$ (c) P_5 (d) $0!$ (e) $_{1000}P_2$

  (f) $_{1000}C_2$ (g) $\dfrac{n!}{(n-1)!}$

4. Let $A = \{1,2\}$, $B = \{3,4,5,\ldots,12\}$, $C = \{1,2,14,16,17\}$. Calculate—
  (a) $n(A \times B)$ (b) $n(A^2)$ (c) $n(A + B)$ (d) $n(A \cup C)$
  (e) $n(A \times B \times C)$

5. (a) Find the number of permutations of the letters in the word

  ENGINEERING

  (b) If the above letters are randomly arranged, what is the probability that the "word" formed begins and ends with E?

6. The Cherry Valley Little League has twenty-five players. How many different nine-player teams can the league form?

7. In the game of *poker dice*, five dice are cast and the resulting combinations of numbers are treated like poker hands. Only two kinds of straights are permitted: 6 5 4 3 2 and 5 4 3 2 1. Five of a kind is the best hand. Rank the possible hands according to the probability of their attainment.

8. In order to test a person's extrasensory perception (ESP), a deck of five different cards are shuffled and the person is asked to state which card is on top. He cannot see the cards. Assume that the person does not have ESP. If the test is repeated ten times, what is the probability that—
  (a) he guesses eight or more correct cards in a row
  (b) he guesses exactly five cards correctly
  (c) he guesses at least five cards correctly?

9. A store sells five different kinds of ice cream.

    (a) How many two-scoop ice-cream cones can be made if—

        (*i*) the flavors must be different

        (*ii*) the flavors do not have to be different?

        Assume that the order of the scoops is unimportant.

    (b) Answer the preceding questions if the order of the scoops is important.

10. Which pattern of cards in a bridge hand is more probable, 5 4 2 2 or 5 4 3 1, if the four digits in the pattern indicate the number of cards in each suit? (For example, 5 4 2 2 means that the hand consists of five cards of one suit, four of a second suit, and two cards from each of the remaining two suits. The pattern 5 4 2 2 is the same as the pattern 2 2 4 5 or 2 5 4 2.)

11. In how many ways is it possible to seat eight people around a round table if—

    (a) there are no restrictions on the seating arrangements

    (b) Sam refuses to sit next to Ralph?

12. A room has six doors. In how many ways is it possible to enter the room by one door and leave by a different door?

13. In a certain Chinese restaurant a family dinner consists of two different items from group A on the menu and two different items from group B. If group A contains twenty items and group B contains ten, how many different family dinners can be ordered?

14. At a certain university a freshman must choose two humanities courses, one physical education course, one social science course, and one science course. If the school offers fourteen humanities courses, seven physical education courses, four social science courses, and six science courses for freshmen, how many different programs can a freshman take?

15. An exam contains eight true-false questions. If a student guesses at random, what is the probability that he or she will get—

    (a) exactly four correct

    (b) at least five correct

    (c) none correct?

16. A store sells red, green, and yellow marbles. In how many ways can a child buy twenty marbles?

17. How many different necklaces can be made from twenty differently colored interlocking beads? (Remember that the necklace can be turned over.)

18. An election is being held to choose a mayor, a council member, and a judge. Suppose that each voter need not vote for candidates to fill all three offices, but must vote for a candidate for at least one office. In how many ways can the ballot be completed if there are three candidates for each office?

19. A man has twenty books but only room for ten of them on his bookshelf. In how many ways can he arrange his shelf if the order of the books is important?

20. If a fair die is tossed 36,000 times, what is the probability that no 2s occur?

21. Expand the following sums:

(a) $\sum_{j=1}^{5} b_j$ (b) $\sum_{k=2}^{4} a_k$ (c) $\sum_{k=1}^{100} 2^k$

22. Expand the following expressions:

(a) $(x + y)^3$ (b) $(x^2 + y)^3$ (c) $(x^2 - 2y^3)^2$

23. Calculate the following sums:

(a) $\sum_{k=1}^{100} k$ (b) $\sum_{k=1}^{100} 2^k$ (c) $\sum_{k=1}^{100} (2k + 3)$

24. Calculate the following sums:

(a) $\sum_{k=2}^{40} \left(\frac{1}{k} - \frac{1}{k + 1}\right)$ (b) $\sum_{k=1}^{50} \frac{1}{5^k}$ (c) $\sum_{n=100}^{200} 2n$ (d) $\sum_{j=1}^{7} 10$

25. Use the binomial expansion to calculate $(1.01)^4$.

26. A die is tossed thirty times. What is the probability that the first 3 will occur—

(a) on the fifth toss

(b) after five or more tosses, but on or before the thirtieth toss

(c) on an even-numbered toss?

Suggested Readings

1. Eisen, Martin. *Elementary Combinatorial Analysis.* New York: Gordon & Breach, Science Publishers, 1970.
2. Niven, Ivan. *Mathematics of Choice.* New York: Random House, 1965.
3. Riordan, J. *An Introduction to Combinatorial Analysis.* New York: John Wiley & Sons, 1958.
4. Shaply, L. S., and Shubik, M. "A Method for Evaluating the Distribution of Power in a Committee System." *American Political Science Review* 48 (1954): 787–92.

Madeleine G. de Varennes

4 Dependence
and
Independence

INTRODUCTION

The picture on the left, taken during the early part of this century, shows a family in which ten boys were born before the first girl was born. What would be the probability that such a distribution of sexes would occur in a randomly chosen family of thirteen children? Any answer to this question would depend on whether the sex of one child influences the sex of the others. Is a boy more likely to have a little brother or a little sister? Or does the sex of the older child have no influence over that of younger siblings?

In general, when do events influence one another and when are they independent? These are the kinds of questions that will be discussed in this chapter. In the process, we will have to give a precise mathematical meaning to the words *dependent* and *independent* and develop a theory of dependent and independent events. We will also discuss how this theory relates to real-life situations like the relationship between the sexes of siblings.

4.1 CONDITIONAL PROBABILITIES

Often, one event depends upon another. For example, a teenager's use of the family car may depend upon his parents' plans. More seriously, a baby's chances of premature birth may depend upon the health and nutrition of its mother. In this section we will discuss the relationship between the probabilities of events where one event depends upon the other. We will begin our discussion of dependent events with an example based on life insurance data.

Example 1   Table 4.1 is an excerpt from the 1958 CSO (Commissioners Standard Ordinary) mortality table.[1] It is based on the experience of ordinary life insurance policies (1950–54) and includes a margin for adverse mortality fluctuations. It is clear from this table that the probability of dying within the next year depends upon age.

Table 4.1	**Probability q_x that a Male Age x**		
Age x	**Dies Before Reaching Age $x + 1$**	**Age x**	q_x
0	0.00708	40	0.00353
5	0.00135	50	0.00832
10	0.00121	60	0.02034
15	0.00146	70	0.04979
20	0.00179	80	0.10998
30	0.00213	90	0.22814

1. For the complete 1958 CSO mortality table, see *Life Insurance Fact Book 1975* (New York: Institute of Life Insurance, 1975), pp. 104–5.

The notation

$$P(B|A)$$

is used to denote the *probability that B occurs, given that A has occurred*. This is read "the *conditional probability of B given A*." The following examples illustrate this concept.

Example 2   Let D be the event "a randomly chosen male dies within the next year" and let A_x be the event "the chosen male is age x." Using Table 4.1, $P(D|A_{20}) = 0.00179$, $P(D|A_{50}) = 0.00832$, and $P(D|A_{90}) = 0.22814$.

Example 3   A quarter is hidden inside one of two boxes, and the first box is opened in an attempt to find it. Let A and B be the events "the coin is in the first box" and "the coin is in the second box," respectively. Then $P(B|A) = 0$, while $P(B|A') = 1$.

Example 4   A student guesses the answers to two true–false questions.

(a) What is the probability that he gets them both correct?

(b) What is the above probability if he knows that at least one of the answers is true?

Solution

(a) A sample space for this situation having equally likely elementary outcomes is

$$\Omega = \{(t,t), (t,f), (f,t), (f,f)\}$$

Let B be the event "both his answers are correct." Applying Laplace's definition, it follows that $P(B) = \frac{1}{4}$.

(b) Since he knows that at least one of the answers is true, the sample space of possible answers has been reduced to

$$\{(t,t), (t,f), (f,t)\}$$

Let A be the event "at least one of the answers is true." It follows that $P(B|A) = \frac{1}{3}$.

In general, suppose that we have a sample space with a finite number of equally likely elementary events. The previous example illustrates how Laplace's definition can be applied to find $P(B|A)$. The total number of cases is $n(A)$ instead of $n(\Omega)$, because we know that A has occurred. The number of favorable cases is $n(AB)$ instead of $n(B)$, because in order for B to occur given that A has occurred, *both* A and B must occur. In other words, we are "reducing" our sample space from Ω to A. It follows that

$$P(B|A) = \frac{n(AB)}{n(A)} = \frac{n(AB)/n(\Omega)}{n(A)/n(\Omega)} = \frac{P(AB)}{P(A)}$$

(Recall that AB is the intersection of the sets A and B. We will be using this notation, instead of $A \cap B$, throughout this chapter.)

Although the preceding equation assumes that we are using Laplace's definition of probability, it motivates the following definition of conditional probability, which holds for all three definitions of probability P.

Definition 4.1   Let A be an event with $P(A) > 0$. The **conditional probability** of B given A is denoted $P(B \mid A)$ and is defined as follows:

$$P(B \mid A) = \frac{P(AB)}{P(A)}$$

The conditional probability $P(B \mid A)$ is not defined if $P(A) = 0$.

The following examples are applications of Definition 4.1.

Example 5   Applying Definition 4.1 to Example 4**(b)** yields

$$P(B \mid A) = \frac{P(AB)}{P(A)} = \frac{\frac{1}{4}}{\frac{3}{4}} = \frac{1}{3}$$

This agrees with our previous result.

Example 6   Suppose $P(A) = 0.10$, $P(B) = 0.20$, and $P(AB) = 0.05$. Then, applying Definition 4.1—

(a) $P(A \mid B) = \dfrac{P(AB)}{P(B)} = \dfrac{0.05}{0.20} = \dfrac{1}{4}$

(b) $P(B \mid A) = \dfrac{P(BA)}{P(A)} = \dfrac{P(AB)}{P(A)} = \dfrac{0.05}{0.10} = \dfrac{1}{2}$

(c) $P(A \mid \Omega) = \dfrac{P(A\Omega)}{P(\Omega)} = \dfrac{P(A)}{P(\Omega)} = \dfrac{P(A)}{1} = 0.1$

(d) $P(A \mid \emptyset)$ is undefined because $P(\emptyset) = 0$.

Theorem 2.3(a) states that if A is any event, then

$$0 \le P(A) \le 1 \tag{1}$$

Let D be an event with $P(D) > 0$. If we replace the probability $P(A)$ in inequality (1) with the conditional probability $P(A \mid D)$, the inequality

$$0 \le P(A \mid D) \le 1 \tag{2}$$

is obtained.

Alternatively, since

$$P(A \mid D) = \frac{P(AD)}{P(D)} \le \frac{P(D)}{P(D)} = 1$$

(see Definition 4.1 and Theorem 2.3(e)), and

$$P(A|D) = \frac{P(AD)}{P(D)} \geq \frac{0}{P(D)} = 0$$

inequality (2) follows.

In general, consider any theorem of Section 2.6. If we replace each probability $P(E)$ by the corresponding conditional probability $P(E|D)$, new statements are obtained. Some of these new statements are listed in Theorem 4.1. They can all be proved by applying Definition 4.1 and Theorems 2.2 and 2.3.

THEOREM 4.1   Let D be an event with $P(D) > 0$. Then—

     (a) $0 \leq P(A|D) \leq 1$
     (b) $P(\Omega|D) = 1$
     (c) $P((A + B)|D) = P(A|D) + P(B|D)$
         if A and B are mutually exclusive
     (d) $P(\varnothing|D) = 0$
     (e) If $A \subseteq B$, then $P(A|D) \leq P(B|D)$.
     (f) $P(A'|D) = 1 - P(A|D)$

The results of Theorem 4.2 are similar to those of Theorems 2.4–2.6. They can be proved from those theorems and Definition 4.1.

THEOREM 4.2   Let D be an event with $P(D) > 0$. Then—

     (a) $P((A + B + C)|D) = P(A|D) + P(B|D) + P(C|D)$
         if A, B, and C are mutually exclusive
     (b) $P((A \cup B)|D) = P(A|D) + P(B|D) - P(AB|D)$
     (c) $P((A \cup B \cup C)|D) = P(A|D) + P(B|D) + P(C|D)$
         $- P(AB|D) - P(BC|D) - P(AC|D) + P(ABC|D)$

The following examples show how Theorems 4.1 and 4.2 can be used to simplify the calculation of conditional probabilities.

Example 7   Suppose that $P(A) = 0.20$, $P(B) = 0.60$, $P(C) = 0.25$, $P(AB) = P(BC) = 0.15$, and A and C are mutually exclusive. Calculate—

     (a) $P(A|C)$ (b) $P(A|B)$ (c) $P(A'|B)$
     (d) $P((A + C)|B)$ (e) $P(\Omega|A)$

Solution

     (a) Applying Definition 4.1,

$$P(A|C) = \frac{P(AC)}{P(C)} = \frac{P(\varnothing)}{P(C)} = 0$$

(b) Applying Definition 4.1,

$$P(A|B) = \frac{P(AB)}{P(B)} = \frac{0.15}{0.60} = \frac{1}{4}$$

(c) Applying Theorem 4.1(f) and part **(b)**,

$$P(A'|B) = 1 - P(A|B) = 1 - \frac{1}{4} = \frac{3}{4}$$

(d) Applying Theorem 4.1(c) and part **(b)**,

$$P((A + C)|B) = P(A|B) + P(C|B) = \frac{1}{4} + \frac{P(CB)}{P(B)}$$

$$= \frac{1}{4} + \frac{0.15}{0.60} = \frac{1}{4} + \frac{1}{4} = \frac{1}{2}$$

(e) Applying Theorem 4.1(b),

$$P(\Omega|A) = 1$$

Example 8 Suppose that $P(C) = 0.60$, $P(AC) = 0.30$, $P(BC) = 0.35$, and $P(ABC) = 0.10$. Calculate (a) $P(AB|C)$ and (b) $P((A \cup B)|C)$.

Solution
(a) Applying Definition 4.1,

$$P(AB|C) = \frac{P(ABC)}{P(C)} = \frac{0.10}{0.60} = \frac{1}{6}$$

(b) Applying Theorem 4.2(b) and part **(a)**,

$$P((A \cup B)|C) = P(A|C) + P(B|C) - P(AB|C)$$

$$= \frac{P(AC)}{P(C)} + \frac{P(BC)}{P(C)} - \frac{1}{6}$$

$$= \frac{0.30}{0.60} + \frac{0.35}{0.60} - \frac{1}{6}$$

$$= \frac{30}{60} + \frac{35}{60} - \frac{10}{60} = \frac{55}{60}$$

Example 9 Al, Bob, Claire, Dick, and Ethel are candidates for school-board president. It is estimated that their probabilities of winning are 0.25, 0.30, 0.20, 0.10, and 0.15, respectively. If Dick withdraws, what is the probability that (a) Al wins, (b) Bob or Claire wins, (c) Al does not win?

Solution Let A, B, C, and E be the events "Al wins," "Bob wins," "Claire wins," and "Ethel wins," respectively. Let N be the event "Dick does not win."

(a) Using Definition 4.1, the required probability is

$$P(A \mid N) = \frac{P(AN)}{P(N)} = \frac{P(A)}{P(A + B + C + E)} = \frac{0.25}{0.25 + 0.30 + 0.20 + 0.15}$$

$$= \frac{0.25}{0.90} = \frac{5}{18}$$

(b) Using Theorem 4.1(c), the required probability is

$$P((B + C) \mid N) = P(B \mid N) + P(C \mid N)$$

$$= \frac{P(BN)}{P(N)} + \frac{P(CN)}{P(N)} = \frac{0.30}{0.90} + \frac{0.20}{0.90} = \frac{5}{9}$$

(c) Using Theorem 4.1(f) and part (a),

$$P(A' \mid N) = 1 - P(A \mid N) = \frac{13}{18}$$

Another useful theorem involving conditional probabilities is the *multiplication theorem of probability* (Theorem 4.3), which follows directly from Definition 4.1. Theorem 4.4 extends this theorem to three events, four events, and so on.

THEOREM 4.3 (Multiplication Theorem of Probability) If $P(A) \neq 0$, then

$$P(AB) = P(A)P(B \mid A)$$

THEOREM 4.4 Let $A, B, C, D, A_1, A_2, \ldots, A_n$ be events with nonzero probability. Then—

(a) $P(ABC) = P(A)P(B \mid A)P(C \mid AB)$

(b) $P(ABCD) = P(A)P(B \mid A)P(C \mid AB)P(D \mid ABC)$

and, in general—

(c) $P(A_1 A_2 \cdots A_n)$
$$= P(A_1)P(A_2 \mid A_1)P(A_3 \mid A_1 A_2) \cdots P(A_n \mid A_1 A_2 \cdots A_{n-1})$$

The remaining examples of this section are applications of Theorems 4.3 and 4.4.

Example 10 A case contains seventeen good bolts and three defective ones.

(a) If two bolts are removed (without replacement), what is the probability that both are defective?

(b) If four bolts are removed, what is the probability that at least one is defective?

Solution

(a) Let A and B be the events "the first bolt is defective" and "the second bolt is defective," respectively. Applying Theorem 4.3, the required probability is $P(AB) = P(A)P(B|A)$. Since there are originally 3 defective bolts in the case of 20, $P(A) = \frac{3}{20}$. If A occurs, one defective bolt has been removed from the case. This leaves 2 defective bolts out of 19. It follows that $P(B|A) = \frac{2}{19}$. Hence,

$$P(AB) = P(A)P(B|A) = \frac{3}{20} \cdot \frac{2}{19} = \frac{3}{190}$$

(b) Let L be the event "at least one bolt is defective," and let G_i $(i = 1,2,3,4)$ be the event "the ith bolt is good." It follows that $L' = G_1 G_2 G_3 G_4$, and so applying Theorem 4.4(b) yields

$$P(L) = 1 - P(L') = 1 - P(G_1 G_2 G_3 G_4)$$

$$= 1 - P(G_1)P(G_2|G_1)P(G_3|G_1 G_2)P(G_4|G_1 G_2 G_3)$$

$$= 1 - (\tfrac{17}{20})(\tfrac{16}{19})(\tfrac{15}{18})(\tfrac{14}{17}) = 0.509$$

Example 11   A secret program is locked into a computer shared by several firms. Five different instructions in the proper order are required to retrieve the program. If a competitor randomly chooses 5 different instructions from a set of 1000 possible instructions, what is the probability that he successfully retrieves the program?

Solution   Let A_i be the event "the ith instruction is correct" $(i = 1,2,3,4,5)$. The required probability is

$$P(A_1 A_2 A_3 A_4 A_5) = \frac{1}{1000} \cdot \frac{1}{999} \cdot \frac{1}{998} \cdot \frac{1}{997} \cdot \frac{1}{996}$$

Using a calculator, this probability is approximately $(1.01)(10^{-15})$.

Example 12   The faculty of a certain university consists of sixty men and forty women. If a faculty president, vice-president, and secretary are randomly chosen, what is the probability that—

(a) all three officeholders are men

(b) exactly one officeholder is a woman?

(Assume that no person can hold more than one position.)

Solution
(a) Let A, B, and C be the events "the president is male," "the vice-president is male," and "the secretary is male," respectively. The required probability is

$$P(ABC) = P(A)P(B|A)P(C|AB) = \frac{60}{100} \cdot \frac{59}{99} \cdot \frac{58}{98} = 0.21$$

(b) Let E be the event "exactly one officeholder is female." Then

$$P(E) = P(ABC') + P(AB'C) + P(A'BC)$$

$$= \frac{60}{100} \cdot \frac{59}{99} \cdot \frac{40}{98} + \frac{60}{100} \cdot \frac{40}{99} \cdot \frac{59}{98} + \frac{40}{100} \cdot \frac{60}{99} \cdot \frac{59}{98}$$

$$= 3\left(\frac{60}{100} \cdot \frac{59}{99} \cdot \frac{40}{98}\right) = 0.44$$

4.1 EXERCISES

1. A large firm is considering using a psychological test to screen applicants for its management-training program. A five-year study of test scores and on-the-job ratings resulted in the following information:

Test Grade	Number Receiving Grade	On-the-job Ratings		
		Excellent (E)	Good (G)	Dismissed (D)
A	300	200	50	50
B	400	200	100	100
C	200	50	100	50
F	100	20	20	60

For example, of the 300 people who received an A on the test, 200 received an excellent rating as manager, 50 received a good rating, and 50 were dismissed. Let A, B, C, F be the events that the employee received an A, B, C, F on the exam, respectively. Calculate—
   (a) $P(A)$ **(b)** $P(F)$ **(c)** $P(A|E)$ **(d)** $P(E|A)$
   (e) $P(D)$ **(f)** $P(E|F)$ **(g)** $P(A|B)$

2. A parent asks her child to hand her a black block from a pile of ten black and ten red blocks. Suppose the child is unable to differentiate colors. What is the probability that he gives his parent a black block ten times in ten attempts if—
   (a) none of the selected blocks are replaced
   (b) the child replaces each block before selecting the next one?

3. A red card is drawn from an ordinary fifty-two card pack. What is the probability that it is (a) a diamond, (b) a club, (c) a king?

4. There are three red balls and two white balls in an urn.
   (a) If a blindfolded individual picks a ball from the urn, replaces it, and then picks another, what is the probability that—
      (i) both balls are red
      (ii) one ball is red and the other white?
   (b) Answer the above questions if the first ball is not replaced.

5. Suppose two cards are picked from the top of a shuffled deck of eight cards containing the four kings and four queens from an ordinary bridge deck. Assuming that the first card is not replaced before the second card is picked, what is the probability that both cards are kings if—
    (a) nothing more is known
    (b) it is known that at least one card is a king
    (c) it is known that one card is the king of hearts
    (d) it is known that at least one card is a black king?

6. Thalassemia is a hereditary disease resulting in anemia. Individuals with genotype MM have severe anemia (*thalassemia major*) and die before reaching adulthood. Those with genotype MN have a milder anemia (*thalassemia minor*) and those with genotype NN are normal.
    (a) What is the probability that an adult whose parents both have thalassemia minor is normal?
    (b) What is the probability that the adult is normal if only one parent has thalassemia minor?

7. A student is taking a ten-question true–false examination. What is the probability that she gets all the right answers if—
    (a) she just guesses
    (b) she is certain of the answers to six of the questions
    (c) she knows that the instructor always puts equal numbers of true and false questions on the test?

8. For each of parts (a)–(c) of Exercise 7, what is the probability that the student answers 60 percent or more of the questions correctly?

9. In the game of bridge, an ordinary fifty-two card deck is used, and thirteen cards are dealt to each of four players. If one player has no hearts, what is the probability that his partner—
    (a) also has no hearts
    (b) has at least one heart?

10. If two dice are rolled and it is known that the sum of the numbers on the faces that turn up is seven, what is the probability that the number showing on one of the dice is six?

11. Answer true or false: $P(B\,|\,A) + P(B\,|\,A') = 1$ if $P(A)$ is neither 0 nor 1. Prove your answer.

12. Suppose that $P(A) = 0.30$, $P(B) = 0.20$, $P(C) = P(AB) = 0.10$, $P(BC) = 0.05$, and $P(ABC) = 0.01$. Calculate—
    (a) $P(A\,|\,B)$ (b) $P(B\,|\,A)$ (c) $P(A'\,|\,B)$ (d) $P(A\,|\,B')$
    (e) $P(C\,|\,B)$ (f) $P(AC\,|\,B)$ (g) $P((A \cup C)\,|\,B)$

13. Suppose $P(A) = 0.1$. Find—
    (a) $P(A\,|\,\varnothing)$ (b) $P(\varnothing\,|\,A)$ (c) $P(A\,|\,\Omega)$
    (d) $P(\Omega\,|\,A)$ (e) $P(A\,|\,A')$

4.2 INDEPENDENT EVENTS

Suppose two fair coins are tossed, and let A and B be the events "the first coin is a head" and "the second coin is a head," respectively. It is intuitively clear that A and B are *independent*; that is, the occurrence or nonoccurrence of A has no influence upon the occurrence or nonoccurrence of B, and vice versa. In other words, we expect $P(B|A)$ to equal $P(B)$. In fact, using Laplace's definition and Definition 4.1,

$$P(B|A) = \frac{P(AB)}{P(A)} = \frac{\frac{1}{4}}{\frac{1}{2}} = \frac{1}{2} = P(B)$$

which verifies our intuitive expectation.

In general, let A and B be any two events with $P(A) > 0$. If we *assume* that $P(B|A) = P(B)$, then the multiplication theorem of probability implies that

$$P(AB) = P(A)P(B|A) = P(A)P(B)$$

This motivates the following definition, which is valid even when $P(A)$ is zero.

Definition 4.2   Two events A and B are **independent events** if and only if

$$P(AB) = P(A)P(B)$$

The following theorems follow from Definition 4.2 and Theorems 4.1 and 4.2.

THEOREM   4.5   Suppose that $P(A) > 0$. Then A and B are independent if and only if

$$P(B|A) = P(B)$$

THEOREM   4.6   If A and B are independent, then so are the pairs A and B', A' and B, and A' and B'.

In most applications we will assume the independence of A and B and use this assumption to compute $P(AB)$. However, in actual practice, extensive testing is required to verify independence. An experiment is performed many times, and the frequencies $f_N(A)$, $f_N(B)$, and $f_N(AB)$ are calculated. It is then determined if $f_N(AB)$ is close enough to $f_N(A) \cdot f_N(B)$ to justify the independence assumption.

It can be argued, for example, that the sex of one child in a family has no effect upon the sex of the next. It would then follow that a study of families of three children would show that the events *bbb*, *bgb*, *ggb*, and so on occur approximately equally often. (The symbol *bgg* means that the oldest child is a boy and the others are girls.) However, the more one considers the reasons that many families have a third child, the less confident one is that these events are independent. It is quite possible that the decision to have three children is influenced by the sex of the first two. It has also been found that some couples are more likely to have children of a particular sex for nongenetic biological reasons.

On the other hand, it would seem obvious that the occurrence or nonoccurrence of rain on a particular day depends on the previous day's weather. However, surveys have shown that this is not the case in some parts of Israel. Thus, the actual dependence or independence of events is not always in agreement with our intuition.

The following examples illustrate the concept of independence.

Example 13   Suppose $P(A) = 0.10$, $P(B) = P(C) = 0.20$, $P(AB) = 0.02$,   and $P(AC) = 0.05$. Which of the following pairs of events are independent?

(a) A and B (b) A and C (c) A' and B (d) A and $\varnothing$

Solution

(a) Applying Definition 4.2, A and B are independent because

$$P(A)P(B) = (0.1)(0.2) = 0.02 = P(AB)$$

(b) Events A and C are not independent, because

$$P(A)P(C) = (0.1)(0.2) = 0.02$$

while $P(AC) = 0.05$.

(c) Events A' and B are independent by Theorem 4.6, because A and B are independent, as shown in part (a).

(d) Applying Definition 4.2, A and $\varnothing$ are independent because

$$P(A\varnothing) = P(\varnothing) = 0 \quad \text{and} \quad P(A)P(\varnothing) = (0.2)(0) = 0$$

Example 14   A ransom note was written on a typewriter of a certain make that had a "3" that was out of line and a scarred "f." A suspect's typewriter has both of these defects. If the probability that a used typewriter of this make has the first defect is $\frac{1}{30}$, while the probability that it has the second defect is $\frac{1}{40}$, what is the probability that a randomly chosen typewriter of this make has both of these defects?

Solution   It seems reasonable to assume that the occurrence of the first defect (event A) has no influence upon the occurrence of the second defect (event B). Hence Definition 4.2 can be applied. The required probability is

$$P(AB) = P(A)P(B) = \frac{1}{30} \cdot \frac{1}{40} = \frac{1}{1200}$$

Example 15 will motivate the extension of Definition 4.2 to three or more events.

Example 15   A certain slot machine has three windows, each of which displays a picture of an apple, banana, or orange when the handle of the slot machine is pulled. However, the machine is made so that the pictures can only appear in certain combinations. If we let a represent apple, b represent banana, and o represent orange, the possible combinations are

$$(a,a,a), \quad (b,b,b), \quad (o,o,o), \quad \text{and} \quad (a,b,o)$$

Each of these four combinations is equally likely to appear when the handle of the slot machine is pulled.

Let A, B, and O be the events that the combination obtained contains an apple, banana, or orange, respectively. Then applying Laplace's definition of probability,

$$P(A) = P(B) = P(O) = \tfrac{1}{2} \quad \text{and} \quad P(AB) = P(AO) = P(BO) = \tfrac{1}{4}$$

Notice that the three events A, B, and O of Example 15 are **pairwise independent events**. That is, every pair of these events satisfies Definition 4.2:

$$P(AB) = P(A)P(B) = \tfrac{1}{4}, \qquad P(AO) = P(A)P(O) = \tfrac{1}{4}$$

and

$$P(BO) = P(B)P(O) = \tfrac{1}{4}$$

However, $P(ABO) \neq P(A)P(B)P(O)$, because

$$P(ABO) = \tfrac{1}{4}$$

while

$$P(A)P(B)P(O) = \tfrac{1}{8}$$

This example motivates the following definition.

Definition 4.3   Three events A, B, and C are **mutually independent events** if and only if

$$P(AB) = P(A)P(B), \qquad P(AC) = P(A)P(C), \qquad P(BC) = P(B)P(C)$$

and

$$P(ABC) = P(A)P(B)P(C)$$

It follows from our definitions that mutual independence implies pairwise independence. However, Example 15 shows that the reverse statement is not true. Definition 4.3 can be generalized to any number n of events.

Definition 4.4   The events $A_1, A_2, \ldots, A_n$ are **mutually independent events** if and only if

$$P(A_{i_1} A_{i_2} \cdots A_{i_k}) = P(A_{i_1})P(A_{i_2}) \cdots P(A_{i_k}) \qquad k = 2, 3, \ldots, n$$

for every possible combination of k events $A_{i_1}, A_{i_2}, \ldots, A_{i_k}$ selected from the n events $A_1, A_2, \ldots, A_n$.

In particular, if $A_1, A_2, \ldots, A_n$ are independent, then

$$P(A_1 A_2 \cdots A_n) = P(A_1)P(A_2) \cdots P(A_n) \tag{3}$$

Theorem 4.6 can also be extended to n events in a natural way.

The following examples are applications of Definitions 4.3 and 4.4 and the generalization of Theorem 4.6.

Example 16 A coin is biased 80 percent in favor of heads; that is, $P(H) = 0.8$. If the coin is tossed three times, what is the probability that (a) exactly three heads appear, (b) exactly two tails occur?

Solution Let A, B, C be the events that a head appears on the first toss, on the second toss, on the third toss, respectively. Then $P(A) = P(B) = P(C) = 0.8$. It has been verified that the events A, B, and C are independent.

(a) Using Definition 4.3, the required probability is

$$P(ABC) = P(A)P(B)P(C) = (0.8)^3 = 0.512$$

(b) The required probability is

$$P = P(A'B'C) + P(AB'C') + P(A'BC')$$

Generalizing from Theorem 4.6, the independence of A, B, and C implies the independence of each of the triples A', B', C; A, B', C'; and A', B, C'. It follows that

$$P = P(A')P(B')P(C) + P(A)P(B')P(C') + P(A')P(B)P(C')$$
$$= 3(0.2)^2(0.8) = 0.096$$

Example 17 Figure 4.1 is the diagram of an electrical circuit. For each switch in the circuit, the probability that the switch is closed is p. If all switches in the circuit close independently, what is the probability that current flows between A and B?

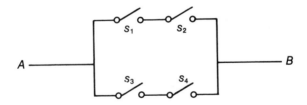

Figure 4.1 Diagram of an electric circuit

Solution Notice that current can flow if s_1 and s_2 are closed, or if s_3 and s_4 are closed, or if all four switches are closed. Therefore, if A_i is the event "switch i is closed" ($i = 1,2,3,4$) and E is the event "current flows between A and B," then $E = (A_1 A_2) \cup (A_3 A_4)$. Hence,

$$P(E) = P((A_1 A_2) \cup (A_3 A_4))$$
$$= P(A_1 A_2) + P(A_3 A_4) - P(A_1 A_2 A_3 A_4)$$
$$= P(A_1)P(A_2) + P(A_3)P(A_4) - P(A_1)P(A_2)P(A_3)P(A_4)$$
$$= p^2 + p^2 - p^4$$
$$= 2p^2 - p^4$$

4.2 EXERCISES

1. Using your intuition, which of the following pairs of events would you consider to be independent?

   (a) Two apples are removed from a barrel of apples. Event A is "first apple is rotten"; event B is "the second apple is rotten."

   (b) Two cards are dealt from a shuffled deck. The first card is not replaced before the second is picked. Event A is "the first card is a king"; event B is "the second card is a king."

   (c) The preceding experiment is performed, but the first card is replaced before the second card is picked.

   (d) Two children from the same family are tested for tuberculosis. Event A is "the first child has the disease"; event B is "the second child has the disease."

2. Assume that $P(A) = 0.2$, $P(AB) = 0.1$, and $P(B) = 0.5$.

   (a) Calculate $(i)\ P(A|B)$, $(ii)\ P(B|A)$, $(iii)\ P(A')$.

   (b) Are A and B independent? Explain.

3. Suppose $P(A) = 0.1$ and $P(B|A) = 0.2$. If A and B are independent, calculate—

   (a) $P(B)$ **(b)** $P(AB)$ **(c)** $P(A|B)$ **(d)** $P(A|B')$ **(e)** $P(A'B')$

4. Let A be any event. Which of the following pairs of events are independent?

   (a) A and $\varnothing$ **(b)** A and A' **(c)** A and Ω

   Which of these pairs of events are mutually exclusive?

5. What is the probability of obtaining ten heads in a row when a fair coin is tossed ten times?

6. What is the probability of getting a head and a 6 when a dime and a die are tossed?

7. In Shakespeare's *Hamlet*, there are approximately 27,000 letters, spaces, and punctuation marks. If a diligent monkey pecks at a typewriter enough times, what is the probability that he will reproduce *Hamlet*? (The typewriter has forty-four keys bearing letters, numbers, and symbols, and it has a space bar. Ignore capitals.)

8. Ninety-eight percent of the genes for a certain trait are dominant, while the remainder are recessive.[2] Remember that traits are determined by a pair of genes, and assume that the inheritance of each gene in the pair is an event that is independent of the inheritance of the other gene. What is the probability that a randomly chosen person has—

   (a) no recessive genes for this trait

   (b) one recessive gene

   (c) two recessive genes?

---

2. See Section 2.5 for a discussion of genes and genetics.

9.   **(a)** What is the probability that a randomly chosen person does not have the trait of Exercise 8?

   (b) What is the probability that he or she does not have the trait but is a carrier of it?

10.   A hunter fires at a moving target until he either hits it or runs out of ammunition. (He has ammunition for twenty shots.) From past experience, he knows that the probability that he will hit a moving target is 0.1. What is the probability that—
   (a) he hits it on the first shot

   (b) he runs out of ammunition before he hits it

   (c) he requires more than two shots?

11.   The following table shows the numbers of various symbols on each of the three dials of a standard slot machine.

Symbol	Dial 1	Dial 2	Dial 3
Bar	1	3	1
Bell	1	3	3
Plum	5	1	5
Orange	3	6	7
Lemon	7	7	0
Cherry	3	0	4

For example, the first dial has 1 bar, 1 bell, 5 plums, etc. After the handle of the slot machine is pulled, the dials spin, and one symbol appears on each dial. If each dial moves independently of the others, what is the probability of getting—
   (a) 3 bars   **(b)** 3 oranges   **(c)** no cherries

   (d) 2 bars and 1 bell?

12.   Suppose that 10 percent of the population has a certain blood type. If twenty people are randomly tested, what is the probability that none of them will have this blood type?

13.   The absence of legs in cattle (amputated cattle) is caused by the presence of two recessive genes. If a normal bull and a normal cow are mated and produce an amputated calf, what is the probability that—
   (a) their next two calves will be amputated

   (b) at least one of their next four calves is healthy?

14.   In Russian roulette, a bullet is placed in one of the six chambers of a gun and the chambers spun. The trigger is then pulled; the gun may or may not fire. The chamber is spun before the trigger is pulled again.
   (a) What is the probability that the gun will fire the first time the trigger is pulled?

   (b) What is the probability that the gun will not fire the first five times the trigger is pulled?

(c) How many times can the trigger be pulled before the probability that the gun will fire becomes greater than or equal to 0.5?

(d) If the gun does not fire the first time the trigger is pulled, what is the probability that it will not fire the second time the trigger is pulled?

15. Answer the questions of Exercise 14 under the assumption that the chamber is not spun before the trigger is pulled again.

16. Suppose that a bullet is placed in one chamber of a gun as described in Exercise 14. Suppose that the chambers are spun and the trigger is pulled ten times.

(a) Is the gun more likely to fire the first or the second time the trigger is pulled?

(b) What is the probability that the gun will not fire any of the ten times that the trigger is pulled?

17. Prove Theorem 4.5.

18. Extend the statement of Theorem 4.6 to three mutually independent events.

19. What is the probability that a man of age forty and his son of age fifteen both live for at least one year? (Use Table 4.1, and assume these two events are independent.)

4.3 THE TOTAL PROBABILITY THEOREM

In many problems calculating a certain probability $P(B)$ may be difficult, but calculating a second probability $P(A)$ and the conditional probabilities $P(B|A)$ and $P(B|A')$ may be relatively simple. In this section we will learn how the *total probability theorem* enables us to calculate $P(B)$ from the other probabilities. The following example will lead to a statement of the theorem.

Example 18   In a certain firm, 40 percent of the male employees earn over 15,000 dollars per year, while only 5 percent of the female employees earn over 15,000 dollars per year. If 30 percent of the employees are female, what is the probability that a randomly chosen employee earns over 15,000 dollars per year?

Solution   Let E be the event "the employee earns over 15,000 dollars per year," and let M be the event "the employee is male." From the information given, we know that

$$P(E|M) = 0.40, \qquad P(E|M') = 0.05, \qquad P(M') = 0.30$$

and

$$P(M) = 1 - 0.30 = 0.70$$

We need to find $P(E)$.

Since $\Omega = M + M'$, it follows that

$$E = \Omega E = (M + M')E = ME + M'E$$

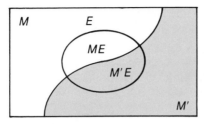

Figure 4.2   Venn diagram for Example 18

by one of the distributive laws for sets. (See Theorem 1.7(a) on page 20.) The relationship between E, M, and M' is shown in Figure 4.2.) Applying the multiplication theorem of probability,

$$P(E) = P(ME) + P(M'E)$$
$$= P(M)P(E\,|\,M) + P(M')P(E\,|\,M')$$
$$= (0.70)(0.4) + (0.30)(0.05) = 0.295$$

In Example 18 we saw that

$$P(E) = P(M)P(E\,|\,M) + P(M')P(E\,|\,M')$$
$$= P(E\,|\,M)P(M) + P(E\,|\,M')P(M')$$

If we replace the particular events E and M in the above equation by any two events B and A (as long as $P(A)$ does not equal 0 or 1), we obtain the total probability theorem. The proof of this theorem is similar to the solution of Example 18.

THEOREM 4.7   (Total Probability Theorem)   If $P(A)$ is not equal to 0 or 1, then $P(B) = P(B\,|\,A)P(A) + P(B\,|\,A')P(A')$

The following examples are applications of Theorem 4.7.

Example 19   A woman goes to a genetic counseling service because her brother is hemophilic, although she and both her parents appear normal. If she contemplates marrying a normal man, what is the probability that her first child is a son with hemophilia?

Solution   The probability that she gives birth to a hemophilic son (event S) clearly depends upon whether or nor she is a carrier (event C). The possible genotypes of her children are given in the following tables:

	X^N	Y
X^h	$X^h X^N$	$X^h Y$
X^N	$X^N X^N$	$X^N Y$

	X^N	Y
X^N	$X^N X^N$	$X^N Y$
X^N	$X^N X^N$	$X^N Y$

If she is a carrier   If she is not a carrier

(See Examples 16 and 17, pages 53–54.)

It follows from the tables that $P(S|C) = \frac{1}{4}$ and $P(S|C') = 0$. Since her father is normal, he must have genotype $X^N Y$; the fact that the mother gave birth to a hemophilic son means that her mother is of genotype $X^N X^h$. Therefore, the table on the left represents the possible genotypes of her parents' children. Using this table, it follows that $P(C) = P(C') = \frac{1}{2}$. Applying the total probability theorem yields

$$P(S) = P(S|C)P(C) + P(S|C')P(C') = (\tfrac{1}{4})(\tfrac{1}{2}) + 0(\tfrac{1}{2}) = \tfrac{1}{8}$$

Example 20 Approximately one-third of all human twins are identical. Find the probability that both children in a set of twins are boys.

Solution Let B be the event that the twins are both boys and let I be the event that they are identical. Applying the total probability theorem yields

$$P(B) = P(B|I)P(I) + P(B|I')P(I') \tag{4}$$

From the information given, we know that

$$P(I) = \tfrac{1}{3} \qquad \text{and} \qquad P(I') = 1 - \tfrac{1}{3} = \tfrac{2}{3}$$

We must still calculate $P(B|I)$ and $P(B|I')$.

If the twins are identical, then they are either both boys or both girls. It follows from Laplace's definition of probability that $P(B|I) = \tfrac{1}{2}$.

If the twins are not identical, there are four equally likely outcomes: bb, bg, gb, and gg. It follows that $P(B|I') = \tfrac{1}{4}$.

Substituting these values in equation (4) yields

$$P(B) = \tfrac{1}{2}(\tfrac{1}{3}) + \tfrac{1}{4}(\tfrac{2}{3}) = \tfrac{1}{3}$$

In each of the preceding examples, the occurrence of an event B depends upon whether or not a second event, A, occurs. Stated differently, the occurrence of B depends upon the occurrence of two mutually exclusive events, A and A', where $A + A' = \Omega$. We have seen that the total probability theorem can be used to find $P(B)$ if we know $P(B|A)$, $P(B|A')$, and $P(A)$.

Sometimes, however, the occurrence of an event B depends upon the occurrence of several mutually exclusive events $A_1, A_2, \ldots, A_n$ where $A_1 + A_2 + \cdots + A_n = \Omega$. The total probability theorem can be extended in a natural way to find $P(B)$ if we know $P(B|A_1)$, $P(B|A_2)$, $\ldots$, $P(B|A_n)$, $P(A_1)$, $\ldots$, $P(A_n)$. The statement of this extended theorem follows. Its proof is similar to the solution of Example 18.

THEOREM 4.8 (The Total Probability Theorem)

If $\Omega = A_1 + A_2 + \cdots + A_n$, $P(A_i) > 0$ $(i = 1, 2, \ldots, n)$, and B is an arbitrary event, then

$$P(B) = P(B|A_1)P(A_1) + P(B|A_2)P(A_2) + \cdots + P(B|A_n)P(A_n)$$

The following examples are applications of this theorem.

Example 21 A company has three factories, F_1, F_2, and F_3, that manufacture lightbulbs. Factories F_2 and F_3 produce an equal number of bulbs, while F_1

produces twice as many bulbs as F_2. It is also known from past experience that 2 percent of the bulbs produced by F_1 and F_2 are defective, while 4 percent of those manufactured by F_3 are defective. All the bulbs produced by the three factories are stored in a common warehouse. If one bulb is chosen from the warehouse at random, what is the probability that it is defective?

Solution   Let B be the event "the bulb is defective" and let A_i be the event "the bulb came from factory F_i." From the facts given, we know that

$$P(A_1) = 2[P(A_2)] \quad \text{and} \quad P(A_2) = P(A_3)$$

It then follows that

$$1 = P(\Omega) = P(A_1 + A_2 + A_3) = P(A_1) + P(A_2) + P(A_3)$$
$$= 2[P(A_2)] + P(A_2) + P(A_2) = 4[P(A_2)]$$

Hence, $P(A_2) = \frac{1}{4} = P(A_3)$, and $P(A_1) = 2[P(A_2)] = \frac{1}{2}$.

The probabilities calculated above are conveniently summarized in the following table:

Event A_i	A_1	A_2	A_3	
$P(B \mid A_i)$	0.02	0.02	0.04	row 1
$P(A_i)$	0.50	0.25	0.25	row 2

For example, the value of $P(B \mid A_3)$ appears in the intersection of a row labeled $P(B \mid A_i)$ and the column labeled A_3: $P(B \mid A_3) = 0.04$.

Applying Theorem 4.8 yields

$$P(B) = P(B \mid A_1)P(A_1) + P(B \mid A_2)P(A_2) + P(B \mid A_3)P(A_3)$$
$$= (0.02)(0.50) + (0.02)(0.25) + (0.04)(0.25) = 0.025$$

Notice that $P(B)$ was obtained by multiplying each entry in row 1 of the table by the corresponding entry in row 2 and then adding the products.

Example 22   In a certain town, 30 percent of the voters are registered Democrats (D), 40 percent are registered Republicans (R), 20 percent are registered Liberals (L), and the remainder are not registered in any party (U). If 90 percent of the Democrats, 10 percent of the Republicans, 50 percent of the Liberals, and 20 percent of the others voted for Smith, what is the probability that a randomly chosen voter in the town voted for Smith?

Solution   Let S be the event "a randomly chosen voter from this town voted for Smith." We must find $P(S)$. The facts given for this example are conveniently summarized in the following table:

Event E	D	R	L	U
$P(S\mid E)$	0.9	0.1	0.5	0.2
$P(E)$	0.3	0.4	0.2	0.1

For example, the probability that a randomly chosen Democrat voted for Smith is $P(S\mid D) = 0.9$, the probability that a randomly chosen voter is a Democrat is $P(D) = 0.3$. Using the total probability theorem,

$$P(S) = P(S\mid D)P(D) + P(S\mid R)P(R) + P(S\mid L)P(L) + P(S\mid U)P(U)$$
$$= (0.9)(0.3) + (0.1)(0.4) + (0.5)(0.2) + (0.2)(0.1) = 0.43$$

It follows that 43 percent of the voters voted for Smith.

Example 23   Three identical treasure chests (A, B, and C) contain two drawers apiece. Each drawer of chest A contains a gold coin, each drawer of chest B contains a silver coin, and chest C has a gold coin in one drawer and a silver coin in the other. If a chest is chosen at random and then a randomly picked drawer of the chosen chest is opened, what is the probability that the opened drawer contains a gold coin?

Solution   Let G be the event "a gold coin is obtained," and let A, B, and C be the events "chest A is opened," "chest B is opened," and "chest C is opened," respectively. Using the total probability theorem, we have

$$P(G) = P(G\mid A)P(A) + P(G\mid B)P(B) + P(G\mid C)P(C)$$
$$= 1(\tfrac{1}{3}) + 0(\tfrac{1}{3}) + \tfrac{1}{2}(\tfrac{1}{3}) = \tfrac{1}{2}$$

This is the answer we might expect, since three out of the six coins are gold.

Example 24   Suppose that the opened drawer of the treasure chest chosen in Example 23 contains a gold coin. What is the probability that the other drawer in the same chest also contains a gold coin?

Solution   Many people "solve" this problem by reasoning like this: Since the drawer contained a gold coin, it is clear that the treasure chest that was chosen was not B. Hence two possibilities remain, A and C (two possible cases). The other coin is gold only if A was chosen (one favorable case). Therefore the required probability is $\tfrac{1}{2}$.

The mistake in this "solution" is deciding that choosing chest A is just as likely as choosing chest C. However, since we know that the chosen chest contains at least one gold coin, and that chest A contains twice as many gold coins as chest B, a moment's reflection makes it clear that it is more likely that chest A was chosen than that chest C was chosen.

To find the correct solution, let us now assume that a gold coin was found. Let A and C be the events "chest A was chosen," and "chest C was chosen," respectively. Notice that chest B could not have been chosen because of our assumption. It follows that $A + C = \Omega$. Let G be the event "the other drawer in the same chest also contains a gold coin." Applying the total probability theorem,

$$P(G) = P(G\,|\,A)P(A) + P(G\,|\,C)P(C) \tag{5}$$

Since there are three gold coins and two of the coins are in chest A,

$$P(A) = \tfrac{2}{3}, \quad\text{and so}\quad P(C) = 1 - P(A) = 1 - \tfrac{2}{3} = \tfrac{1}{3}$$

Moreover, since chest A contains two gold coins while chest C contains one gold coin,

$$P(G\,|\,A) = 1 \quad\text{and}\quad P(G\,|\,C) = 0$$

Substituting these values in equation (5) yields

$$P(G) = 1(\tfrac{2}{3}) + 0(\tfrac{1}{3}) = \tfrac{2}{3}$$

4.3 EXERCISES

1. One of three urns is selected at random. Urn A contains five red and two green balls, urn B contains six red and seven green balls, while urn C contains five red and five green balls. If a ball is randomly chosen from the selected urn, what is the probability that the ball is (*a*) green, (*b*) red?

2. What is the probability that the first daughter of the woman described in Example 19 is a carrier?

3. Brown eye color is dominant over blue. Assume that one-third of the brown-eyed people have genotype *BB*, while two-thirds have genotype *Bb*. A person with blue eyes has genotype *bb*. If a brown-eyed woman marries a blue-eyed man, what is the probability that their child is (*a*) brown-eyed, (*b*) blue-eyed?

4. From past experience it is known that a certain machine works properly 90 percent of the time. When it is working well, only 0.5 percent of the screws produced are defective. When it is working poorly, 9 percent of the screws produced are defective.
(a) What is the probability that a given screw is defective?
(b) If a screw is chosen at random and found to be defective, what is the probability that the machine is not working properly?

5. A commando operation is tentatively scheduled to begin at 12:00 midnight. The commandos must decide now whether or not to postpone the operation. According to the weather forecast, the probability of heavy rain at that time is 0.6, the probability of showers is 0.3, while the probability of no rain at all is 0.1. The commandos feel that their probability of success is 0.9 if there is a storm, 0.6 if it rains lightly, and 0.3 if it does not rain. They do not want to schedule the operation unless the probability of success is more than 0.7. Should they postpone the operation? Why?

6.   A mixture of corn seeds is made up of 50 percent type T_1 seeds, 40 percent type T_2, and 10 percent type T_3. For seed types T_1, T_2, and T_3, the probabilities that a randomly selected ear of corn has more than 100 kernels are 0.75, 0.40, and 0.25, respectively. If a farmer plants this seed mixture, what is the probability that an ear of corn randomly chosen from the resulting crop has more than 100 kernels?

7.   A woman has three identical purses. One contains three pennies, a nickel, and a dime, another contains five pennies and two dimes, and the third contains three pennies. A purse is randomly chosen and a coin is removed at random.

   (a) Find the probability that the coin is—

       (*i*) a penny   (*ii*) a dime   (*iii*) a nickel   (*iv*) a quarter

   (b) What is the sum of the probabilities calculated in part **(a)**?

8.   A student has three cards. The first card is white on both sides, the second is red on both sides, and the third is white on one side and red on the other. A card is randomly chosen and the side that faces up is white. What is the probability that the other side is also white?

9.   Construct the cards described in Exercise 8, and see if you can convince a friend to accept even odds that the other side is a different color than the first. Repeat the experiment ten times, betting even money each time. Compare your results to the probabilities calculated in Exercise 8.

10.   Records show that about 5 percent of men are color-blind, but only 0.2 percent of women are color-blind. What is the probability that a randomly selected person is color-blind? (Assume that 50 percent of the population is male.)

11.   Suppose it is known that only 40 percent of the population of a certain age is male. What is the probability that a randomly selected person of this age is color-blind? (See Exercise 10.)

12.   Twenty percent of the women and 5 percent of the men in a certain firm earn over 20,000 dollars per year. If 70 percent of the employees are women, what is the probability that a randomly picked employee earns over 20,000 dollars per year?

13.   Jane has five playing cards, including three aces. Suppose she discards the two cards that are not aces. What is the probability that she draws another ace if—

   (a) she draws two cards from the deck containing all the other cards

   (b) she draws two cards, but her opponent is holding three cards she cannot draw

   (c) she draws two cards, but she has two opponents holding five cards each?

14.   Suppose that a survey results in the following data: Twenty percent of the people surveyed are heavy smokers, but only 10 percent of the women are heavy smokers. Forty percent of the heavy smokers have heart disease, but only 20 percent of those who are not heavy smokers have heart disease. Fifteen percent of the women who smoke heavily have heart disease, but only 5 percent of the remaining women have heart disease. Based on these statistics—

(a) If a person is randomly chosen, what is the probability that he or she has heart disease?

(b) If a woman is randomly chosen, what is the probability that she has heart disease? (Hint: Let D be the event "the woman has heart disease," let H be the event "the woman is a heavy smoker," and then use the total probability theorem to find $P(D)$.)

4.4 BAYES' THEOREM

Suppose that A and B are events and that

$$P(B \mid A), \quad P(B \mid A'), \quad \text{and} \quad P(A)$$

are known. In the last section we learned how to calculate $P(B)$ using these probabilities and the total probability theorem. In this section we will see how these probabilities also enable us to calculate $P(A \mid B)$. The following example shows how this can be done.

Example 25   Suppose that a nickel biased 70 percent in favor of heads (i.e., $P(H) = 0.7$) accidentally gets mixed up with five fair nickels. If, in an attempt to find the biased coin, you toss one of the coins six times and its lands heads up each time, what is the probability that the nickel you've tossed is the biased one?

Solution   Let B and H be the events "the coin is the biased one" and "the coin lands heads up six times in a row," respectively. Then, applying Definition 4.1, Theorem 4.3, and Theorem 4.7,

$$P(B \mid H) = \frac{P(BH)}{P(H)} = \frac{P(B)P(H \mid B)}{P(H \mid B)P(B) + P(H \mid B')P(B')}$$

$$= \frac{(\frac{1}{6})(0.7)^6}{(0.7)^6(\frac{1}{6}) + (0.5)^6(\frac{5}{6})} = 0.60$$

The method used to solve Example 25 can also be used to prove the following theorem.

THEOREM 4.9   (Bayes' Theorem)   Let A and B be events, where $P(A)$ is not equal to 0 or 1. Then

$$P(A \mid B) = \frac{P(B \mid A)P(A)}{P(B \mid A)P(A) + P(B \mid A')P(A')} \tag{6}$$

Bayes' theorem enables us to calculate $P(A \mid B)$ if we know $P(B \mid A)$ and several other probabilities. Notice that the denominator of the expression for $P(A \mid B)$ in equation (6) is equal to $P(B)$. That is, $P(B) = P(B|A)P(A) + P(B|A')P(A')$, by Theorem 4.7. Therefore, equation (6) is equivalent to

$$P(A|B) = \frac{P(B|A)P(A)}{P(B)}$$

Theorem 4.9, like Theorem 4.7, can be extended in a natural way to n mutually exclusive events $A_1, A_2, \ldots, A_n$ where $A_1 + A_2 + \cdots + A_n = \Omega$. The statement of this extended theorem follows. Its proof is similar to the solution of Example 25.

THEOREM 4.10 (Bayes' Theorem) Suppose $A_1, A_2, \ldots, A_n$, and B are events. If $P(A_i) > 0$ ($i = 1, 2, \ldots, n$), and $\Omega = A_1 + A_2 + \cdots + A_n$, then

$$P(A_i|B) = \frac{P(B|A_i)P(A_i)}{P(B|A_1)P(A_1) + P(B|A_2)P(A_2) + \cdots + P(B|A_n)P(A_n)} \qquad (7)$$

Note that the denominator of the expression for $P(A_i|B)$ in equation (7) is equal to $P(B)$.

The usefulness of Theorems 4.9 and 4.10 is illustrated in the following examples.

Example 26 Suppose that 10 percent of the women and 5 percent of the men at a certain university are on the dean's list. Moreover, 70 percent of the students are male. If a randomly chosen student is on the dean's list, what is the probability that the student is male?

Solution Let M and D be the events " the student is male " and " the student is on the dean's list," respectively. From the facts given, we know that

$$P(D|M') = 0.10, \qquad P(D|M) = 0.05, \qquad P(M) = 0.70$$

and

$$P(M') = 1 - P(M) = 0.30$$

We need to calculate $P(M|D)$. Applying Theorem 4.9,

$$P(M|D) = \frac{P(D|M)P(M)}{P(D|M)P(M) + P(D|M')P(M')}$$

$$= \frac{(0.05)(0.70)}{(0.05)(0.70) + (0.10)(0.30)} = \frac{35}{65}$$

Example 27 At a certain university, 20 percent of the freshmen (F), 10 percent of the sophomores (S), 5 percent of the juniors (J), and 3 percent of the seniors (L) are on probation. Moreover, 35 percent, 30 percent, 20 percent, 15 percent of the students in the school are freshmen, sophomores, juniors, seniors, respectively. If a student on probation (N) is randomly chosen, what is the probability that he or she is a freshman?

Solution The information given in this example is summarized in the following table:

Event E	Column 1 F	2 S	3 J	4 L	
$P(N\|E)$	0.20	0.10	0.05	0.03	row 1
$P(E)$	0.35	0.30	0.20	0.15	row 2

The first row of the table gives the values of $P(N|F)$, $P(N|S)$, $P(N|J)$, and $P(N|L)$. The second row gives the values of $P(F)$, $P(S)$, $P(J)$, and $P(L)$.

Applying Theorem 4.10,

$$P(F|N) = \frac{P(N|F)P(F)}{P(N|F)P(F) + P(N|S)P(S) + P(N|J)P(J) + P(N|L)P(L)}$$

$$= \frac{(0.20)(0.35)}{(0.20)(0.35) + (0.10)(0.30) + (0.05)(0.2) + (0.03)(0.15)}$$

$$= \frac{700}{1145}$$

Notice that the denominator of this probability was found by multiplying each entry in row 1 of the table by the corresponding entry in row 2, and then adding the products. The numerator was calculated by multiplying the entry in row 1, column 1, by the corresponding entry in row 2, column 1.

Example 28 It is estimated that one percent of the people in a certain neighborhood have undiagnosed tuberculosis. Free chest x rays are given to the people in this area to detect the disease. From past experience, it is known that, if a person does not have tuberculosis, the probability that the test will be positive is 0.1. If the person does have the disease, the probability that the test will be positive is 0.99.

(a) If a person's test is positive, what is the probability that the person does not have tuberculosis?

(b) If the test is negative, what is the probability that the person has tuberculosis?

Solution

(a) Let T be the event "the person has tuberculosis," and let N be the event "the test is negative." We know from the information given that

$$P(T) = 0.01, \qquad P(T') = 0.99$$

$$P(N'|T) = 0.99, \qquad \text{and} \qquad P(N'|T') = 0.10$$

Applying Theorem 4.9, the required probability is

$$P(T'|N') = \frac{P(N'|T')P(T')}{P(N'|T')P(T') + P(N'|T)P(T)}$$

$$= \frac{(0.10)(0.99)}{(0.10)(0.99) + (0.99)(0.01)} = 0.909$$

Therefore, even if a person's test is positive, there is still a very large probability that the person does not have the disease.

(b) On the other hand,

$$P(T\,|\,N) = \frac{P(N\,|\,T)P(T)}{P(N\,|\,T)P(T) + P(N\,|\,T')P(T')}$$

The unknown quantities, $P(N\,|\,T)$ and $P(N\,|\,T')$, can be easily calculated by applying Theorem 4.1(f), page 126.

$$P(N\,|\,T) = 1 - P(N'\,|\,T) = 1 - 0.99 = 0.01$$

$$P(N\,|\,T') = 1 - P(N'\,|\,T') = 1 - 0.10 = 0.90$$

Hence,

$$P(T\,|\,N) = \frac{(0.01)(0.01)}{(0.01)(0.01) + (0.90)(0.99)} = \frac{1}{8911} = 0.00011$$

Therefore, if a person's test is negative, there is only a very, very small probability that the person has the disease. The person has good reason, therefore, to assume that he or she does not have tuberculosis.

One may argue that the test of Example 28 is not a good one because the overwhelming majority (90.9 percent) of the people who fail the test are healthy. However, if the test were made less stringent, more sick people would escape notice. (See Exercise 1 following this section.) Even worse, more sick people would have a doctor's assurance that they are healthy and would then ignore relevant symptoms. It is much better to require that healthy people undergo further testing than to allow sick people to remain untreated.

Example 29   Let Ω be the set of adults, and let M, W, D, and S be the sets of married, widowed, divorced, and single (i.e., never married) adults. Moreover, let H be the set of all adults that are happy in their own opinion. Suppose that a survey of adult residents in a certain city results in the following statistics:

Event E	M	W	D	S
$P(H\,|\,E)$	0.40	0.30	0.50	0.35
$P(E)$	0.50	0.15	0.25	0.10

For example, the probability that a married person is happy is $P(H\,|\,M) = 0.40$.

(a) What is the probability that a randomly chosen person is happy?

(b) If a randomly chosen person is found to be happy, what is the probability that he or she is widowed?

Solution

(a) We need to calculate $P(H)$. Applying the total probability theorem,

$$P(H) = P(H \mid M)P(M) + P(H \mid W)P(W) + P(H \mid D)P(D) + P(H \mid S)P(S)$$

$$= (0.40)(0.50) + (0.30)(0.15) + (0.50)(0.25) + (0.35)(0.10) = 0.405$$

(b) Applying Bayes' Theorem, the total probability theorem, and part **(a)**, the required probability is

$$P(W \mid H) = \frac{P(H \mid W)P(W)}{P(H \mid M)P(M) + P(H \mid W)P(W) + P(H \mid D)P(D) + P(H \mid S)P(S)}$$

$$= \frac{(0.3)(0.15)}{P(H)} = \frac{0.045}{0.405} = 0.111$$

4.4 EXERCISES

1. Suppose, because of objections from healthy people and the cost of retesting, the test described in Example 28 is made less rigorous. Only 1 percent of the healthy people who take the new test fail it (i.e., obtain a positive reaction). Unfortunately, the less stringent requirements also mean that 10 percent of the sick people who take the test pass it.

    (a) If a person fails the test, what is the probability that he or she is healthy?

    (b) If a person passes the test, what is the probability that he or she is sick?

2. In a certain university, 40 percent of the students are women, 10 percent of the women are science majors, and 30 percent of the men are science majors. If a student chosen at random is found to be a science major, what is the probability that the student is male?

3. In the same university, a professor grades 30 percent of the papers in all his classes, and his graduate assistant grades the rest. A student insists, from long experience with this teacher, that if the professor grades her paper, her probability of passing the test is 0.9, while if the assistant grades it, her probability of passing is only 0.6. If she passes the test, what is the probability that the professor graded her paper?

4. The probabilities that a male has blood genotype OA, AA, or AB are 0.352, 0.069, and 0.015, respectively. If a child of an OO mother has genotype OA, what is the probability that its father has blood type AB?

5. If the child of a blue-eyed mother has brown eyes, find the probability that the father's genotype is—

    (a) BB **(b)** bB **(c)** bb

Use the facts given in Exercise 3, Section 4.3, page 143.

6. In a certain city, 35 percent of the people live in houses, 50 percent live in apartments, and 15 percent live in high-rise condominiums. Transportation statistics show that 10 percent of the house dwellers in the city, 5 percent of the apartment dwellers, and 1 percent of the high-rise dwellers belong to car pools.

(a) What is the probability that a city resident belongs to a car pool?

(b) If a resident of the city belongs to a car pool, what is the probability that he or she lives in a house?

7.   Suppose that 50 percent of the burglaries in Newton are committed by residents of the town, 30 percent are committed by residents of a neighboring city, Roosevelt, and the rest are committed by residents of neighboring suburbs. If the method used in a certain burglary is the one used by 30 percent of the burglars in Newton, 60 percent of those in Roosevelt, and 20 percent of those in the neighboring suburbs, what is the probability that the burglar came from—

(a) Newton

(b) Roosevelt

(c) the neighboring suburbs?

Key Word Review

Each key word is followed by the number of the section in which it is defined.

conditional probability   4.1   **pairwise independent events   4.2**
independent events   4.2   **mutually independent events   4.2**

CHAPTER 4   **REVIEW EXERCISES**

1.   Define each of the words in the Key Word Review.

2.   What does the symbol $P(B|A)$ mean?

3.   Suppose that $P(A) = 0.30$, $P(B) = 0.20$, $P(AB) = 0.10$, $P(C) = 0.40$, $P(BC) = 0.15$, and $P(ABC) = 0.05$. Calculate the following probabilities:
     (a) $P(A|B)$ **(b)** $P(B|A)$ **(c)** $P(A'|B)$
     (d) $P(AC|B)$ **(e)** $P((A \cup C)|B)$ **(f)** $P(\varnothing|A)$
     (g) $P(A|\varnothing)$ **(h)** $P(A|\Omega)$ **(i)** $P(\Omega|A)$

4.   Are the events A and B of Exercise 3 independent? Are they mutually exclusive? Explain.

5.   Suppose that two dice are tossed. What is the probability that the sum of the dots showing on their faces is 8 if—
     (a) nothing more is known
     (b) the first die shows a 5
     (c) both dice show the same number
     (d) the total is even
     (e) the second die shows a 1?

6.   A ten-year study of patients of a Veteran's Administration hospital resulted in the following statistics:

	Suffered a fatal heart attack	Did not suffer a fatal heart attack
Polyunsaturated diet	48	376
Conventional diet	70	352

Based on these data,[3] what is the probability that—
     (a) a patient chosen at random (without regard to diet) suffers a fatal heart attack during the next ten years

3. Reported by Donal Drake in the *Philadephia Inquirer*, April 18, 1971, based on an article by Dr. M. Pearce and Dr. S. Dayton in the medical journal *Lancet*.

(b) a patient on a conventional diet suffers a fatal heart attack during the next ten years

(c) a patient on a polyunsaturated diet suffers a fatal heart attack during the next ten years?

7. Let C be the event "the patient is on a conventional diet" and let F be the event "the patient suffers a fatal heart attack during the next ten years." Using the data from Exercise 6—

    (a) calculate—

        (*i*) $P(C)$ (*ii*) $P(F)$ (*iii*) $P(CF)$ (*iv*) $P(F \mid C)$ (*v*) $P(C \mid F)$

    (b) are the events C and F independent? Explain.

8. The following table gives the number of freshmen (F), sophomores (S), juniors (J), and seniors (A) who are on probation (R), not on probation but not on the dean's list (B), or on the dean's list (D) at a certain university.

Class	On Probation (R)	Not on Probation, Not on Dean's List (B)	On Dean's List (D)
Freshmen (F)	70	200	30
Sophomore (S)	55	175	30
Junior (J)	15	190	35
Senior (A)	5	150	45

    (a) What is the probability that a randomly chosen student is (*i*) on the dean's list, (*ii*) a freshman on the dean's list?

    (b) What is the probability that a randomly chosen sophomore is (*i*) on probation, (*ii*) not on the dean's list?

    (c) Calculate—

        (*i*) $P(F)$ (*ii*) $P(F \mid R)$ (*iii*) $P(FR)$ (*iv*) $P(D \mid R)$

9. Let A be any set. Prove that A and $\varnothing$ are independent events.

10. Suppose that 20 percent of the students in a certain school take math, 15 percent take physics, and 10 percent take both. A student is randomly chosen and is found to be taking math. What is the probability that this student is not taking physics?

11. Approximately one out of every thirty Jews is a carrier of Tay-Sachs disease, a recessive hereditary disease that causes severe mental retardation and early death.

    (a) If two Jews marry, what is the probability that—

        (*i*) they are both carriers

        (*ii*) their first child has Tay-Sachs disease?

(b) If their first child has Tay-Sachs disease, what is the probability that—

    (*i*) their second child has Tay-Sachs disease

    (*ii*) their second and third child have Tay-Sachs disease

    (*iii*) at least one of their next three children is completely normal (does not have Tay-Sachs disease and is not a carrier)?

12. Suppose that two Jews have a child who does not have Tay-Sachs disease. What is the probability that the parents are both carriers?

13. Assume that 20 percent of the science majors, 10 percent of the liberal arts majors, and 5 percent of the business majors in a certain university continue their education after college. If 20 percent of the students at this college are science majors, 45 percent are liberal arts majors, and 35 percent are business majors, what is the probability that—

(a) a randomly chosen student will continue his or her education after college

(b) a randomly chosen graduate who has continued his or her education after college was a business major?

Suggested Readings

1. Eisen, Martin. *Introduction to Mathematical Probability Theory.* Englewood Cliffs, N.J.: Prentice-Hall, 1969.
2. Eisen, Martin, and Eisen, Carole. *Probability and Its Applications.* New York: Quantum Press, 1975.
3. Hodges, J. L., Jr., and Lehmann, E. L. *Elements of Finite Probability.* 2d ed. San Francisco: Holden-Day, 1970.

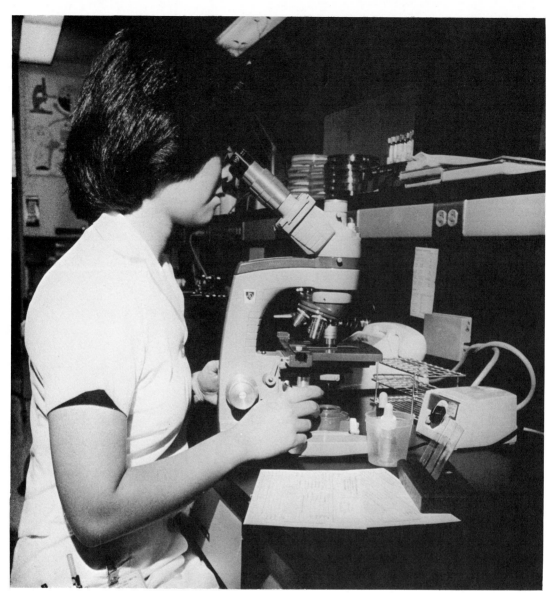

Hospital of the Good Samaritan

INTRODUCTION

The medical technologist in the photograph on the left is examining samples of blood to determine their type. If the samples were taken from randomly selected volunteers, what is the most likely number of slides she will have to examine before finding two samples of blood type A? What is the probability that she will not find three samples of blood type A until she examines the fifth slide?

We can answer these questions by using the probability theory that we have learned and the fact that 40 percent of all Americans have blood type A. However, the answers are not obvious. Students often complain that they do not know how to begin solving probability problems because each problem seems different. In this chapter we will see that problems with different descriptions sometimes have similar mathematical characteristics. These mathematical characteristics will enable us to place many (but not all) probability problems into one of several categories, and to use the same rules to solve all problems in each category.

The questions that we have asked about blood typing, for example, belong to a category of problems called *Pascal*. Other problems in this category include die-tossing, poll-taking, and personnel problems, in addition to many others. By learning the general rules for solving a Pascal problem, we will learn how to solve each of the seemingly different problems in this category.

5.1 THE CONCEPT OF FUNCTION

The classification of probability problems requires an understanding of the concept of a *function*. The following example will lead us to a definition of this word.

Example 1 The final grades given at a certain school are 1, 2, 3, 4, 5, P, and F. If there are five students in a freshman mathematics class, the instructor might post the following grades:

Student	*Grade*
Brown	2
Cinelli	1
Levy	2
Martin	1
Lewis	F

Notice that each element in the set *A* of students corresponds to *one and only one* element in the set *B* of final grades. This correspondence is an example of a function.

Definition 5.1   A **function** f from the set A into the set B is a correspondence that associates each element a of A with one and only one element b of B.

This correspondence is sometimes written $f(a) = b$. The set A is called the **domain** of f. The set $f(A) = \{b \mid f(a) = b$ for some $a \in A\}$ is the **range**, or *image*, of f. The following examples illustrate this definition.

Example 2   Let f be the function of Example 1.

   (a) The domain of f is {Brown, Cinelli, Levy, Martin, Lewis}.

   (b) The range of f is {1,2,F}. Notice that 3 is not in the range, although 3 is a grade given by the school.

   (c) $f(\text{Brown}) = 2$, $f(\text{Cinelli}) = 1$, $f(\text{Levy}) = 2$, $f(\text{Martin}) = 1$,   and $f(\text{Lewis}) = \text{F}$.

Example 3   Let $A = \{1,2,4\}$ and $B = \{1,2,3\}$. The following table defines a function f from A into B.

a	1	2	4
$f(a)$	1	3	3

   (a) The domain of f is A.

   (b) The range of f is {1,3}. Note that 2 is not in the range, although $2 \in B$.

   (c) $f(1) = 1, f(2) = 3$, and $f(4) = 3$.

Example 4   Let A be the set of integers. The correspondence

$$f(x) = x^2 + 1 \tag{1}$$

is a function from A into A. For example, $f(1)$ is found by replacing x by 1 in equation (1):

$$f(1) = 1^2 + 1 = 2$$

Similarly,

$$f(3) = 3^2 + 1 = 10 \quad \text{and} \quad f(-3) = (-3)^2 + 1 = 10$$

The range of f is {1,2,5,10,17,26,...}.

The function f of Example 1 is represented by the following diagram:

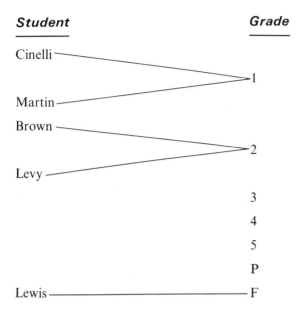

From the diagram, we see that—

(a) every student received a grade

(b) no person received two grades

(c) there are grades that were not received by anyone

(d) some grades were received by more than one person.

Using Definition 5.1, the reader can verify that statements similar to (a) and (b) can be made about every function. Statements similar to (c) and (d) are true about some functions, but not about others.

Most of the functions that we will study in this book are *real-valued functions*; that is, their range is a subset of the real numbers.[1] The functions of Examples 3 and 4 are examples of real-valued functions. The function f of Example 1, on the other hand, is not real valued because the grade F is in the range of f, but F is not a number.

Real-valued functions are similar in many ways to numbers. We know, for example, that numbers can be combined by the operations of addition, subtraction, multiplication, and division. Real-valued functions can be combined by these operations also. Each of the operations with real-valued functions is defined in terms of the corresponding operation with numbers.

---

1. The reader has probably already studied real numbers in high school. To review briefly, the *rational numbers* are the numbers which can be written in the form P/Q where P and Q are integers. Examples of rational numbers are $\frac{1}{2}$, $-\frac{7}{102}$, $\frac{2}{1} = 2$, and $0.3 = \frac{3}{10}$. Numbers such as $\sqrt{2}$ and $1 - \pi$ cannot be written in this way. These numbers are called *irrational numbers*. The set consisting of all rational numbers *and* all irrational numbers is called the set of *real numbers*.

Definition 5.2   Let f and g be real-valued functions with domains A and B, respectively. The *sum* $(f + g)$, *difference* $(f - g)$, *product* $(f \cdot g)$, and *quotient* (f/g) of these functions are defined as follows:

(a) $(f + g)(x) = f(x) + g(x)$ The domain of $f + g$ is $A \cap B$

(b) $(f - g)(x) = f(x) - g(x)$ The domain of $f - g$ is $A \cap B$

(c) $(f \cdot g)(x) = [f(x)] \cdot [g(x)]$ The domain of $f \cdot g$ is $A \cap B$

(d) $\dfrac{f}{g}(x) = \dfrac{f(x)}{g(x)}$ if $g(x) \neq 0$ The domain of $\dfrac{f}{g}$ is $\{x \mid x \in A \cap B, \, g(x) \neq 0\}$

Notice that the value assigned to a number x by the sum, difference, product, or quotient function depends entirely upon the values assigned to the number by f and g. However, f and g assign values only to numbers that are in their domains. Therefore, the sum, difference, product, and quotient functions only assign values to numbers that are in the domains of both f and g. This is the reason that the domains of $f + g, f - g$, and $f \cdot g$ are the intersection of the domains of f and g, $A \cap B$. Furthermore, the domain of the quotient function cannot include any number x for which $g(x)$ is zero, because division by zero is not defined. This is why the domain of f/g is $\{x \mid x \in A \cap B, \, g(x) \neq 0\}$.

The following example shows how Definition 5.2 applies to particular functions. For simplicity, in this example and throughout the remainder of the book, we will frequently not state the domain and range of the functions we discuss. When the domain and range are not stated, the domain will be assumed to be the set under discussion, or the largest set of real numbers for which the formula defining the function yields a real number. For example, the real-valued function $f(x) = 1/x$ yields a real value for every real number x except zero. Therefore, the domain of f is the set of all real numbers except zero.

Example 5   Let

$$f(x) = x^2 \quad \text{and} \quad g(x) = 2x$$

Then—

(a) $(f + g)(x) = f(x) + g(x) = x^2 + 2x$. In particular,

$(f + g)(1) = 1^2 + 2(1) = 3$

(b) $(f - g)(x) = f(x) - g(x) = x^2 - 2x$. In particular,

$(f - g)(2) = 2^2 - 2(2) = 0$

(c) $(f \cdot g)(x) = x^2(2x) = 2x^3$. In particular,

$(f \cdot g)(3) = 2(3^3) = 54$

(d) $\dfrac{f}{g}(x) = \dfrac{x^2}{2x} = \dfrac{x}{2}$ if $x \neq 0$. In particular,

$\dfrac{f}{g}(-2) = \dfrac{-2}{2} = -1$

In this case, the domain of f/g may be extended in a natural way to include 0 by defining

$$\frac{f}{g}(0) = \frac{0}{2} = 0$$

5.1 EXERCISES

1. Let f be the function defined by the following table:

x	1	2	3	4
$f(x)$	3	1	5	1

Find (a) the domain of f, (b) the range of f, (c) $f(3)$.

2. State the domain of the following functions.

   (a) $f(x) = 2x + 1$ (b) $g(x) = \dfrac{2}{x}$ (c) $f(x) = \dfrac{1}{2x + 1}$

3. Let $f(x) = x^4$. Calculate—
   (a) $f(1)$ (b) $f(-3)$ (c) $f(5)$ (d) $f(k)$

4. Let $f(x) = x^3$ and $g(x) = x^2$. Calculate—

   (a) $(f + g)(x)$ (b) $\dfrac{f}{g}(x)$ (c) $(f \cdot g)(x)$

5. Let $f(x) = 2x$ and $g(x) = x^2$. Calculate

   (a) $f(2)$,   (b) $\dfrac{f}{g}(2)$,   and   (c) $(f - g)(2)$.

6. Let $f(x)$ be the largest integer that is less than or equal to x; $f(x)$ is written $[x]$. For example, $[3.5] = 3$ and $[-2.1] = -3$. Calculate—
   (a) $[\frac{3}{2}]$ (b) $[\pi]$ (c) $[-9.8]$ (d) $[-7]$

7. Let $f(x) = x!$, $x = 1, 2, 3, \ldots$. Find—
   (a) $f(3)$ (b) $f(6)$ (c) $f(0)$

5.2 THE CONCEPT OF RANDOM VARIABLES

In this section we will study a particular type of function called a *random variable*. In the remainder of this chapter we will see that these functions make the classification of probability problems possible. This classification allows us to use the same formula or method to solve many apparently dissimilar problems.

We saw in Section 5.1 that every function has a domain and a range. In particular, if the domain of the function is a sample space Ω and its range is a subset of the real numbers, then the function is called a *random variable*.

Random variables will be denoted by capital letters, usually those from the end of the alphabet (e.g., $R, S, T, \ldots$). Corresponding lowercase letters (i.e., $r, s, t, \ldots$) will denote values assumed by the random variable.

The function described in the following example is a random variable.

Example 6   Let S be the number of heads that are obtained when a fair coin is tossed three times. This function is defined in the following table:

ω	(h,h,h)	(h,h,t)	(h,t,t)	(h,t,h)	(t,h,h)	(t,t,h)	(t,h,t)	(t,t,t)
$S(\omega)$	3	2	1	2	2	1	1	0

The domain of S is the sample space $\Omega = \{(h,h,h),(h,h,t),\ldots,(t,t,t)\}$; its range is $\{0,1,2,3\}$, which is a subset of the set of real numbers. It follows that S is a random variable.

Frequently, random variables are discussed without specifically describing their domains. The following examples illustrate this procedure.

Example 7

(a) A lightbulb is tested. The function X defined by setting $X = 1$ if the bulb works and $X = 0$ if it does not is a random variable whose range is $\{0,1\}$.

(b) A radioactive source is emitting alpha particles. The emission of these particles is observed on a counting device during a specified period of time. The number N of particles observed is a random variable whose range is the nonnegative integers.

(c) The temperature T at noon in New York is a random variable.

(d) The number of students in a class who wear glasses is a random variable.

(e) A machine produces 1000 bolts per hour. The number of defective bolts that are produced in an hour is a random variable with range $\{0,1,2,\ldots,1000\}$.

(f) The number of children in a randomly selected family is a random variable.

We have said that any function whose domain is a sample space Ω and whose range is a subset of the real numbers is a random variable. The following example will show why it is important to extend this definition.

Example 8   Suppose a coin is tossed until the first head appears. Let G be the number of tosses up to and including the appearance of the first head. The domain of G is a sample space Ω. (Several elements of Ω are $(h,h,h,\ldots)$, $(h,t,h,t,h,\ldots)$, and $(h,h,t,t,t,\ldots)$.)

Solution   Let us now discuss the range of G. Since it is clear that 1, 2, 3, ... tosses may be required, the range of G includes all the positive integers. However, it is also possible that we may never get a head; that is, the event $\{(t,t,t,...)\}$ may occur. (We will see in Section 5.9 that the probability of this event is 0.) If this event occurs, we would have to keep on tossing forever. It follows that the set $\{1,2,3,...\}$ does not describe everything that can happen. We need an additional symbol to represent $G((t,t,t,...))$. Since we will be tossing indefinitely, it is convenient to let $G((t,t,...))$ be ∞ (this is the symbol for infinity). The range of G is therefore $\{\infty,1,2,3,...\}$.

Let $\mathbf{R}$ be the set of real numbers and let $\mathbf{R}^* = \mathbf{R} \cup \{-\infty,\infty\}$. The set $\mathbf{R}^*$ is called the set of **extended real numbers**. The function G of Example 8 is similar to the random variables that we have discussed previously. Its domain is a sample space Ω. However, its range is not a subset of $\mathbf{R}$ (i.e., G is not a real-valued function). Instead, its range is a subset of $\mathbf{R}^*$. A function whose range is a subset of $\mathbf{R}^*$ is called an *extended-real-valued function*. Since many important situations can be described by functions like G, it is convenient to extend the definition of a random variable to include these functions.

Definition 5.3   A **random variable** is a function whose domain is a sample space Ω and whose range is a subset of $\mathbf{R}^*$, the set of extended real numbers ($\mathbf{R}^* = \mathbf{R} \cup \{-\infty,\infty\}$ where $\mathbf{R}$ is the set of all real numbers).

In particular, the function G of Example 8 is a random variable.

We saw in Section 5.1 that real-valued functions can be added, subtracted, multiplied, and divided. These operations can be extended in a natural way to extended-real-valued functions. It follows that random variables can be added, subtracted, and so on. The following example illustrates this.

Example 9   Suppose that two dice are tossed. Let U and V be the numbers of dots on the first and second dies, respectively. Then $T = U + V$ is the total number of dots on both dice, and $D = U - V$ is the number of dots on the first die minus the number of dots on the second die. The functions U, V, T, and D are all random variables.

5.2 EXERCISES

1.   A coin is tossed three times. Let the random variable X be the number of the toss on which the first tail occurs; if no tails occurs, let $X = 0$.
   (a) What is the range of X?
   (b) Calculate (i) $X(h,h,h)$, (ii) $X(t,t,h)$, (iii) $X(h,t,t)$.

2.   A fair die is rolled twice. Let X and Y be the number of spots on the first and second rolls, respectively, and let $S = X + Y$. Evaluate—
   (a) $X(1,2)$ (b) $Y(1,2)$ (c) $S(1,2)$

3.   Find the range of the following random variables.

    (a) Townspeople are randomly interviewed until the first person who enjoys Health-Crisp Cereal is found, or until 100 people have been interviewed. Let the random variable N be the number of people interviewed.

    (b) Townspeople are randomly interviewed until the first person who enjoys Health-Crisp Cereal is found. The same person may be interviewed more than once. Let the random variable G be the number of people interviewed.

5.3 DENSITY AND DISTRIBUTION FUNCTIONS

In this section we will see that every random variable is associated with two functions: its density function and its distribution function. These functions simplify the calculations in many probability problems.

    Let X be any random variable. We willl find it convenient to use the notation

$$(X = x)$$

to represent the event "X is x occurs." (Remember that the capital letter, X, represents a random variable; the corresponding lowercase letter, x, represents a number, ∞, or $-\infty$.) For example, if N is the number of students absent from school on a particular day, then

$$(N = 5)$$

is the event "5 children are absent from school," while $(N = 7)$ is the event "7 children are absent from school." Similarly,

$$(N \leq 7) \quad \text{and} \quad (N > 4)$$

are the events "at most 7 children are absent from school" and "more than 4 children are absent from school," respectively.

    Many probability problems can be solved if we know $P(X = x)$ or $P(X \leq x)$. The following definition is therefore useful.

Definition 5.4   Let X be a random variable.

    (a) The function f defined by

$$f(x) = P(X = x)$$

      is called the **density function** of X.

    (b) The function F defined by

$$F(x) = P(X \leq x)$$

      is called the **distribution function** of X.

    In this definition x can be any real number. It can also be ∞ or $-\infty$ if these symbols are in the range of X.

The following examples illustrate Definition 5.4.

Example 10 Let X be the number of television sets sold by a certain store in a day and let f be its density function. From past records, it is known that the following table represents the values of $f(x)$.

x	0	1	2	3	4
$f(x)$	0.25	0.30	0.20	0.15	0.10

The unlisted values of $f(x)$ are all 0. What is the probability that the store will sell (a) exactly 3, (b) at most 2, (c) exactly 5 television sets tomorrow?

Solution

(a) Using the table and Definition 5.4, $P(X = 3) = f(3) = 0.15$.

(b) Similarly,

$$P(X \leq 2) = P(X = 0) + P(X = 1) + P(X = 2)$$
$$= f(0) + f(1) + f(2)$$
$$= 0.25 + 0.30 + 0.20 = 0.75$$

(c) Since no value of $f(5)$ is listed, $f(5) = 0$. It follows that $P(X = 5) = 0$.

Example 11 Let X be the random variable of Example 10 and let F be its distribution function. Calculate—

(a) $F(-1)$ (b) $F(0)$ (c) $F(1)$ (d) $F(2.5)$ (e) $F(5)$

Solution

(a) $F(-1) = P(X \leq -1) = P(\varnothing) = 0$. Notice that $F(-2) = F(-3) = F(-2.5) = 0$ also. In fact, $F(x) = 0$, if $x < 0$.

(b) $F(0) = P(X \leq 0) = P(X = 0) = f(0) = 0.25$

(c) $F(1) = P(X \leq 1) = f(0) + f(1) = 0.25 + 0.30 = 0.55$

(d) $F(2.5) = P(X \leq 2.5) = P(X = 0) + P(X = 1) + P(X = 2) = 0.25 + 0.30 + 0.20 = 0.75$

(e) $F(5) = P(X \leq 5) = 0.25 + 0.30 + 0.20 + 0.15 + 0.10 = 1$. Notice that $F(x) = 1$ if $x \geq 4$.

We saw in Example 10 that the values $f(x)$ of the density function f of a random variable may be zero for many values of x. In defining f, it is therefore conventional to list only the nonzero values of $f(x)$. (This was done in Example 10.) It is then understood that the unlisted values of $f(x)$ are zero.

Let X be the random variable of Example 10 and let F be its distribution function. Although $F(x)$ is defined for every value of x, we need only list the values of $F(0)$, $F(1)$, $F(2)$, $F(3)$, and $F(4)$. These are the values of x for which $f(x) \neq 0$. These values are given in the following table.

x	0	1	2	3	4
$F(x)$	0.25	0.55	0.75	0.90	1.00

The remaining values of $F(x)$ can now be easily calculated from the table, Definition 5.4, and the fact that $f(x) = 0$ if $x \neq 0, 1, 2, 3, 4$. For example,

$$F(3.2) = P(X \leq 3.2) = P(X \leq 3) = F(3) = 0.9$$
$$F(7) = P(X \leq 7) = P(X \leq 4) = 1.0$$

This approach will also be used in determining the values of other distribution functions.

Consider the table of Example 10. The values $f(x)$ that appear in the second row of that table are somewhat arbitrary. For example, $f(0)$ could be 0.1, 0.4, or even 1 depending upon the store's business. However, since $f(0)$ is the probability that no television sets are sold, it is clear that $f(0)$ cannot be 7, 4, or $-\frac{1}{2}$. In fact $f(0)$ (and every other entry in row 2) must be a number between 0 and 1, inclusive.

In general, if f is any density function and x is any number, then

$$0 \leq f(x) \leq 1 \tag{2}$$

by Definition 5.4 and Theorem 2.3. Consider the table of Example 10 again. We have said that $f(0)$ can be any number between 0 and 1. However, if we change $f(0)$ to 1, it is clear that we must change all the other entries in row 2 to 0, because the sum of all the probabilities must be 1. This fact is conveniently symbolized by

$$\sum_x f(x) = 1 \tag{3}$$

where $\sum\limits_x$ means that the summation is taken over all values in the range of X.

Statements (2) and (3) are true of any density function f and are listed in Theorem 5.1. Additional properties of the density and distribution functions of a random variable are given in Theorems 5.2 and 5.3.

THEOREM 5.1 Let X be a random variable and let f be its density function. Then—

(a) $0 \leq f(x) \leq 1$ for all x
(b) $\sum\limits_x f(x) = 1$

Conversely, if f is a function satisfying statements (a) and (b) of Theorem 5.1, then there is a random variable X that has f as its density function.

The values of many distribution functions f are given in tables. One such table is the one on page 164. Other tables of values for distribution functions appear throughout the remainder of this book. Theorem 5.2 will enable us to use these tables to solve many probability problems. This theorem is easily proved by applying the definition of a distribution function and the properties of probability.

THEOREM 5.2   Let F be the distribution function of a random variable X, and let a and b be real numbers such that $a < b$. Then—

(a) $P(a < X \leq b) = F(b) - F(a)$
(b) $P(X > a) = 1 - F(a)$

Additional properties of the distribution function F that are important in calculating probabilities are listed in Theorem 5.3. These properties are best understood by using an example.

Let X be the random variable of Example 10. The values of its distribution function F for which $f(x) \neq 0$ are given in the following table:

x	0	1	2	3	4
$F(x)$	0.25	0.55	0.75	0.90	1.00

On page 164, we saw how to calculate the values of $F(x)$ if x is not equal to 0, 1, 2, 3 or 4. From the table of values for F and the method we used to calculate $F(x)$ for values of x that do not appear in the table, we see that—

(a) $F(x)$ is defined for every real number x
(b) $F(x)$ is a number at least equal to 0 and at most equal to 1
(c) the values of $F(x)$ increase as x gets larger $(0.25 \leq 0.55 \leq 0.75 \leq 0.90 \leq 1.00)$
(d) if x is large (i.e., if $x \geq 4$), then $F(x) = 1$
(e) if x is small (i.e., if $x < 0$), then $F(x) = 0$.

Statements (a), (b), and (c) are true for every distribution function F. Statements (d) and (e) are true for many distribution functions—in particular, the ones that we shall consider in this book—but they are not true for all distribution functions. For some distribution functions, the values of $F(x)$ get closer and closer to 1 as x gets larger and larger, but they never actually achieve the value 1. For example, $F(5)$ may equal 0.9, $F(6)$ may equal 0.99, $F(7)$ may equal 0.999, and so on. Similarly, for some distribution functions, the values of $F(x)$ get closer and closer to 0 as x gets smaller and smaller, but they never actually achieve the value 0. Theorem 5.3 lists the five properties of a distribution function that correspond to statements (a)–(e).

THEOREM 5.3   A distribution function F has the following properties:

(a) It is defined for every real number x.
(b) $0 \leq F(x) \leq 1$
(c) It is an increasing function of x; that is, if $a < b$, then $F(a) \leq F(b)$.

(d) Either   (*i*) $F(x) = 1$, if x is sufficiently large,

       or   (*ii*) $F(x)$ gets closer and closer to 1 as x gets larger and larger.

(e) Either   (*i*) $F(x) = 0$, if x is sufficiently small,

       or   (*ii*) $F(x)$ gets closer and closer to 0 as x gets smaller and smaller.

The following examples show how to use Theorems 5.1–5.3.

Example 12   Let X be a random variable with distribution function F defined by

$$F(x) = \frac{x^2}{100} \qquad x = 1, 2, 3, \ldots, 10$$

Find—

(a) $P(X \le 4)$ **(b)** $P(X > 7)$ **(c)** $P(2 < X \le 5)$ **(d)** $P(X = 3)$

Solution

(a) Applying Definition 5.4,

$$P(X \le 4) = F(4) = \frac{4^2}{100} = 0.16$$

(b) Applying Theorem 5.2(b),

$$P(X > 7) = 1 - F(7) = 1 - \frac{7^2}{100} = 0.51$$

(c) Applying Theorem 5.2(a),

$$P(2 < X \le 5) = F(5) - F(2) = \frac{5^2}{100} - \frac{2^2}{100} = 0.21$$

(d) Since $f(x) = 0$ if $x \ne 1, 2, 3, \ldots, 10,$

$$F(3) = f(1) + f(2) + f(3)$$

and

$$F(2) = f(1) + f(2)$$

It then follows that

$$P(X = 3) = f(3) = F(3) - F(2) = \frac{3^2}{100} - \frac{2^2}{100} = 0.05$$

Example 13   Let X be a random variable with density function f defined by

$$f(x) = kx \qquad x = 1, 2, \ldots, 10$$

Find k.

Solution   Applying Theorem 5.1(b),

$$1 = f(1) + f(2) + \cdots + f(10) = k + 2k + 3k + \cdots + 10k$$

$$= k(1 + 2 + \cdots + 10) = k\,\frac{(10)(11)}{2} = 55k$$

because

$$1 + 2 + \cdots + 10 = \frac{(10)(11)}{2}$$

by formula (15) of Chapter 3. It follows that $k = \frac{1}{55}$.

5.3 EXERCISES

1.   The following table represents the number X of students absent from a certain class.

x	0	1	2	3	4	5
$f(x)$	0.1	0.2	0.3	0.2	0.1	0.1

What is the probability that (a) exactly 4, (b) exactly 6, (c) at most 2, (d) at least 3 students will be absent tomorrow?

2.   Let X be the random variable of Exercise 1, and let f and F be its density and distribution functions, respectively. Calculate—
(a) $f(3)$ (b) $f(5)$ (c) $F(3)$ (d) $F(2)$
(e) $P(X = 1)$ (f) $P(1 < X)$ (g) $P(2 < X \leq 4)$

3.   Let X be the random variable of Exercise 1 and let F be its distribution function. Make a table defining $F(x)$.

4.   Groups of five children in a certain school are given vision tests. Let X be the number of children in a group who fail the test. The distribution function F of X is given below.

x	0	1	2	3	4	5
$F(x)$	0.3	0.5	0.65	0.8	0.9	1.0

Using the table, find the probability that—
(a) at most four children fail the test
(b) all the children pass the test
(c) exactly three children fail the test.

5.   Let X be the random variable of Exercise 4 and let f be its density function. Make a table giving all nonzero values of $f(x)$.

6.   The density function f of a random variable X is defined by

$$f(x) = \frac{20}{49x} \qquad x = 1, 2, \ldots, 6$$

Let F be the distribution function of X. Find—
(a) $P(X = 3)$ **(b)** $P(X = 11)$ **(c)** $P(X \neq 4)$
(d) $P(X < 1)$ **(e)** $F(2)$ **(f)** $F(10)$

5.4 MEAN, MODE, VARIANCE, AND STANDARD DEVIATION

The density function of a random variable X tells us the value of $P(X = x)$ for any number x. Since a random variable may have many nonzero probabilities associated with it, its density function may give us too much information to be useful in a particular situation. Therefore, the question of how all this information can be summarized naturally arises.

Suppose, for example, that an anthropologist constructs a density function for the heights of pygmies and a second density function for the heights of Americans. He might summarize his data by giving the average height of pygmies and the average height of Americans. A second anthropologist might summarize these data by giving the most likely height of pygmies and the most likely height of Americans. These two methods of summarizing data, the average value, or *mean*, and the most likely value, or *mode*, will be studied in this section.

Since the average is a single number, it is obvious that it cannot adequately represent all the information given by the density function. For example, if the average age of employees in a certain firm is forty-five, does this mean that most of the employees are approximately forty-five years old, or does it mean that half the employees are twenty and the other half are seventy? Two measures of the degree to which the values of the random variable vary from the mean, the *standard deviation* and the *variance*, will also be studied in this section.

The following example shows how the average is found, and will help us formulate a mathematical definition of this concept.

Example 14   Find the average number of heads that appear when three coins are tossed.

Solution   A sample space for this experiment in which the elementary events are equally likely is

$$\Omega = \{(h,h,h),(h,h,t),(h,t,h),(h,t,t),(t,h,h),(t,h,t),(t,t,h),(t,t,t)\}$$

Let X be the number of heads that appear. Using Ω and Laplace's definition of probability, the reader can verify that Table 5.1 gives the values of the density function of X.

Table 5.1

x	0	1	2	3
$f(x)$	$\frac{1}{8}$	$\frac{3}{8}$	$\frac{3}{8}$	$\frac{1}{8}$

Let us see what happens if this experiment is repeated a large number of times, say 800 times. Although we do not know what will happen on any particular toss, we do know that 0 heads (event E_0) will occur approximately $\frac{1}{8} \times 800 = 100$ times. Therefore, the total number of heads that will result from the occurrence of E_0 is $0 \cdot 100 = 0$. Similarly, since 1 head (event E_1) will occur approximately $\frac{3}{8} \times 800 = 300$ times, the total number of heads that will result from the occurrence of E_1 is $300 \cdot 1 = 300$. This procedure is summarized in Table 5.2.

Table 5.2

1 x	2 $f(x)$	3 Approximate Number of Occurrences (Column 2) $\times$ 800	4 Approximate Number of Heads (Column 3) $\times$ (Column 1)
0	$\frac{1}{8}$	100	0
1	$\frac{3}{8}$	300	300
2	$\frac{3}{8}$	300	600
3	$\frac{1}{8}$	100	300
		800	1200

After tossing 800 times, we will have gotten approximately 1200 heads. It follows that the

average number of heads per toss

$$= \frac{\text{total number of heads}}{\text{number of tosses}} \approx \frac{1200}{800} = 1.5$$

(Recall that the symbol $\approx$ means "approximately.") The reader can check that the result is the same if the coins are tossed N times where N is any large number. We therefore say that the *average number*, or *expected number*, of heads is 1.5.

In order to generalize the concept illustrated in Example 14, let us review how the average number of heads was obtained. The sum in column 4 of Table 5.2 is the approximate total number of heads that will result from 800 tosses. This sum was found by multiplying each entry in column 3 by the corresponding entry in column 1 and then adding the products:

approximate total number of heads

$$= 0(100) + 1(300) + 2(300) + 3(100)$$

It follows that the

approximate average number
of heads per toss

$$= \frac{\text{approximate total number of heads}}{800}$$

$$= \frac{0(100) + 1(300) + 2(300) + 3(100)}{800}$$

$$= 0\left(\frac{100}{800}\right) + 1\left(\frac{300}{800}\right) + 2\left(\frac{300}{800}\right) + 3\left(\frac{100}{800}\right)$$

$$= 0 \cdot f(0) + 1 \cdot f(1) + 2 \cdot f(2) + 3 \cdot f(3) \qquad (4)$$

In other words, the average number of heads could have been found quickly by multiplying each entry in row 1 of Table 5.1 by the corresponding entry in row 2 and then adding the products.

The last sum of equation (4) is denoted $E(X)$ and is called the *expected value* of the random variable X. In general, the following definition is used.

Definition 5.5 Let X be a random variable with range $\{x_1, x_2, \ldots, x_n\}$ and density function f defined by Table 5.3.

Table 5.3

x	x_1	x_2	x_3	$\cdots$	x_n
$f(x)$	$f(x_1)$	$f(x_2)$	$f(x_3)$	$\cdots$	$f(x_n)$

The **expected value** of X (or *mathematical expectation* of X), denoted $E(X)$ or EX, is defined by

$$E(X) = x_1 \cdot f(x_1) + x_2 \cdot f(x_2) + \cdots + x_n \cdot f(x_n) = \sum_{i=1}^{n} x_i \cdot f(x_i)$$

A similar definition of $E(X)$ is used if the range of X is $\{x_1, x_2, \ldots\}$.

Notice that $E(X)$ can be calculated by multiplying each entry of row 1 of Table 5.3 by the corresponding entry of row 2 and then adding the products.

The number $E(X)$ is also called the **mean** or *average* of X and is often denoted μ or μ_X (μ is the Greek letter *mu*). The number $E(X)$ is also called the mean of the density function f.

Notice that $E(X)$ gives the *average value* of X. If X is the number of boys in families with five children, $E(X)$ is the average number of boys in such families. If X is the amount of rainfall per day in New York, $E(X)$ is the average amount of rainfall.

The following examples show how to use Definition 5.5.

Example 15 Let N be the number of accidents in a certain industrial plant in one day. The following table defines the density function of N.

n	0	1	2	3	4
$f(n)$	0.40	0.30	0.15	0.10	0.05

What is the average number of accidents that will occur in one day?

Solution   Using Definition 5.5,

$$E(N) = 0(0.40) + 1(0.30) + 2(0.15) + 3(0.10) + 4(0.05) = 1.10$$

The average number of accidents per day is therefore 1.1.

Notice that $E(N)$ of Example 15 was found by multiplying each entry in row 1 of the table by the corresponding entry in row 2 and then adding the products.

Example 16   The Keepkleene Company is preparing a bid for a contract to supply the state prisons with linens. They are considering two alternate bids: a high bid that will yield a 20,000-dollar profit, and a low bid that will yield a 10,000-dollar profit. From past experience, they realize that if they bid high, the probability of being awarded the contract is 0.2, while if they bid low, the probability of being awarded the contract is 0.5. Should they bid high or low?

Solution   The company naturally wants to maximize its profits. Therefore, let X be the profit.

Expected Profit with High Bid. In this case X is either 20,000 or 0, depending upon whether or not the company is awarded the contract. The density function of X is given below.

x	20,000	0
$f(x)$	0.2	0.8

Multiplying each entry of row 1 by the corresponding entry of row 2 and then adding the products yields

$$E(X) = 20,000(0.2) + 0(0.8) = 4000 \text{ dollars}$$

Expected Profit with Low Bid. In this case X is either 10,000 or 0. The density function of X is given below.

x	10,000	0
$f(x)$	0.5	0.5

Multiplying each entry of row 1 by the corresponding entry of row 2 and then adding yields

$$E(X) = 10,000(0.5) + 0(0.5) = 5000 \text{ dollars}$$

The company should therefore bid low.

The preceding examples show how to find the expected value of a random variable X, that is, they show how to find $E(X)$. However, it is sometimes necessary to find the expected value of a function of X; for example, $E(X^2)$, $E(2X)$, $E(3X + 4)$, and so on. The next example shows how this is done.

Example 17   Suppose a die is tossed. If 1 dot appears, you receive $1^2 = 1$ dollars, if 2 dots appear, you receive $2^2 = 4$ dollars, and, in general, if x dots appear you receive x^2 dollars. Find the average number of dots that appear and the average amount of money that you receive.

Solution   Let X be the number of dots that appear and let Y be the amount of money that you receive. Notice that $Y = X^2$. We must calculate $E(X)$ and $E(Y) = E(X^2)$.
   The density function of X is given below

x	1	2	3	4	5	6
$f(x)$	$\frac{1}{6}$	$\frac{1}{6}$	$\frac{1}{6}$	$\frac{1}{6}$	$\frac{1}{6}$	$\frac{1}{6}$

Applying Definition 5.5,

$$E(X) = 1(\tfrac{1}{6}) + 2(\tfrac{1}{6}) + 3(\tfrac{1}{6}) + 4(\tfrac{1}{6}) + 5(\tfrac{1}{6}) + 6(\tfrac{1}{6}) = 3.5$$

Since $Y = X^2$ and the range of X is $\{1,2,3,4,5,6\}$, it follows that the range of Y is $\{1,4,9,16,25,36\}$. We must now find the corresponding probabilities. Let g be the density function of Y. (We cannot use f to denote this density function because f is the density function of X.) Since $Y = X^2$, Y will be 1 if and only if X is 1. Hence,

$$g(1) = P(Y = 1) = P(X = 1) = f(1) = \tfrac{1}{6}$$

Similarly, Y will be 4 if and only if X is 2. Hence,

$$g(4) = P(Y = 4) = P(X = 2) = f(2) = \tfrac{1}{6}$$

and so on. The density function of Y is tabulated below.

x	1	2	3	4	5	6
$y = x^2$	1	4	9	16	25	36
$g(y) = f(x)$	$\frac{1}{6}$	$\frac{1}{6}$	$\frac{1}{6}$	$\frac{1}{6}$	$\frac{1}{6}$	$\frac{1}{6}$

Applying Definition 5.5,

$$E(X^2) = E(Y) = \sum_y y \cdot g(y) = \sum_x x^2 \cdot f(x)$$

$$= 1(\tfrac{1}{6}) + 4(\tfrac{1}{6}) + 9(\tfrac{1}{6}) + 16(\tfrac{1}{6}) + 25(\tfrac{1}{6}) + 36(\tfrac{1}{6})$$

$$= \tfrac{91}{6}$$

Notice that in Example 17

$$E(X^2) = \sum x^2 \cdot f(x)$$

(For simplicity, we have written $\sum$ instead of $\sum_x$ because the letter under the $\sum$ is obvious.) By similar reasoning, we can prove that

$$E(2X) = \sum 2x \cdot f(x), \qquad E(X^3) = \sum x^3 \cdot f(x),$$
$$E(4X^2 - 1) = \sum (4x^2 - 1) \cdot f(x)$$

and so on. Moreover, these equations hold for any random variable X.

Let a and b be any real numbers and let X be any random variables. It follows from the above comments and the laws of arithmetic that

$$
\begin{aligned}
E(aX + b) &= \sum (ax + b)f(x) = \sum [ax \cdot f(x) + b \cdot f(x)] \\
&= \sum ax \cdot f(x) + \sum b \cdot f(x) \\
&= a \sum x \cdot f(x) + b \sum f(x) \\
&= a[E(X)] + b
\end{aligned}
$$

Hence, if a and b are any two numbers and X is any random variable, then

$$E(aX + b) = a[E(X)] + b \tag{5}$$

For example, if $E(X) = 10$, then

$$E(2X) = 2[E(X)] = 2(10) = 20, \qquad E(4X + 5) = 4[E(X)] + 5 = 45$$

and

$$E(X - 1) = 9$$

Although the mean gives us important information about a random variable, Example 18 shows that additional information is useful.

Example 18   Let X and Y be the ages of the children in two different kindergartens. The density functions for X and Y—f and g, respectively—are defined by the following tables:

y	5
$f(y)$	1

z	4	5	6
$g(z)$	0.25	0.50	0.25

Calculate the average age in each class.

Solution   Applying Definition 5.5, the average age in the first class is

$$E(Y) = 5 \cdot 1 = 5$$

while the average age in the second class is

$$E(Z) = 4(0.25) + 5(0.50) + 6(0.25) = 5$$

The random variables Y and Z of Examples 18 both have the same mean, 5. However, the values of Z are more spread out from the mean than the values of X. As a measure of this spread, we might consider using the average deviation of a random variable X from its mean μ; that is, we might try to use $E(X - \mu)$. However, it follows from equation (5) that

$$E(X - \mu) = E(X) - \mu = \mu - \mu = 0$$

To avoid the cancellation of negative and positive deviations, we could consider using $E(|X - \mu|)$ or $E[(X - \mu)^2]$.[2] The quantity $E[(X - \mu)^2]$ is called the *variance* and is more frequently used as a measure of spread because of its simpler algebraic properties. The greater the probability that the values of X are spread out about the mean, the larger is the variance. Equivalently, the smaller the variance, the smaller is the probability that the values of X are spread out about the mean.

Definition 5.6   Let X be a random variable with mean μ.

(a)   The **variance** of X is denoted $\mathrm{Var}(X)$, $\sigma^2(X)$, or σ^2. It is defined by

$$\mathrm{Var}(X) = E[(X - \mu)^2]$$

(b)   The square root of the variance is called the **standard deviation** of X and is denoted σ or σ_X.

The calculation of the variance is simplified by the following theorem.

THEOREM   5.4   Let X be a random variable with mean μ. Then

$$\mathrm{Var}(X) = E(X^2) - \mu^2$$

where $E(X^2) = \sum_x x^2 f(x)$.

---

2. $|X - \mu|$ is the absolute value of $X - \mu$. If x is a real number, $|x|$, the *absolute value of x*, is defined as follows: $|x| = x$ if $x \geq 0$, $|x| = -x$ if $x < 0$. For example, $|2| = 2$, $|0| = 0$, $|-3| = 3$.

Proof   Applying Definition 5.6(a) and equation (5), we have

$$\text{Var}(X) = E[(X - \mu)^2] = E(X^2 - 2X\mu + \mu^2)$$
$$= E((X^2 - 2X\mu) + \mu^2)$$
$$= E(X^2 - 2X\mu) + \mu^2$$
$$= E(X^2) + E(-2X\mu) + \mu^2$$
$$= E(X^2) - 2\mu[E(X)] + \mu^2$$
$$= E(X^2) - 2\mu^2 + \mu^2$$
$$= E(X^2) - \mu^2$$

Example 19 shows how to apply Theorem 5.4 and illustrates the importance of the variance of a random variable.

Example 19   Let Y and Z be the random variables of Example 18. Calculate $\text{Var}(Y)$ and $\text{Var}(Z)$.

Solution   Applying Theorem 5.4,

$$\text{Var}(Y) = E(Y^2) - \mu_Y^2 \quad \text{and} \quad \text{Var}(Z) = E(Z^2) = \mu_Z^2$$

The means μ_Y and μ_Z were calculated in Example 18:

$$\mu_Y = \mu_Z = 5$$

The quantities $E(Y^2) = \sum_y y^2 f(y)$ and $E(Z^2) = \sum_z z^2 f(z)$ can be easily calculated with the help of the following tables based on the tables for f and g given in Example 18:

Table 5.4

y	5
y^2	25
$f(y)$	1

Table 5.5

z	4	5	6
z^2	16	25	36
$g(z)$	0.25	0.50	0.25

Multiplying each entry of row 2 of Table 5.4 by the corresponding entry in row 3 and then adding the product yields

$$E(Y^2) = 25 \cdot 1 = 25$$

Similarly,

$$E(Z^2) = 16(0.25) + 25(0.50) + 36(0.25) = 25.5$$

Hence,

$$\text{Var}(Y) = E(Y^2) - \mu_Y^2 = 25 - 5^2 = 0$$

and
$$\text{Var}(Z) = E(Z^2) - \mu_Z^2 = 25.5 - 5^2 = 0.5$$

Notice that the random variable Y is not at all spread out about the mean, so its variance is 0. The random variable Z is slightly spread out about the mean, so its variance is 0.5.

If the variance is large, the mean is a poor characterization of the random variable. Suppose that each of the nine employees of a firm earns 10,000 dollars while the owner earns 200,000 dollars. The mean salary, 29,000 dollars, is a poor reflection of the salaries paid by the firm. The value X of a randomly selected salary paid by the firm is better characterized by 10,000 dollars, its most probable value.

Definition 5.7   A **mode** of a random variable X is a most probable value of X.

In other words, a mode x_o is any value of the random variable X that makes the density function largest; that is, $f(x_o) \geq f(x)$ for all x. The following example shows that a random variable may have more than one mode.

Example 20   Find the mode or modes of the following random variables.

x	0	1	2	3
$f(x)$	0.1	0.4	0.3.	0.2

y	0	1	2	3
$g(y)$	0.1	0.3	0.3	0.3

Solution   Since the largest value of $f(x)$ is $f(1) = 0.4$, the mode of X is 1. The random variable Y has three modes: 1, 2, and 3.

Example 21 reviews all the concepts studied in this section.

Example 21   Let X be a random variable with density function defined by the following table:

x	-2	-1	0	1	2
$f(x)$	0.1	0.2	0.2	0.3	0.2

Find—
(a) $E(X)$ (b) $\text{Var}(X)$ (c) the mode or modes of X
(d) μ (e) σ^2 (f) σ (g) the standard deviation of X

Solution

(a) Multiplying each entry of the first row of the table by the corresponding entry in the second row and then adding the products yields

$$E(X) = -2(0.1) + (-1)(0.2) + 0(0.2) + 1(0.3) + 2(0.2) = 0.3$$

(b) The following table will help us calculate Var(X).

x	-2	-1	0	1	2
x^2	4	1	0	1	4
$f(x)$	0.1	0.2	0.2	0.3	0.2

Multiplying each entry of row 2 by the corresponding entry in row 3 and adding the products yields

$$E(X^2) = 4(0.1) + 1(0.2) + 0(0.2) + 1(0.3) + 4(0.2) = 1.7$$

It then follows that

$$\text{Var}(X) = E(X^2) - \mu^2 = E(X^2) - [E(X)]^2 = 1.7 - (0.3)^2 = 1.61$$

(c) Since the largest value of $f(x)$ is $f(1) = 0.3$, the mode of X is 1.
(d) $\mu = E(X) = 0.3$
(e) Using part **(b)**, $\sigma^2 = \text{Var}(X) = 1.61$.
(f) Using part **(e)**, $\sigma = \sqrt{\sigma^2} = \sqrt{1.61} = 1.27$.
(g) The standard deviation is $\sigma = 1.27$.

5.4 EXERCISES

1.   The density function of the random variable X is defined by the following table:

x	-1	0	1	2
$f(x)$	$\frac{1}{2}$	$\frac{1}{4}$	$\frac{1}{8}$	$\frac{1}{8}$

Calculate—
(a) $E(X)$ **(b)** Var(X)   **(c)** the mode or modes of X **(d)** σ

2.   From long experience, a woman feels that the following table represents the size of her catch after a day's fishing.

Number X of fish caught	0	1	2	3	4
$P(X = x)$	0.20	0.30	0.30	0.15	0.05

(a) What is the average number of fish that she catches?
(b) What is the most likely number of fish that she will catch today?

(c) If a fishing trip costs 4 dollars and every fish caught is worth 2 dollars, is fishing profitable?

3.   A fair coin is tossed four times. Each time a head occurs, you receive 2 dollars. If you pay 4 dollars to play the game, what is your average profit?

4.   The following tables define the density functions of the random variables X and Y.

x	-1	0	1
$f(x)$	$\frac{1}{2}$	0	$\frac{1}{2}$

y	-4	0	4
$g(y)$	$\frac{1}{2}$	0	$\frac{1}{2}$

(a) Without calculating the variances, which random variable would you expect to have the larger variance?

(b) Show that $E(X) = E(Y)$, but $\text{Var}(X) \neq \text{Var}(Y)$.

5.   The blood samples of ten randomly chosen people will be pooled and tested together. If the test is positive (i.e., shows evidence for a disease), then each of the ten must be tested separately and, in all, eleven tests are required. Assume that the probability that a person has the disease being tested for is 0.1.

(a) Let N be the number of tests required. Calculate $P(N = 1)$ and $P(N = 11)$.

(b) Find the mean number of tests required.

6.   The weather forecast is for a low of -10 degrees Celsius. A fruit shipper has found from long and careful observation that, under these conditions, the chance of encountering a temperature of -12 degrees Celsius or below is $\frac{1}{5}$. He also knows that a temperature of -12 degrees or below will result in a total loss of 15 percent of his fruit. To protect the fruit from the lower temperature involves an increased protection cost of 4 cents on every dollar of profit. Should the shipper protect his fruit?

7.   Find (a) the mean, (b) the mode or modes, (c) the variance, and (d) the standard deviation of the following density function:

$$f(x) = \frac{12}{13x}, \qquad x = 2, 3, 4.$$

8.   A friend offers to play one of the following games with you, all based on tossing a fair coin once.

	If heads occurs he will pay you—	If tails occurs he will pay you—
Game 1	$0	$2
Game 2	$-$ $1	$3
Game 3	$-$ $1	$5
Game 4	$-$ $20,000	$40,000

(The entry " $-\$1$ " means that you pay him 1 dollar.)

(a) What are the expected values of the above games?

(b) Which game would you choose to play? (The game will be played only once.)

9. Suppose that your friend agrees to play one of the games of Exercise 8 two hundred times. The bet will be collected after all the games are played. Which game would you choose?

5.5 THE UNIFORM RANDOM VARIABLE

In this section we will begin classifying random variables by studying the uniform variable U. We will see that many apparently dissimilar probability problems can be quickly solved once we have derived formulas for the density function of U, the distribution function of U, the mean of U, and so on.

Let X be the number of dots that appear when a die is tossed. Since each number from 1 to 6 is equally likely to appear, it is clear that

$$P(X = 1) = P(X = 2) = P(X = 3)$$
$$= P(X = 4) = P(X = 5) = P(X = 6) = \tfrac{1}{6}$$

In other words, X assumes all its values with equal probability. A random variable that has this characteristic is called a **uniform random variable**.

Examples of uniform random variables are given next.

Example 22

(a) A firm employs 100 people, each identified by a different number from 1 to 100. Suppose that an employee is randomly chosen from this firm and let U be the employee's number. Since U assumes all its values with equal probability, U is a uniform random variable. Its range is $\{1,2,\ldots,100\}$.

(b) A restaurant sells ten different lunches; each lunch is identified by a number from 1 to 10. Suppose that a lunch is randomly chosen and let V be the number of the lunch. Then V is a uniform random variable. The range of V is $\{1,2,\ldots,10\}$.

Theorem 5.5 lists the important characteristics of the uniform random variable. Several of these characteristics are also listed in Table A.4 of the Appendix.

THEOREM 5.5 Let U be a uniform random variable with range $\{1,2,\ldots,n\}$. Some properties of U are—

(a) Density function $\qquad f(u) = \dfrac{1}{n} \qquad u = 1, 2, \ldots, n$

(b) Distribution function $\qquad F(u) = \dfrac{u}{n} \qquad u = 1, 2, \ldots, n$

(c)   Mean $$\mu = \frac{n+1}{2}$$

(d)   Variance $$\sigma^2 = \frac{n^2 - 1}{12}$$

(e)   Modes $1, 2, \ldots, n$

Proof   We shall prove parts (a)–(c). The proofs of the remaining parts are left for the exercises.

(a) By definition of a uniform random variable,

$$f(1) = f(2) = \cdots = f(n) \tag{6}$$

Moreover, since all the probabilities must add to 1,

$$f(1) + f(2) + \cdots + f(n) = 1 \tag{7}$$

Substituting equation (6) into equation (7) and simplifying yields

$$f(1) + f(1) + \cdots + f(1) = nf(1) = 1$$

Hence,

$$f(1) = \frac{1}{n} = f(2) = \cdots = f(n)$$

and so

$$f(u) = \frac{1}{n} \qquad u = 1, 2, \ldots, n$$

(b) The density function of U is written in tabular form as follows:

Table 5.6

u	1	2	$\cdots$	n
$f(u)$	$\frac{1}{n}$	$\frac{1}{n}$		$\frac{1}{n}$

Using this table and the definition of a distribution function.,

$$F(1) = \frac{1}{n}$$

$$F(2) = \frac{1}{n} + \frac{1}{n} = \frac{2}{n}$$

and, in general,

$$F(u) = \frac{u}{n} \qquad u = 1, 2, 3, \ldots, n$$

(c) Multiplying each entry in row 1 of Table 5.6 by the corresponding entry in row 2 and adding the products yields

$$E(U) = 1\left(\frac{1}{n}\right) + 2\left(\frac{1}{n}\right) + \cdots + n\left(\frac{1}{n}\right)$$

$$= \frac{1}{n}[1 + 2 + \cdots + n] = \frac{1}{n}\left[\frac{n(n+1)}{2}\right] = \frac{n+1}{2}$$

The following example shows how Theorem 5.5 is applied.

Example 23 Twenty raffle tickets numbered 1–20 are placed in a hat and one ticket is randomly selected. Find—

(a) the probability that ticket number 2 is drawn

(b) the probability that a ticket bearing a number less than 5 is drawn

(c) the average number of the ticket drawn.

Solution Let U be the number on the selected ticket. Then U is a uniform random variable because each ticket is equally likely to be selected. The range of U is $\{1, 2, \ldots, 20\}$.

(a) Using Theorem 5.5(a), the required probability is

$$P(U = 2) = f(2) = \tfrac{1}{20}$$

(b) Using Theorem 5.5(b), the required probability is

$$P(U < 5) = P(U \le 4) = F(4) = \tfrac{4}{20} = \tfrac{1}{5}$$

(c) Using Theorem 5.5(c), the average number drawn is

$$\mu = \frac{21}{2} = 10.5$$

5.5 EXERCISES

1. What is the average number of dots that appear when a die is tossed?

2. Suppose that you receive 1 dollar for each dot that appears when a die is tossed. Find—
   (a) the most likely amount that you will receive
   (b) the average amount that you will receive
   (c) the probability that you will receive 4 dollars
   (d) the probability that you receive at most 5 dollars.

3. A roulette wheel is divided into thirty equal compartments each bearing a different number from 1 to 30. The wheel is spun and the ball lands on one of the compartments. What is the probability that the ball lands on a compartment bearing a number greater than seven?

4.   Let U be a uniform random variable with range $\{2,4,6,8,10\}$. Find—

   (a) $P(U = 2)$ **(b)** $P(U \le 6)$ **(c)** $E(U)$

5.   Let U be a uniform random variable with range $\{1,2,\ldots,50\}$. Find—

   (a) $E(U)$ **(b)** $\text{Var}(U)$ **(c)** $P(U = 12)$ **(d)** $P(U \le 15)$

6.   Finish the proof of Theorem 5.5. (Hint: Use the fact that

$$1^2 + 2^2 + \cdots + n^2 = \frac{(n + 1)(2n + 1)}{6}$$

and show that $E(U^2) = \dfrac{(n + 1)(2n + 1)}{6}$.)

5.6 THE BINOMIAL RANDOM VARIABLE

In this section we will study a second kind of random variable—the binomial random variable—that has a large number of important applications.

Suppose a fair coin is tossed n times. This experiment has the following characteristics:

Characteristics of a Sequence of Bernoulli Trials

(1)   An experiment T (e.g., tossing a coin) is repeated n times. Each repetition of T is called a *trial*.

(2)   Exactly one of two mutually exclusive events, S (success) or F (failure), occurs on each trial. (In the coin-tossing experiment, S may be heads and F tails, or vice versa.)

(3)   The outcome of each trial is independent of the outcomes of the other trials.

(4)   The probabilities $P(S) = p$ and $P(F) = q = 1 - p$ are the same for each trial. (In the coin-tossing experiment, $p = q = \frac{1}{2}$.)

A sequence of trials having these four characteristics is called a **sequence of Bernoulli trials.**

Definition 5.8   Suppose a sequence of Bernoulli trials consists of n repetitions of an experiment T. The **binomial random variable** S_n is the number of times S (success) occurs in the trials. The range of S_n is $\{0,1,2,\ldots,n\}$.

Examples of the binomial random variable are the number of heads in fourteen tosses of a coin, the number of black guinea pigs in a litter of six guinea pigs, and the number of voters who favor Smith in a sample of forty voters.

The next example illustrates some of the distinctions between binomial and nonbinomial random variables.

Example 24

(a) *Sampling with replacement.* A card is removed from a deck containing fifty-two cards. The card is then replaced, the deck shuffled, and a second card removed. Let X be the total number of kings obtained. Then X is a binomial random variable because this process satisfies the four characteristics of a sequence of Bernoulli trials with $n = 2$.

(b) *Sampling without replacement.* Two cards are again removed from a deck containing fifty-two cards. However, this time, the first card is *not* replaced before the second card is chosen. Let Y be the total number of kings obtained. Although X and Y are similar, Y is *not* a binomial random variable because characteristic 4 of a sequence of Bernoulli trials is not satisfied. Clearly, the probability of getting a king on the second pick depends upon whether or not a king was obtained on the first pick. In fact, if we let K_1 and K_2 be the events that a king is selected on the first and second picks, respectively, then $P(K_2 | K_1) = \frac{3}{51}$, while $P(K_2 | K_1') = \frac{4}{51}$.

Sampling with replacement, which is a binomial situation, will be studied in this section. Sampling without replacement is a *hypergeometric* situation and will be studied in the next section.

The following example will lead us to a formula for the density function of the binomial random variable S_n.

Example 25

About 40 percent of the people in the United States belong to blood group A. If six people are randomly chosen, what is the probability that—

(a) exactly two have blood type A

(b) exactly five have blood type A

(c) exactly x have blood type A?

Solution   The number S_6 of people having blood type A is a binomial random variable because: (1) the experiment is repeated $n = 6$ times; (2) each trial results in a success S (the person has blood type A) or failure F (the person does not have blood type A); (3) the trials are independent; and (4) $P(S) = 0.4$ is the same on each trial.

(a) We need to find $P(S_6 = 2)$. Let S_i be the event "the ith person has blood type A" and let $F_i = S_i'$.[3] Suppose that $E_1, E_2, \ldots, E_k$ are all of the events which are intersections of two S_i's and four F_i's. For example, E_1 might be $S_1 S_2 F_3 F_4 F_5 F_6$ and E_2 might be $S_1 F_2 S_3 F_4 F_5 F_6$. The event $(S_6 = 2)$ will occur if E_1 occurs, or E_2 occurs, and so on. Because the outcomes of the trials are independent,

$$P(E_1) = P(S_1 S_2 F_3 F_4 F_5 F_6)$$
$$= P(S_1)P(S_2)P(F_3)P(F_4)P(F_5)P(F_6)$$
$$= (0.4)^2(0.6)^4$$

---

3. The symbol S_i is used here to label both the binomial random variable and the event "the ith person has blood type A." It will always be clear from the context which is meant.

Similarly, $P(E_2) = (0.4)^2(0.6)^4$ and, in general,

$$P(E_j) = (0.4)^2(0.6)^4 \qquad j = 1, 2, \ldots, k$$

It follows that

$$
\begin{aligned}
P(S_6 = 2) &= P(E_1 + E_2 + \cdots + E_k) \\
&= P(E_1) + P(E_2) + \cdots + P(E_k) \\
&= (0.4)^2(0.6)^4 + (0.4)^2(0.6)^4 + \cdots + (0.4)^2(0.6)^4 \\
&\qquad \qquad (k \text{ summands}) \\
&= k(0.4)^2(0.6)^4
\end{aligned}
$$

We must now find k, the number of events E_j. Each E_j has the form $E_j = ABCDEF$, where exactly two of the six events A, B, C, D, E, and F are S_i's and the remainder are F_i's. It follows that k is the number of ways of selecting two out of six events to be S_i's; that is, $k = {_6}C_2$. The required probability is therefore

$$P(S_6 = 2) = {_6}C_2(0.4)^2(0.6)^4$$

(b) Similarly, $P(S_6 = 5) = {_6}C_5(0.4)^5(0.6)$.

(c) In general,

$$f(x) = P(S_6 = x) = {_6}C_x(0.4)^x(0.6)^{6-x} \qquad x = 0, 1, 2, \ldots, 6$$

The following theorem can be proved in a similar way.

THEOREM   5.6   If S_n is any binomial random variable and f is its density function, then

$$f(x) = P(S_n = x) = {_n}C_x p^x q^{n-x} \qquad x = 0, 1, 2, \ldots, n$$

where n is the number of trials, $p = P(S)$, and $q = P(F)$.

Notice that $f(x) = P(S_n = x)$ is the probability of obtaining exactly x successes in n trials.

The following example shows that Theorem 5.6 can be applied to a large number of different situations. Notice that the underlying structure of each of these situations is the same.

Example 26

(a) If two cream guinea pigs are crossed, the probability that a particular offspring is cream is 0.5. What is the probability that fifteen offspring of such matings include exactly eight cream guinea pigs?

(b) If a coin is tossed fifteen times, what is the probability of obtaining exactly eight heads?

(c) Fifty percent of all lightbulbs of a certain make last more than 200 hours. What is the probability that exactly eight out of fifteen lightbulbs will last more than 200 hours?

(d) The probability that a new automobile of a certain make has a leaky windshield is 0.5. What is the probability that eight out of fifteen new cars of this make will have leaky windshields?

(e) A certain die is biased so that 6 occurs on 50 percent of the tosses. If it is tossed fifteen times, what is the probability that exactly eight 6s occur?

Solution Applying Theorem 5.6, the answer to each of these problems is

$$_{15}C_8(0.5)^8(0.5)^7 = {}_{15}C_8(0.5)^{15}$$

Although the solution to Example 26 is correct, it is not very satisfactory in a practical sense because it is not a numerical answer—we do not even known whether this probability is large or small. However, certain numerical values of the binomial density and distribution functions appear in Tables A.1 and A.2 of the Appendix. The binomial density $P(S_n = x) = {}_nC_x p^x q^{n-x}$ $(x = 0,1,2,\ldots,n)$ and binomial distribution $P(S_n \leq x)$ are denoted

$$b(x;n,p) = P(S_n = x) \quad \text{and} \quad B(x;n,p) = P(S_n \leq x)$$

respectively. This notation indicates that x is the variable while n and p are fixed constants called *parameters*. Note that we do not have to specify the value of q because $q = 1 - p$. We can find the numerical answer to Example 26 by looking up $b(8;15,0.5)$ in Table A.1. We first look for the section labeled $n = 15$. (This is the last section of the table.) The value for $b(8;15,0.5)$ appears in this section in the intersection of the row labeled $x = 8$ and the column labeled $p = 0.50$. Using the table, the required probability is 0.19638.

The following examples illustrate the use of Tables A.1 and A.2.

Example 27 Find the numerical values of—

(a) $b(2;15,0.2)$ **(b)** $b(10;10,0.06)$ **(c)** $B(10;10,0.2)$

Solution

(a) The value of $b(2;15,0.2)$ appears in the section of Table A.1 labeled $n = 15$ and in the intersection of the row labeled $x = 2$ and the column labeled $p = 0.20$. The value of $b(2;15,0.2)$ is 0.23090.

(b) Similarly, $b(10;10,0.06) = 0$.

(c) The value of $B(10;10,0.2)$ appears in the section of Table A.2 labeled $n = 10$ and in the intersection of the row labeled $x = 10$ and the column labeled $p = 0.20$. The value of $B(10;10,0.2)$ is 1.0000.

Example 28 If $p = 0.06$, find—

(a) $P(S_{10} = 1)$ **(b)** $P(S_5 \leq 2)$ **(c)** $P(S_5 > 3)$

Solution

(a) $P(S_{10} = 1) = b(1;10,0.06) = 0.34380$

(b) $P(S_5 \leq 2) = B(2;5,0.06)$. However, the values for $B(x;n,p)$ are not found in Table A.2 for $p = 0.06$. Using Table A.1,

$$P(S_5 \leq 2) = P(S_5 = 0) + P(S_5 = 1) + P(S_5 = 2)$$

$$= b(5;0,0.06) + b(5;1,0.06) + b(5;2,0.06)$$

$$= 0.73390 + 0.23422 + 0.02990$$

$$= 0.99802$$

(c) $P(S_5 > 3) = P(S_5 = 4) + P(S_5 = 5) = b(5;4,0.06) + b(5;5,0.06) = 0.00006$
$+ 0.00000 = 0.00006$

Example 29   If $p = 0.25$, find—

    (a) $P(S_{15} \leq 6)$ **(b)** $P(S_5 = 4)$ **(c)** $P(S_{10} > 2)$ **(d)** $P(S_{15} < 12)$

Solution

    (a) Using Table A.2, $P(S_{15} \leq 6) = B(6;15,0.25) = 0.94338$.

    (b) Table A.1 cannot be used because values of $b(x;n,p)$ do not appear for $p = 0.25$. From Table A.2,

$$P(S_5 = 4) = P(S_5 \leq 4) - P(S_5 \leq 3)$$

$$= B(5;4,0.25) - B(5;3,0.25)$$

$$= 0.99902 - 0.98438 = 0.01464$$

    (c) $P(S_{10} > 2) = 1 - P(S_{10} \leq 2) = 1 - B(10;2,0.25) = 1 - 0.52559 = 0.47441$

    (d) $P(S_{15} < 12) = P(S_{15} \leq 11) = B(15;11,0.25) = 0.99999$

Example 30   A company has a passing rate of 70 percent in its training program. If five trainees start the program, what is the probability that exactly two pass? What is the probability that exactly four pass?

Solution   All of the conditions for a sequence of Bernoulli trials are satisfied. There are five independent trials, each consisting of an individual's participating in the training program. If we let success be the event "the trainee passes," then $p = P(S) = 0.7$, which does not appear in the top heading of Table A.1. Hence, let S be the event "the trainee fails." Since "two pass" means "three fail," and "four pass" means "one fails," the required answers are $b(3;5,0.3) = 0.13230$ and $b(1;5,0.3) = 0.36015$.

     We can also find the required probabilities from Table A.1 by using the column headings on the bottom of the table and row headings on the right of the table. The required probability $P = P(\text{exactly two pass})$ is found by locating the section labeled $n = 5$, the column which is labeled $p = 0.7$ on the bottom, and the row labeled $x = 2$ on the right. Therefore, $P = 0.13230$, as we calculated before.

The mean and variance of a binomial random variable are given in the following theorem. They are also listed in Table A.4 of the Appendix.

THEOREM 5.7   Let S_n be a binomial random variable with $p = P(S)$ and $q = P(F)$. The mean and variance of S_n are:

Mean $\mu = np$

Variance $\sigma^2 = npq$

Example 31 shows how to use Theorem 5.7.

Example 31   A deck of five cards, each bearing a different symbol, are used in a certain ESP (extrasensory perception) experiment. The deck is shuffled, one card is removed, and the person being tested is asked to name the chosen card without looking at it. This card is then *returned to the deck* and the experiment repeated. Suppose the experiment is repeated ten times. If the person being tested does not have ESP (and hence just guesses), find—

(a) the average number of correct answers
(b) $\text{Var}(X)$ where X is the number of correct answers.

Solution   This is an example of *sampling with replacement*, and thus the binomial random variable applies. Using Theorem 5.7 with $n = 10$, $p = \frac{1}{5}$, and $q = 1 - \frac{1}{5} = \frac{4}{5}$ yields—

(a) $\mu = np = 10(\frac{1}{5}) = 2$
(b) $\text{Var}(X) = npq = 10(\frac{1}{5})(\frac{4}{5}) = \frac{8}{5}$

The reader may well wonder why S_n is called a *binomial* random variable. The name binomial arises from the fact that the $(x + 1)$st term in the binomial expansion

$$(q + p)^n = q^n + {}_nC_1 q^{n-1}p + {}_nC_2 q^{n-2}p^2 + \cdots + {}_nC_x q^{n-x}p^x + \cdots + p^n$$

is ${}_nC_x p^x q^{n-x} = P(S_n = x)$. This is the term in which the exponent of p is x. The following example shows how this fact can be used to calculate probabilities.

Example 32   If a die is tossed four times, what is the probability of obtaining zero 6s, one 6, two 6s, and so on?

Solution   Since

$$(q + p)^4 = q^4 + 4q^3p + 6q^2p^2 + 4qp^3 + p^4$$

it follows that the probability of zero 6s is $q^4 = (\frac{5}{6})^4$, the probability of one 6 is $4q^3p = 4(\frac{5}{6})^3(\frac{1}{6})$, the probability of two 6s is $6q^2p^2 = 6(\frac{5}{6})^2(\frac{1}{6})^2$, and so on. Alternatively, Theorem 5.6 may be applied directly.

5.6 EXERCISES

Use Tables A.1 and A.2 wherever possible.

1. Find the numerical value of—
   (a) $b(3;5,0.2)$ (b) $B(3;5,0.2)$ (c) $b(8;15,0.25)$
   (d) $B(10;15,0.08)$ (e) $b(8;10,0.7)$ (f) $B(3;5,0.8)$

2. If $p = 0.01$, find—
   (a) $P(S_{10} = 2)$ (b) $P(S_{10} \leq 3)$ (c) $P(S_{10} > 2)$
   (d) $P(1 < S_{10} < 4)$

3. If $p = 0.25$, calculate—
   (a) $P(S_{15} = 4)$ (b) $P(S_{15} \leq 6)$ (c) $P(S_{15} < 5)$
   (d) $P(S_{15} > 4)$ (e) $P(1 \leq S_{15} \leq 10)$

4. If $p = 0.8$, find—
   (a) $P(S_5 = 4)$ (b) $P(S_5 \leq 3)$ (c) $P(S_5 > 2)$

5. Suppose 2 percent of the trucks sent out by a trucking firm have accidents.
   (a) If three trucks are sent out, what is the probability that none will have an accident?
   (b) If ten trucks are sent out, what is the probability that—
      (i) at least one will have an accident
      (ii) more than three will have an accident?

6. Victor, Stephen, and Jim are the only three employees in an office. One is randomly selected each day to work late. If the office is open five days a week, what is the probability that Victor will work late—
   (a) every day next week
   (b) at least three days next week
   (c) no days next week?

7. Suppose Victor of Exercise 6 has been selected to work late nine out of the previous ten nights, and he complains that somebody has cheated. Assume that the selection was really random.
   (a) What is the probability that Victor is selected nine or more days out of ten?
   (b) What is the average number of days out of ten that Victor would be required to stay late?

8. According to newspaper accounts, 1 out of every 1000 cans of soup sold by a certain company is spoiled. If a family consumes 300 cans of soup a year and never checks the cans for bulges and other defects, what is the probability that at least one can they consume is spoiled?

9. To test a dolphin's sonar, two fish were placed in the dolphin's tank at feeding time. One fish was simply placed in the tank; the other was placed behind a sheet of glass. The glass could not be seen, but it could be detected by sonar. In 202 tests, the dolphin never tried to reach the fish behind the glass. Assume that the dolphin just used his eyes and so was equally likely to

choose either fish; that is, assume that the dolphin did not use sonar. What is the probability of obtaining the given results? What does this probability mean?

10. A college has a 30-percent dropout rate in its freshman class. Suppose that 1000 freshmen are currently enrolled and let X be the number of students who drop out. Find—
    (a) $P(X = 300)$ **(b)** $P(X = 500)$ **(c)** $P(X \geq 300)$
    (d) $\text{Var}(X)$ **(e)** $E(X)$ **(f)** σ

11. Write the answers to Exercises 10**(a)**–**(c)** using the symbols $b(x;n,p)$ and $B(x;n,p)$.

12. A man shoots at a target six times. If the probability that he hits the bull's-eye is 0.10, what is the probability that he gets—
    (a) at least one hit
    (b) exactly four hits
    (c) more than three hits
    (d) at least four hits?

13. A roulette wheel contains thirty-eight compartments, each one bearing a different number from 1 to 36 or a blank (there are two blanks). The wheel is spun, and the ball is equally likely to stop at each compartment. John decides to play five times, betting on two different numbers each time. What is the probability that he will win—
    (a) at least once
    (b) exactly three times
    (c) more than two times?

5.7 THE HYPERGEOMETRIC RANDOM VARIABLE

In Section 5.6 we learned that if two cards are removed from a deck with replacement (i.e., the first card is replaced before the second is removed), the number S_2 of kings that are selected is a binomial random variable. However, if the two cards are removed *without replacement*, the number H of kings is not a binomial random variable. In fact, H is a *hypergeometric random variable*, which will be studied in this section.

In general, consider any collection of two kinds of objects (e.g., kings and other cards; men and women; doctors and lawyers, etc.). For convenience, let us represent the two kinds of objects by b black balls and w white balls in an urn. If n $(n \leq b + w)$ balls are randomly selected from the urn *without replacement*, then H, the number of black balls in the sample, is a **hypergeometric random variable**. The range of H is $\{M, M + 1, M + 2, \ldots, m\}$, where the number m is the smaller of n and b and the number M is the larger of 0 and $n - w$.

Remember that the black balls and white balls in the preceding paragraph can stand for any two kinds of objects.

The following example will lead us to a formula for the density function of H.

Example 33 Four grapefruit are randomly chosen from a bin containing three pink grapefruit and nine yellow grapefruit. Let H be the number of pink grapefruit in the sample and let f be the density function of H. Calculate—

(a) $f(1)$ **(b)** $f(3)$ **(c)** $f(x)$

Solution The random variable H is hypergeometric with pink and yellow grapefruit taking the place of black and white balls.

(a) Laplace's definition of probability can be used. Since there are $_{12}C_4$ ways of selecting 4 grapefruit from 12, the total number of cases is $_{12}C_4$. Since there are $_3C_1$ ways of selecting 1 pink grapefruit from 3 pink grapefruit and $_9C_3$ ways of selecting 3 white grapefruit from 9 white grapefruit, the number of favorable cases is $_3C_1 \; _9C_3$. It follows that

$$f(1) = P(H = 1) = \frac{_3C_1 \cdot _9C_3}{_{12}C_4}$$

(b) Similarly,

$$f(3) = P(H = 3) = \frac{_3C_3 \cdot _9C_1}{_{12}C_4}$$

(c) In general,

$$f(x) = P(H = x) = \frac{_3C_x \cdot _9C_{4-x}}{_{12}C_4} \qquad x = 0, 1, 2, 3$$

The following theorem is proved using the same method we used to solve Example 33.

THEOREM 5.8 The density function of the hypergeometric random variable H is given by the formula

$$f(x) = \frac{_bC_x \; _wC_{n-x}}{_{b+w}C_n} \quad \begin{cases} x = M, \, M + 1, \, M + 2, \, \ldots, \, m \\ m = \text{the smaller of } n \text{ and } b \\ M = \text{the larger of } 0 \text{ and } n - w \end{cases}$$

where b is the number of black balls, w is the number of white balls, n is the number of balls in the sample ($n \leq b + w$), and x is the number of black balls in the sample.

Remember that the black balls and white balls represent any two kinds of objects (e.g., freshmen and sophomores, hearts and cards which are not hearts, children and adults, etc.). For example, the density function of Example 33 can be calculated from Theorem 5.8 by setting $b = 3$, $w = 9$, and $n = 4$.

Although the answers to Examples 33(**a**) and (**b**) are correct, numerical answers would be much more useful. If the total number of objects, $b + w$, is small, as in

Example 33 ($b + w = 12$), the probabilities can be easily calculated using a hand calculator. Using this method, the probabilities Example 33**(a)** and **(b)** are

$$f(1) = \frac{252}{495} = 0.509 \quad \text{and} \quad f(3) = \frac{9}{495} = 0.018.$$

In addition, tables of values of the hypergeometric density function are available if the total number of objects is less than or equal to 50.

Suppose, however, that the total number of objects is larger than 50. We have seen that the hypergeometric situation (sampling without replacement) is very similar to the binomial situation (sampling with replacement). Intuitively, one feels that, if the total number of objects, $b + w$, is large in comparison to the sample size n, then it does not make much difference whether or not each object is replaced before the next one is chosen. The next example shows that this intuitive feeling is quite true.

Example 34   A certain club contains 100 members—20 Republicans and 80 Democrats. Suppose that 5 members of this club are randomly selected one at a time and asked their opinions about the next election. Find the probability that the 5 members selected include exactly 2 Republicans if—

(a) a member may be questioned twice (sampling with replacement)

(b) a member may not be questioned twice (sampling without replacement).

Solution   Let X be the number of Republicans who are questioned.

(a) Since a member may be questioned twice, the probability of choosing a Republican does not depend upon whether or not a Republican was previously chosen. It follows that the trials (selecting a member) are independent and X is a binomial random variable with $n \doteq 5$, $p = \frac{20}{100} = 0.2$, and $x = 2$. Using Table A.1, the required probability is $b(2;5,0.2) = 0.20480$.

(b) Since a member may not be questioned twice, the probability of choosing a Republican depends upon whether a Republican was chosen previously. Therefore, X is not a binomial random variable. However, the characteristics of a hypergeometric random variable are satisfied, with Republicans and Democrats replacing black and white balls. Hence, X is a hypergeometric random variable with $b = 20$, $w = 80$ $n = 5$, and $x = 2$. Using Theorem 5.8 and a hand calculator, the required probability is

$$f(2) = \frac{{}_{20}C_2 \cdot {}_{80}C_3}{{}_{100}C_5} = 0.20734$$

This is approximately equal to $b(2;5,0.2)$.

We saw in Example 34 that the hypergeometric probability of part **(b)** is approximately the same as the corresponding binomial probability of part **(a)**. This means that the hypergeometric probability could have been approximated by the value of the binomial density function. The advantage of this procedure is that tables for the binomial density function are readily available, and therefore tedious hand calculations can be avoided.

Let H be a hypergeometric random variable, and let $h(x;n,b,w)$ be its density function. The notation $h(x;n,b,w)$ indicates that the hypergeometric random variable H depends on three parameters: n, the number of balls in the sample; b, the number of black balls in the urn; and w, the number of white balls in the urn. This notation for the hypergeometric density function is similar to the notation $b(x;n,p)$ used for the binomial density function. The following theorem gives a formula for approximating the values of the hypergeometric density function by the values of the binomial density function.

THEOREM 5.9   Let $h(x;n,b,w)$ be a hypergeometric density function. Then if

$$\frac{n}{b+w} < 0.1,$$

$$h(x;n,b,w) \approx b(x;n,p) \quad \text{where} \quad p = \frac{b}{b+w}$$

Notice that the parameter p of the corresponding binomial density function $b(x;n,p)$ is

$$p = \frac{b}{b+w}$$

This is the probability of removing a black ball from the urn.

Since tables of the hypergeometric density are available if $b + w \le 50$, the approximation of Theorem 5.9 is generally used only if $b + w > 50$.

Whenever the binomial density function is used to approximate the hypergeometric density function, an error is introduced. Formulas for finding the size of this error are given in more advanced books. In a practical situation, this error must be calculated before the binomial approximation is used. It can then be determined whether the error is small enough to meet the precision required by the particular problem being studied. We will not, however, consider the error in this book, because its calculation is difficult. We shall simply use the approximation when $n/(b + w) < 0.1$ and $b + w > 50$.

The following example shows how to apply Theorem 5.9.

Example 35   The freshman class of 1000 students at a certain university contains 200 mathematics majors. If a committee of 15 freshmen is randomly chosen, what is the probability that the committee contains—

(a) exactly 2 math majors
(b) at most 4 math majors
(c) at least 3 math majors?

Solution   The number H of math majors in the committee is a hypergeometric random variable with $b = 200$, $w = 800$, and $n = 15$. Since

$$\frac{n}{b+w} = \frac{15}{1000} = 0.015 < 0.1$$

the binomial approximation can be used, with

$$p = \frac{b}{b+w} = \frac{200}{1000} = 0.2 \quad \text{and} \quad n = 15$$

Using Theorem 5.9 and Tables A.1 and A.2,

(a) $P(H = 2) \approx b(2;15,0.2) = 0.23090$

(b) $P(H \le 4) = \sum_{k=0}^{4} P(H = x) \approx \sum_{x=0}^{4} b(x;15,0.2) = B(4;15,0.2) = 0.83577$

(c) $P(H \ge 3) = 1 - P(H \le 2) \approx 1 - B(2;15,0.2) = 1 - 0.39802 = 0.60198$

Since the hypergeometric density function $h(x;n,b,w)$ is approximated (under certain conditions) by the binomial density function $b(x;n,p)$ with $p = \dfrac{b}{b + w}$, and the mean of the binomial random variable is np, we might suspect that the mean of the hypergeometric random variable is also np where $p = \dfrac{b}{b + w}$. Similarly, we might suspect that the variance of H is npq where $p = \dfrac{b}{b + w}$ is the probability of selecting a black ball from the urn and $q = \dfrac{w}{b + w}$ is the probability of selecting a white ball from the urn. The following theorem indicates that our first suspicion is correct and our second suspicion is almost correct.

THEOREM 5.10 Let H be a hypergeometric random variable. The mean and variance of H are:

Mean $\qquad\qquad \mu = np$

Variance $\qquad\quad \sigma^2 = npq\left(\dfrac{b + w - n}{b + w - 1}\right)$

$$\text{where } p = \frac{b}{b + w} \text{ and } q = \frac{w}{b + w}.$$

5.7 EXERCISES

1. Small electric motors are shipped in lots of 100. Before a shipment is accepted, 10 randomly chosen motors are inspected. If none are defective, the lot is accepted; otherwise, the entire lot is inspected. Suppose that there are 5 defective motors among the 100. What is the probability that the entire lot will have to be inspected?

2. What is the probability that a five-card poker hand will contain (a) exactly two clubs, (b) at most one spade?

3. A child eats five cookies from a package of twenty cookies. If six of the cookies are spoiled and eating at least three spoiled cookies will make the child ill, what is the probability that the child gets sick?

4. The faculty of a certain school consists of eighty men and thirty women. What is the probability that a randomly chosen committee of ten faculty members contains exactly five women?

5. Twenty dogs and four cats are entered in a children's pet show. Suppose that five prizes are awarded randomly, and let X be the number of cats that receive prizes.
 (a) What kind of random variable is X?
 (b) What is the range of X?
 (c) Find the density function of X.
 (d) Calculate $P(X = 3)$, $E(X)$, and $\text{Var}(X)$.

6. A bridge hand (thirteen cards) is randomly selected from an ordinary fifty-two card deck. Let X and Y be the numbers of aces and diamonds in the hand, respectively.
 (a) Find the probability of—
  (i) $(X = 3)$ (ii) $(Y = 3)$ (iii) $(X = 5)$ (iv) $(Y = 5)$
 (b) What is the density function of X? What is the density function of Y?

7. Let H be a hypergeometric random variable with $b = 100$, $w = 900$, and $n = 5$.
 (a) Approximate (i) $P(H = 2)$, (ii) $P(H > 2)$, and (iii) $P(1 < H < 5)$.
 (b) State why the approximation can be used.
 (c) Which value of H is the most probable?

5.8 THE POISSON RANDOM VARIABLE

We saw in Section 5.6 that exact numerical values of the binomial density function can be found by using Table A.1. This table can also be used to find approximate numerical values of the hypergeometric random variable if

$$\frac{n}{b + w} < 0.1 \quad \text{and} \quad b + w > 50$$

However, complete tables of the binomial density function are not available for large values of n. If n is large, p is small, and the product $\mu = np$ is of moderate size (e.g., $p = 0.002$, $n = 5000$, and $\mu = 10$), the binomial density function may be approximated by the density function of a random variable called the *Poisson* random variable.

The following example introduces the Poisson random variable.

Example 36 (Poisson density function) It can be proved theoretically and verified statistically that the number X of calls that are received during a fixed interval of time at a telephone switchboard is a random variable with density function

$$f(x) = \frac{e^{-\mu}\mu^x}{x!} \quad x = 0, 1, 2, 3, \dots \tag{8}$$

where e is an irrational number approximately equal to 2.71828 and μ is the average number of calls in a certain time period.

For example, suppose that the average number of telephone calls is 2 per minute. The probability that there will be 3 calls during the next minute is found by substituting $\mu = 2$ and $x = 3$ in equation (8). The required probability is

$$f(3) = \frac{e^{-2}2^3}{3!}$$

The probability that there will be 8 calls in a *five-minute* interval is found by substituting $\mu = 2 \cdot 5 = 10$ and $x = 8$ in equation (8), because the average number of calls in a five-minute interval is 10. This yields

$$f(8) = \frac{e^{-10}(10)^8}{8!}$$

The density function given in equation (8) is called a *Poisson density function*. In general, any random variable X whose density function is given by

$$f(x) = \frac{e^{-\mu}\mu^x}{x!} \qquad x = 0, 1, 2, \ldots$$

where μ is a fixed positive constant, is called a **Poisson random variable**. Its density function is called a *Poisson density function*. Examples of Poisson random variables are: the number of defects per linear unit of wire, the number of strands in a cross section of thread, the number of defective teeth in an individual, and the frequency of earthquakes.

Notice that the Poisson density function depends not only on x, but also on a fixed positive constant μ. We will therefore let

$$p(x;\mu) = \frac{e^{-\mu}\mu^x}{x!} \qquad x = 0, 1, 2, \ldots$$

denote the Poisson density function. This notation means that x is a variable while μ is a parameter. (Recall that similar notations, $b(x;n,p)$ and $h(x;n,b,w)$, are used to denote binomial and hypergeometric random variables.)

Values of $p(x;\mu)$ appear in Table A.3 of the Appendix. For example, the probability $f(3) = p(3;2)$ of Example 36 is found in the intersection of the column of Table A.3 labeled $\mu = 2$ and the row labeled $x = 3$. The required probability is $p(3;2) = 0.1804$. Similarly, the probability $f(8) = p(8;10)$ of Example 36 is found in the intersection of the column of Table A.3 labeled $\mu = 10$ and the row labeled $x = 8$; the value is $f(8) = 0.1126$.

The important characteristics of a Poisson density function are listed below. They are also listed in Table A.4 of the Appendix.

THEOREM 5.11   Let X be a Poisson random variable with parameter $\mu > 0$. The density function, mean, and variance of X are:

Density function $\quad p(x;\mu) = \dfrac{e^{-\mu}\mu^x}{x!} \qquad x = 0, 1, 2, \ldots$

Mean $\qquad\qquad\qquad \mu$

Variance $\qquad\qquad\qquad \mu$

Notice that the mean and variance of the Poisson random variable are both μ. The following examples are applications of Theorem 5.11.

Example 37 The number X of misprints per page in a certain Sunday newspaper is a Poisson random variable. The average number of misprints per page is one. What is the probability that the three-page sports section will contain—

(a) exactly four misprints
(b) at most two misprints
(c) more than five misprints?

Solution Let N be the number of misprints in the sports section. We must first find the parameter μ, the average number of misprints in a three-page section. Since the average number of misprints per page is 1, it follows that $\mu = 3 \cdot 1 = 3$. Using Theorem 5.11 and Table A.3—

(a) $P(N = 4) = p(4;3) = 0.1680$
(b) $P(N \leq 2) = p(0;3) + p(1;3) + p(2;3)$
$\qquad = 0.0498 + 0.1494 + 0.2240 = 0.4232$
(c) $P(N > 5) = p(6;3) + p(7;3) + \cdots = 0.0504$
$\qquad\qquad + 0.0216 + 0.0081 + 0.0027 + 0.0008$
$\qquad\qquad + 0.0002 + 0.0001 = 0.0839$

Notice that $p(13;3)$ does not appear in the table. This means that $p(13;3)$ is 0 to at least four decimal places.

Example 38 Let N be the number of misprints in the three-page sports section of Example 37. Find the mean and variance of N.

Solution Applying Theorem 5.11 and using the results of Example 37, the mean and variance of N are both $\mu = 3$.

We have seen that the Poisson density function is important because it can be applied to many different practical situations. As mentioned at the beginning of this section, it is also important because it can be used as an approximation to the binomial density function. To be specific, if n is large, p is small, and $\mu = np$ is of moderate size, it can be shown that

$$b(x;n,p) \approx p(x;\mu) \qquad \text{where} \quad \mu = np$$

This approximation is illustrated in Table 5.7 and Example 39.

It is clear from Table 5.7 that an error is introduced whenever the Poisson density function is used to approximate the binomial density function. Formulas for finding the size of this error are given in more advanced books, and this error must be calculated before the Poisson approximation is applied to a practical situation. If the error is small enough to meet the precision required by the particular problem being studied, the approximation can be used. As in the case of the binomial approximation, we will not consider the error in this book because its calculation is difficult.

Table 5.7   Examples of the Poisson Approximation

x	$b(x; 30, 0.01)$	$p(x; 0.3)$	$b(x; 30, 0.02)$	$p(x; 0.6)$
0	0.73970	0.74082	0.54548	0.54881
1	0.22415	0.22225	0.33397	0.32929
2	0.03283	0.03334	0.09883	0.09878
3	0.00310	0.00333	0.01882	0.01976
4	0.00021	0.00025	0.00259	0.00296
5	0.00001	0.00002	0.00028	0.00036
6	0.00000	0.00000	0.00002	0.00004

Example 39   According to the 1958 CSO mortality table, the probability that a male of age forty-two will die within one year is approximately 0.004. If an insurance company has 1000 policy holders of that sex and age, find the probability that there will be (*a*) no claims, (*b*) one claim, (*c*) at least two claims next year.

Solution   The number of deaths S_{1000} is a binomial random variable. Since $p = 0.004$ is small, $n = 1000$ is large, and $\mu = np$ is moderate, the Poisson density function may be used to approximate the binomial density function. The required probabilities are—

(**a**) $b(0;1000,0.004) \approx p(0;4) = 0.0183$
(**b**) $b(1;1000,0.004) \approx p(1;4) = 0.0733$
(**c**) $P(S_{1000} \geq 2) = 1 - P(S_{1000} < 2)$
$$= 1 - [b(0;1000,0.004) + b(1;1000,0.004)]$$
$$\approx 1 - [0.0183 + 0.0733] = 0.9084$$

5.8 EXERCISES

1.   Find numerical values for the following probabilities:
(**a**) $p(5;3)$ (**b**) $p(3;5)$ (**c**) $b(5;1000,0.01)$
(**d**) $B(3;100,0.02)$ (**e**) $h(2;5,30,70)$

2.   In an electronic computer, the number of transistors that fail in any one-hour period is a Poisson random variable. Assume that the machine becomes inoperative if two or more transistors fail and that, on the average, there is one transistor failure every ten hours. A computation requiring ten hours of computing time is started. What is the probability that it can be successfully completed without a breakdown?

3.   The number of alpha particles emitted from a radioactive source during a fixed time t is a Poisson random variable. If the average number of particles emitted and recorded is two per second, find the probability that—
(**a**) twelve particles will be recorded in a five-second interval
(**b**) at most five particles will be recorded in a two-second interval.

4.   The number of white blood corpuscles on a microscope slide is a Poisson random variable. If the average number of white corpuscles per slide is two, what is the probability that two slides will contain—

(a) at least three white corpuscles

(b) at most six white corpuscles?

5. Let S_{1000} be a binomial random variable with $p = 0.001$. Approximate the following probabilities:

(a) $P(S_{1000} = 2)$ (b) $P(S_{1000} \neq 1)$ (c) $P(S_{1000} < 1)$

6. The chance of triplets in a human birth is approximately 0.0001. Assuming independence, what is the probability of observing at least five sets of triplets in a record of 10,000 human births?

7. If 1 percent of the lightbulbs manufactured by a certain company are defective, use the Poisson approximation to find the probability that a case of 100 bulbs will contain—

(a) at least one defective bulb

(b) exactly 3 defective bulbs.

Use a calculator to find the exact answers to these problems.

8. The number of accidents in a certain factory is a Poisson random variable. Assume that the average number of accidents per year is 52, and let X be the number of accidents that occur in a week.

(a) Find $E(X)$ and $\text{Var}(X)$.

(b) What is the most likely number of accidents that will occur next week?

5.9 THE GEOMETRIC AND PASCAL RANDOM VARIABLES

Suppose a coin is tossed ten times and let S_{10} be the number of heads that appear. We saw in Section 5.6 that X is a binomial random variable. Now suppose that the coin is not tossed a definite number of times (e.g., ten times), but instead is tossed *until the third head is obtained.* Let N_3 be the number of times that the coin is tossed; N_3 is an example of a *Pascal random variable.* If we replace the particular experiment "a coin is tossed" by any experiment having similar characteristics, and the outcome "a head occurs" by the general outcome S ("a success occurs"), and if we replace 3 heads by k successes, then we have the general definition of a Pascal random variable. This definition follows.

Definition 5.9   Suppose that an experiment T has the following characteristics:

(1) T has two mutually exclusive outcomes, S (success) and F (failure), and exactly one of these outcomes occurs each time T is repeated.

(2) T is repeated until the kth success occurs. Each repetition of T is called a *trial.*

(3) The outcome of each trial is independent of the outcomes of the other trials.

(4) $P(S) = p$ and $P(F) = q = 1 - p$ are the same for each trial.

The **Pascal random variable** N_k is the number of times T is repeated. The range of N_k is $\{\infty, k, k + 1, k + 2, \ldots\}$ where $\{\infty\}$ is the event "the kth success never occurs." In particular, the **geometric random variable** G is defined to be N_1. The range of G is $\{\infty, 1, 2, 3, \ldots\}$. (See Example 8 in Section 5.1.)

Notice that the geometric random variable G is the number of times that T is repeated until the *first* success occurs.

It is important that the reader clearly understand the difference between the binomial and Pascal random variables: The binomial random variable S_n is the number of successes in n trials; the Pascal random variable N_k is the number of trials until the kth success.

Examples of the Pascal random variable follow.

Example 40

(a) People are questioned until four people are found who favor Brown for mayor. No person is questioned more than once. The number N_4 of people who are questioned is a Pascal random variable.

(b) Children are given vision tests until the second child with unsatisfactory vision is found. The number N_2 of children who are tested is a Pascal random variable.

(c) Bob shoots a target until he hits a bull's-eye. The number N_1 of shots that he takes is a Pascal random variable. Since $k = 1$, N_1 is also a geometric random variable.

The following examples will lead us to a formula for the density function of a Pascal random variable.

Example 41   A die is tossed until three 6s appear. What is the probability that the third 6 will appear on the fifth toss, on the eighth toss, on the xth toss $(x = 3,4,5,\ldots)$?

Solution   The number N_3 of the toss on which the third 6 appears is a Pascal random variable. Let S_i be the event "a 6 appears on the ith toss," and let $F_i = S_i'$. The event $(N_3 = 5)$ will occur if the event $A_1 = S_1 S_2 F_3 F_4 S_5$ occurs, or if the event $A_2 = F_1 F_2 S_3 S_4 S_5$ occurs, and so on. Hence,

$$(N_3 = 5) = A_1 + A_2 + \cdots + A_m$$

where each $A_i = ABCDS_5$ and exactly two of the events A, B, C, D are S_i's and two are F_i's.

Using independence (characteristic (3) of Definition 5.9),

$$P(A_1) = P(S_1 S_2 F_3 F_4 S_5) = P(S_1)P(S_2)P(F_3)P(F_4)P(S_5)$$
$$= (\tfrac{1}{6})^3 (\tfrac{5}{6})^2$$

Similarly,

$$P(A_2) = (\tfrac{1}{6})^3 (\tfrac{5}{6})^2$$

and, in general,

$$P(A_k) = (\tfrac{1}{6})^3 (\tfrac{5}{6})^2$$

It follows that

$$P(N_3 = 5) = P(A_1 + A_2 + \cdots + A_m) = P(A_1) + P(A_2) + \cdots + P(A_m)$$
$$= m(\tfrac{1}{6})^3(\tfrac{5}{6})^2$$

The coefficient m is the number of ways of selecting 2 out of the 4 events $A, B, C,$ D to be S_i's; that is, $m = {}_4C_2$. It follows that

$$P(N_3 = 5) = {}_4C_2(\tfrac{1}{6})^3(\tfrac{5}{6})^2$$

Similarly,

$$P(N_3 = 8) = {}_7C_2(\tfrac{1}{6})^3(\tfrac{5}{6})^5$$

and, in general,

$$f(x) = P(N_3 = x) = {}_{x-1}C_2(\tfrac{1}{6})^3(\tfrac{5}{6})^{x-3} \qquad x = 3, 4, 5, 6, \ldots$$

Example 42   Referring to the die-tossing experiment of Example 41, what is the probability that the third 6 will *never* appear?

Solution   We need to find $P(N_3 = \infty)$. The probabilities $P(N_3 = x)$, $x = 1, 2, 3,$ $\ldots$, can be found by using the results of Example 41. Several of these values are listed in the following table:

x	$P(N_3 = x)$	
11	${}_{10}C_2(\tfrac{1}{6})^3(\tfrac{5}{6})^8$	$= 0.05$
19	${}_{18}C_2(\tfrac{1}{6})^3(\tfrac{5}{6})^{16}$	$= 0.04$
35	${}_{34}C_2(\tfrac{1}{6})^3(\tfrac{5}{6})^{32}$	$= 0.008$
67	${}_{66}C_2(\tfrac{1}{6})^3(\tfrac{5}{6})^{64}$	$= 0.00008$
131	${}_{130}C_2(\tfrac{1}{6})^3(\tfrac{5}{6})^{128}$	$= 0.0000000$

We see from this table that the probability that we would have to toss the die 35 times before obtaining three 6s is very small; the probability that we would have to toss it 67 times is even smaller. The probability that we would have to toss it 131 times is zero to at least seven decimal places. It is clear from the table that, as x gets larger and larger, $P(N_3 = x)$ gets closer and closer to zero.[4] For this reason, we say that the probability that three 6s are *never* obtained is zero. In other words, $P(N_3 = \infty) = 0$.

The following theorem can be proved by the method used in Examples 41 and 42.

THEOREM   5.12   The density function of the Pascal random variable N_k is

$$f(x) = {}_{x-1}C_{k-1}p^k q^{x-k} \qquad x = k, k+1, k+2, \ldots$$

---

4. This statement is only true when $x > 12$.

Remember that $f(x)$ *is the probability that the kth success will be obtained on the xth trial.* Remember also that, since the value of $f(\infty)$ is not listed in the theorem, $f(\infty) = 0$.

Substituting $k = 1$ in Theorem 5.12 yields the following formula for the density function of the geometric random variable:

THEOREM   5.13   The density function of the geometric random variable G is

$$f(x) = pq^{x-1} \qquad x = 1, 2, 3, \ldots$$

Remember that $f(x)$ *is the probability that the first success will be obtained on the xth trial.* Remember also that $f(\infty) = 0$, so $f(\infty)$ is not listed in the theorem.

In the coin-tossing experiment of Example 8 the random variable G was defined as the number of tosses until the first head appears. Since G is a geometric random variable with density function f, Theorem 5.13 tells us that the probability that a head will never appear is $f(\infty) = 0$.

Let G be a geometric random variable. The density function of G is tabulated as follows:

x	1	2	3	4	5	$\cdots$
$f(x)$	p	pq	pq^2	pq^3	pq^4	$\cdots$

Notice that the sum of the entries in row 2, $p + pq + pq^2 + pq^3 + pq^4 + \cdots$, forms a geometric series. This is the reason that G is called a geometric random variable.

The following examples are applications of Theorems 5.12 and 5.13.

Example 43   Blood is needed for a student with blood type A. American volunteers who are not related to the student are tested for blood type. If 40 percent of the people in the United States have blood type A, what is the probability that—

(a) the fifth person tested is the first person with this blood type

(b) the seventh person tested is the first person tested with this blood type

(c) no one will ever be found with blood type A?

Solution   The number G of persons tested to find the first person with blood type A is a geometric random variable with $p = 0.4$ and $q = 1 - 0.4 = 0.6$.

(a) Applying Theorem 5.13 with $x = 5$ yields the required probability:

$$f(5) = (0.4)(0.6)^4 = 0.05$$

(b) Applying Theorem 5.13 with $x = 7$ yields

$$f(7) = (0.4)(0.6)^6 = 0.02$$

(c) The required probability is $f(\infty)$. The fact that $f(\infty)$ is not listed in Theorem 5.13 indicates that $f(\infty) = 0$.

Example 44   From past experience, a company's personnel officers know that 20 percent of the college graduates they interview will be qualified to work for their firm. They wish to hire four people. What is the probability that they must interview at least seven college graduates to find four qualified people?

Solution   Assume that the company interviews college graduates one at a time until they find four qualified people and let N_4 be the number of people interviewed. Then N_4 is a Pascal random variable with $p = 0.2$ and $q = 1 - 0.2 = 0.8$. Applying Theorem 5.12 and using a hand calculator, the required probability is

$$P(N_4 \geq 7) = 1 - [P(N_4 = 4) + P(N_4 = 5) + P(N_4 = 6)]$$
$$= 1 - [{}_3 C_3 (0.2)^4 (0.8)^0 + {}_4 C_3 (0.2)^4 (0.8)^1 + {}_5 C_3 (0.2)^4 (0.8)^2]$$
$$= 0.98304$$

The following example will lead us to formulas for the mode and distribution function of a geometric random variable.

Example 45   A couple is told that, if they follow certain recommended procedures, the probability that they will have a daughter is 0.8. Suppose the couple keeps on having children until they have a daughter.

(a) What is the most likely number of children that they will have?

(b) What is the probability that they will have at most two children, at most four children, at most x children?

Solution   The number G of children that they will have is a geometric random variable with $p = 0.8$.

(a) We wish to find the mode of G. Applying Theorem 5.13,

$$f(1) = 0.8, \qquad f(2) = (0.8)(0.2) = f(1)(0.2)$$
$$f(3) = (0.8)(0.2)^2 = f(2)(0.2)$$

and, in general,

$$f(x + 1) = (0.8)(0.2)^x = [(0.8)(0.2)^{x-1}](0.2) = f(x)(0.2)$$

This means that $f(x + 1)$ is less than $f(x)$ because 0.2 times any number is less than that number. Therefore,

$$f(1) > f(2) > f(3) > \cdots$$

which implies that the mode is 1.

(b) We need to calculate $F(2)$, $F(4)$, and $F(x)$.

$$F(2) = P(G \leq 2) = 1 - P(G > 2)$$

The event $(G > 2)$ will occur if and only if the couple's first two children are boys. Let B_i be the event "the ith child is a boy." Then

$$F(2) = 1 - P(G > 2) = 1 - P(B_1 B_2) = 1 - P(B_1)P(B_2) = 1 - (0.2)(0.2)$$
$$= 1 - (0.2)^2$$

Similarly,

$$F(4) = P(G \leq 4) = 1 - P(G > 4) = 1 - P(B_1 B_2 B_3 B_4) = 1 - (0.2)^4$$

and, in general,

$$F(x) = 1 - (0.2)^x \qquad x = 1, 2, 3, 4, \ldots$$

Using a similar method, we can easily prove that the mode of any geometric random variable is 1 and its distribution function is $F(x) = 1 - q^x$. These and other important characteristics of a geometric random variable are listed in Theorem 5.14. Theorem 5.15 lists the important characteristics of a Pascal random variable.

THEOREM 5.14 The density function, distribution function, mean, variance, and mode of the geometric random variable G are:

Density function	$f(x) = pq^{x-1}$	$x = 1, 2, 3, 4, \ldots$
Distribution function	$F(x) = 1 - q^x$	$x = 1, 2, 3, 4, \ldots$
Mean	$\mu = \dfrac{1}{p}$	
Variance	$\sigma^2 = \dfrac{q}{p^2}$	
Mode	1	

THEOREM 5.15 The density function, mean, and variance of the Pascal random variable N_k are:

Density function	$f(x) = {}_{x-1}C_{k-1}p^k q^{x-k}$	$x = k, k + 1, k + 2, \ldots$
Mean	$\mu = \dfrac{k}{p}$	
Variance	$\sigma^2 = \dfrac{kq}{p^2}$	

The following examples show how to apply Theorems 5.14 and 5.15.

Example 46 A doctor knows that 10 percent of his patients will have an allergic reaction to a certain vaccination. What is the probability that—

(a) the third person he injects will be the first person to get a reaction

(b) the fifth person he injects will be the second person to have a reaction

(c) the first person who has a reaction will be among the first four people he injects?

Solution   Suppose that the doctor injects patients one at a time until he finds one (two) of them who have a reaction. Let G (N_2) be the number of patients that he injects. Then G and N_2 are geometric and Pascal random variables, respectively, with $p = 0.1$ and $q = 1 - 0.1 = 0.9$.

(a) Applying Theorem 5.14, $P(G = 3) = f(3) = (0.1)(0.9)^2 = 0.081$.
(b) Applying Theorem 5.15, $P(N_2 = 5) = f(5) = {}_4C_1(0.1)^2(0.9)^3 = 0.029$.
(c) Applying Theorem 5.14, $P(G \leq 4) = F(4) = 1 - (0.9)^4 = 0.3439$.

Example 47   Referring to Example 46, find—

(a) the average number of patients the doctor will inject before finding one who has a reaction
(b) average number of patients he will inject before finding two who have a reaction
(c) the most likely number of patients that he will inject before finding one who has a reaction.

Solution

(a) Using the random variables defined in Example 46 and applying Theorem 5.14, the average is

$$E(G) = \mu = \frac{1}{p} = \frac{1}{0.1} = 10$$

(b) Applying Theorem 5.15, the average is

$$E(N_2) = \mu = \frac{2}{0.1} = 20$$

(c) Applying Theorem 5.14, the most likely number is the mode, 1.

5.9 EXERCISES

1.   A coin is biased 60 percent in favor of heads. If the coin is tossed repeatedly, what is the probability that—
(a) the first head will appear on the fifth toss
(b) the first head will appear on or before the fourth toss
(c) the third head will appear on the fifth toss?

2.   If 10 percent of the people in a certain area show a positive reaction to a test for tuberculosis, what is the probability that fewer than five negative reactions occur before the first positive one?

3. Experience has shown that 90 percent of the wells drilled in a certain area will have water.

    (a) What is the probability that four wells will have to be drilled to find water in two wells?

    (b) What is the most likely number of wells that will have to be drilled to find water in two wells?

4. A doctor estimates that the probability that a certain couple will have a boy is 0.8, provided that they follow certain recommended procedures. The couple decide to keep having children until they have a boy. (Assume that there are no multiple births.)

    (a) Find the probability that they have (*i*) three children, (*ii*) seven children, (*iii*) *x* children.

    (b) What is the most likely number of children they will have?

    (c) What is the probability that they have at most five children?

5. Answer the questions of Exercise 4 under the assumption that the couple decide to keep on having children until they have two sons.

6. Ruth's batting average is 0.400. What is the probability that her first hit will occur—

    (a) on her third turn at bat

    (b) on her sixth turn at bat or later

    (c) on or before her fourth turn at bat?

7. A travel agent keeps calling motels until he finds one that has a room available for the night his client needs it.

    (a) If 50 percent of the motels have rooms available on that night, how many calls should he expect to make?

    (b) What is the most likely number of calls that he will make?

5.10 APPLYING RANDOM VARIABLES

A number of different random variables have been discussed in this chapter: the binomial, hypergeometric, Poisson, Pascal, and geometric random variables. The important characteristics of these random variables are listed in Table A.4 of the Appendix and in the theorems of this chapter. Before solving a probability problem, one must decide which (if any) of these random variables apply. The following two examples point out the differences in applying the various random variables we have studied.

Example 48   The probability that a certain type of antiaircraft missile will hit its target is 0.7. Moreover, the accuracy of each missile is unaffected by the success or failure of the others.

    (a) If ten missiles are launched, what is the probability that exactly six will be successful?

    (b) If the missiles are launched one at a time, what is the probability that the fourth missile will be the first to hit its target?

(c) If three airplanes are attacking and the missiles are launched one at a time, what is the probability that six launchings will be necessary to destroy all three planes? (Assume for the sake of simplicity that no plane is hit by more than one missile.)

Solution   The binomial random variable S_n is the number of successes in n trials. The geometric random variable G is the number of trials until the first success. The Pascal random variable N_k is the number of trials until the kth success. In this example a success occurs when a missile hits its target. It follows that the binomial random variable S_{10} is the number of hits in 10 launchings; the geometric random variable G is the number of launchings until the first hit; and the Pascal random variable N_3 is the number of launchings until the third hit.

Therefore, parts **(a)**–**(c)** can be solved by applying Tables A.1 and A.4 with $p = 0.7$ and $q = 0.3$. The answers are as follows:

(a) Using the binomial density function with $n = 10$ and $x = 6$, we have

$$P(S_{10} = 6) = {}_{10}C_6(0.7)^6(0.3)^4 = b(6;10,0.7) = 0.20012$$

(b) Using the geometric density function with $x = 4$, we have

$$P(G = 4) = (0.3)^3(0.7) = 0.0189$$

(c) Using the Pascal density function with $k = 3$ and $x = 6$, we have

$$P(N_3 = 6) = {}_5C_2(0.7)^3(0.3)^3 = 0.09261$$

Example 49   Suppose that two of our planes are engaged in close combat with ten enemy planes. Moreover, suppose that each of our antiaircraft missiles is certain to hit one of the twelve planes (maybe one of ours). However, because of their close proximity, we are unable to predict which plane will be hit. If we assume for the sake of simplicity that each missile will hit a different plane, what is the largest number of rockets that can be launched if the probability of missing all of our planes is to be greater than 0.5?

Solution   Let H_n be the number of our planes shot down if n missiles are launched, and let $p_n = P(H_n = 0)$. Then H_n is a hypergeometric random variable with our planes represented by black balls and enemy planes represented by white balls, and p_n is the probability that 0 (i.e., none) of our planes will be shot down if n missiles are launched. Using Table A.4 with $b = 2$ and $q = 10$, we see that

$$p_n = \frac{{}_2C_0 \cdot {}_{10}C_n}{{}_{12}C_n} = \frac{{}_{10}C_n}{{}_{12}C_n}$$

We wish to find the largest n for which $p_n > 0.5$. Since $p_3 = \frac{6}{11}$, $p_4 = \frac{14}{33}$, and the probability decreases as n increases, the answer is 3 rockets.

Key Word Review

Each key word is followed by the number of the section in which it is defined.

function 5.1	standard deviation 5.4
domain 5.1	mode 5.4
range 5.1	uniform random variable 5.5
extended real numbers 5.2	sequence of Bernoulli trials 5.6
random variable 5.2	binomial random variable 5.6
density function 5.3	hypergeometric random variable 5.7
distribution function 5.3	Poisson random variable 5.8
expected value 5.4	Pascal random variable 5.9
mean 5.4	geometric random variable 5.9
variance 5.4	

CHAPTER 5   REVIEW EXERCISES

1. Define each word in the Key Word Review.

2. What do the following symbols mean?
   (a) $P(X \leq k)$ (b) $\text{Var}(X)$ (c) μ (d) σ (e) $b(x;n,p)$
   (f) $B(x;n,p)$ (g) $p(x;\mu)$ (h) σ^2 (i) $E(X)$ (j) S_n

3. Let $f(x) = x^2$ and $g(x) = 2x$. Calculate—
   (a) $f(3)$ (b) $(f+g)(x)$ (c) $f(k)$ (d) $\dfrac{f}{g}(x)$

4. Find numerical values for the following probabilities:
   (a) $b(3;10,0.02)$ (b) $B(5;10,0.25)$ (c) $p(6;7)$
   (d) $b(5;10,0.96)$ (e) $B(12;20,0.75)$

5. Let S_{10} be a binomial random variable with $p = 0.08$. Calculate—
   (a) $P(S_{10} = 3)$ (b) $P(S_{10} \leq 4)$ (c) $P(1 < S_{10} < 5)$
   (d) $P(S_{10} \geq 3)$

6. Let S_{20} be a binomial random variable with $p = 0.25$. Calculate—
   (a) $P(S_{20} \leq 7)$ (b) $P(S_{20} > 9)$ (c) $P(S_{20} = 8)$

7. Let X be a random variable with density function f defined as follows:

x	0	1	2	3	4
$f(x)$	0.1	0.2	0.25	0.25	0.2

Calculate—
   (a) $P(X = 1)$ (b) $P(X > 4)$ (c) $E(X)$ (d) $\text{Var}(X)$
   (e) σ (f) the mode or modes of X

8. A supermarket buys a certain type of cake for 50 cents each and sells them for 1 dollar each. Unsold cakes are a total loss. From past experience, the density function of X—the number of cakes that the market sells—has the values tabulated in the following chart:

x	0	1	2	3	4	5
$P(X=x)$	0.05	0.15	0.25	0.30	0.20	0.05

Assume that, if only k cakes are available and a customer wants more than k cakes, he will buy the k cakes instead of going elsewhere.

(a) If the market buys five cakes, what is the most probable number of cakes that it will sell?

(b) If the market buys five cakes, what is the probability that it will sell at least two of them?

(c) How many cakes can the market expect to sell if it buys three cakes?

(d) What is the expected profit if the market buys two cakes?

(e) How many cakes should the market buy to maximize the expected profit?

9. The probability that a male of age fifty will die during the next year is 0.0083. Suppose that an insurance company has 1000 male policyholders of this age.

(a) What is the average number of these men who will die within the next year?

(b) What is the most likely number of these men who will die within the next year?

(c) What is the probability that at least 16 of these men will die within the next year?

10. The number of babies born at a certain hospital is a Poisson random variable, and the average number of babies born in that hospital per day is two. What is the probability that—

(a) no children will be born at the hospital in the next three days

(b) at least one child will be born tomorrow

(c) at least one child will be born each day next week?

11. Ten students are randomly chosen from the student body at a college at which 10 percent of the students are married.

(a) What is the probability that exactly two married students will be chosen?

(b) What is the probability that at most five married students will be chosen?

(c) What is the average number of married students that will be chosen?

(d) What is the most likely number of married students that will be chosen?

12. A certain type of radio uses tubes with a yearly failure rate of 0.80. That is, the probability that a tube will stop working before the end of one year is 0.80.

What is the probability that a ten-tube radio will have—

(a) at least one defective tube by the end of the year

(b) exactly two defective tubes

(c) more than four defective tubes?

13. A certain baseball team wins 60 percent of the games it plays. What is the probability that the team will have a losing streak before its first win of—

(a) exactly three games

(b) at least six games

(c) at most four games?

14. (a) What is the average number of games that the team of Exercise 13 will lose before its first win?

(b) What is the most likely number of games that the team will lose before its first win?

15. What is the probability that the team of Exercises 13 and 14 will have lost exactly five games by the time it has won four? That is, what is the probability that the team will win exactly five out of nine games, where the last game is a win?

16. The faculty of a certain school consists of thirty tenured and seventy non-tenured teachers. If a committee of five teachers is randomly selected to evaluate the administration, what is the probability that the committee contains—

(a) only tenured teachers

(b) exactly three nontenured teachers

(c) at least one nontenured teacher?

Give exact answers.

17. (a) Approximate the answers to the questions of Exercise 16.

(b) State why the approximations may be used.

Suggested Readings

1. Derman, Cyrus; Glaser, Leslie; and Olkin, Ingram. *A Guide to Probability Theory and Its Applications*. New York: Holt, Rinehart & Winston, 1973.
2. Eisen, Martin, and Eisen, Carole. *Probability and Its Applications*. New York: Quantum Press, 1975.
3. Feller, William. *An Introduction to Probability Theory and Its Applications*. Vol. 1. 3d ed. New York, John Wiley & Sons, 1968. Vol. 2. 2d ed. 1971.
4. Parzen, Emanuel. *Modern Probability Theory and Its Applications*. New York: John Wiley & Sons, 1960.
5. Selby, Samuel, and Sweet, Leonard. *Sets, Relations, Functions*. 2d. ed. New York: McGraw-Hill Book Co., 1969.

Jean Claude Lejeune/Stockmarket, Los Angeles

6 Vectors,
Matrices,
and
Linear Equations

INTRODUCTION

The pigeonholes in the picture on the facing page are arranged in a rectangular array with six rows and five columns. In this chapter we will find that rectangular arrays of numbers have many important mathematical uses. Numerical information is frequently displayed in rectangular arrays like Table 6.1. An array of numbers with m rows and n columns is called a *matrix*. (Table 6.1 has three rows and three columns.) If the matrix has only one row or one column, it is called a *vector*.

Vectors and matrices are used to organize information and solve problems in the fields of statistics, probability, sociology, physics, economics, and biology. In this chapter we will learn how to add and multiply matrices and how to apply these operations to practical problems. We will also study systems of linear equations and learn how to use matrices to solve them. In Chapters 7 and 8 we will explore further applications of matrices.

6.1 VECTORS AND MATRICES

The following example illustrates how tables of numerical data can be used in a practical setting.

Example 1 The salaries of employees at a certain firm are given in Table 6.1.

Table 6.1

	Less than $10,000	$10,000–$20,000	More than $20,000
Managers	20	30	40
Salespersons	80	150	100
Factory Workers	200	150	100

(a) The number N of managers earning between 10,000 and 20,000 dollars per year is found at the intersection of the row labeled managers (row 1) and the column labeled 10,000–20,000 dollars (column 2); therefore, $N = 30$.

(b) From the table, 80 salespersons earn less than 10,000 dollars per year.

(c) The firm employs $20 + 30 + 40 = 90$ managers, $80 + 150 + 100 = 330$ salespersons, and 450 factory workers.

The rectangular array of numbers in Table 6.1 is called a *matrix*. It is often convenient to omit the headings of a matrix and enclose the array in parentheses. The matrix of Example 1 can then be written

$$\mathbf{B} = \begin{pmatrix} 20 & 30 & 40 \\ 80 & 150 & 100 \\ 200 & 150 & 100 \end{pmatrix} \tag{1}$$

Array **B** is called a 3×3 (read "3 by 3") matrix because it has 3 rows and 3 columns. Other 3×3 matrices are

$$\mathbf{C} = \begin{pmatrix} 1 & 2 & 3 \\ -1 & 0 & 1 \\ 2 & 3 & 4 \end{pmatrix}, \qquad \mathbf{D} = \begin{pmatrix} 2 & 1 & 0 \\ 3 & -1 & \frac{1}{2} \\ 5 & 7 & 0.3 \end{pmatrix}$$

$$\mathbf{E} = \begin{pmatrix} 0.1 & 0.8 & 0.1 \\ 0.2 & 0.8 & 0 \\ 1 & 0 & 0 \end{pmatrix}$$

In general, we can let

$$\mathbf{A} = \begin{pmatrix} a_{11} & a_{12} & a_{13} \\ a_{21} & a_{22} & a_{23} \\ a_{31} & a_{32} & a_{33} \end{pmatrix}$$

represent any 3×3 matrix. Each of the entries $a_{11}, a_{12}, a_{13}, \dots$ represents a real number. The double subscript denotes the row and column in which the entry appears. The first digit in the double subscript denotes the row; the second digit denotes the column. For example, a_{11} appears in the first row, first column of **A**, a_{12} appears in the first row, second column of **A**, and a_{21} appears in the second row, first column of **A**.

These concepts can be extended to rectangular arrays having m rows and n columns.

Definition 6.1   An $m \times n$ **matrix A** is a rectangular array of real numbers arranged in m rows and n columns.

$$\mathbf{A} = \begin{pmatrix} a_{11} & a_{12} & \cdots & a_{1j} & \cdots & a_{1n} \\ a_{21} & a_{22} & \cdots & a_{2j} & \cdots & a_{2n} \\ \vdots & \vdots & & \vdots & & \vdots \\ a_{i1} & a_{i2} & \cdots & a_{ij} & \cdots & a_{in} \\ \vdots & \vdots & & \vdots & & \vdots \\ a_{m1} & a_{m2} & \cdots & a_{mj} & \cdots & a_{mn} \end{pmatrix}$$

The entry in the ith row, jth column is denoted a_{ij}; the matrix **A** is sometimes denoted by $\mathbf{A} = (a_{ij})$.

Matrices are usually denoted by boldface capital letters (e.g., **A**, **B**, **C**). The entries of the matrices are real numbers, which in this context are frequently called *scalars*; these scalars are usually denoted by lowercase letters (e.g., a, b, c, a_{11}, a_{ij}).

Note that an $m \times n$ matrix has m rows and n columns. A 2×3 matrix has 2 rows and 3 columns; a 3×2 matrix has 3 rows and 2 columns. The matrix **B** in equation (1) is a 3×3 matrix. Matrices like **B** that have the same number of rows and columns are called *square* matrices.

Definition 6.2

(a)   An $m \times m$ matrix is called a **square matrix**.

(b)   A $1 \times n$ matrix is called a **row vector**.

(c)   An $m \times 1$ matrix is called a **column vector**.

A row vector is a matrix with just one row; a column vector is a matrix with just one column. The entries of a vector are called **components**; the entries of a row vector are usually separated by commas. For example, (1,0,0) and (0.1,0.2,0.3) are row vectors. The first component of (1,0,0) is 1; the second and third components are both 0.

Example 2

$$A = \begin{pmatrix} 1 & 3 \\ 0 & 1 \end{pmatrix}, \qquad B = \begin{pmatrix} 0.1 \\ 0.9 \end{pmatrix}, \qquad C = (2,3,-1)$$

$$D = \begin{pmatrix} 1 & -4 & -5 \\ 0 & 1 & -3 \end{pmatrix}, \qquad E = \begin{pmatrix} \frac{1}{10} \\ \frac{9}{10} \end{pmatrix}, \qquad F = (0.1,0.9)$$

are all matrices. Matrices **B**, **C**, **E**, and **F** are the only vectors; **B** and **E** are column vectors, while **C** and **F** are row vectors. Matrix **A** is the only square matrix.

Example 3

(a) The rectangular array

$$A = \begin{pmatrix} 2 & 4 & 24 \\ 6 & 5 & -4 \end{pmatrix}$$

is a matrix having 2 rows and 3 columns. For this reason, it is called a 2×3 matrix. Matrix **A** is also denoted (a_{ij}). This means that the entry in the intersection of the ith row, jth column is denoted a_{ij}. For example, a_{11} is the entry in the intersection of the first row, first column, so $a_{11} = 2$; a_{12} is the entry in the intersection of the first row, second column, so $a_{12} = 4$.

(b) The rectangular array

$$B = \begin{pmatrix} 1 & -1 \\ 2 & 3 \\ 3 & 4 \end{pmatrix}$$

is a 3×2 matrix $B = (b_{ij})$; b_{21} is the entry in row 2, column 1, so $b_{21} = 2$, and b_{12} is the entry in row 1, column 2, so $b_{12} = -1$.

(c) The ordered pairs (1,2), (4,6), and (-5,3) are all row vectors. In general, any ordered pair (x,y) is a row vector.

A number can be written in many different ways. For example, 4, $\frac{8}{2}$, $6 - 2$, and 2^2 all identify the same number; all are equal. Matrices can also be written in different ways. Matrices **B** and **E** of Example 2, for example, are both 2×1 matrices and their corresponding entries are equal. Matrices like these are called *equal matrices*. The matrices **B** and **F** of Example 2 list the same numbers but are not equal, because **B** is 2×1 and **F** is 1×2.

Definition 6.3 Suppose $A = (a_{ij})$ and $B = (b_{ij})$ are two $m \times n$ matrices. Then **A** and **B** are **equal matrices** if and only if $a_{ij} = b_{ij}$ for each $i = 1, 2, \ldots, m$ and $j = 1, 2, \ldots, n$.

In other words, $\mathbf{A} = \mathbf{B}$ if the entry in the ith row, jth column of $\mathbf{A}$ is the same as the entry in the ith row, jth column of $\mathbf{B}$ for all values of i and j. Note that $\mathbf{A}$ and $\mathbf{B}$ cannot be equal unless they have the same number of rows and columns.

Example 4   Find x if—

(a) $\begin{pmatrix} 1 & 2 \\ 4 & -1 \end{pmatrix} = \begin{pmatrix} 1 & 2 \\ 4 & x \end{pmatrix}$

(b) $\begin{pmatrix} 1 & 2 \\ 4 & -1 \end{pmatrix} = \begin{pmatrix} 1 & 2 \\ 2x & -1 \end{pmatrix}$

Solution

(a) Since the two matrices are equal, corresponding entries must be equal. It follows that $x = -1$.

(b) Since $2x = 4$, $x = 2$.

Example 5   Find x, y, and z if

$$\begin{pmatrix} x & y - x \\ x + z & 3 \end{pmatrix} = \begin{pmatrix} 0 & 1 \\ 2 & 3 \end{pmatrix}$$

Solution   Equating corresponding entries yields

$$x = 0 \tag{2}$$

$$y - x = 1 \tag{3}$$

$$x + z = 2 \tag{4}$$

Substituting equation (2) into equations (3) and (4) yields $y = 1$ and $z = 2$. It follows that $x = 0$, $y = 1$, and $z = 2$.

Example 6   Let

$$\mathbf{A} = (a_{ij}) = \begin{pmatrix} 1 & 2 & 3 \\ 3 & 1 & -1 \end{pmatrix} \quad \text{and} \quad \mathbf{B} = (b_{ij}) = \begin{pmatrix} 0 & 1 & -1 \\ 2 & -1 & 3 \end{pmatrix}$$

Find—

(a) $(a_{ij} + 1)$ **(b)** $(2a_{ij})$ **(c)** $(a_{ij} + b_{ij})$

Solution

(a) Matrix $(a_{ij} + 1)$ is the matrix whose entry in the ith row, jth column is $a_{ij} + 1$; that is, one more than the entry in the ith row, jth column of $\mathbf{A}$. In other words, we must add 1 to each entry of $\mathbf{A}$. Hence,

$$(a_{ij} + 1) = \begin{pmatrix} 1 + 1 & 2 + 1 & 3 + 1 \\ 3 + 1 & 1 + 1 & -1 + 1 \end{pmatrix} = \begin{pmatrix} 2 & 3 & 4 \\ 4 & 2 & 0 \end{pmatrix}$$

(b) The matrix $(2a_{ij})$ is formed by multiplying each entry of **A** by 2.

$$(2a_{ij}) = \begin{pmatrix} 2 \cdot 1 & 2 \cdot 2 & 2 \cdot 3 \\ 2 \cdot 3 & 2 \cdot 1 & 2(-1) \end{pmatrix} = \begin{pmatrix} 2 & 4 & 6 \\ 6 & 2 & -2 \end{pmatrix}$$

(c) The matrix $(a_{ij} + b_{ij})$ is formed by adding each entry of **A** to the corresponding entry of **B**.

$$(a_{ij} + b_{ij}) = \begin{pmatrix} 1+0 & 2+1 & 3-1 \\ 3+2 & 1-1 & -1+3 \end{pmatrix} = \begin{pmatrix} 1 & 3 & 2 \\ 5 & 0 & 2 \end{pmatrix}$$

6.1 EXERCISES

1.   Which of the following matrices are equal?

$$\mathbf{A} = (1,3,5,7), \qquad \mathbf{B} = \begin{pmatrix} 1 & 3 \\ 5 & 7 \end{pmatrix}, \qquad \mathbf{C} = \begin{pmatrix} 1 \\ 3 \\ 5 \\ 7 \end{pmatrix}$$

$$\mathbf{D} = \begin{pmatrix} 3 & 1 \\ 5 & 7 \end{pmatrix}, \qquad \mathbf{E} = (1,3,5,7), \qquad \mathbf{F} = \begin{pmatrix} 1.0 & \frac{3}{1} \\ 5.0 & 7 \end{pmatrix}$$

2.   Which of the matrices of Exercise 1 are (a) vectors, (b) column vectors, (c) square, (d) 1×4, (e) 4×1?

3.   Let

$$\mathbf{A} = (a_{ij}) = \begin{pmatrix} 1 & 2 & -1 & -5 \\ 3 & 4 & 0 & -8 \\ 4 & 1 & 0 & 5 \end{pmatrix}$$

Find—
(a) a_{11} **(b)** a_{13} **(c)** a_{31} **(d)** a_{34} **(e)** a_{43} **(f)** a_{33}

4.   Let

$$\mathbf{A} = (a_{ij}) = \begin{pmatrix} 1 & 2 & 3 \\ 4 & 0 & 1 \end{pmatrix} \quad \text{and} \quad \mathbf{B} = (b_{ij}) = \begin{pmatrix} 0 & 1 & 3 \\ 2 & 0 & 1 \end{pmatrix}$$

Find—
(a) $(a_{ij} + 2)$ **(b)** $(2a_{ij})$ **(c)** $(3a_{ij} - 5)$ **(d)** $(a_{ij} + b_{ij})$
(e) $(2a_{ij} - 3b_{ij})$

5.   Let $\mathbf{A} = (a_{ij})$ be a 2×3 matrix. Find **A** if—
(a) $a_{ij} = 1$; $i = 1, 2, j = 1, 2, 3$ (i.e., every entry is 1)
(b) $a_{ij} = i + j$ (e.g., $a_{12} = 1 + 2$, $a_{23} = 2 + 3$)
(c) $a_{ij} = i$ (e.g., $a_{23} = 2$, $a_{12} = 1$).

6.   The snack bar in a theater lobby sold three popsicles, ten candy bars, and fifteen sodas during the first feature, and six popsicles, five candy bars, and twenty sodas during the second feature. Write a matrix that displays this information.

7. Find x and y if—

(a) $\begin{pmatrix} 1 & 2 & 3 \\ 4 & 5 & 6 \end{pmatrix} = \begin{pmatrix} x & 2 & 3 \\ 4 & 5 & y \end{pmatrix}$

(b) $\begin{pmatrix} 1 & 2 & 3 \\ 4 & 5 & 6 \end{pmatrix} = \begin{pmatrix} x & 2 & y - x \\ 4 & 5 & 6 \end{pmatrix}$

8. Write the answers to Exercise 7 as row vectors (x,y).

9. Let S, C, E, M, and W denote salespersons, clerks, executives, men, and women, respectively. The following matrix gives the number of employees of a certain firm by sex and position.

$$\begin{array}{c} \\ M \\ W \end{array} \begin{array}{ccc} S & C & E \\ \begin{pmatrix} 20 & 10 & 5 \\ 30 & 25 & 10 \end{pmatrix} \end{array}$$

How many employees of the firm are (a) male executives, (b) female clerks, (c) men, (d) executives?

10. Consider the firm of Exercise 9.
(a) What percentage of the men are executives?
(b) What percentage of the women are executives?
(c) What percentage of the employees are clerks?
(d) What percentage of the salespersons are women?
(e) What percentage of the employees are women?
(f) What percentage of the employees are male salespersons?

6.2 MATRIX ADDITION AND SCALAR MULTIPLICATION

There are three operations which are defined for matrices: matrix addition, matrix multiplication, and scalar multiplication. Addition and scalar multiplication will be studied in this section; matrix multiplication will be studied in Section 6.3. We will see that addition of matrices has many of the same properties as addition of numbers.

Example 7 Let T, R, W, J, and F denote television sets, refrigerators, washing machines, January, and February, respectively. The number of items sold by the two branches of the EZ Kredit Company last year are given in the following matrices:

$$\mathbf{A} = \begin{array}{c} \\ J \\ F \end{array} \begin{array}{ccc} T & R & W \\ \begin{pmatrix} 14 & 12 & 13 \\ 13 & 11 & 14 \end{pmatrix} \end{array} \qquad \mathbf{B} = \begin{array}{c} \\ J \\ F \end{array} \begin{array}{ccc} T & R & W \\ \begin{pmatrix} 15 & 13 & 12 \\ 16 & 12 & 10 \end{pmatrix} \end{array}$$

Branch A   Branch B

For example, branch A sold thirteen washing machines in January and eleven refrigerators in February. The total number of units of each item sold by the firm each month can be found by adding each entry of **A** to the corresponding entry of **B**. These sales are given in the following matrix:

$$\mathbf{C} = \begin{pmatrix} 14 + 15 & 12 + 13 & 13 + 12 \\ 13 + 16 & 11 + 12 & 14 + 10 \end{pmatrix} = \begin{pmatrix} 29 & 25 & 25 \\ 29 & 23 & 24 \end{pmatrix}$$

Total sales

The entry in the first row, first column of **C** is 29, for example, because branch A sold fourteen television sets in January and branch B sold fifteen.

Example 8 The EZ Kredit Company expects this year's sales at branch A to be twice last year's. (See Example 7.) Therefore, branch A's expected sales can be found by multiplying each entry of matrix **A** of Example 7 by 2. This yields the matrix

$$\mathbf{D} = \begin{pmatrix} 2 \cdot 14 & 2 \cdot 12 & 2 \cdot 13 \\ 2 \cdot 13 & 2 \cdot 11 & 2 \cdot 14 \end{pmatrix} = \begin{pmatrix} 28 & 24 & 26 \\ 26 & 22 & 28 \end{pmatrix}$$

Expected sales at Branch A

We saw in Examples 7 and 8 that it is often necessary to form a new matrix by adding corresponding entries of two matrices or by multiplying each entry of a matrix by a real number (a scalar). The first operation is called *matrix addition*; the second is called *scalar multiplication*.

Definition 6.4 **(matrix addition)** Let $\mathbf{A} = (a_{ij})$ and $\mathbf{B} = (b_{ij})$ be two $m \times n$ matrices. The *sum* $\mathbf{A} + \mathbf{B}$ of **A** and **B** is the matrix $(a_{ij} + b_{ij})$.

In other words, two matrices are added by adding their corresponding entries. Symbolically, if (a_{ij}) and (b_{ij}) are both $m \times n$ matrices, then

$$(a_{ij}) + (b_{ij}) = (a_{ij} + b_{ij})$$

Note that matrix addition is not defined if the matrices to be added do not have the same numbers of rows and columns.

Definition 6.5 **(scalar multiplication)** Let $\mathbf{A} = (a_{ij})$ be a matrix and k be a real number. The *scalar product* $k\mathbf{A}$ of k and **A** is the matrix (ka_{ij}).

In other words, a matrix **A** is multiplied by a real number k by multiplying each entry of **A** by k. Symbolically,

$$k(a_{ij}) = (ka_{ij})$$

The operation defined in Definition 6.5 is called scalar multiplication because it defines the product of a scalar (real number) with a matrix. *Matrix multiplication* (finding the product of two matrices) will be defined in Section 6.3.

Example 9 illustrates Definitions 6.4 and 6.5.

Example 9   Let

$$A = \begin{pmatrix} 1 & 2 \\ 0 & 1 \\ 2 & 3 \end{pmatrix}, \quad B = \begin{pmatrix} 1 & 1 \\ 0 & 1 \\ -1 & 0 \end{pmatrix}, \quad C = \begin{pmatrix} 1 & 0 \\ 2 & 1 \end{pmatrix}$$

$$D = \begin{pmatrix} 0 & 0 \\ 0 & 0 \\ 0 & 0 \end{pmatrix}, \quad \text{and} \quad k = 3$$

If possible, calculate—

(a) $A + B$ (b) $B + A$ (c) $B + C$ (d) $A + D$ (e) kC

(f) $2A + 5B$ (g) $2A + (-5)B$

Solution

(a)
$$A + B = \begin{pmatrix} 1 & 2 \\ 0 & 1 \\ 2 & 3 \end{pmatrix} + \begin{pmatrix} 1 & 1 \\ 0 & 1 \\ -1 & 0 \end{pmatrix} = \begin{pmatrix} 1+1 & 2+1 \\ 0+0 & 1+1 \\ 2-1 & 3+0 \end{pmatrix} = \begin{pmatrix} 2 & 3 \\ 0 & 2 \\ 1 & 3 \end{pmatrix}$$

· (b)
$$B + A = \begin{pmatrix} 1 & 1 \\ 0 & 1 \\ -1 & 0 \end{pmatrix} + \begin{pmatrix} 1 & 2 \\ 0 & 1 \\ 2 & 3 \end{pmatrix} = \begin{pmatrix} 1+1 & 1+2 \\ 0+0 & 1+1 \\ -1+2 & 0+3 \end{pmatrix}$$

$$= \begin{pmatrix} 2 & 3 \\ 0 & 2 \\ 1 & 3 \end{pmatrix} = A + B$$

(c) The sum $B + C$ is not defined because B and C do not have the same number of rows.

(d)
$$A + D = \begin{pmatrix} 1+0 & 2+0 \\ 0+0 & 1+0 \\ 2+0 & 3+0 \end{pmatrix} = \begin{pmatrix} 1 & 2 \\ 0 & 1 \\ 2 & 3 \end{pmatrix} = A$$

(e)
$$kC = 3\begin{pmatrix} 1 & 0 \\ 2 & 1 \end{pmatrix} = \begin{pmatrix} 3 \cdot 1 & 3 \cdot 0 \\ 3 \cdot 2 & 3 \cdot 1 \end{pmatrix} = \begin{pmatrix} 3 & 0 \\ 6 & 3 \end{pmatrix}$$

(f)
$$2A + 5B = 2\begin{pmatrix} 1 & 2 \\ 0 & 1 \\ 2 & 3 \end{pmatrix} + 5\begin{pmatrix} 1 & 1 \\ 0 & 1 \\ -1 & 0 \end{pmatrix}$$

$$= \begin{pmatrix} 2 & 4 \\ 0 & 2 \\ 4 & 6 \end{pmatrix} + \begin{pmatrix} 5 & 5 \\ 0 & 5 \\ -5 & 0 \end{pmatrix} = \begin{pmatrix} 7 & 9 \\ 0 & 7 \\ -1 & 6 \end{pmatrix}$$

(g)
$$2A + (-5)B = \begin{pmatrix} 2 & 4 \\ 0 & 2 \\ 4 & 6 \end{pmatrix} + \begin{pmatrix} -5 & -5 \\ 0 & -5 \\ 5 & 0 \end{pmatrix} = \begin{pmatrix} -3 & -1 \\ 0 & -3 \\ 9 & 6 \end{pmatrix}$$

Every entry of the matrix D of Example 9 is 0. Matrix D is therefore called a *zero matrix*. When D is added to the matrix A in Example 9(d), the result is A. A similar proof shows that this is true for every matrix A and every zero matrix Φ, as long as A and Φ have the same number of rows and columns.

Definition 6.6 A **zero matrix**, denoted Φ, is a matrix that has 0 for every entry.

The matrices

$$\begin{pmatrix} 0 & 0 & 0 \\ 0 & 0 & 0 \end{pmatrix}, \quad \begin{pmatrix} 0 \\ 0 \end{pmatrix}, \quad \text{and} \quad (0,0)$$

are all zero matrices.

THEOREM 6.1 Let $\mathbf{A}$ be an $m \times n$ matrix and Φ be an $m \times n$ zero matrix. Then

$$\mathbf{A} + \Phi = \mathbf{A} \tag{5}$$

Proof

$$\mathbf{A} + \Phi = (a_{ij}) + (0) = (a_{ij} + 0) = (a_{ij}) = \mathbf{A}$$

Note that the sum in equation (5) is not defined unless $\mathbf{A}$ and Φ have the same number of rows and columns. This requirement was stated in Theorem 6.1. It will be assumed in the remainder of the book.

Consider equation (5). If we replace the matrix $\mathbf{A}$ by the number a and the matrix Φ by the number 0, equation (5) becomes the arithmetic law $a + 0 = a$. In other words, the zero matrix Φ acts in matrix addition like the number 0 in the addition of numbers. Certain other matrix laws can also be remembered by using analogies with ordinary arithmetic. In Table 6.2, arithmetic laws for real numbers appear in the second column. If we replace numbers by matrices and 0 by Φ, we obtain the matrix laws in the right column. Note that the matrices must be chosen so that matrix addition is defined.

Table 6.2

Property	Arithmetic Law	Matric Law A, B, C and Φ are $m \times n$ matrices
Commutativity	$a + b = b + a$	$A + B = B + A$
Associativity	$(a + b) + c = a + (b + c)$	$(A + B) + C = A + (B + C)$
Existence of an identity	$a + 0 = a$	$A + \Phi = A$

Suppose $\mathbf{A}$, $\mathbf{B}$, and $\mathbf{C}$ are $m \times n$ matrices. Since $(\mathbf{A} + \mathbf{B}) + \mathbf{C} = \mathbf{A} + (\mathbf{B} + \mathbf{C})$, we may drop the parentheses and define the sum $\mathbf{A} + \mathbf{B} + \mathbf{C}$ of $\mathbf{A}$, $\mathbf{B}$, and $\mathbf{C}$ to be either $(\mathbf{A} + \mathbf{B}) + \mathbf{C}$ or $\mathbf{A} + (\mathbf{B} + \mathbf{C})$. This is analogous to the sum of three numbers. For example, if

$$\mathbf{A} = \begin{pmatrix} 1 & 1 \\ 2 & 0 \end{pmatrix}, \quad \mathbf{B} = \begin{pmatrix} 1 & -1 \\ -1 & 2 \end{pmatrix}, \quad \text{and} \quad \mathbf{C} = \begin{pmatrix} 2 & -1 \\ 1 & 3 \end{pmatrix}$$

then

$$\mathbf{A} + \mathbf{B} + \mathbf{C} = (\mathbf{A} + \mathbf{B}) + \mathbf{C} = \begin{pmatrix} 2 & 0 \\ 1 & 2 \end{pmatrix} + \begin{pmatrix} 2 & -1 \\ 1 & 3 \end{pmatrix} = \begin{pmatrix} 4 & -1 \\ 2 & 5 \end{pmatrix}$$

A similar definition holds for the sum of four or more matrices.

Addition properties of matrices are given in the following theorem.

THEOREM   6.2   Let $\mathbf{A} = (a_{ij})$ be any matrix and k be any real number. Then—

(a) $0 \cdot \mathbf{A} = \Phi$

(b) $k\Phi = \Phi$

Proof   We will prove statement (a). Statement (b) can be proved similarly. Using the definition of scalar multiplication,

$$0 \cdot \mathbf{A} = 0 \cdot (a_{ij}) = (0 \cdot a_{ij}) = (0) = \Phi$$

Note that Theorem 6.2(a) states that the product of the *number* 0 and the matrix $\mathbf{A}$ is the *matrix* Φ.

The difference of two matrices is defined exactly like the difference of two numbers.

Definition 6.7   Let $\mathbf{A}$ and $\mathbf{B}$ be two $m \times n$ matrices. The *difference* $\mathbf{A}$ *minus* $\mathbf{B}$ is defined by

$$\mathbf{A} - \mathbf{B} = \mathbf{A} + (-1)\mathbf{B}$$

Example 10 shows how to apply Definition 6.7 and Theorems 6.1 and 6.2 to particular calculations.

Example 10   Let

$$\mathbf{A} = \begin{pmatrix} 1 & 2 & 3 \\ 4 & 2 & 0 \end{pmatrix} \quad \text{and} \quad \mathbf{B} = \begin{pmatrix} 2 & 3 & 1 \\ 4 & 5 & 1 \end{pmatrix}$$

Calculate—

(a) $\mathbf{A} + \Phi$ (b) $\mathbf{A} - \mathbf{B}$ (c) $\mathbf{B} - \mathbf{A}$ (d) $\mathbf{A} - \Phi$

(e) $0 \cdot \mathbf{A}$ (f) 2Φ (g) $3\mathbf{A} - 2\mathbf{B}$

Solution

(a) We do not have to specify which zero matrix is meant, because $\mathbf{A} + \Phi$ is not defined unless Φ and $\mathbf{A}$ have the same number of rows and columns. Hence, in this problem, Φ is a 2×3 matrix. Using Theorem 6.1, $\mathbf{A} + \Phi = \mathbf{A}$.

(b)
$$\mathbf{A} - \mathbf{B} = \mathbf{A} + (-1)\mathbf{B} = \begin{pmatrix} 1 & 2 & 3 \\ 4 & 2 & 0 \end{pmatrix} + \begin{pmatrix} -2 & -3 & -1 \\ -4 & -5 & -1 \end{pmatrix}$$

$$= \begin{pmatrix} 1-2 & 2-3 & 3-1 \\ 4-4 & 2-5 & 0-1 \end{pmatrix} = \begin{pmatrix} -1 & -1 & 2 \\ 0 & -3 & -1 \end{pmatrix}$$

Note that $\mathbf{A} - \mathbf{B}$ is formed by subtracting every entry of $\mathbf{B}$ from the corresponding entry of $\mathbf{A}$.

(c)
$$\mathbf{B} - \mathbf{A} = \begin{pmatrix} 2-1 & 3-2 & 1-3 \\ 4-4 & 5-2 & 1-0 \end{pmatrix} = \begin{pmatrix} 1 & 1 & -2 \\ 0 & 3 & 1 \end{pmatrix}$$

(d) $\mathbf{A} - \Phi = \mathbf{A} + (-1)\Phi = \mathbf{A} + \Phi = \mathbf{A}$

(e) Applying Theorem 6.2(a), $0 \cdot \mathbf{A} = \Phi$.

(f) Applying Theorem 6.2(b), $2\Phi = \Phi$.

(g)
$$3\mathbf{A} - 2\mathbf{B} = 3\begin{pmatrix} 1 & 2 & 3 \\ 4 & 2 & 0 \end{pmatrix} - 2\begin{pmatrix} 2 & 3 & 1 \\ 4 & 5 & 1 \end{pmatrix}$$
$$= \begin{pmatrix} 3 & 6 & 9 \\ 12 & 6 & 0 \end{pmatrix} - \begin{pmatrix} 4 & 6 & 2 \\ 8 & 10 & 2 \end{pmatrix}$$
$$= \begin{pmatrix} -1 & 0 & 7 \\ 4 & -4 & -2 \end{pmatrix}$$

6.2 EXERCISES

1. Let

$$\mathbf{A} = \begin{pmatrix} 1 & 7 & 5 \\ 2 & 0 & 3 \end{pmatrix}, \qquad \mathbf{B} = \begin{pmatrix} 4 & 7 & 0 \\ 1 & 5 & 6 \end{pmatrix}, \qquad \mathbf{C} = \begin{pmatrix} 0 & -5 \\ 2 & 3 \end{pmatrix},$$

and

$$\mathbf{D} = \begin{pmatrix} 1 & -1 & 0 \\ -2 & 4 & 2 \end{pmatrix}$$

If possible, calculate—

(a) $\mathbf{A} + \mathbf{B}$ **(b)** $\mathbf{B} + \mathbf{A}$ **(c)** $\mathbf{A} + \mathbf{C}$ **(d)** $(-1)\mathbf{C}$

(e) $\mathbf{A} - \mathbf{B}$ **(f)** $\mathbf{C} - \mathbf{A}$ **(g)** $\mathbf{A} + \mathbf{B} + \mathbf{D}$ **(h)** $\mathbf{D} + \mathbf{A} + \mathbf{B}$

(i) $\mathbf{A} + \mathbf{B} + \mathbf{A}$ **(j)** $\mathbf{B} + 2\mathbf{A}$

2. Let

$$\mathbf{A} = \begin{pmatrix} 2 & 4 & 6 \\ 1 & 0 & 4 \end{pmatrix}, \qquad \mathbf{B} = \begin{pmatrix} 3 & 5 & 9 \\ 0 & -4 & 1 \end{pmatrix},$$

and

$$\mathbf{C} = \begin{pmatrix} 2 & 4 & 0 \\ 0 & 3 & 1 \end{pmatrix}$$

Show that—

(a) $\mathbf{A} + \mathbf{B} = \mathbf{B} + \mathbf{A}$ (Hint: Calculate $\mathbf{A} + \mathbf{B}$ and $\mathbf{B} + \mathbf{A}$ and show that they are equal.)

(b) $\mathbf{A} + (\mathbf{B} + \mathbf{C}) = (\mathbf{A} + \mathbf{B}) + \mathbf{C}$

3.   Let Φ be a 2×5 zero matrix and let k be a real number.
(a) Write out Φ.
(b) Show that $k\Phi = \Phi$.

4.   Prove Theorem 6.2(b)

5.   Calculate the following, if possible. If the calculation is not possible, explain why.

(a) $(1,2,3) + 4(2,1,5)$

(b) $2(3,5,7) - 4(1,5,6)$

(c) $\begin{pmatrix} 1 & 0 & 1 \\ 2 & 4 & 8 \end{pmatrix} + \Phi$

(d) $\begin{pmatrix} 1 & 2 & -1 \\ 3 & 4 & 7 \end{pmatrix} - \begin{pmatrix} 2 & 1 \\ 3 & 5 \end{pmatrix}$

(e) $3\Phi - \Phi$

6.   Find $\mathbf{X}$ if $(2,3,5) + \mathbf{X} = (1,5,9)$.

7.   Find x and y such that—

(a) $2\begin{pmatrix} 2 & y \\ 1 & 3 \end{pmatrix} = \begin{pmatrix} 4 & -2 \\ 2 & -6x \end{pmatrix}$

(b) $\begin{pmatrix} 2x & 1 & 2 \\ 1 & 0 & y \end{pmatrix} + \begin{pmatrix} 1 & 0 & 1 \\ 2 & 4 & 1 \end{pmatrix} = \begin{pmatrix} 5 & 1 & 3 \\ 3 & 4 & 0 \end{pmatrix}$

(c) $\begin{pmatrix} x & y \\ y & x \end{pmatrix} + \Phi = \begin{pmatrix} 2 & 4 \\ 4 & 2 \end{pmatrix}$

Write each answer as a vector (x,y).

8.   Complete the following equations.
(a) $\mathbf{A} - \mathbf{A} = $ ____   (b) $\mathbf{A} - \Phi = $ ____   (c) $\Phi - \mathbf{A} = $ ____

9.   The costs per house (in dollars) of material (M) and labor (L) for planting trees (T), bushes (B), and grass (G) at two different locations are given in the following matrices:

$$
\begin{array}{c}
\begin{array}{ccc} T & B & G \end{array} \\
\begin{array}{c} M \\ L \end{array} \begin{pmatrix} 200 & 100 & 50 \\ 150 & 100 & 25 \end{pmatrix}
\end{array}
\qquad
\begin{array}{c}
\begin{array}{ccc} T & B & G \end{array} \\
\begin{array}{c} M \\ L \end{array} \begin{pmatrix} 100 & 150 & 80 \\ 80 & 150 & 40 \end{pmatrix}
\end{array}
$$

Location A   Location B

(a) Write a single matrix that represents the total itemized costs of landscaping one house at location A and one at location B.

(b) A builder has five houses at location A and two at location B. Display the itemized cost of landscaping all seven houses in matrix form.

(c) What is the total cost of the materials needed to landscape a house at location A?

10.   The monthly production (in numbers of cases) of screws (S), nuts (N), bolts (B), and washers (W) at factories X, Y, and Z are given in the following matrices:

$$
\begin{array}{c}
\begin{array}{cccc} S & N & B & W \end{array} \\
\begin{array}{c} X \\ Y \\ Z \end{array}
\begin{pmatrix}
1000 & 5000 & 5000 & 4000 \\
2000 & 4000 & 4000 & 6000 \\
500 & 1000 & 2000 & 1000
\end{pmatrix}
\end{array}
\qquad
\begin{array}{c}
\begin{array}{cccc} S & N & B & W \end{array} \\
\begin{array}{c} X \\ Y \\ Z \end{array}
\begin{pmatrix}
1500 & 6000 & 5500 & 3500 \\
2400 & 4800 & 4200 & 7000 \\
1000 & 800 & 1600 & 1200
\end{pmatrix}
\end{array}
$$

January   February

Write a matrix that displays the increase in production from January to February. (An increase of -500 means a decrease of 500.)

6.3 MATRIX MULTIPLICATION

In Section 6.2 the product of a real number and a matrix (scalar multiplication) was defined. In this section we will define the product of two matrices (matrix multiplication). The following example will illustrate the need for matrix multiplication.

Example 11   A bakery produces three types of rolls—kaiser rolls (K), bagels (B), and twists (T)—in two sizes, small (S) and large (L). The cost in dollars of producing a dozen rolls is given in the vector **C**. Yesterday's production of rolls (in dozens) is given in matrix **A**.

$$
\mathbf{C} = \begin{array}{c} \begin{array}{cc} S & L \end{array} \\ (0.50,\ 0.75), \end{array}
\qquad
\mathbf{A} = \begin{array}{c} \begin{array}{ccc} K & B & T \end{array} \\ \begin{array}{c} S \\ L \end{array} \begin{pmatrix} 10 & 20 & 20 \\ 40 & 40 & 20 \end{pmatrix} \end{array}
$$

Find the total cost in dollars of producing each of the three types of rolls.

Solution   From the two matrices, we see that the bakery produced 10-dozen small kaiser rolls at a cost of 50 cents a dozen and 40-dozen large kaiser rolls at a cost of 75 cents a dozen. The total cost of producing the kaiser rolls is therefore

$$(0.50)(10) + (0.75)(40) = 35 \text{ dollars} \tag{6}$$

Similarly, the total cost of producing the bagels is

$$(0.50)(20) + (0.75)(40) = 40 \text{ dollars} \tag{7}$$

and the total cost of producing the twists is

$$(0.50)(20) + (0.75)(20) = 25 \text{ dollars} \tag{8}$$

It follows from equations (6)–(8) that the total cost in dollars of producing each type of roll is given in matrix **D**:

$$\mathbf{D} = \begin{array}{c} \begin{array}{ccc} K & B & T \end{array} \\ (35, 40, 25) \end{array}$$

Let us review how matrix $\mathbf{D}$ of Example 11 was calculated. This will lead us to a definition of matrix multiplication. Suppose we write

$$\mathbf{C} \quad \times \quad \mathbf{A} \quad = \quad \mathbf{D}$$

$$(0.50, 0.75)\begin{pmatrix} 10 & 20 & 20 \\ 40 & 40 & 20 \end{pmatrix} = (35, 40, 25)$$

and carefully consider equation (6):

$$(0.50)(10) + (0.75)(40) = 35$$

We see that the entry in the *first row, first column* of $\mathbf{D}$ (35) was found by multiplying each entry in the *first row* of $\mathbf{C}$ (0.50, 0.75) by the corresponding entry in the *first column* of $\mathbf{A}$ (10, 40), and then adding the products.

Similarly, from equation (7) we see that the entry in the *first row, second column* of $\mathbf{D}$ was formed by multiplying the entries in the *first row* of $\mathbf{C}$ by the corresponding entries in the *second column* of $\mathbf{A}$, and then adding the products:

$$\mathbf{C} \quad \times \quad \mathbf{A} \quad = \quad \mathbf{D}$$

$$(0.50, 0.75)\begin{pmatrix} 10 & 20 & 20 \\ 40 & 40 & 20 \end{pmatrix} = (35, 40, 25)$$

$$(0.50)(20) + (0.75)(40) = 40$$

From equation (8) we see that the entry in the *first row, third column* of $\mathbf{D}$ was formed by multiplying the entries in the *first row* of $\mathbf{C}$ by the corresponding entries in the *third column* of $\mathbf{A}$, and then adding the products:

$$\mathbf{C} \quad \times \quad \mathbf{A} \quad = \quad \mathbf{D}$$

$$(0.50, 0.75)\begin{pmatrix} 10 & 20 & 20 \\ 40 & 40 & 20 \end{pmatrix} = (35, 40, 25)$$

$$(0.50)(20) + (0.75)(20) = 25$$

In other words, the entry in the *ith row, jth column* of $\mathbf{D}$ was found by multiplying each entry of the *ith row* of $\mathbf{C}$ by the corresponding entry of the *jth column* of $\mathbf{A}$ and then adding the products. This suggests the following definition.

Definition 6.8   (matrix multiplication)   Let $\mathbf{A} = (a_{ij})$ be an $m \times n$ matrix and $\mathbf{B} = (b_{ij})$ be an $n \times s$ matrix. The *matrix product* $\mathbf{C} = (c_{ij})$ of $\mathbf{A}$ and $\mathbf{B}$, denoted $\mathbf{AB}$, is an $m \times s$ matrix. The entry c_{ij} in the *ith row, jth column* of $\mathbf{C}$ is obtained by multiplying the entries of the *ith row* of $\mathbf{A}$ by the corresponding entries of the *jth column* of $\mathbf{B}$ and then adding the products. That is,

$$\mathbf{AB} = \begin{pmatrix} a_{11} & a_{12} & \cdots & a_{1n} \\ \vdots & \vdots & & \vdots \\ a_{i1} & a_{i2} & \cdots & a_{in} \\ \vdots & \vdots & & \vdots \\ a_{m1} & a_{m2} & \cdots & a_{mn} \end{pmatrix} \begin{pmatrix} b_{11} & \cdots & b_{1j} & \cdots & b_{1s} \\ \vdots & & \vdots & & \vdots \\ b_{i1} & \cdots & b_{ij} & \cdots & b_{is} \\ \vdots & & \vdots & & \vdots \\ b_{n1} & \cdots & b_{nj} & \cdots & b_{ns} \end{pmatrix} = \begin{pmatrix} c_{11} & \cdots & c_{1j} & \cdots & c_{1s} \\ \vdots & & \vdots & & \vdots \\ c_{i1} & \cdots & c_{ij} & \cdots & c_{is} \\ \vdots & & \vdots & & \vdots \\ c_{m1} & \cdots & c_{mj} & \cdots & c_{ms} \end{pmatrix}$$

where

$$c_{ij} = a_{i1}b_{1j} + a_{i2}b_{2j} + \cdots + a_{in}b_{nj}$$

Note that the product **AB** *is not defined unless the number of columns of the first matrix* **A** *equals the number of rows of the second matrix* **B**. If **A** is a 2 × 3 matrix and **AB** is defined, **B** must have **3** rows. If **BA** is defined, **B** must have 2 columns.

Suppose **A** is any matrix. The product **AA** is not defined unless the number of columns of **A** equals the number of rows of **A**; that is, unless **A** is a square matrix. If **A** is a square matrix, the product **AA** is denoted A^2. In general, the *powers* of **A** are defined as follows:

$$A^3 = A^2A, \qquad A^4 = A^3A, \ldots$$

It can be shown that $A^iA^j = A^{i+j} = A^jA^i$. It follows, for example, that

$$A^4 = A^2A^2 = AA^3 = A^3A$$

Examples 12–15 illustrate matrix multiplication.

Example 12   Let

$$A = \begin{pmatrix} 1 & 2 & 3 \\ 2 & 1 & 4 \end{pmatrix} \quad \text{and} \quad B = \begin{pmatrix} 1 & 2 \\ 2 & 3 \end{pmatrix}$$

Which of the following products are defined?

(a) **AB**   (b) **BA**   (c) **AA**   (d) **BB**

Solution

(a) The product **AB** is not defined unless the number of columns of the first matrix, **A**, equals the number of rows of the second matrix, **B**. Therefore, **AB** is not defined, because **A** has 3 columns but **B** has only 2 rows.

(b) The product **BA** is defined because the number of columns of the first matrix, **B**, equals the number of rows of the second matrix, **A**. (Matrix **B** has 2 columns, and **A** has 2 rows.)

(c) The product **AA** is not defined because **A** has 3 columns but only 2 rows.

(d) The product **BB** is defined because **B** has 2 columns and 2 rows.

Example 13   Let **A** and **B** be the matrices of Example 12. Calculate—

(a) **BA**   (b) B^2

Solution

(a) Since **B** is a 2 × 2 matrix and **A** is a 2 × 3 matrix, the product **BA** is a 2 × 3 matrix $C = (c_{ij})$.

The entry c_{11} in the *first row, first column* of **C** is found by multiplying the entries in the *first row* of **B** by the corresponding entries in the *first column* of **A** and then adding the products:

$$\begin{pmatrix} 1 & 2 \\ 2 & 3 \end{pmatrix}\begin{pmatrix} 1 & 2 & 3 \\ 2 & 1 & 4 \end{pmatrix} = \begin{pmatrix} 1(1) + 2(2) & x & x \\ x & x & x \end{pmatrix} = \begin{pmatrix} 5 & x & x \\ x & x & x \end{pmatrix}$$

(The x's in the above matrices denote the entries which are not being discussed. These entries may not all be the same.)

The entry c_{12} in the *first row, second column* of **C** is found by multiplying the entries in the *first row* of **B** by the corresponding entries of the *second column* of **A** and adding the products:

$$\begin{pmatrix} 1 & 2 \\ 2 & 3 \end{pmatrix}\begin{pmatrix} 1 & 2 & 3 \\ 2 & 1 & 4 \end{pmatrix} = \begin{pmatrix} x & 1(2)+2(1) & x \\ x & x & x \end{pmatrix} = \begin{pmatrix} x & 4 & x \\ x & x & x \end{pmatrix}$$

The remaining entries of **C** are calculated similarly.

$$\begin{pmatrix} 1 & 2 \\ 2 & 3 \end{pmatrix}\begin{pmatrix} 1 & 2 & 3 \\ 2 & 1 & 4 \end{pmatrix} = \begin{pmatrix} 5 & 4 & 1(3)+2(4) \\ 2(1)+3(2) & 2(2)+3(1) & 2(3)+3(4) \end{pmatrix}$$

$$= \begin{pmatrix} 5 & 4 & 11 \\ 8 & 7 & 18 \end{pmatrix}$$

(b) The product $\mathbf{B}^2 = \mathbf{BB}$ is 2×2 because **B** is $\mathbf{2 \times 2}$.

$$\mathbf{B}^2 = \begin{pmatrix} 1 & 2 \\ 2 & 3 \end{pmatrix}\begin{pmatrix} 1 & 2 \\ 2 & 3 \end{pmatrix} = \begin{pmatrix} 1(1)+2(2) & 1(2)+2(3) \\ 2(1)+3(2) & 2(2)+3(3) \end{pmatrix} = \begin{pmatrix} 5 & 8 \\ 8 & 13 \end{pmatrix}$$

Example 14

(a) Let

$$\mathbf{A} = \begin{pmatrix} 1 & 2 & 0 \\ 2 & 1 & 3 \end{pmatrix} \quad \text{and} \quad \mathbf{B} = \begin{pmatrix} 0 & 1 & 2 \\ -1 & 0 & 1 \\ 2 & 1 & 0 \end{pmatrix}$$

Then

$$\mathbf{AB} = \begin{pmatrix} 1 & 2 & 0 \\ 2 & 1 & 3 \end{pmatrix}\begin{pmatrix} 0 & 1 & 2 \\ -1 & 0 & 1 \\ 2 & 1 & 0 \end{pmatrix} = \begin{pmatrix} -2 & 1 & 4 \\ 5 & 5 & 5 \end{pmatrix}$$

However, **BA** is not defined, because **B** has 3 columns while **A** has only 2 rows. Note that **AB** *is defined, but* **BA** *is not defined.*

(b) Let

$$\mathbf{A} = \begin{pmatrix} 1 & 0 & 1 \\ 0 & 2 & -1 \end{pmatrix} \quad \text{and} \quad \mathbf{B} = \begin{pmatrix} 2 & 1 \\ -1 & 0 \\ 3 & 2 \end{pmatrix}$$

Then

$$\mathbf{AB} = \begin{pmatrix} 1 & 0 & 1 \\ 0 & 2 & -1 \end{pmatrix}\begin{pmatrix} 2 & 1 \\ -1 & 0 \\ 3 & 2 \end{pmatrix} = \begin{pmatrix} 5 & 3 \\ -5 & -2 \end{pmatrix}$$

while

$$\mathbf{BA} = \begin{pmatrix} 2 & 1 \\ -1 & 0 \\ 3 & 2 \end{pmatrix}\begin{pmatrix} 1 & 0 & 1 \\ 0 & 2 & -1 \end{pmatrix} = \begin{pmatrix} 2 & 2 & 1 \\ -1 & 0 & -1 \\ 3 & 4 & 1 \end{pmatrix}$$

Note that **AB** *and* **BA** *are both defined, but* $\mathbf{AB} \neq \mathbf{BA}$.

(c) Let

$$\mathbf{A} = \begin{pmatrix} 1 & 2 \\ 3 & 4 \end{pmatrix} \quad \text{and} \quad \mathbf{B} = \begin{pmatrix} 2 & 0 \\ 0 & 2 \end{pmatrix}$$

Then

$$\mathbf{AB} = \begin{pmatrix} 1 & 2 \\ 3 & 4 \end{pmatrix}\begin{pmatrix} 2 & 0 \\ 0 & 2 \end{pmatrix} = \begin{pmatrix} 2 & 4 \\ 6 & 8 \end{pmatrix}$$

and

$$\mathbf{BA} = \begin{pmatrix} 2 & 0 \\ 0 & 2 \end{pmatrix}\begin{pmatrix} 1 & 2 \\ 3 & 4 \end{pmatrix} = \begin{pmatrix} 2 & 4 \\ 6 & 8 \end{pmatrix} = \mathbf{AB}$$

In this case, **AB** *and* **BA** *are both defined, and* $\mathbf{AB} = \mathbf{BA}$.

If two *numbers* are multiplied, the order of multiplication is unimportant. In other words, if a and b are numbers, then $ab = ba$. This is called the *commutative law* for multiplication of numbers. Example 14 shows that the corresponding law does *not* hold for matrices. If **AB** is defined, then **BA** may not even be defined. If **BA** *is* defined, then **BA** may equal **AB**, or **BA** may be different from **AB**.

Example 15 Let **A** be any 2×2 matrix, and let

$$\mathbf{I}_2 = \begin{pmatrix} 1 & 0 \\ 0 & 1 \end{pmatrix}$$

Find $\mathbf{I}_2 \mathbf{A}$ and $\mathbf{A}\mathbf{I}_2$.

Solution We may let

$$\mathbf{A} = \begin{pmatrix} a & b \\ c & d \end{pmatrix}$$

where a, b, c, and d denote any four real numbers. (Other letters may be used in place of a, b, c, and d.) Then

$$\mathbf{I}_2 \mathbf{A} = \begin{pmatrix} 1 & 0 \\ 0 & 1 \end{pmatrix}\begin{pmatrix} a & b \\ c & d \end{pmatrix} = \begin{pmatrix} 1 \cdot a + 0 \cdot c & 1 \cdot b + 0 \cdot d \\ 0 \cdot a + 1 \cdot c & 0 \cdot b + 1 \cdot d \end{pmatrix}$$

$$= \begin{pmatrix} a & b \\ c & d \end{pmatrix} = \mathbf{A}$$

Similarly,

$$\mathbf{A}\mathbf{I}_2 = \mathbf{A}$$

It follows that

$$\mathbf{I}_2 \mathbf{A} = \mathbf{A}\mathbf{I}_2 = \mathbf{A}$$

Let

$$A = \begin{pmatrix} a_{11} & a_{12} & \cdots & a_{1n} \\ a_{21} & a_{22} & \cdots & a_{2n} \\ \vdots & \vdots & & \vdots \\ a_{n1} & a_{n2} & \cdots & a_{nn} \end{pmatrix}$$

be any $n \times n$ matrix. The numbers $a_{11}, a_{22}, \ldots, a_{nn}$ that appear in the first row, first column of A, second row, second column of A, and so on, to the nth row, nth column of A, are called the *diagonal entries* of A. For example, the diagonal entries of I_2 of Example 15 are all 1; all other entries of this matrix are 0.

We saw in Example 15 that, if I_2 is multiplied by any 2×2 matrix A, the result is A. The matrix I_2 is therefore called an *identity matrix*. Definition 6.9 extends this concept to matrices with n rows and n columns.

Definition 6.9   The **identity matrix** I_n is a matrix having the following characteristics:

(a)   It has n rows and n columns.
(b)   Every diagonal entry is 1.
(c)   Every other entry is 0.

$$I_n = \begin{pmatrix} 1 & 0 & 0 & 0 & \cdots & 0 \\ 0 & 1 & 0 & 0 & \cdots & 0 \\ 0 & 0 & 1 & 0 & \cdots & 0 \\ \vdots & \vdots & \vdots & \vdots & & \vdots \\ 0 & 0 & 0 & 0 & \cdots & 1 \end{pmatrix}$$

For example,

$$I_1 = (1), \qquad I_3 = \begin{pmatrix} 1 & 0 & 0 \\ 0 & 1 & 0 \\ 0 & 0 & 1 \end{pmatrix},$$

and

$$I_4 = \begin{pmatrix} 1 & 0 & 0 & 0 \\ 0 & 1 & 0 & 0 \\ 0 & 0 & 1 & 0 \\ 0 & 0 & 0 & 1 \end{pmatrix}$$

are all identity matrices.

In Example 15 we saw that, if A is any 2×2 matrix, $AI_2 = I_2 A = A$. A similar proof shows that if A is any $n \times n$ matrix,

$$AI_n = I_n A = A \tag{9}$$

The reader may well wonder if equation (9) holds if A does not have n rows and n columns. The following example shows that A *must* be $n \times n$ in equation (9).

Example 16   Suppose equation (9) holds for some matrix A. Prove that A is $n \times n$.

Solution   Since $\mathbf{AI}_n$ is defined, the number of columns of $\mathbf{A}$ must equal the number n of rows of $\mathbf{I}_n$; that is, $\mathbf{A}$ must have n columns. Similarly, since $\mathbf{I}_n\mathbf{A}$ is defined, $\mathbf{A}$ must have n rows.

Consider equation (9). If we replace the arbitrary $n \times n$ matrix $\mathbf{A}$ by an arbitrary number a, the matrix $\mathbf{I}_n$ by the number 1, and matrix multiplication by the multiplication of numbers, equation (9) becomes the arithmetic identity

$$a(1) = 1(a) = a$$

In other words, the identity matrix $\mathbf{I}_n$ and the number 1 have similar properties.

Additional arithmetic properties of matrices are given in Table 6.3. As in Table 6.2, the arithmetic laws for real numbers appear in the second column of the table. If we replace numbers by matrices and 1 by $\mathbf{I}_n$, we obtain the matrix laws in the right-hand column. Note that the matrices must be chosen so that matrix multiplication and addition are defined.

Table 6.3

Property	Arithmetic Law	Matrix Law	
Associativity	$a(bc) = (ab)c$	$\mathbf{A(BC)} = \mathbf{(AB)C}$	$\mathbf{A}$ is $m \times n$, $\mathbf{B}$ is $n \times s$, $\mathbf{C}$ is $s \times t$
Existence of an identity	$a \cdot 1 = a = 1 \cdot a$	$\mathbf{AI}_n = \mathbf{A} = \mathbf{I}_n\mathbf{A}$	$\mathbf{A}$ is $n \times n$
Distributivity	$a(b + c) = ab + ac$	$\mathbf{A(B + C)} = \mathbf{AB} + \mathbf{AC}$	$\mathbf{A}$ is $m \times n$, $\mathbf{B}$ and $\mathbf{C}$ are $n \times s$
	$(b + c)a = ba + ca$	$\mathbf{(B + C)A} = \mathbf{BA} + \mathbf{CA}$	$\mathbf{A}$ is $m \times n$, $\mathbf{B}$ and $\mathbf{C}$ are $s \times m$

Table 6.3 shows that there are many similarities between matrix multiplication and the multiplication of numbers. However, matrix multiplication is not commutative, as we saw in Example 14.

Suppose $\mathbf{A}$ is $m \times n$, $\mathbf{B}$ is $n \times s$, and $\mathbf{C}$ is $s \times t$. Since $\mathbf{A(BC)} = \mathbf{(AB)C}$, we may drop the parentheses and define the product $\mathbf{ABC}$ as either $\mathbf{A(BC)}$ or $\mathbf{(AB)C}$. For example, if

$$\mathbf{A} = \begin{pmatrix} 1 & 1 \\ 2 & 1 \end{pmatrix}, \qquad \mathbf{B} = \begin{pmatrix} 2 \\ 1 \end{pmatrix}, \qquad \text{and} \qquad \mathbf{C} = (1,3)$$

then

$$\mathbf{ABC} = \mathbf{(AB)C} = \left[\begin{pmatrix} 1 & 1 \\ 2 & 1 \end{pmatrix}\begin{pmatrix} 2 \\ 1 \end{pmatrix}\right](1,3) = \begin{pmatrix} 3 \\ 5 \end{pmatrix}(1,3) = \begin{pmatrix} 3 & 9 \\ 5 & 15 \end{pmatrix}$$

The product of four or more matrices is defined similarly.

6.3 EXERCISES

1. Let

$$A = \begin{pmatrix} 1 & 2 \\ 3 & 4 \\ 5 & 6 \end{pmatrix} \quad \text{and} \quad B = \begin{pmatrix} 1 & 2 & 0 \\ 4 & 5 & 6 \end{pmatrix}$$

Which of the following products are defined?

$$\mathbf{AB}, \quad \mathbf{BA}, \quad \mathbf{B}^2, \quad \mathbf{A}^2, \quad \mathbf{AI}_2, \quad \mathbf{I}_2\mathbf{A}, \quad \mathbf{A\Phi}$$

Explain your answers.

2. Write the matrix $\mathbf{I}_5$.

3. Suppose

$$\begin{pmatrix} 2 & 3 \\ x & x \end{pmatrix}\begin{pmatrix} 4 & x \\ 1 & x \end{pmatrix} = \begin{pmatrix} z & x \\ x & x \end{pmatrix}$$

Find z. (The x's stand for unknown numbers—they are not necessarily equal.)

4. Let

$$A = \begin{pmatrix} 1 & 2 \\ 2 & 1 \end{pmatrix}, \quad B = \begin{pmatrix} 0 & 0 \\ 1 & 0 \end{pmatrix}, \quad \text{and} \quad C = \begin{pmatrix} 1 & 2 \\ -1 & 1 \end{pmatrix}$$

Calculate—
(a) **AB**   (b) **BA**   (c) **A + B**   (d) **B + A**   (e) $(\mathbf{A}^2)\mathbf{B}$
(f) **ABC**   (g) **CBA**   (h) **C(BA)**

5. Let

$$A = \begin{pmatrix} 1 & 0 \\ 3 & 0 \end{pmatrix}, \quad B = \begin{pmatrix} 1 & 0 \\ 0 & 2 \end{pmatrix}, \quad \text{and} \quad C = \begin{pmatrix} 0 & 1 \\ 2 & 0 \end{pmatrix}$$

Show that—
(a) $\mathbf{A(BC)} = \mathbf{AB(C)}$ (b) $\mathbf{A(B + C)} = \mathbf{AB + AC}$
(c) $\mathbf{(B + C)A} = \mathbf{BA + CA}$

6. Let

$$A = \begin{pmatrix} 2 & 0 \\ 0 & 0 \end{pmatrix}$$

Calculate—
(a) $\mathbf{A}^2$ (b) $\mathbf{A}^3$ (c) $\mathbf{A}^n$, $n = 1, 2, 3, \ldots$

7. Let **A** be any matrix. Calculate **AΦ**.

8. A child chose three pairs of pants, two vests, four blouses, and one jacket at a clothing store. Her younger sister chose four pairs of pants, one vest, three

blouses, and two jackets. The pants, vests, blouses, and jackets cost 8, 6, 5, and 9 dollars each, respectively.

(a) Display the clothing choice made by the two children in a 2×4 matrix.

(b) Display the costs of the various types of clothing in a column vector.

(c) Use matrix multiplication to find the total cost of clothing for each child.

(d) How much did the children's mother spend on the oldest child's clothes if she paid for all purchases at the store?

9.   A garment manufacturer makes a certain shirt in three sizes: infants' (I), children's (C), and adults' (A). The shirts require fabric (F), buttons (B), trimming (T), and labor (L), and are sold in ten-shirt lots. The number of units of each item required to make ten shirts is given in the matrix X. The cost in dollars of a unit of each item is given in the matrix Y.

$$
\begin{array}{c}
 \\
 \\
X = \begin{array}{c} I \\ C \\ A \end{array}
\end{array}
\begin{array}{cccc}
F & B & T & L \\
\left(\begin{array}{cccc}
5 & 30 & 5 & 10 \\
10 & 50 & 8 & 10 \\
20 & 70 & 10 & 15
\end{array} \right)
\end{array}
\qquad
\begin{array}{c}
\text{Cost} \\
Y = \begin{array}{c} F \\ B \\ T \\ L \end{array}
\left(\begin{array}{c}
3 \\
0.1 \\
0.3 \\
4
\end{array} \right)
\end{array}
$$

(a) Calculate XY.

(b) What information is displayed in the matrix XY?

(c) What is the production cost of—

(i) ten infant shirts

(ii) thirty adult shirts?

10.   A store wishes to purchase 50 infants', 100 children's, and 80 adults' shirts from the manufacturer of Exercise 9.

(a) Write the purchase order (in number of ten-shirt lots) as a row vector Z.

(b) Use matrix multiplication to find the total number of units of fabrics, buttons, and so on required to fill the order.

(c) How many units of fabric does the company need to fill the order?

11.   Let X, Y, and Z be the matrices of Exercises 9 and 10. Use matrix multiplication to find the total production cost of filling the store's order. (Hint: Calculate $(ZX)Y$.)

12.   The probabilities that Nick (N), Debby (D), and Lewis (L) will make 10, 20, or 50 points when shooting darts are given in the following matrix:

$$
A = \begin{array}{c} 10 \\ 20 \\ 50 \end{array}
\begin{array}{ccc}
N & D & L \\
\left(\begin{array}{ccc}
0.5 & 0.2 & 0.3 \\
0.3 & 0.6 & 0.4 \\
0.2 & 0.2 & 0.3
\end{array} \right)
\end{array}
$$

(a) Use matrix multiplication to find the expected number of points each player will get. (Hint: Let $X = (10,20,50)$ and calculate XA.)

(b) Which player has the largest expected number of points?

6.4 LINEAR EQUATIONS AND THEIR SOLUTIONS

Many practical problems require solving a system of two or more linear equations. In Section 6.6 we will see how matrices can be used to find the solution to such a system, if it exists. In this section we will study single linear equations and their solutions.

Before we can study linear equations, we must consider equations in general. Equations such as

$$x^2 + y^2 = 5, \qquad x - y^2 = 3, \qquad \text{and} \qquad x + y = 2$$

are called equations in *two variables*. The letters x and y have been used to designate the variables in each of these equations. However, any other pair of letters—such as t and v, r and s, or u and v—may be used.

The equations

$$x^2 + y^2 + z^2 = 5, \qquad r = x + 2s, \qquad \text{and} \qquad x_1 = x_2 x_3 - 5x_1$$

are all equations in three variables.

Consider the equation

$$2x - y = 5 \tag{10}$$

It is clear that (10) is a true statement for some values of x and y, but not for others. If, for example, we replace x by 4 and y by 3, then (10) is true; that is,

$$2(4) - 3 = 5$$

is a true statement. Therefore, the vector $\mathbf{v} = (4,3)$ is called a *solution* of equation (10). (Note that the first component of $\mathbf{v}$ replaces x; the second component replaces y.) On the other hand, the vector (3,4) is not a solution of equation (10), because replacing x by 3 and y by 4 yields

$$2(3) - 4 = 5$$

which is not a true statement.

Definition 6.10 generalizes these ideas.

Definition 6.10

(a) Let E be any equation in two variables x and y. The vector (x_o, y_o) is a **solution** of E (or *satisfies* E) if and only if the substitution of x_o for x and y_o for y in E yields a true statement.

(b) Let E_1 and E_2 be a system (set) of two equations in two variables x and y. The vector (x_o, y_o) is a **solution of the system** (or *satisfies* the system) if and only if (x_o, y_o) is a solution of both E_1 and E_2.

A similar definition holds if the variables are denoted r and s, or u and v, and so on, instead of x and y. Definition 6.10 can also be easily extended to equations in three variables, four variables, and so on.

The following examples illustrate Definition 6.10.

Example 17

(a) The vector $(1,3)$ is a solution of

$$x + y^2 = 10 \tag{11}$$

because the substitution of $x = 1$ and $y = 3$ in equation (11) yields the true statement

$$1 + 3^2 = 10$$

(b) However, the vector $(3,1)$ is not a solution of equation (11), because the substitution of $x = 3$ and $y = 1$ in equation (11) yields

$$3 + 1^2 = 10$$

which is not a true statement.

Example 18   If E is an equation in three variables x_1, x_2, and x_3, then the solutions of E are denoted by ordered triples (x_1,x_2,x_3). Which of the following vectors are solutions of

$$2x_1 + 3x_2 - 4x_3 = -2? \tag{12}$$

(a) $(3,0,2)$ (b) $(1,2,\frac{5}{2})$ (c) $(0,3,2)$

Solution

(a) The vector $(3,0,2)$ is a solution of equation (12) because substituting $x_1 = 3$, $x_2 = 0$, and $x_3 = 2$ into equation (12) yields the true statement

$$2(3) + 3(0) - 4(2) = -2$$

(b) Similarly, $(1,2,\frac{5}{2})$ is a solution of equation (12) because

$$2(1) + 3(2) - 4(\tfrac{5}{2}) = -2$$

is a true statement.

(c) On the other hand, $(0,3,2)$ is not a solution of equation (12), because

$$2(0) + 3(3) - 4(2) \neq -2$$

In general, if E is any equation in n variables $x_1, x_2, \ldots, x_n$, the solutions of E are denoted by ordered n-tuples $(x_1,x_2,\ldots,x_n)$.

Example 19

(a) The vector $(2,4)$ is a solution of the system of equations

$$x + 2y = 10 \tag{13}$$

$$4x - y = 4 \tag{14}$$

because $(2,4)$ satisfies both equation (13) and equation (14).

(b) On the other hand, the vector $(10,0)$ is not a solution of the above system because $(10,0)$ does not satisfy equation (14). (Notice, however, that $(10,0)$ does satisfy equation (13).)

We already know that a number, set, or matrix can be written in many different ways. For example, the sets

$$A = \{1,2,3\} \quad \text{and} \quad B = \{3,1,2\}$$

are written differently, but they contain the same elements.

An equation can also be written in many ways. Consider the equation

$$x^2 + 2y^2 = 9 \tag{15}$$

If we multiply both sides of equation (15) by 5, we obtain the equation

$$5x^2 + 10y^2 = 45 \tag{16}$$

We know from algebra that every solution of equation (15) is a solution of equation (16), and vice versa. We can therefore say that equation (16) is just another way of writing equation (15). Equations that have the same set of solutions, as equations (15) and (16) do, are called **equivalent equations**.

We are now able to define a linear equation.

Definition 6.11   Any equation that can be written in the form

$$ax + by = c \tag{17}$$

where a, b, and c are fixed numbers and a and b are not both zero, is called a **linear equation** *in two variables* (or *two unknowns*).

In equation (17), the variables have been denoted by the letters x and y. Other pairs of letters, such as r and s, may also be used. However, *if S is a system of two or more equations in two variables, the same two letters must be used for the variables in each equation in S.*

Definition 6.11 can be easily extended to equations having three variables, four variables, and so on. For example, any equation that can be written in the form

$$ax + by + cz = d \tag{18}$$

where a, b, c, and d are fixed numbers and a, b, and c are not all zero, is called a *linear equation in three variables*.

The following examples illustrate Definition 6.11.

Example 20

(a) The equations

$$3x + y = 10, \quad 2x - 8y = 5, \quad \text{and} \quad 6x + 5y = 0$$

are all linear equations in two variables, because they each have the form of equation (17).

(b) The equations

$$34r - 2s = 5, \qquad 3v - 3w = 10, \qquad \text{and} \qquad 4x_1 + 63x_2 = 9$$

are also linear equations in two variables. In these equations the variables have been denoted by r and s, v and w, and x_1 and x_2, respectively.

(c) The equations

$$5x = 9 - 7y, \qquad 7y = 9 - 5x, \qquad \text{and} \qquad 5x + 7y - 9 = 0$$

are all linear equations in two variables, because each of these equations can be written in the equivalent form

$$5x + 7y = 9$$

(d) On the other hand, none of the equations

$$x^2 + y^2 = 9, \qquad x - y^2 = 10, \qquad \text{or} \qquad xy = 5$$

are linear, because none of these equations can be written in the form of equation (17).

Example 21   The set of equations

$$x - y + 7z = 1$$
$$2x + y + 8z = 2$$
$$3x + y - z = 0$$

is a system of three linear equations in three variables. Notice that the same three letters, x, y, and z, have been used to denote the variables in each of these equations.

In the remainder of this chapter, we will learn how to find all solutions (if any exist) of n linear equations in n variables. The solution of m linear equations in n variables where m is not necessarily equal to n will be considered at the end of Section 6.7.

Example 22 will show that a system of two linear equations in two variables can have a unique solution, an infinite number of solutions, or no solution. Similar statements can be made for systems of three linear equations in three variables, four linear equations in four variables, and so on.

Example 22

(a) Let (x,y) be any solution of the following system of linear equations:

$$x + y = 1 \tag{19}$$
$$y = 2 \tag{20}$$

It follows from equation (20) that y must be 2. Substituting this value in equation (19) yields $x + 2 = 1$, or $x = 1 - 2 = -1$. The system therefore has exactly one solution, $(-1,2)$. The reader can check this solution by showing that $(-1,2)$ satisfies both equation (19) and equation (20).

(b) Let (x,y) be any solution of the following system of linear equations:

$$x + y = 1 \tag{21}$$

$$x + y = 2 \tag{22}$$

and let $z = x + y$. Then

$$z = 1 \quad \text{and} \quad z = 2$$

Since this is impossible, the system of equations (21) and (22) is *inconsistent*; that is, the system has no solution.

(c) The system of equations

$$x + y = 1$$

$$2x + 2y = 2$$

has an infinite number of solutions: $(0,1)$, $(1,0)$, $(2,-1)$, $(3,-2)$, and so on.

The three systems of equations discussed in Example 22 were fairly simple to solve. In the next section we will study two methods for solving more difficult systems.

6.4 EXERCISES

1.   Which of the following equations are linear?
     (a) $x + 4y = 7$ **(b)** $x^2 - 2y = 3$ **(c)** $2xy = 5$
     (d) $x_1 + 2x_2 = 5$ **(e)** $x + y^2 - 2z = 10$

2.   Which of the following equations have $(2,1)$ as a solution?
     (a) $x + y = 3$ **(b)** $2x - 7y = 3$ **(c)** $4x - 6y = 2$
     (d) $2x + 5y = 12$

3.   Which of the equations of Exercise 2 have $(1,2)$ as a solution?

4.   Which of the following equations are equivalent to $y^2 + 7x = 5$?
     (a) $y^2 + 7x - 5 = 0$ **(b)** $2y^2 + 14x = 5$ **(c)** $y^2 = 5 - 7x$
     (d) $x = \dfrac{5 - y^2}{7}$

5.   Determine which of the following vectors are solutions of the system

$$x + 2y = 3$$

$$3x + 4y = 9$$

     (a) $(0,3)$ **(b)** $(3,0)$ **(c)** $(1,1)$ **(d)** $(1,\tfrac{3}{2})$

6. Robert receives a weekly salary of 20 dollars plus 50-cent commission on every subscription he sells. Let x be the number of subscriptions that he sells during the week and let y be his total salary.

(a) Show that $y = 0.5x + 20$.

(b) Is the equation of part **(a)** linear?

(c) Find his total earnings if he sells 50, 80, or 100 subscriptions during the week.

7. Which of the following systems have (i) exactly one solution, (ii) an infinite number of solutions, (iii) no solution?

(a) $\quad\begin{aligned} y &= 5 \\ 2x + 3y &= 80 \end{aligned}$ **(b)** $\quad\begin{aligned} x + y &= 5 \\ 2x + 2y &= 10 \end{aligned}$ **(c)** $\quad\begin{aligned} x + y &= 5 \\ 2x + 2y &= 8 \end{aligned}$

(d) $\quad\begin{aligned} x &= 4 \\ 2x + y &= 10 \end{aligned}$

Give the solution if the solution is unique. Give three solutions if there are an infinite number of solutions.

6.5 SOLUTIONS OF TWO LINEAR EQUATIONS IN TWO VARIABLES

Example 22 shows that a system of two linear equations in two variables may have one solution, no solution, or an infinite number of solutions. It can be proved that these are the only possible cases. In this section we will study systems of equations that have a unique solution, and we will give two methods for finding this solution. In Section 6.6 we will see how matrices can be used to solve systems of equations. Inconsistent systems and systems having an infinite number of solutions will be studied in Section 6.7.

Consider the system of linear equations

$$2x + 3y = 8 \tag{23}$$

$$4x - 2y = 0 \tag{24}$$

If we multiply equation (23) by 3, a new set of equations is formed:

$$6x + 9y = 24 \tag{25}$$

$$4x - 2y = 0 \tag{26}$$

We know from algebra that any solution of the first set of equations, (23) and (24), is a solution of the second set, (25) and (26), and vice versa. Two systems of equations like (23)–(24) and (25)–(26) that have the same set of solutions are called **equivalent systems**.

Similarly, if equation (24) is replaced by the sum of equation (24) and 3 times equation (23), the result is the equivalent system

$$2x + 3y = 8$$

$$10x + 7y = 24$$

In general, if any of the following operations is performed upon a set of linear equations, the result is an equivalent set of equations.

Operations That Transform a System of Linear Equations into an Equivalent System

Type A: Interchanging two equations

Type B: Multiplying or dividing an equation by a nonzero constant

Type C: Replacing one equation by the sum of that equation and a nonzero constant times another equation in the system.

In particular, one equation in the system may be replaced by the sum or difference of itself and another equation in the system.

Let S be the following system of linear equations:

$$ax + by = c$$

$$dx + ey = f$$

If a, b, d, or e is 0, it is easy to decide whether a solution exists. Moreover, if there is a solution, it is easy to find. (See Example 22(a).) We will therefore only study systems in which a, b, d, and e are all nonzero.

Let S be such a system and assume that it has a unique solution $z = (x_o, y_o)$. By repeatedly performing the above three operations, S can be systematically transformed into an equivalent system in which the solution is obvious. Two methods of doing this are illustrated in the following examples. The first method may be easier to apply in some cases; the second method may be simpler in others. In Section 6.6 we will see how the second method can be simplified by using matrices.

Example 23 Two weights weighing 20 kilograms and 36 kilograms, respectively, are hung at the ends of a 14-meter weightless pole. (See Figure 6.1.) Let x

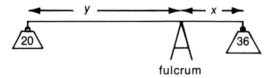

fulcrum

Figure 6.1

be the distance between the fulcrum and the heavier weight, and let y be the distance between the fulcrum and the lighter weight. The pole will be balanced if x and y satisfy the following equations:

$$x + \quad y = 14 \tag{27}$$

$$36x - 20y = \quad 0 \tag{28}$$

Find x and y.

Solution To find the value of y, we will transform equations (27) and (28) into an equivalent system in which the value of y is obvious. To do this, we follow these steps:

(1) *Transform the given system into one in which the coefficients of x in the two equations are equal.* Multiplying both sides of equation (27) by 36 (a type B operation) yields the equivalent system

$$36x + 36y = 504 \tag{29}$$

$$36x - 20y = \quad 0 \tag{30}$$

(2) *Subtract equation (30) from equation (29).* This is a type C operation; the nonzero constant is -1. This yields the equivalent system

$$56y = 504 \tag{31}$$

$$36x - 20y = \quad 0$$

(3) *Divide equation (31) by the coefficient of y.* This type B operation yields the equivalent system

$$y = 9 \tag{32}$$

$$36x - 20y = 0 \tag{33}$$

Since we have used only permissible operations, equations (27) and (28) have the same solutions as equations (32) and (33). It is clear from equation (32) that y must be 9.

The value of x can now be found:

(4) *Substitute equation (32) into equation (27) and solve for x.* Substituting $y = 9$ into equation (27) yields $x + 9 = 14$. It follows that $x = 5$. Notice that no other value of x is possible.

The reader can verify that $(5,9)$ is a solution of the system of equations (27)–(28) by showing that this vector satisfies both equation (27) and equation (28). This shows that the pole will be balanced if the fulcrum is 5 meters from the heavier weight and 9 meters from the lighter weight.

The following example shows how to apply the method of Example 23 to another system of linear equations.

Example 24   Solve the following system of linear equations:

$$4x - 2y = 2 \tag{34}$$

$$5x - 3y = 1 \tag{35}$$

Solution   We will use the method of Example 23.

(1) Multiplying both sides of equation (34) by 5 and both sides of equation (35) by 4 yields

$$20x - 10y = 10 \tag{36}$$

$$20x - 12y = \quad 4 \tag{37}$$

(2) Subtracting equation (37) from equation (36) yields

$$2y = 6 \tag{38}$$

(3) Dividing equation (38) by 2 yields

$$y = 3 \tag{39}$$

(4) Substituting equation (39) into equation (34) yields

$$4x - 6 = 2$$

Therefore, $4x = 8$, and so $x = 2$.

It follows that this system has a unique solution, $(2,3)$.

Example 25 gives a second method for solving a system of two linear equations in two variables whose coefficients a, b, c, and d are all nonzero.

Example 25   Solve the following system of linear equations.

$$2x + 3y = 8 \tag{40}$$
$$4x - 2y = 0 \tag{41}$$

Solution   Equations (40) and (41) can be solved by using the following four steps:

(1) *Transform equation (40) into an equivalent equation with x coefficient equal to 1.* Dividing both sides of equation (40) by 2, a type B operation,[1] yields the equivalent system

$$x + \tfrac{3}{2}y = 4 \tag{42}$$
$$4x - 2y = 0 \tag{43}$$

(2) *Eliminate x from equation (43).* Multiplying equation (42) by -4 and adding the result to equation (43), a type C operation, yields the equivalent system

$$x + \tfrac{3}{2}y = \quad 4 \tag{44}$$
$$-8y = -16 \tag{45}$$

(3) *Transform equation (45) into an equivalent equation with y coefficient equal to 1.* Dividing equation (45) by -8 yields

$$x + \tfrac{3}{2}y = 4 \tag{46}$$
$$y = 2 \tag{47}$$

---

1. If the coefficient of x in the second equation is 1, it is easier to begin by interchanging the two equations, a type A operation.

(4) *Eliminate y from equation* (46). Multiplying equation (47) by $-\frac{3}{2}$ and adding the result to equation (47) yields

$$x \quad = 1 \tag{48}$$

$$y = 2 \tag{49}$$

It is now obvious that equations (40) and (41) have exactly one solution, (1,2).

The method of Example 25 can easily be modified to yield the solution (if exactly one solution exists) of any two linear equations in two variables whose coefficients *a*, *b*, *d*, and *e* are all nonzero. The next section will present a simplification of this method using matrices.

6.5 EXERCISES

1. Determine which of the following systems of linear equations are equivalent to

$$4x + 7y = 9$$
$$2x - 3y = 8$$

(a) $2x - 3y = 8$ **(b)** $4x - 6y = 16$ **(c)** $2x - 3y = 8$
 $4x + 7y = 9$ $4x + 7y = 9$ $4x - 6y = 16$

(d) $2x - 3y = 8$
 $13y = -7$

2. Solve the following systems of linear equations.

(a) $2x + 3y = 13$ **(b)** $3x + 4y = 11$ **(c)** $4x - 9y = -1$
 $4x - 7y = -13$ $2x - 8y = -14$ $8x - 6y = 10$

3. The Chocolate Treat Company makes two varieties of chocolate milk. Regular chocolate milk uses 1 gallon of syrup to 12 gallons of milk; extra-chocolaty uses 2 gallons of syrup to 19 gallons of milk. The company has 260 gallons of syrup and 2720 gallons of milk. How many gallons of each type of chocolate milk should the company produce if it wishes to use up all of its ingredients? (Hint: Let x and y be the amounts of regular and extra-chocolaty milk, respectively, that the company should produce. Show that $x + 2y = 260$ and $12x + 19y = 2720$.)

6.6 LINEAR EQUATIONS AND MATRICES

In Section 6.5 we learned two methods for solving a system of two linear equations in two variables. In this section we will see how the method of Example 25 can be simplified by using matrices. Section 6.7 will show how this method can be extended to the solution of n linear equations in n variables.

Let

$$ax + by = c$$
$$dx + ey = f$$

be any system of two linear equations in two variables. This system can be represented by the matrix

$$\mathbf{K} = \begin{pmatrix} a & b & c \\ d & e & f \end{pmatrix}$$

called the **corresponding matrix**.

Example 26

(a) The corresponding matrix of the system

$$4x - 5y = 7$$
$$2x + 3y = 8$$

is

$$\begin{pmatrix} 4 & -5 & 7 \\ 2 & 3 & 8 \end{pmatrix}$$

(b) The set of linear equations represented by the matrix

$$\begin{pmatrix} 2 & 4 & -\frac{7}{2} \\ 9 & 3 & 8 \end{pmatrix}$$

is

$$2x + 4y = -\tfrac{7}{2}$$
$$9x + 3y = 8$$

We will now see how matrices can be used to solve the system of linear equations (40) and (41) of Example 25. These equations (repeated here) are:

$$2x + 3y = 8 \tag{40}$$
$$4x - 2y = 0 \tag{41}$$

Equations (40) and (41) can be represented by the corresponding matrix

$$\mathbf{K} = \begin{pmatrix} 2 & 3 & 8 \\ 4 & -2 & 0 \end{pmatrix} \tag{50}$$

The operations performed in Example 25 on this system of equations can now be translated into corresponding operations on the corresponding matrix.

(1) The first step in Example 25 was to transform equation (40) into an equivalent equation with x coefficient equal to 1. This was done by dividing both sides of

equation (40) by 2. Since the first row of **K** corresponds to equation (40), the corresponding operation is to divide the entries in that row by 2. This yields

$$\begin{pmatrix} 1 & \frac{3}{2} & 4 \\ 4 & -2 & 0 \end{pmatrix} \tag{51}$$

Note that this matrix corresponds to the system of equations (42) and (43) on page 240.

(2) The second step in Example 25 was to eliminate x from equation (43). This was done by multiplying equation (42) by -4 and adding the result to equation (43). Since the coefficients of equation (42) appear in the first row of matrix (51), and the coefficients of equation (43) appear in the second row, the corresponding operation is to multiply the first row of matrix (51) by -4 and add the result to the second row. This results in the matrix

$$\begin{pmatrix} 1 & \frac{3}{2} & 4 \\ 0 & -8 & -16 \end{pmatrix} \tag{52}$$

which corresponds to equations (44) and (45) of Example 25.

(3) The third step in Example 25 was to transform equation (45) into an equivalent equation with y coefficient equal to 1. This was done by dividing equation (45) by -8. If we perform the same operation on the entries in the second row of matrix (52)—the row corresponding to equation (45)—we have

$$\begin{pmatrix} 1 & \frac{3}{2} & 4 \\ 0 & 1 & 2 \end{pmatrix} \tag{53}$$

(4) The final step in Example 25 was to eliminate y from equation (46) by multiplying equation (47) by $-\frac{3}{2}$ and adding the result to equation (46). The corresponding operation on matrix (53) is to multiply the second row by $-\frac{3}{2}$ and add the result to the first row. The resulting matrix is

$$\mathbf{A} = \begin{pmatrix} 1 & 0 & 1 \\ 0 & 1 & 2 \end{pmatrix}$$

Matrix **A** corresponds to the system of equations $x = 1$, $y = 2$ that yields the required solution. Notice that if the last column is removed from **A**, the identity matrix $\mathbf{I}_2$ is formed.

$$\begin{pmatrix} 1 & 0 & 1 \\ 0 & 1 & 2 \end{pmatrix} \rightarrow \begin{pmatrix} 1 & 0 \\ 0 & 1 \end{pmatrix} = \mathbf{I}_2$$

Matrix **A** is called a *solution matrix*. The solution is given in the last column.

Definition 6.12   Let **A** be a 2×3 matrix. If the matrix formed from **A** by deleting the last column of **A** is $\mathbf{I}_2$, then **A** is a **solution matrix**.

$$\mathbf{A} = \begin{pmatrix} 1 & 0 & k \\ 0 & 1 & m \end{pmatrix} \qquad \text{where } k \text{ and } m \text{ are any real numbers}$$

The matrix method used to solve equations (40)–(41) can be applied to any system of two linear equations in two variables. As we have seen, this method involves operating upon the rows of the matrix **K** that represents the given system of equations. These row operations, called **elementary row operations**, correspond to the operations upon equations that yield equivalent systems. Elementary row operations are classified into three types, as follows:

Elementary Row Operations

Type A: Interchanging two rows
Type B: Multiplying or dividing a row by a nonzero constant
Type C: Replacing one row by the sum of that row and a nonzero constant times another row

The above elementary row operations correspond to the operations of types A, B, and C that may be performed on systems of linear equations without changing the set of solutions. (See page 238.)

The matrix method for solving two linear equations in two variables can be summarized as follows:

Method for Solving a System of Two Linear Equations in Two Variables

Step 1 Replace the system by the corresponding matrix **K**.
Step 2 If possible, reduce **K** to a solution matrix by means of elementary row operations.

Definition 6.13 If a matrix **A** can be transformed into a matrix **B** by a series of elementary row operations, then **A** and **B** are **equivalent matrices**. This is denoted

$$\mathbf{A} \sim \mathbf{B}$$

It can be shown that if $\mathbf{A} \sim \mathbf{B}$, then $\mathbf{B} \sim \mathbf{A}$. Moreover, **A** and **B** are equivalent if and only if the systems of linear equations that they represent are also equivalent.

Using Definition 6.13, Step 2 of the method for solving a system of two linear equations in two variables can be rewritten as follows:

Step 2′ If possible, reduce **K** to an equivalent solution matrix.

We shall now review how matrix (50) was reduced to an equivalent solution matrix. This will lead us to a general form for the reduction of any matrix **K** corresponding to a system of two equations in two unknowns.

Let **K** be the matrix corresponding to equations (40) and (41). Then

$$\mathbf{K} = \begin{pmatrix} 2 & 3 & 8 \\ 4 & -2 & 0 \end{pmatrix}$$

The matrix $\mathbf{K}$ was reduced to an equivalent solution matrix

$$\mathbf{A} = \begin{pmatrix} 1 & 0 & 1 \\ 0 & 1 & 2 \end{pmatrix}$$

by the four steps described on pages 242–243.

In step 1, $\mathbf{K}$ was reduced to the equivalent matrix

$$\begin{pmatrix} 1 & \frac{3}{2} & 4 \\ 4 & -2 & 0 \end{pmatrix}$$

by dividing the first row of $\mathbf{K}$ by 2. This is a type B operation. This step can be symbolized by

$$\mathbf{K} = \begin{pmatrix} 2 & 3 & 8 \\ 4 & -2 & 0 \end{pmatrix} \overset{B}{\underset{\sim}{}} \begin{pmatrix} 1 & \frac{3}{2} & 4 \\ 4 & -2 & 0 \end{pmatrix}$$

The letter over the equivalence sign indicates the type of elementary row operation that was used.

The complete set of matrices that were used to reduce $\mathbf{K}$ to $\mathbf{A}$ is as follows:

$$\begin{aligned}
\mathbf{K} = \begin{pmatrix} 2 & 3 & 8 \\ 4 & -2 & 0 \end{pmatrix} &\overset{B}{\underset{\sim}{}} \begin{pmatrix} 1 & \frac{3}{2} & 4 \\ 4 & -2 & 0 \end{pmatrix} \\
\overset{C}{\underset{\sim}{}} \begin{pmatrix} 1 & \frac{3}{2} & 4 \\ 0 & -8 & -16 \end{pmatrix} &\overset{B}{\underset{\sim}{}} \begin{pmatrix} 1 & \frac{3}{2} & 4 \\ 0 & 1 & 2 \end{pmatrix} \\
\overset{C}{\underset{\sim}{}} \begin{pmatrix} 1 & 0 & 1 \\ 0 & 1 & 2 \end{pmatrix} &= \mathbf{A}
\end{aligned}$$

(54)

In general, suppose that the system of linear equations

$$\begin{aligned} ax + by &= c \\ dx + ey &= f \end{aligned} \qquad a, b, c, d \neq 0$$

has a unique solution. This solution can be found by applying the method for solving two linear equations in two variables:

Step 1   The corresponding matrix $\mathbf{K}$ is

$$\mathbf{K} = \begin{pmatrix} a & b & c \\ d & e & f \end{pmatrix}$$

Step 2   We must now try to reduce $\mathbf{K}$ to a solution matrix by using elementary row operations. This can be done using matrices patterned after the matrices of equation (54). Applying this method yields the following sequence of matrices:

General Form for the Reduction of K to a Solution Matrix

$$\mathbf{K} = \begin{pmatrix} a & b & c \\ d & e & f \end{pmatrix} \underset{\sim}{\overset{B}{}} \begin{pmatrix} 1 & g & h \\ d & e & f \end{pmatrix}$$

$$\underset{\sim}{\overset{C}{}} \begin{pmatrix} 1 & g & h \\ 0 & i & j \end{pmatrix} \underset{\sim}{\overset{B}{}} \begin{pmatrix} 1 & g & h \\ 0 & 1 & k \end{pmatrix} \tag{55}$$

$$\underset{\sim}{\overset{C}{}} \begin{pmatrix} 1 & 0 & m \\ 0 & 1 & k \end{pmatrix}$$

where g, h, i, j, k, m are numbers obtained by applying the indicated row operations. The solution is $x = m$, $y = k$.

If $d = 1$, it is simpler to begin by interchanging rows 1 and 2 (a type A operation).

If the sequence of operations listed in equation (55) is not possible, the system of equations either has no solution, or it has an infinite number of solutions. These cases will be studied in the next section.

The following example is an application of the solution method to a real-life situation.

Example 27   The Hi-Price Pharmaceutical Company produces drugs A and B in packets of ten capsules. One packet of drug A requires 5 hours of compounding and 20 hours of distillation; one packet of drug B requires 4 hours of compounding and 15 hours of distillation. The company has 210 hours of compounding time and 800 hours of distillation time available next week. How many capsules of each drug should it produce next week if the company wishes to utilize all of its time?

Solution   Let x and y be the numbers of packets of drugs A and B, respectively, that the company should produce. Since one packet of drug A requires 5 hours of compounding, $5x$ hours will be spent compounding drug A next week. Similarly, $4y$ hours will be spent compounding drug B. It follows that the total amount of compounding time that will be spent is $5x + 4y$. Since the company wishes to utilize all 210 hours of available compounding time,

$$5x + 4y = 210 \tag{56}$$

Similarly, since the company wishes to use all available distillation time,

$$20x + 15y = 800 \tag{57}$$

Equations (56)–(57) can be solved by the matrix method for solving two equations in two unknowns:

Step 1   *Replace equations (56)–(57) by the corresponding matrix* **K**.

$$\mathbf{K} = \begin{pmatrix} 5 & 4 & 210 \\ 20 & 15 & 800 \end{pmatrix}$$

Step 2   *If possible, reduce* **K** *to a solution matrix by means of elementary row operations.* The general form for the reduction of **K** is given in equation (55). Using this guide, we obtain

$$
\mathbf{K} = \begin{pmatrix} 5 & 4 & 210 \\ 20 & 15 & 800 \end{pmatrix} \overset{\mathbf{B}}{\underset{\sim}{}} \begin{pmatrix} 1 & \frac{4}{5} & 42 \\ 20 & 15 & 800 \end{pmatrix}
$$

$$
\overset{\mathbf{C}}{\underset{\sim}{}} \begin{pmatrix} 1 & \frac{4}{5} & 42 \\ 0 & -1 & -40 \end{pmatrix} \overset{\mathbf{B}}{\underset{\sim}{}} \begin{pmatrix} 1 & \frac{4}{5} & 42 \\ 0 & 1 & 40 \end{pmatrix}
$$

$$
\overset{\mathbf{C}}{\underset{\sim}{}} \begin{pmatrix} 1 & 0 & 10 \\ 0 & 1 & 40 \end{pmatrix}
$$

If we delete the last column from the last matrix, we have the identity matrix I_2. It follows that the last matrix in Step 2 is a solution matrix; the solution appears in its last column. Using this matrix, the company should produce 10 packets of drug A and 40 packets of drug B. In other words, the company should produce 100 capsules of A and 400 capsules of B. This solution may be verified by showing that (10,40) satisfies equations (56) and (57).

6.6 EXERCISES

1.   Write the matrices that correspond to the following systems of linear equations.

   (a) $2x - (0.3)y = 8$ **(b)** $14x + 3y = 12$ **(c)** $4r + 3s = 6$

   $4x - \quad 90y = 7$ $6x - 3y = 15$ $7s + 2r = 4$

2.   **(a)** Write the systems of equations that correspond to the following matrices.

   (i) $\begin{pmatrix} 3 & 4 & 0 \\ -1 & 2 & 7 \end{pmatrix}$ *(ii)* $\begin{pmatrix} 40 & -5 & \frac{1}{2} \\ 7 & 9 & -6 \end{pmatrix}$

   (iii) $\begin{pmatrix} \pi & 0.5 & -9 \\ 2 & 6 & 3\pi \end{pmatrix}$ *(iv)* $\begin{pmatrix} 1 & 0 & 4 \\ 0 & 1 & 2 \end{pmatrix}$

   (b) Which of the matrices in part **(a)** are solution matrices?

3.   Perform the following elementary row operations on

   $$\mathbf{A} = \begin{pmatrix} 1 & 3 & 5 \\ 4 & 7 & 2 \end{pmatrix}$$

   (a) Multiply row 1 by 4.
   (b) Interchange rows 1 and 2.
   (c) Replace row 2 by the sum of row 2 and -4 times row 1.

4.   **(a)** Write the equations that correspond to matrix **A** of Exercise 3.
   (b) Write the equations that correspond to the matrices obtained in parts **(a)**, **(b)**, and **(c)** of Exercise 3.

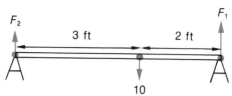

Figure 6.2

5. A 10-pound weight is suspended 2 feet from one end of a weightless 5-foot pole. The ends of the pole are resting on two benches. (See Figure 6.2.) The first bench exerts a force of F_1 pounds, and the second bench exerts a force of F_2 pounds. These forces can be found by solving the following equations:

$$F_1 + F_2 = 10$$
$$2F_1 - 3F_2 = 0$$

Find F_1 and F_2.

6. Use the matrix method to solve the following systems of linear equations.

   (a) $3x + 2y = 9$ **(b)** $5x - 3y = 14$

           $4x + 3y = 13$ $2x + 5y = 18$

7. Find $\begin{pmatrix} x \\ y \end{pmatrix}$ if—

   (a) $\begin{pmatrix} 1 & 2 \\ 3 & 4 \end{pmatrix} \begin{pmatrix} x \\ y \end{pmatrix} = \begin{pmatrix} 5 \\ 13 \end{pmatrix}$ **(b)** $\begin{pmatrix} 2 & 3 \\ -1 & 1 \end{pmatrix} \begin{pmatrix} x \\ y \end{pmatrix} = \begin{pmatrix} 16 \\ 2 \end{pmatrix}$

8. Jan has invested 1000 dollars in two savings certificates: one pays 6-percent interest per year, the other pays 7-percent interest. Find the amount of each investment if the total interest is 64 dollars. (Hint: Let x and y be the amounts invested at 6 and 7 percent, respectively. The interest from the first certificate is $0.06x$. Show that $0.06x + 0.07y = 64$ and $x + y = 1000$.)

9. A dietician has ascertained that patients on a certain diet require five units of protein (p) and four units of vitamins (v) for lunch. The nutritional values of a sandwich (s) and a piece of fruit (f) are given in the following matrix:

$$\begin{array}{cc} & \begin{array}{cc} s & f \end{array} \\ \begin{array}{c} p \\ v \end{array} & \begin{pmatrix} 2 & 1 \\ 1 & 2 \end{pmatrix} \end{array}$$

What should be served for lunch if the nutritional needs of the patients are to be met exactly?

6.7 SOLUTIONS OF n LINEAR EQUATIONS IN n VARIABLES

In Section 6.6 we learned how to use matrices to solve a system of two linear equations in two variables. In this section we will see how to extend this method to solving n linear equations in n variables.

The following definition and theorem are needed in order to solve systems of n linear equations in n variables. Definition 6.14 is a generalization of Definition 6.12.

Definition 6.14   An $n \times (n + 1)$ matrix **A** is a **solution matrix** if and only if the matrix formed by deleting the last column of **A** is $\mathbf{I}_n$.

THEOREM   6.3   Consider the system of linear equations

$$a_{11}x_1 + a_{12}x_2 + \cdots + a_{1n}x_n = c_1$$

$$a_{21}x_1 + a_{22}x_2 + \cdots + a_{2n}x_n = c_2$$

$$\vdots \qquad \vdots \qquad \vdots \qquad \vdots$$

$$a_{n1}x_1 + a_{n2}x_2 + \cdots + a_{nn}x_n = c_n$$

and let

$$\mathbf{K} = \begin{pmatrix} a_{11} & a_{12} & \cdots & a_{1n} & c_1 \\ a_{21} & a_{22} & \cdots & a_{2n} & c_2 \\ \vdots & \vdots & & \vdots & \vdots \\ a_{n1} & a_{n2} & \cdots & a_{nn} & c_n \end{pmatrix}$$

be the corresponding matrix.

(a) If **K** is equivalent to a solution matrix

$$\begin{pmatrix} 1 & 0 & 0 & \cdots & 0 & k_1 \\ 0 & 1 & 0 & \cdots & 0 & k_2 \\ \vdots & \vdots & \vdots & & \vdots & \vdots \\ 0 & 0 & 0 & \cdots & 1 & k_n \end{pmatrix}$$

then the system has exactly *one* solution, $(k_1, k_2, \ldots, k_n)$.

(b) If **K** is equivalent to a matrix having at least one zero row (i.e., each entry of the row is 0), then the system has an *infinite* number of solutions.

(c) If **K** is equivalent to a matrix having a row

$$0 \quad 0 \quad 0 \quad \cdots \quad 0 \quad 1$$

then the system has *no* solution.

Moreover, **K** is equivalent to exactly one of these three types of matrices.

The following examples show how to use Theorem 6.3.

Example 28   Reports from our secret agents indicate that an enemy munitions factory received 30 tons of aluminum, 30 tons of steel, and 36 tons of copper last year. The reports also state that the entire amount of material was used to make three types of weapons, A, B, and C. It is known that each weapon A requires 1 ton of aluminum, 2 tons of steel, and 1 ton of copper. Each weapon B requires 2 tons of aluminum, 1 ton of steel, and 3 tons of copper. Each weapon C requires 2 tons of aluminum, 1 ton of steel, and 2 tons of copper. How many weapons of each type were produced by the factory last year?

Solution   Let x, y, and z be the numbers of weapons A, B, and C produced. Since 30 tons of aluminum were used and A, B, and C require 1, 2, and 2 tons of aluminum, respectively, it follows that

$$1x + 2y + 2z = 30 \tag{58}$$

Similarly, analysis of the steel usage implies that

$$2x + 1y + 1z = 30 \tag{59}$$

and consideration of the copper usage implies that

$$1x + 3y + 2z = 36 \tag{60}$$

Applying Theorem 6.3, the system of equations (58)–(60) can be solved by replacing the system by the corresponding matrix and reducing the matrix (if possible) to an equivalent solution matrix.

Step 1   The corresponding matrix is

$$\mathbf{K} = \begin{pmatrix} 1 & 2 & 2 & 30 \\ 2 & 1 & 1 & 30 \\ 1 & 3 & 2 & 36 \end{pmatrix}$$

Step 2   We must now try to reduce $\mathbf{K}$ to an equivalent solution matrix. We will extend the method used in Section 6.6 to systems of n linear equations in n variables as follows:

(a) We transform $\mathbf{K}$ into an equivalent matrix $\mathbf{L}$ having 1 in row 1, column 1.

$$\mathbf{L} = \begin{pmatrix} 1 & x & x & x \\ x & x & x & x \\ x & x & x & x \end{pmatrix}$$

(b) We then transform $\mathbf{L}$ into an equivalent matrix $\mathbf{M}$ having 0 in every position in the first column except the first.

$$\mathbf{M} = \begin{pmatrix} 1 & x & x & x \\ 0 & x & x & x \\ 0 & x & x & x \end{pmatrix}$$

(c) Using two steps similar to steps (a) and (b), we transform $\mathbf{M}$ into the equivalent matrix

$$\mathbf{N} = \begin{pmatrix} 1 & 0 & x & x \\ 0 & 1 & x & x \\ 0 & 0 & x & x \end{pmatrix}$$

(d) Using steps similar to steps (a) and (b), we transform $\mathbf{N}$ into a solution matrix.

Following this plan,

$$\mathbf{K} = \begin{pmatrix} 1 & 2 & 2 & 30 \\ 2 & 1 & 1 & 30 \\ 1 & 3 & 2 & 36 \end{pmatrix} \overset{C}{\sim} \begin{pmatrix} 1 & 2 & 2 & 30 \\ 0 & -3 & -3 & -30 \\ 1 & 3 & 2 & 36 \end{pmatrix}$$

$$\overset{C}{\sim} \begin{pmatrix} 1 & 2 & 2 & 30 \\ 0 & -3 & -3 & -30 \\ 0 & 1 & 0 & 6 \end{pmatrix} \overset{B}{\sim} \begin{pmatrix} 1 & 2 & 2 & 30 \\ 0 & 1 & 1 & 10 \\ 0 & 1 & 0 & 6 \end{pmatrix}$$

$$\overset{C}{\sim} \begin{pmatrix} 1 & 2 & 2 & 30 \\ 0 & 1 & 1 & 10 \\ 0 & 0 & -1 & -4 \end{pmatrix} \overset{C}{\sim} \begin{pmatrix} 1 & 0 & 0 & 10 \\ 0 & 1 & 1 & 10 \\ 0 & 0 & -1 & -4 \end{pmatrix}$$

$$\overset{B}{\sim} \begin{pmatrix} 1 & 0 & 0 & 10 \\ 0 & 1 & 1 & 10 \\ 0 & 0 & 1 & 4 \end{pmatrix} \overset{C}{\sim} \begin{pmatrix} 1 & 0 & 0 & 10 \\ 0 & 1 & 0 & 6 \\ 0 & 0 & 1 & 4 \end{pmatrix}$$

If we delete the last column from the last matrix, we obtain the identity matrix I_3. It follows that this last matrix is a solution matrix; the solution (10,6,4) appears in the last column. The factory therefore produced 10 weapons of type A, 6 weapons of type B, and 4 weapons of type C.

Example 29   Further intelligence reports indicate that a second factory produces weapons E, F, and G. The total amount (in tons) of aluminum (*A*), steel (*S*), and copper (*C*) required by each weapon is given in matrix **R**. The total amount (in tons) of each material received by the factory last year is given in matrix **M**.

$$\mathbf{R} = \begin{matrix} A \\ S \\ C \end{matrix} \begin{matrix} E & F & G \\ \begin{pmatrix} 1 & 1 & 3 \\ 2 & 2 & 6 \\ 2 & 3 & 8 \end{pmatrix} \end{matrix} \qquad \mathbf{M} = \begin{matrix} A \\ S \\ C \end{matrix} \begin{pmatrix} 30 \\ 60 \\ 80 \end{pmatrix}$$

How many weapons of each type were produced by the factory if the entire supply of the three materials was used to produce the three types of weapons?

Solution   Let *x*, *y*, and *z* be the numbers of weapons E, F, and G produced, respectively. Then it follows from the facts given that

$$x + y + 3z = 30 \tag{61}$$

$$2x + 2y + 6z = 60 \tag{62}$$

$$2x + 3y + 8z = 80 \tag{63}$$

(See the solution of Example 28.) The corresponding matrix is **K**, where

$$\mathbf{K} = \begin{pmatrix} 1 & 1 & 3 & 30 \\ 2 & 2 & 6 & 60 \\ 2 & 3 & 8 & 80 \end{pmatrix} \sim \begin{pmatrix} 1 & 1 & 3 & 30 \\ 0 & 0 & 0 & 0 \\ 0 & 1 & 2 & 20 \end{pmatrix} = \mathbf{L}$$

By Theorem 6.3(b), the zero row of **L** tells us that the system has an infinite number of solutions. In order to find these solutions, we can simplify the matrix further using elementary row operations.

$$\mathbf{L} = \begin{pmatrix} 1 & 1 & 3 & 30 \\ 0 & 0 & 0 & 0 \\ 0 & 1 & 2 & 20 \end{pmatrix} \sim \begin{pmatrix} 1 & 1 & 3 & 30 \\ 0 & 1 & 2 & 20 \\ 0 & 0 & 0 & 0 \end{pmatrix}$$

$$\sim \begin{pmatrix} 1 & 0 & 1 & 10 \\ 0 & 1 & 2 & 20 \\ 0 & 0 & 0 & 0 \end{pmatrix}$$

The *three* equations (61)–(63) are therefore equivalent to the *two* equations

$$x + \qquad z = 10 \tag{64}$$

$$y + 2z = 20 \tag{65}$$

(Obviously, any solution (x,y,z) will satisfy $0x + 0y + 0z = 0$. This equation can be omitted, just as we may omit all terms of an equation whose coefficients are 0.)

The solutions of equations (64)–(65) can be determined by arbitrarily fixing z and letting $x = 10 - z$ and $y = 20 - 2z$. Although the system has an infinite number of solutions, we are not interested in solutions such as $(11,22,-1)$, $(-1,-2,11)$, or $(0.5,1,9.5)$, because we know from the original problem that x, y, and z must be nonnegative integers. It is also clear from equations (61)–(63) that, even if only weapon G is produced, at most 10 weapons of type G can be made from the available materials. It follows that z can only be 0, 1, 2, ..., or 10. Therefore, the only possible solutions to the problem are $(10,20,0)$, $(9,18,1)$, ..., $(0,0,10)$. The factory may have produced 10, 20, 0 weapons of type A, B, C, respectively, or 9, 18, 1 weapons of type A, B, C, and so on. Since each of these answers fits the data, our spies have given us insufficient information to answer the question.

Look again at the system of equations (61)–(63). We can easily see that equation (62) is superfluous, because it is twice equation (61). Any solution of equation (62) must be a solution of equation (61), and vice versa. Therefore, equation (62) adds no new information to the system and may be omitted. It follows that the system of three equations, (61)–(63), can be reduced to a system of two equations, (61) and (63). This reduction is reflected in the zero row of matrix **L** of Example 29. In general, *a zero row in a matrix shows that the original system of n linear equations can be reduced to a system of n − 1 linear equations.* In Exercise 10, following this section, the reader will reduce a system of four linear equations in three variables to a system of three linear equations in three variables and then solve the system by the methods of this section.

Example 30   Intelligence information indicates that the solution of the following system of linear equations will reveal the quantities of various weapons manufactured at an enemy factory:

$$x + y + z = 3$$
$$x + y + z = 4$$
$$x + y - z = 1$$

If possible, solve the system.

Solution   The corresponding matrix

$$\mathbf{K} = \begin{pmatrix} 1 & 1 & 1 & 3 \\ 1 & 1 & 1 & 4 \\ 1 & 1 & -1 & 1 \end{pmatrix} \overset{C}{\sim} \begin{pmatrix} 1 & 1 & 1 & 3 \\ 0 & 0 & 0 & 1 \\ 1 & 1 & -1 & 1 \end{pmatrix} = \mathbf{L}$$

It follows from Theorem 6.3(c) that no solution exists. Notice that row 2 of $\mathbf{L}$ represents the equation

$$0x + 0y + 0z = 1$$

This is clearly false for any values of x, y, and z. The information given by our spies is therefore inconsistent.

Thus far, we have only considered the solution of n linear equations in n variables. Let us briefly consider the general case, the solution of m linear equations in n variables.

(a) If $m < n$ The system either has an infinite number of solutions (see equations (64)–(65)), or has no solution (see Exercise 11(d) following this section).

(b) If $m > n$ The system may either be reduced to a system of n linear equations in n variables (see Exercise 10), or it has no solution (see Exercise 11(c)).

(c) If $m = n$ Theorem 6.3 applies.

6.7 EXERCISES

1.   The matrix

$$\mathbf{K} = \begin{pmatrix} 1 & 2 & 3 \\ 0 & 0 & 0 \\ 2 & 3 & 5 \end{pmatrix}$$

corresponds to a system of three linear equations in three variables. Does this system have one solution, no solution, or an infinite number of solutions? Explain your answer.

2.   Redo Exercise 1 for the matrix

$$\mathbf{K} = \begin{pmatrix} 1 & 2 & 3 \\ 1 & 4 & 6 \\ 0 & 0 & 1 \end{pmatrix}$$

3.   Let

$$K = \begin{pmatrix} 2 & 3 & 1 & 10 \\ 1 & -3 & 2 & -1 \\ 2 & 1 & 1 & 6 \end{pmatrix}, \qquad L = \begin{pmatrix} 1 & -3 & 2 & -1 \\ 2 & 3 & 1 & 10 \\ 2 & 1 & 1 & 6 \end{pmatrix}$$

$$M = \begin{pmatrix} 1 & -3 & 2 & -1 \\ 0 & 9 & -3 & 12 \\ 0 & 7 & -3 & 8 \end{pmatrix}, \qquad N = \begin{pmatrix} 1 & -3 & 2 & -1 \\ 0 & 1 & -\frac{1}{3} & \frac{4}{3} \\ 0 & 7 & -3 & 8 \end{pmatrix}$$

$$P = \begin{pmatrix} 1 & 0 & 1 & 3 \\ 0 & 1 & -\frac{1}{3} & \frac{4}{3} \\ 0 & 0 & -\frac{2}{3} & -\frac{4}{3} \end{pmatrix}$$

Why are the following matrices equivalent?

(a) **K** and **L**   (b) **L** and **M**   (c) **M** and **N**
(d) **N** and **P**   (e) **K** and **P**

4.   Are the matrices of Exercise 3 equivalent to a solution matrix? Explain your answer.

5.   Find the matrices that correspond to the following systems of linear equations.

(a) $x + y + 2z = 3$ (b) $x + y + 2z = 3$
      $2x - y + 3z = 3$ $x - y + 4z = 4$
      $x + 3y - 4z = 5$ $3x - y + 10z = 11$
(c) $x + 2y + 3z = 0$
      $x - y + z = 1$
      $4x + 5y + 10z = 2$

6.   If possible, solve the systems of linear equations given in Exercise 5. Give one solution if the solution is unique. Give three solutions if there are infinitely many solutions.

7.   Find $\begin{pmatrix} x \\ y \\ z \end{pmatrix}$ if

$$\begin{pmatrix} 1 & 2 & 3 \\ 0 & 1 & 2 \\ 4 & 1 & 3 \end{pmatrix} \begin{pmatrix} x \\ y \\ z \end{pmatrix} = \begin{pmatrix} 4 \\ 2 \\ 7 \end{pmatrix}$$

8.   Find (x,y,z) if

$$\begin{pmatrix} x & 2x \\ y & 0 \end{pmatrix} + \begin{pmatrix} y & y-z \\ 2z & 3 \end{pmatrix} = \begin{pmatrix} 3 & 3 \\ 3 & 3 \end{pmatrix}$$

9.   If possible, solve the following system of linear equations. Give one solution if the solution is unique. Give three solutions if there are infinitely many solutions.

$$x + y + z + w = 4$$
$$x - y + z - w = 0$$
$$2x + y - z + 2w = 3$$
$$x - y + 2z - w = 1$$

10.   **(a)** Which equation can be omitted from the following system?

$$x + 2y + z = 6$$
$$2x + 3y + z = 9$$
$$3x + y - z = 4$$
$$3x + 6y + 3z = 18$$

   Explain your answer.
   (b) Solve the system.

11.   Which of the following systems are inconsistent (i.e., have no solution)?

(a) $x + y = 4$ **(b)** $x + y + z = 3$ **(c)** $x + y = 5$
     $x + y = 5$ $x + y - z = 2$ $2x - y = 7$
                     $2x + 2y = 6$ $x + 2y = 3$

(d) $x + y + z = 5$
     $2x + 2y + 2z = 8$

Explain your answers.

12.   A parent is preparing three types of cake—spice (S), vanilla (V), and fudge (F)—for a PTA cake sale. The number of units of the basic ingredients required for each cake are given in matrix **A**. The number of units of these ingredients that are available are given in vector **C**.

$$
\mathbf{A} = \begin{matrix} & S & V & F \\ \text{Eggs} \\ \text{Flour} \\ \text{Sugar} \end{matrix} \begin{pmatrix} 2 & 3 & 4 \\ 2 & 2 & 3 \\ \frac{1}{2} & 1 & 1 \end{pmatrix} \qquad \mathbf{C} = \begin{matrix} \text{Eggs} \\ \text{Flour} \\ \text{Sugar} \end{matrix} \begin{pmatrix} 19 \\ 15 \\ 5 \end{pmatrix}
$$

For example, spice cake requires 2 eggs, 2 cups of flour, and $\frac{1}{2}$ cup of sugar. The parent wishes to know how many cakes of each kind he should bake if he must use up all the available ingredients.
(a) Let

$$\mathbf{X} = \begin{pmatrix} x \\ y \\ z \end{pmatrix}$$

Show that the number of cakes of each kind that should be baked can be found by solving the matrix equation $AX = C$ (i.e., show that the problem can be solved by finding X such that $AX = C$ is satisfied).

(b) Solve the matrix equation $AX = C$.

(c) How many spice cakes should the parent bake?

13. A store sells peanuts at $1.20 per pound, raisins at $2.00 per pound, and chocolate chips at $2.40 per pound. The storekeeper would like to make a 100-pound mixture, half chocolate chips, that would sell for $1.96 per pound. How many pounds of each item should he put in the mixture?

14. A chemist has three solutions of nitric acid—a 10-percent solution, a 15-percent solution, and a 20-percent solution. How can these solutions be mixed in order to obtain 100 cubic centimeters of a 12-percent solution? (Hint: Show that the following system of linear equations must be solved:

$$x + \quad y + \quad z = 100$$
$$(0.10)x + (0.15)y + (0.20)z = \quad 12)$$

15. Redo Exercise 14 under the assumption that—
(a) the chemist does not have any 20-percent solution left
(b) the chemist does not have any 10-percent solution left.

16. The results of an experiment indicate that the resistances R_1, R_2, and R_3 (in ohms) of three electrical resistors satisfy the following equations:

$$4R_1 + 2R_2 + \quad R_3 = 27$$
$$2R_1 + 3R_2 + 2R_3 = 27$$
$$2R_1 + \quad R_2 + 2R_3 = 21$$

Find the resistances.

17. Let A be any matrix. Prove that $A \sim A$. (Hint: Find an elementary row operation that will transform A into A.)

Key Word Review

Each key word is followed by the number of the section in which it is defined.

matrix   6.1
square matrix   6.1
row vector   6.1
column vector   6.1
components   6.1
equal matrices   6.1
matrix addition   6.2
scalar multiplication   6.2
zero matrix   6.2
matrix multiplication   6.3

identity matrix   6.3
solution   6.4
solution of a system   6.4
equivalent equations   6.4
linear equation   6.4
equivalent systems   6.5
corresponding matrix   6.6
solution matrix   6.6, 6.7
elementary row operations   6.6
equivalent matrices   6.6

CHAPTER 6   **REVIEW EXERCISES**

1. Define each word in the Key Word Review.

2. What do the following symbols represent?
    (a) I_n (b) $A \sim B$ (c) Φ

3. Let

$$A = \begin{pmatrix} 1 & 2 \\ 3 & 0 \end{pmatrix}, \qquad B = \begin{pmatrix} 1 & 5 \\ 2 & 4 \end{pmatrix}, \qquad C = (1,2)$$

If possible, calculate—
    (a) $A + B$ (b) $B + A$ (c) AB (d) BA (e) AC
    (f) CA (g) $2A$ (h) A^2 (i) C^2

4. Using matrices **A**, **B**, and **C** of Exercise 3, calculate (if possible)—
    (a) $2A + 3B$ (b) $A - B^2$ (c) AI_2 (d) $A\Phi$
    (e) $A + I_2$ (f) $A + \Phi$

5. Find (x,y) if—

(a) $\begin{pmatrix} 1 & 2 \\ 3 & x \end{pmatrix} = \begin{pmatrix} y & 2 \\ 3 & 1 \end{pmatrix}$

(b) $\begin{pmatrix} 1 & 2 \\ x & 1 \end{pmatrix} + \begin{pmatrix} 2 & x \\ y & 0 \end{pmatrix} = \begin{pmatrix} 3 & 4 \\ 5 & 1 \end{pmatrix}$

(c) $\begin{pmatrix} 1 & 2 \\ 3 & 4 \end{pmatrix} \begin{pmatrix} x \\ y \end{pmatrix} = \begin{pmatrix} 7 \\ 17 \end{pmatrix}$

6. A hardware store has two branches. Branch A has forty screwdrivers, fifty wrenches, twenty pliers, and seventy hammers in stock. Branch B has twenty screwdrivers, twenty wrenches, ten pliers, and fifty hammers in stock. The screwdrivers, wrenches, pliers, and hammers cost the store $2.00, $2.50, $1.50, and $2.25 each, respectively.

(a) Display the store's inventory at the two branches as a 2 × 4 matrix **X**.

(b) Display the store's cost per item as a column vector **Y**.

(c) Use matrix multiplication to find the total cost of these items for each branch.

7. The store of Exercise 6 wishes to charge its customers twice its cost for each item. Write a matrix that will display the prices that customers will be charged.

8. Martin wishes to buy bread (B), cheese (C), milk (M), and eggs (E). From newspaper ads, he learns that the prices per unit of these items at stores F, G, and H are those displayed in matrix **A**. The numbers of units of these items that he requires are given in matrix **Y**.

$$
\mathbf{A} = \begin{array}{c} \\ F \\ G \\ H \end{array}
\begin{array}{cccc} B & C & M & E \\ \begin{pmatrix} 0.50 & 0.83 & 1.20 & 0.70 \\ 0.48 & 0.85 & 1.15 & 0.65 \\ 0.51 & 0.90 & 1.25 & 0.60 \end{pmatrix} \end{array}
\qquad
\mathbf{Y} = \begin{array}{c} B \\ C \\ M \\ E \end{array}
\begin{pmatrix} 2 \\ 1 \\ 3 \\ 2 \end{pmatrix}
$$

(a) Use matrix multiplication to find the total cost of shopping at each store.

(b) Which store should he shop at if he wishes to minimize his food bill?

(c) What will his total food bill be if he shops in the store that minimizes his total food bill?

9. (a) Display the minimum cost for each item of Exercise 8 in a 1 × 4 matrix.

(b) Use matrix multiplication to find Martin's total food bill if he buys each item at the store that sells it at the lowest price.

10. Which of the following equations are linear?

(a) $x + 2y = 7$ (b) $xy = 5$ (c) $x^2 - 2y + 3z = 10$

(d) $x + 3y = 5z + 2w$ (e) $r = s + t$

11. Determine which of the following vectors are solutions of the equation

$$2x + 3y = 10$$

(a) $(4,0)$ (b) $(5,0)$ (c) $(0,5)$ (d) $(2,2)$

12. Determine which of the following vectors are solutions of the system of equations

$$2x + 3y = 5$$
$$4x - \; y = 3$$

(a) $(3,0)$ (b) $(\tfrac{5}{2},0)$ (c) $(1,1)$

13. Write the matrices that correspond to the following systems of linear equations.

(a) $2x + 5y = 12$ 　　**(b)** $x + y - z = 3$ 　　**(c)** $x + 2y = 3$

　　　$3x - 4y = -5$ 　　　　　$2x - 2y + z = -2$ 　　　　$2x + 4y = 5$

　　　　　　　　　　　　　　　　$x + 2y + 3z = 5$

(d) $x + y + 2z = 1$

　　　$2x + 3y - 4z = 2$

　　　$5x + 6y + 2z = 5$

14. If possible, solve the systems of linear equations of Exercise 13. Give the solution if it is unique. Give three solutions if there are infinitely many solutions.

15. A zoo is able to purchase three brands of animal food: A, B, and C. The numbers of units of protein (p), vitamins (v), and minerals (m) contained in one pound of each brand are given in matrix **N**. The daily nutritional requirements of a certain animal are given in matrix **R**.

$$\mathbf{N} = \begin{array}{c} \\ A \\ B \\ C \end{array} \begin{array}{ccc} p & v & m \\ \begin{pmatrix} 1 & 2 & 3 \\ 2 & 1 & 2 \\ 3 & 3 & 1 \end{pmatrix} \end{array}, \qquad \mathbf{R} = \begin{array}{c} p \\ v \\ m \end{array} \begin{pmatrix} 8 \\ 6 \\ 0 \end{pmatrix}$$

How many pounds of each brand of food should the zookeepers give the animal if they wish to meet its nutritional requirements exactly?

Suggested Readings

1. Batschelet, E. *Introduction to Mathematics for Life Scientists*. Edited by K. Krickeberg et al. Biomathematics, vol. 2. New York: Springer-Verlag, New York, 1971.
2. Bavelas, A. "A Mathematical Model for Group Structure." *Applied Antropology*, 1948, pp. 16–30.
3. Festinger, L. "The Analysis of Sociograms Using Matrix Algebra." *Human Relations* 2 (1949):153–58.
4. Forsyth, E., and Katz, L. "A Matrix Approach to the Analysis of Sociometric Data: Preliminary Report." *Sociometry* 9 (1946):340–47.
5. Ijiri, Yuji. *Management Goals and Accounting for Control*. Chicago: Rand McNally & Co., 1965.

CHAPTER 7 Markov
Chains

INTRODUCTION

The rat in the photograph on the facing page is learning to find his way through a maze. After many practice trials, the rat will be able to find his way through the maze without false turns. However, the first time he is placed in an unfamiliar maze, the rat's behavior will probably be random: if he has several choices, he will be equally likely to make each of them. It will probably take many of these random moves before the rat chances upon the exit from the maze.

  The situation in which the inexperienced rat finds himself is one of a whole class of situations that can be mathematically described by a branch of probability called *Markov chain theory*. A *Markov chain* is a mathematical device for studying certain kinds of processes that move through a series of stages. In the case of the rat, the stages are the series of choices that the rat must make to find his way out. In this chapter we will apply our knowledge of matrices to this theory. We will also see how Markov chain theory can be used to predict the weather and to study the inheritance of genetic traits over several generations.

7.1 MARKOV CHAINS AND TRANSITION MATRICES

To introduce the concept of a Markov chain, we will consider a simple example. If a rat is placed in a compartment of the maze diagrammed in Figure 7.1, he can move into one of the other compartments. In studying the rat's movements in the maze, we might ask the following question: What is the probability that the rat will be in compartment 4 after two moves if he is originally placed in compartment 1? In Example 1 we begin an analysis that will eventually allow us to answer this question.

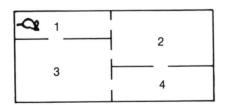

Figure 7.1

Example 1 Suppose a rat is placed in compartment 1 of the maze of Figure 7.1. We will call the initial placement of the rat the **initial trial**. Let E_i be the event "the rat is in compartment i." Then the **initial outcome** (i.e., the outcome of the initial trial) is E_1.

(1) From compartment 1, the rat can move into compartment 2 or compartment 3. (There is no entrance between compartments 1 and 4.) Let us arbitrarily assume that he moves into compartment 2. This movement is called trial 1; its outcome is E_2.

(2)   From compartment 2, he can move into compartment 1, 3, or 4. Let us
           suppose he moves into compartment 4. Then the outcome of trial 2 is E_4.

(3)   From compartment 4, he can go into compartment 2 or 3. We will suppose
           that he goes into compartment 2. Then the outcome of trial 3 is E_2.

This process can continue indefinitely. The sequence of the rat's movements
has the following characteristics:

(a) Each movement (trial) can only result in one of $m = 4$ possible outcomes:
    E_1, E_2, E_3, or E_4.[1]
(b) The compartment the rat enters next depends only on the compartment he
    is in now. It does not depend on the compartment he was in two moves
    ago (or three moves ago, etc.), and it does not depend on whether this is his
    first move, his second move, or some other move.

In general, consider a sequence of consecutive trials having the following
characteristics:

(a) Each trial can only result in one of m possible outcomes $E_1, E_2, \ldots, E_m$. These
    outcomes are called *states*.
(b) The probability that E_i occurs on the nth trial depends only on the outcome of
    the $(n - 1)$st trial. Knowledge of the outcomes of the trials preceding the
    $(n - 1)$st trial does not affect this probability; nor does this probability
    depend on the number n of the trial.

A sequence of consecutive trials with these characteristics forms a *homogeneous
Markov chain*.[2]

Characteristic **b** of the preceding informal definition is intuitive. A formal
definition will be given after we define and discuss a certain random variable X_n.
This random variable will also greatly simplify our description of sequences of trials
like the rat's movements in Example 1.

Let X_n denote the outcome of the nth trial as follows:

$$X_n = i \quad \text{if and only if } E_i \text{ occurs on the } n\text{th trial}$$

In particular, let X_o denote the outcome of the initial trial. The following examples
will clarify this definition.

Example 2   Consider the sequence of movements of a rat placed in a maze, as
     described in Example 1.

(a) Using our new notation, $(X_7 = 3)$ is the event "the rat chooses compart-
    ment 3 on his seventh move"; $(X_8 = 2, X_9 = 3)$ is the event "the rat
    chooses compartment 2 on his eighth move and compartment 3 on his
    ninth move."

---

1. These are the only possible outcomes; however, some of these outcomes may not be
   possible on a particular trial.

2. In a *nonhomogeneous Markov chain*, the probability that E_i occurs on the nth trial depends
   on the result of the $(n - 1)$st trial, and *also* depends on n, the number of the trial.

(b) The rat's first few movements, which were verbally described in Example 1, can be more concisely described as $(X_o = 1)$, $(X_1 = 2)$, $(X_2 = 4)$, and $(X_3 = 2)$, because the outcomes of the initial, first, second, and third moves are E_1, E_2, E_4, and E_2, respectively.

(c) $P(X_7 = 4 | X_6 = 1) = 0$ because compartments 1 and 4 are not connected.

Example 3   Suppose a second rat is initially placed in compartment 4 of the maze of Example 1. Furthermore, assume that the rat is equally likely to enter any connecting compartment.

(a) If the rat chose compartments 2, 4, 3, 1 on his first four moves (in that order), what is the probability that he chooses compartment 2 on his fifth move?

(b) If the rat is in compartment 1 on his eighth move, what is the probability that he will be in compartment 2 on his ninth move?

Solution

(a) We need to calculate $P(X_5 = 2 | X_4 = 1, X_3 = 3, X_2 = 4, X_1 = 2, X_o = 4)$. However, since the compartment the rat enters on his fifth move depends only on the compartment he entered on his fourth move,

$$P(X_5 = 2 | X_4 = 1, X_3 = 3, X_2 = 4, X_1 = 2, X_o = 4)$$
$$= P(X_5 = 2 | X_4 = 1) \qquad (1)$$

Since there are two compartments connecting with compartment 1, and the rat is equally likely to choose either one of them, $P(X_5 = 2 | X_4 = 1) = \frac{1}{2}$.

(b) The probability that the rat is in compartment 2 next depends only on where he is now, not on the number n of the move. Hence,

$$P(X_9 = 2 | X_8 = 1) = P(X_5 = 2 | X_4 = 1) = \frac{1}{2} \qquad (2)$$

Equations similar to equations (1) and (2) are true for any sequence of choices of compartments. If the rat was first placed in compartment i_o ($i_o = 1,2,3,$ or 4), and then chooses compartments $i_1, i_2, \ldots, i_{n-1}$ (in that order), the probability that he chooses compartment i_n on the nth trial is

$$P(X_n = i_n | X_{n-1} = i_{n-1}, \ldots, X_1 = i_1, X_o = i_o)$$
$$= P(X_n = i_n | X_{n-1} = i_{n-1})$$

Moreover, since the choice of compartments does not depend on the number n of the move,

$$P(X_1 = j | X_o = i) = P(X_2 = j | X_1 = i) = \cdots$$
$$= P(X_n = j | X_{n-1} = i) = \cdots$$

We are now in a position to formally define a Markov chain.

Definition 7.1   Consider a sequence of consecutive trials such that—

(a) Each trial can only result in m possible outcomes $E_1, E_2, \ldots, E_m$. These outcomes are called **states.**

Let $X_n = i$ if and only if E_i occurs on the nth trial. If—

(b) $P(X_n = i_n | X_{n-1} = i_{n-1}, \ldots, X_1 = i_1, X_o = i_o) = P(X_n = i_n | X_{n-1} = i_{n-1})$
then the sequence of trials, or the associated sequence of random variables $X_1, X_2, \ldots$, forms a *Markov chain M*. If, in addition—

(c) $P(X_1 = j | X_o = i) = P(X_2 = j | X_1 = i) = \cdots = P(X_n = j | X_{n-1} = i) = \cdots$
then M is a **homogeneous Markov chain.**

Since we will study only homogeneous Markov chains in this book, the adjective "homogeneous" will usually be omitted.

Definition 7.1 does not rely on our intuitive ideas of what a Markov chain should be. If (a), (b), and (c) of Definition 7.1 are satisfied, then M is a (homogeneous) Markov chain even if there is no intuitive basis for believing this. (See Example 8 and Excercise 2 at the end of this section.)

Let M be a Markov chain. If E_i occurs on the nth trial (i.e., if $X_n = i$), we will say that M is in state E_i (on the nth trial). If a transition is made from state E_i to state E_j, we will say that M moves from E_i to E_j.

Let

$$p_{ij} = P(X_1 = j | X_o = i) = P(X_2 = j | X_1 = i)$$
$$= P(X_3 = j | X_2 = i) = \cdots$$

denote the probability that a transition is made from state E_i to state E_j. The conditional probability p_{ij} is called a **transition probability.** The transition probabilities can be conveniently displayed in a square matrix called a *transition matrix*.

Definition 7.2   Suppose M is a Markov chain with m states $E_1, E_2, \ldots, E_m$. The **transition matrix T** of M is the $m \times m$ matrix (p_{ij}). That is, the entry in the ith row, jth column of **T** is the transition probability p_{ij}.

$$
\mathbf{T} = \begin{array}{c} \\ E_1 \\ E_2 \\ \vdots \\ E_m \end{array}
\begin{array}{cccc}
E_1 & E_2 & \cdots & E_m \\
\left(\begin{array}{cccc}
p_{11} & p_{12} & \cdots & p_{1m} \\
p_{21} & p_{22} & \cdots & p_{2m} \\
\vdots & \vdots & & \vdots \\
p_{m1} & p_{m2} & \cdots & p_{mm}
\end{array} \right)
\end{array}
$$

The transition probability p_{21}, for example, is in the interaction of the second row and first column. This is the row labeled E_2 and the column labeled E_1. The row and column headings remind us that p_{21} is the probability that M moves from state E_2 to state E_1. The headings are sometimes omitted.

The following examples will illustrate the definitions of a Markov chain and transition matrix.

Example 4 Consider the Markov chain formed by the rat's movements in Example 1. The transition matrix for this Markov chain is:

$$\mathbf{T} = \begin{matrix} & \begin{matrix} E_1 & E_2 & E_3 & E_4 \end{matrix} \\ \begin{matrix} E_1 \\ E_2 \\ E_3 \\ E_4 \end{matrix} & \begin{pmatrix} 0 & \frac{1}{2} & \frac{1}{2} & 0 \\ \frac{1}{3} & 0 & \frac{1}{3} & \frac{1}{3} \\ \frac{1}{3} & \frac{1}{3} & 0 & \frac{1}{3} \\ 0 & \frac{1}{2} & \frac{1}{2} & 0 \end{pmatrix} \end{matrix}$$

For example, if the rat is in compartment 2 (state E_2), he can move into compartment 1, 3, or 4. (See Figure 7.1.) Since we assumed in Example 3 that the rat is equally likely to enter any connecting compartment, $p_{21} = p_{23} = p_{24} = \frac{1}{3}$. These probabilities are displayed in the second row of $\mathbf{T}$. Notice that $p_{22} = 0$. This is because we are only interested in the rat's movement from one compartment to another.

Similarly, if the rat is in compartment 1, he can move into compartment 2 or compartment 3, but not into compartment 4. Therefore, $p_{12} = p_{13} = \frac{1}{2}$, and $p_{14} = p_{11} = 0$. These values appear in the first row of $\mathbf{T}$.

Example 5 Let M be a Markov chain with two states E_1 and E_2 and transition probabilities $p_{11} = 0.2$, $p_{12} = 0.8$, $p_{21} = 0.3$, and $p_{22} = 0.7$. Find the transition matrix $\mathbf{T}$ of M.

Solution Since M has two states, $\mathbf{T}$ has two rows and two columns. The entry in row 1, column 1 is $p_{11} = 0.2$; the entry in row 1, column 2 is $p_{12} = 0.8$. Similarly, the entries in the second row are $p_{21} = 0.3$ and $p_{22} = 0.7$, in that order. Thus,

$$\mathbf{T} = \begin{pmatrix} 0.2 & 0.8 \\ 0.3 & 0.7 \end{pmatrix}$$

Example 6 Al and Bob are evenly matched on the first game that they play. However, every time Al wins a game, he gains confidence and the probability that he wins the next game is 0.6. On the other hand, every time Al loses, he becomes very discouraged, and the probability that he wins the next game is 0.3. Find the transition matrix of this Markov chain.

Solution This sequence of games is a Markov chain with $m = 2$ states: A (Al wins) and B (Bob wins). Since M has two states, its transition matrix $\mathbf{T} = (p_{ij})$ has two rows and 2 columns:

$$\mathbf{T} = \begin{matrix} & \begin{matrix} A & B \end{matrix} \\ \begin{matrix} A \\ B \end{matrix} & \begin{pmatrix} p_{11} & p_{12} \\ p_{21} & p_{22} \end{pmatrix} \end{matrix}$$

We must therefore calculate the transition probabilities p_{11}, p_{12}, p_{21}, and p_{22}.

The transition probability p_{11} is the probability that a transition from the first state (Al wins) to the first state (Al wins) occurs. In other words, p_{11} is the

probability that Al wins the next game if he wins this game. From the information given, $p_{11} = 0.6$.

The transition probability p_{12} is the probability that a transition from the
first state (Al wins) to the second state (Bob wins) occurs. The probability p_{12}
that Bob wins the next game if Al wins this game is $1 - 0.6 = 0.4$.

The remaining transition probabilities, $p_{21} = 0.3$ and $p_{22} = 0.7$, are calculated similarly. It follows that

$$\mathbf{T} = \begin{array}{c} \\ A \\ B \end{array} \begin{array}{c} A \quad\quad B \\ \begin{pmatrix} 0.6 & 0.4 \\ 0.3 & 0.7 \end{pmatrix} \end{array}$$

Example 7   What is the probability that Al wins the third game of Example 6?

Solution   The tree diagram in Figure 7.2 conveniently shows the different ways
that the three games can be played. Since Al and Bob are equally likely to win

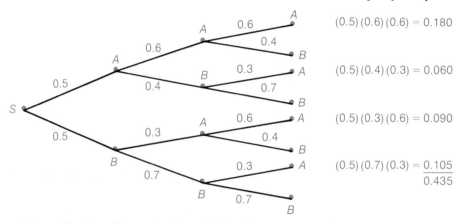

Figure 7.2   Tree diagram for Example 7

the first game, 0.5 is placed on the two lines beginning at S (start). Every
time Al wins a game, the probability that he wins the next game is $p_{11} = 0.6$.
(See the transition matrix of Example 6.) For this reason, 0.6 is placed on
each line joining state A of one trial (game) with state A of the next trial. On
the other hand, if Al wins the game, the probability that Bob wins the
following game is $p_{12} = 0.4$. For this reason, 0.4 is placed on all lines
connecting state A of one trial with state B of the next. The remainder of the
tree diagram is completed similarly.

Let $X_n = 1$ if Al wins the nth game, and let $X_n = 2$ if he loses it. The
topmost sequence of line segments in Figure 7.2 corresponds to the event
$(X_1 = 1, X_2 = 1, X_3 = 1)$. Applying the multiplication theorem of probability
(Theorem 4.3), the probability of this event is $(0.5)(0.6)(0.6) = 0.180$. This
number is written to the right of the sequence of line segments in Figure 7.2.
The probabilities of the other three sequences that result in Al's winning the
third game are calculated similarly, and are also written on the right. The sum
of these probabilities is the required probability P; it follows that $P = 0.435$.

Example 8   In 1962 two researchers found that a two-state Markov chain provides a good description of the occurrence of wet and dry days in Tel Aviv during the rainy period (December–February). The data, gathered over a period of twenty-seven years, were as follows:

Previous Day	Present Day	
	Dry	Wet
Dry	1049	350
Wet	351	687

Letting dry be state 1 and wet be state 2, the estimated transition probabilities (using relative frequency) are

$$p_{11} = \frac{1049}{1049 + 350} = 0.750, \qquad p_{12} = \frac{350}{1049 + 350} = 0.250, \qquad \text{etc.}$$

Hence, the transition matrix is

$$\mathbf{T} = \begin{array}{c} \\ \text{Dry} \\ \text{Wet} \end{array} \begin{array}{cc} \text{Dry} & \text{Wet} \\ \begin{pmatrix} 0.750 & 0.250 \\ 0.338 & 0.662 \end{pmatrix} \end{array}$$

Let $\mathbf{T}$ be the transition matrix of Example 8. The sum of the entries in the first row of $\mathbf{T}$ is $0.750 + 0.250 = 1$; the sum of the entries in the second row is $0.338 + 0.662 = 1$. The same observation is true of the transition matrices of Examples 4, 5, and 6. In general, the ith row of any transition matrix is

$$p_{i1}, p_{i2}, \ldots, p_{im}$$

and gives the probabilities that the chain makes a transition from state E_i to states $E_1, E_2, \ldots, E_m$, respectively. Since it is certain that there will be a transition from state E_i to some state E_j,[3] it follows that the sum of the entries in any row is 1. Moreover, since the entries are all probabilities, each entry is nonnegative. A matrix with the preceding two characteristics is called a *stochastic matrix*.

Definition 7.3   The matrix $\mathbf{S}$ is a **stochastic matrix** if—

(a)   every entry of $\mathbf{S}$ is nonnegative

(b)   the sum of the entries in every row of $\mathbf{S}$ is 1.

In particular, a $1 \times m$ stochastic matrix such as

$$(1,0,0), \quad (\tfrac{1}{2},\tfrac{1}{2}), \quad \text{or} \quad (0.3,0,0.3,0,0.4)$$

---

3. State E_j may be E_i. In this case, M stays in state E_i; the corresponding transition probability is p_{ii}.

is called a **probability vector**. Column vectors such as

$$\begin{pmatrix} 0 \\ 1 \\ 0 \end{pmatrix}, \quad \begin{pmatrix} \frac{1}{2} \\ \frac{1}{2} \end{pmatrix}, \quad \text{and} \quad \begin{pmatrix} 0 \\ 0.7 \\ 0.1 \\ 0 \\ 0.2 \end{pmatrix}$$

which have nonnegative components that sum to 1 are also called probability vectors. Notice that a probability vector has either one row or one column.

Example 9 illustrates these concepts.

Example 9   Which of the following matrices are stochastic matrices? Which are probability vectors?

$$\mathbf{A} = \begin{pmatrix} 1 & 2 & 3 \\ 4 & 5 & 6 \end{pmatrix}, \quad \mathbf{B} = \begin{pmatrix} 0.1 & 0.9 \\ 0.2 & 0.8 \end{pmatrix}, \quad \mathbf{C} = (1,0)$$

$$\mathbf{D} = \begin{pmatrix} 1 \\ 0 \end{pmatrix}, \quad \mathbf{E} = \begin{pmatrix} -1 & 2 \\ 1 & 0 \end{pmatrix}$$

Solution   The matrices **B** and **C** are stochastic matrices because—

(a) every entry is nonnegative
(b) the sum of the entries in each row is 1.

On the other hand, **A** and **D** are not stochastic matrices because the sum of the entries in the second row is not 1. Matrix **E** is also not stochastic because it has a negative entry, -1.

The only probability vectors are **C** and **D**.

Notice that—

(a) every row of a stochastic matrix is a probability vector
(b) every transition matrix is stochastic.

The converse of **(b)**, however, is not true, because a transition matrix must be square. For example, the matrix **C** of Example 9 is stochastic, but is not a transition matrix because it is not square. However, if a stochastic matrix **T** is square, we can let M be a Markov chain whose transition probabilities are the entries of **T**. Then **T** is the transition matrix of M. It follows that *every square stochastic matrix is a transition matrix*.

The transition probabilities of a Markov chain are often represented graphically by a *transition diagram* like those in Figure 7.3. A transition from the state E_i to the state E_j is denoted by an arrow from the point labeled E_i to the point labeled E_j with the transition probability p_{ij} written beside the arrow. The transition diagrams for the Markov chains of Examples 8 and 1 appear in Figures 7.3(a) and 7.3(b), respectively.

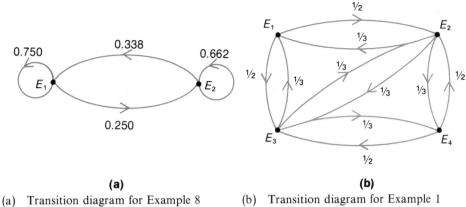

(a)   **(b)**

(a) Transition diagram for Example 8   (b)   Transition diagram for Example 1

Figure 7.3

7.1 EXERCISES

1.   Which of the following processes are Markov chains? State your reasons.

(a) Three boys—Al, Bob, and Carl—are playing catch. Al is just as likely to throw the ball to Bob as to Carl; Bob always throws the ball to Al; Carl is just as likely to throw the ball to Al as to Bob.

(b) An urn initially contains three red and three black balls. A ball is randomly selected and replaced. After each even-numbered draw, a black ball is added to the contents of the urn; after each odd-numbered draw, a red ball is added.

(c) An urn contains five red and five black balls. A ball is randomly selected and replaced; then another ball is drawn and replaced; and so on.

(d) The process described in part (c) is carried out; however, after each selection, an additional ball of the same color as the selected ball is added to the urn.

2.   From statistical data, it was found that the vowels and consonants in a Samoan word form a Markov chain. A consonant is never followed by a consonant, and there is a probability of 0.51 that a vowel will be followed by a vowel.

(a) Find the transition matrix.

(b) If the first letter of a Samoan word is a consonant, what is the probability that—

(*i*) the third letter in the word is a consonant
(*ii*) the third letter is a consonant and the fourth is a vowel?
(Hint: Draw a tree diagram like the one for Example 7.)

3.   Customers in a certain city are continually switching the brand of soap they use. If a customer is now using brand A, the probability that he will use brand A next week is 0.5; the probability that he will switch to brand B is 0.3, while the probability that he will switch to brand C is 0.2. On the other hand, if he now uses brand B, the probability that he continues to use brand B is 0.6, while the probability that he switches to brand C is 0.4; he will not switch to brand A. If he now uses brand C, the probability that he continues to use brand C is 0.4, the probability that he switches to brand A is 0.2, and the

probability that he switches to B is 0.4. Let $X_n = 1, 2, 3$, respectively, if the customer uses brand A, B, C in week n. (Today is week 0.) Suppose the customer is now using brand B.

(a) State the verbal equivalent of the following events:

(i) $(X_3 = 1)$

(ii) $(X_2 = 1, X_3 = 2)$

(iii) $(X_3 = 1 | X_2 = 1)$

(b) Calculate the following probabilities:

(i) $P(X_o = 2)$

(ii) p_{12}

(iii) $P(X_2 = 2 | X_1 = 1)$

(iv) $P(X_3 = 2 | X_2 = 1)$

(v) $P(X_5 = 3 | X_4 = 1, X_3 = 1)$

4.   Suppose the customer of Exercise 3 is now using brand A. Draw a tree diagram and use it to answer the following questions.

(a) What is the probability that he will still be using brand A in two weeks?

(b) What brand of soap is he most likely to be using in two weeks?

(c) What is the probability of the event of part (b)? (That is, what is the probability that in two weeks the customer will be using the most likely brand of soap?)

5.   Let M be a Markov chain with transition matrix

$$\mathbf{T} = \begin{pmatrix} 0.1 & 0.2 & 0.7 \\ 0.3 & 0.7 & 0 \\ 0.5 & 0 & 0.5 \end{pmatrix}$$

(a) Calculate the following probabilities:

(i) p_{12} (ii) p_{21}

(iii) p_{33} (iv) $P(X_2 = 1 | X_1 = 2)$

(b) Draw the transition diagram for $\mathbf{T}$.

(c) Calculate the probability that—

(i) a transition is made from E_1 to E_3

(ii) M will be in state E_2 next if M is now in E_3

(iii) M moves from E_2 to E_1.

6.   Which of the following matrices are (a) stochastic matrices, (b) probability vectors, (c) transition matrices?

$$A = (1,0,0), \qquad B = \begin{pmatrix} 0.1 & 0.9 \\ 0.5 & 0.5 \end{pmatrix}, \qquad C = \begin{pmatrix} -1 & 2 \\ 1 & 0 \end{pmatrix}$$

$$D = \begin{pmatrix} 0.1 & 0.3 & 0.6 \\ 0.3 & 0.5 & 0.2 \end{pmatrix}, \qquad E = (0.4, 0.5, 0.2)$$

7.   Suppose $(x, 2x, 3x)$ is a probability vector. Find x.

8.   Figure 7.4 is the transition diagram of a Markov chain M.

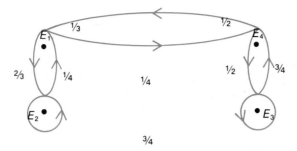

Figure 7.4

(a) Use the transition diagram to calculate the following probabilities:

(i) p_{12} (ii) p_{22}

(iii) $P(X_3 = 2 \mid X_2 = 1)$ (iv) $P(X_3 = 2 \mid X_1 = 1)$

(b) Find the transition matrix of M.

9.   A psychologist studying the behavior of mice found that 70 percent of the mice who completed a maze in two minutes or less on one trial were then able to finish the maze in two minutes or less on the next trial. However, only 20 percent of the mice who took more than two minutes to finish the maze were able to finish it in two minutes or less on the next trial.

(a) Find the transition matrix of this Markov chain.

(b) If a certain mouse finished the maze in two minutes or less on his first attempt, what is the probability that he will finish the maze in two minutes or less on his third attempt?

10.   A chimpanzee is being trained to read the word BANANA. A series of words are flashed on the screen. If the chimpanzee pushes a button when the word BANANA appears (a correct response), he receives a banana. If he does not push the button when a different word appears (a correct reponse), the instructor compliments him. If the chimpanzee fails to push the button when BANANA appears or pushes it at any other time (an incorrect response), he receives a mild shock. Suppose the probability that a correct response is followed by an incorrect response is 0.1, while the probability that an incorrect response is followed by a correct one is 0.3. If the probability that the chimpanzee's first response is correct is 0.1, what is the probability that—

(a) his second response is correct

(b) his third response is correct

(c) his second and third responses are both correct?

7.2 ABSOLUTE AND HIGHER TRANSITION PROBABILITIES

In Example 8 we saw that the weather in Tel Aviv forms a Markov chain M. The following are reasonable questions to ask about Tel Aviv's weather:

(a) What is the probability of rain tomorrow if it rains today?

(b) What is the probability of rain in three days if it rains today?

(c) What is the probability of rain in three days if there is a 50-percent chance of rain today?

The probability p_{22} of rain tomorrow if it rains today is a *transition probability*. This is the probability of a transition from state E_2 (rain) to state E_2 in one step (i.e., it is the probability that M moves from E_2 to E_2 in one step). These probabilities were discussed in Section 7.1.

The probability $p_{22}(3)$ of rain in 3 days if it rains today is called a *3-step transition probability*. It is the probability of a transition from state E_2 to state E_2 in 3 steps. The probability of rain in 3 days if there is a 50-percent chance of rain today is called an *absolute probability at time n = 3*. It is the probability of a transition to state E_2 in 3 steps if the initial conditions are known. Absolute and n-step transition probabilities will be studied in this section.

Definition 7.4   Let M be a Markov chain with m states.

The *absolute probabilities* at time n, denoted $p_j(n)$, are defined by

$$p_j(n) = P(X_n = j) \qquad j = 1,2,\ldots,m$$

The vector

$$\mathbf{p}(n) = (p_1(n),p_2(n),\ldots,p_m(n))$$

is the **absolute probability vector** at time n. In particular, $\mathbf{p}(0)$ is called the **initial probability vector**.

Example 10   Let M be any Markov chain with states $E_1,E_3,\ldots,E_m$.

(a) The absolute probability $p_1(3)$ is the probability that M is in state E_1 at time $n = 3$ (after 3 steps).

(b) The absolute probability $p_3(1)$ is the probability that M is in state E_3 at time $n = 1$ (after 1 step).

(c) The initial probability vector $\mathbf{p}(0) = (p_1(0),p_2(0),\ldots,p_m(0))$ gives the probabilities that M is initially in states $E_1,E_2,\ldots,E_m$, respectively.

Example 11   Suppose the probability of rain in Tel Aviv today is 0.1. (See Example 8.) Find—
(a) the initial probability vector
(b) $p_1(2)$
(c) $\mathbf{p}(2)$

Solution

(a) The initial probability vector is $\mathbf{p}(0) = (p_1(0), p_2(0))$. The first component of this probability vector, $p_1(0)$, is that the Markov chain M is initially in state E_1 (dry); the second component, $p_2(0)$, is the probability that M is initially in state E_2 (wet). From the given facts, $p_2(0) = 0.1$. It follows that $p_1(0) = 1 - p_2(0) = 1 - 0.1 = 0.9$, and so $\mathbf{p}(0) = (0.9,0.1)$.

(b) The absolute probability $p_1(2)$ is the probability that M is in state E_1 (dry) at time $n = 2$; that is, $p_1(2)$ is the probability that it will not rain 2 days from now. This probability can be found by using the tree diagram in Figure 7.5.

Let $X_n = 1$ if the weather is dry on the nth day and let $X_n = 2$ if the weather is wet. Using the tree diagram, we see that

$$p_1(2) = P(X_o = 1, X_1 = 1, X_2 = 1) + P(X_o = 1, X_1 = 2, X_2 = 1)$$
$$+ P(X_o = 2, X_1 = 1, X_2 = 1) + P(X_o = 2, X_1 = 2, X_2 = 1)$$

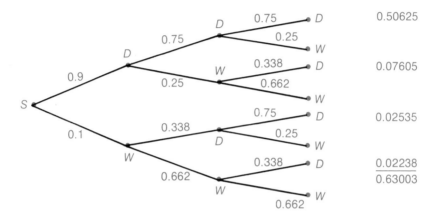

Figure 7.5   Tree diagram for determining $p_1(2)$ in Example 11

By the multiplication theorem of probability,

$$P(X_o = 1, X_1 = 1, X_2 = 1) = (0.9)(0.75)(0.75) = 0.50625$$

This probability is written to the right of the sequence of line segments representing this event on the tree diagram. Similarly,

$$P(X_o = 1, X_1 = 2, X_2 = 1) = 0.07605$$

$$P(X_o = 2, X_1 = 1, X_2 = 1) = 0.02535$$

$$P(X_o = 2, X_1 = 2, X_2 = 1) = 0.02238$$

Summing these probabilities yields

$$p_1(2) = 0.63$$

(c) The absolute probability $p_1(2)$ is the probability that it will not rain 2 days from now; $p_2(2)$ is the probability that it *will* rain 2 days from now. It follows that $p_2(2) = 1 - p_1(2) = 1 - 0.63 = 0.37$. Hence

$$\mathbf{p}(2) = (p_1(2), p_2(2)) = (0.63, 0.37).$$

We will now define and discuss n-step transition probabilities.

Definition 7.5   Let M be a Markov chain.

(a) The *n-step transition probabilities*, $p_{ij}(n)$, are defined by

$$p_{ij}(n) = P(X_n = j \mid X_o = i)$$

(b) The matrix

$$\mathbf{T}^{(n)} = (p_{ij}(n))$$

is called the *n*-**step transition matrix**.

In other words, $p_{ij}(n)$ is the probability that the system moves from state E_i to state E_j in exactly n steps. In particular, $p_{ij}(1)$ is the probability that M moves from E_i to E_j in 1 step; that is, $p_{ij}(1) = p_{ij}$.

Example 12   Let M be any Markov chain with states $E_1, E_2, \ldots, E_m$.

(a) The 3-step transition probability $p_{12}(3)$ is the probability that M moves from state E_1 to state E_2 in 3 steps.

(b) The 2-step transition probability is the probability that M moves from state E_3 to E_1 in 2 steps.

(c) The 1-step transition probability $p_{23}(1)$ is the probability of a transition from E_2 to E_3 in 1 step; therefore, $p_{23}(1) = p_{23}$.

Example 13   The daily traffic on a certain bus route is a Markov chain with 3 states: E_1 (light), E_2 (average), and E_3 (heavy).

(a) The 3-step transition probability $p_{21}(3)$ is the probability that traffic will be light in 3 days if it is average today.

(b) The 3-step transition probability $p_{12}(3)$ is the probability that traffic will be average in 3 days if it is light today.

(c) The 1-step transition probability $p_{32}(1) = p_{32}$ is the probability that traffic will be average tomorrow if it is heavy today.

Consider the traffic situation in Example 13, and suppose today is Monday. It follows from Definition 7.1(c) that the probability that traffic is heavy on Tuesday, given that it is light on Monday, is the same as the probability that traffic is heavy on Wednesday, given that it is light on Tuesday; and so on. Symbolically,

$$P(X_1 = 3 \mid X_o = 1) = P(X_2 = 3 \mid X_1 = 1) = \cdots$$

Similarly, it can be proved that the probability that traffic is heavy on Thursday, given that it is light on Monday, is the same as the probability that traffic is heavy on Friday, given that it is light on Tuesday; and so on. Symbolically,

$$P(X_3 = 3 \mid X_o = 1) = P(X_4 = 3 \mid X_1 = 1) = \cdots$$

In general, if M is any homogeneous Markov chain, it can be shown that

$$P(X_n = j \mid X_o = i) = P(X_{n+1} = j \mid X_1 = i)$$
$$= P(X_{n+2} = j \mid X_2 = i) = \cdots \tag{3}$$

In other words, the probability of going from state E_i to state E_j in *exactly n steps* does not depend upon the number of the trial; symbolically,

$$P(X_{n+t} = j \mid X_t = i)$$

does not depend on t.

Tree diagrams can be used to calculate both absolute and n-step transition probabilities. (See Example 11(**b**).) However, this method is clearly impractical if either the number n of steps or the number m of states is large. In such cases, we can use matrix multiplication to calculate these probabilities by applying the following theorem.

THEOREM 7.1   Let **T** be the transition matrix of a Markov chain M and let $\mathbf{p}(0)$ be the initial probability vector. Then—

(a) $\mathbf{T}^{(n)} = \mathbf{T}^n$

(b) $\mathbf{p}(n) = \mathbf{p}(0)\mathbf{T}^n$

In words, Theorem 7.1 says that—

(**a**) the n-step transition matrix of M is the nth power of the transition matrix **T**

(**b**) The absolute probability vector $\mathbf{p}(n)$ is the product of the initial probability vector $\mathbf{p}(0)$ and the n-step transition matrix $\mathbf{T}^n$.

The following examples illustrate the usefulness of Theorem 7.1.

Example 14   If the rat of Example 1 is placed in compartment 1, what is the probability that he will be in compartment 4 after two moves?

Solution   This is the question that was raised at the beginning of Section 7.1. To answer it, we need to calculate $p_{14}(2)$. We can do this by calculating $\mathbf{T}^{(2)}$, because $p_{14}(2)$ will be the entry in row 1, column 4 of $\mathbf{T}^{(2)}$. By Theorem 7.1, $\mathbf{T}^{(2)} = \mathbf{T}^2$. Therefore, using the results of Example 4, we have:

$$\mathbf{T}^{(2)} = \mathbf{T}^2 = \begin{pmatrix} 0 & \frac{1}{2} & \frac{1}{2} & 0 \\ \frac{1}{3} & 0 & \frac{1}{3} & \frac{1}{3} \\ \frac{1}{3} & \frac{1}{3} & 0 & \frac{1}{3} \\ 0 & \frac{1}{2} & \frac{1}{2} & 0 \end{pmatrix} \begin{pmatrix} 0 & \frac{1}{2} & \frac{1}{2} & 0 \\ \frac{1}{3} & 0 & \frac{1}{3} & \frac{1}{3} \\ \frac{1}{3} & \frac{1}{3} & 0 & \frac{1}{3} \\ 0 & \frac{1}{2} & \frac{1}{2} & 0 \end{pmatrix}$$

$$= \begin{pmatrix} \frac{1}{3} & \frac{1}{6} & \frac{1}{6} & \frac{1}{3} \\ \frac{1}{9} & \frac{4}{9} & \frac{1}{3} & \frac{1}{9} \\ \frac{1}{9} & \frac{1}{3} & \frac{4}{9} & \frac{1}{9} \\ \frac{1}{3} & \frac{1}{6} & \frac{1}{6} & \frac{1}{3} \end{pmatrix}$$

The entry in row 1, column 4 of this matrix is $\frac{1}{3}$. Therefore, if the rat is placed in compartment 1, the probability that he will be in compartment 4 after two moves is $p_{14}(2) = \frac{1}{3}$.

Example 15   A study by a psychiatrist indicates that the emotional stability of individuals depends upon the emotional stability of their parents. When parents and their children were classified to be above-average (O), average (A), and below-average (B) in emotional stability, the following data were obtained:

$$
\begin{array}{cc}
 & \text{Child} \\
\text{Parent} \quad O \quad A \quad B
\end{array}
$$

$$
\mathbf{T} = \begin{array}{c} O \\ A \\ B \end{array}
\begin{pmatrix}
0.6 & 0.3 & 0.1 \\
0.4 & 0.5 & 0.1 \\
0.3 & 0.4 & 0.3
\end{pmatrix}
$$

For example, the probabilities that children of an above-average parent are above-average, average, below-average, respectively, in emotional stability are 0.6, 0.3, 0.1. Assume that the emotional stability of individuals from generation to generation forms a Markov chain. What is the probability that—

(a) the child of an average individual is below-average in emotional stability
(b) the grandchild of an average individual is below-average in emotional stability
(c) the child and great-grandchild of an above-average individual are both below-average in emotional stability?

Solution

(a) The required probability p_{23} is found in the intersection of row 2, column 3 of $\mathbf{T}$; thus, $p_{23} = 0.1$.

(b) We must calculate the probability $p_{23}(2)$ that the Markov chain moves from state A to state B in 2 steps. Applying Theorem 7.1, this probability is found in the intersection of row 2, column 3 of $\mathbf{T}^{(2)} = \mathbf{T}^2$. Since

$$
\mathbf{T}^2 = \begin{pmatrix}
0.6 & 0.3 & 0.1 \\
0.4 & 0.5 & 0.1 \\
0.3 & 0.4 & 0.3
\end{pmatrix} \times \begin{pmatrix}
0.6 & 0.3 & 0.1 \\
0.4 & 0.5 & 0.1 \\
0.3 & 0.4 & 0.3
\end{pmatrix}
$$

$$
= \begin{pmatrix}
0.51 & 0.37 & 0.12 \\
0.47 & 0.41 & 0.12 \\
0.43 & 0.41 & 0.16
\end{pmatrix}
$$

it follows that $p_{23}(2) = 0.12$.

(c) Let C, G be the events that the child, great-grandchild, respectively, of an above-average person is below-average in emotional stability. Applying the multiplication theorem of probability (Theorem 4.3), the required probability is

$$
P(CG) = P(C)P(G \,|\, C) \tag{4}
$$

The value of $P(C) = p_{13}$ is found in the intersection of row 1, column 3 of $\mathbf{T}$; therefore, $P(C) = 0.1$. It follows from equation (3) that the probability that the great-grandchild is below-average, given that the child is below-average, is the same as the probability that the grandchild is below-average, given that the individual is below-average. This is because both probabilities are the probability $p_{33}(2)$ that a transition is made from state B to state B in 2 steps. This probability is found in the intersection of row 3, column 3 of $\mathbf{T}^2$; thus, $P(G \,|\, C) = 0.16$.

Substituting these values in equation (4) yields

$$P(CG) = (0.1)(0.16) = 0.016$$

Example 16   Suppose that in a certain town 30 percent, 50 percent, 20 percent of all adults between the ages of twenty and thirty-five are above-average, average, below-average, respectively, in emotional stability.

(a) If the children of this group are tested fifteen years later, what percentage of the children will be above-average in emotional stability?

(b) If the grandchildren of this group are tested thirty years later, what is the probability that a randomly chosen grandchild will be below-average in emotional stability?

Solution   It follows from the given information that the initial probability vector is $\mathbf{p}(0) = (0.3, 0.5, 0.2)$.

(a) We need to calculate $p_1(1)$. Applying Theorem 7.1,

$$\mathbf{p}(1) = (p_1(1), p_2(1), p_3(1))$$

$$= \mathbf{p}(0)\mathbf{T} = (0,3,0.5,0.2) \begin{pmatrix} 0.6 & 0.3 & 0.1 \\ 0.4 & 0.5 & 0.1 \\ 0.3 & 0.4 & 0.3 \end{pmatrix}$$

$$= (0.44, 0.42, 0.14)$$

Hence, $p_1(1) = 0.44$, and so 44 percent of the children will be above-average in emotional stability.

(b) We must calculate $p_3(2)$. Applying Theorem 7.1 and the results of Example 15(b),

$$\mathbf{p}(2) = (p_1(2), p_2(2), p_3(2))$$

$$= \mathbf{p}(0)\mathbf{T}^2 = (0.3, 0.5, 0.2) \begin{pmatrix} 0.51 & 0.37 & 0.12 \\ 0.47 & 0.41 & 0.12 \\ 0.43 & 0.41 & 0.16 \end{pmatrix}$$

$$= (0.474, 0.398, 0.128)$$

It follows that the required probability is 0.128.

The matrices $\mathbf{T}^2$, $\mathbf{p}(0)\mathbf{T}$, and $\mathbf{p}(0)\mathbf{T}^2$ of Examples 15 and 16 are products of stochastic matrices and are stochastic. In general, the following theorem holds.

THEOREM   7.2   The product of two stochastic matrices is stochastic.

In particular, each n-step transition matrix $\mathbf{T}^{(n)} = \mathbf{T}^n$ is stochastic. Theorem 7.2 can therefore be used to check the calculation of $\mathbf{T}^n$, because this theorem tells us that the sum of the entries in each row of $\mathbf{T}^n$ must be 1.

The following examples review all the concepts studied in this chapter so far.

Example 17   A student noticed that, if her teacher gives a surprise quiz one day, the probability that she gives a surprise quiz the next day is 0.2. However, if she does not give a surprise quiz, the probability that she gives a quiz the following day is 0.6. Suppose that the teacher gave a test today and that the teacher's testing procedure forms a Markov chain.

(a) How many states does this Markov chain have?

(b) What is the transition matrix $\mathbf{T}$?

(c) What is the initial probability vector?

(d) State in words the meaning of $p_2(0)$, $p_1(2)$, $p_{12}(4)$, and p_{12}.

Solution

(a) There are two states: E_1, a quiz is given, and E_2, no quiz is given.

(b)

$$
\begin{array}{cc}
& \begin{array}{cc} E_1 & E_2 \end{array} \\
\mathbf{T} = \begin{array}{c} E_1 \\ E_2 \end{array} & \begin{pmatrix} 0.2 & 0.8 \\ 0.6 & 0.4 \end{pmatrix}
\end{array}
$$

(c) The value of $p_1(0)$ is the probability that the teacher gave a quiz to-day. According to the information given, $p_1(0) = 1$. Then $p_2(0) = 1 - p_1(0) = 1 - 1 = 0$. So the initial probability vector is $\mathbf{p}(0) = (p_1(0), p_2(0)) = (1,0)$.

(d) In general, $p_i(n)$ is the probability that the Markov chain is in state E_i at time n. Hence, $p_2(0)$ is the probability that the Markov chain is initially (time $n = 0$) in state E_2. In this example, $p_2(0)$ is the probability that the teacher gives no quiz today.

The value of $p_1(2)$ is the probability that M is in state E_1 at time $n = 2$. Thus, in this case, $p_1(2)$ is the probability the teacher gives a quiz two days from today.

In general, $p_{ij}(n)$ is the probability that the Markov chain makes a transition from E_i to E_j in n steps. In this example, $p_{12}(4)$ is the probability that the teacher does not give a quiz four days from now if she gave a quiz today.

In general, p_{ij} is the probability that M makes a transition from E_i to E_j in one step. In this example, p_{12} is the probability that the teacher does not give a quiz tomorrow if she gave a quiz today.

Example 18   What is the probability that the teacher of Example 17 gives a test in two days if—

(a) she gave a test today

(b) she did not give a test today?

Solution   For part **(a)**, we must calculate the probability that the Markov chain made a transition from state E_1 to state E_1 in 2 moves; for part **(b)**, we must calculate the probability that the Markov chain made a transition from state E_2 to state E_1 in 2 moves. In other words, we must calculate $p_{11}(2)$ and $p_{21}(2)$. By

Theorem 7.1(a), these probabilities are entries of $\mathbf{T}^2$, where $\mathbf{T}$ is the transition matrix of Example 17(b).

$$\mathbf{T}^2 = \begin{pmatrix} 0.2 & 0.8 \\ 0.6 & 0.4 \end{pmatrix} \begin{pmatrix} 0.2 & 0.8 \\ 0.6 & 0.4 \end{pmatrix}$$

$$= \begin{pmatrix} 0.52 & 0.48 \\ 0.36 & 0.64 \end{pmatrix} = \begin{pmatrix} p_{11}(2) & p_{12}(2) \\ p_{21}(2) & p_{22}(2) \end{pmatrix}$$

Note that $\mathbf{T}^2$ is stochastic. (This fact may be used to check the calculation of $\mathbf{T}^2$.) The required probabilities are—

(a) $p_{11}(2) = 0.52$
(b) $p_{21}(2) = 0.36$

Example 19   If the probability that the teacher of Example 17 gives a quiz on Monday is 0.4, what is the probability that she gives a quiz on Wednesday?

Solution   We need to calculate the probability that M will be in state E_1 (quiz) in $n = 2$ days. In other words, we must calculate $p_1(2)$. By Theorem 7.1,

$$\mathbf{p}(2) = (p_1(2), p_2(2)) = \mathbf{p}(0)\mathbf{T}^2$$

where $\mathbf{T}$ is the transition matrix of Example 17. The matrix $\mathbf{T}^2$ is calculated in Example 18. Since the probability that M is initially in state E_1 (quiz) is 0.4, the probability that M is initially in state E_2 (no quiz) is $1 - 0.4 = 0.6$. Therefore, the initial probability vector is $\mathbf{p}(0) = (0.4, 0.6)$. Hence,

$$\mathbf{p}(2) = (0.4, 0.6) \begin{pmatrix} 0.52 & 0.48 \\ 0.36 & 0.64 \end{pmatrix} = (0.424, 0.576) = (p_1(2), p_2(2))$$

The required probability is $p_1(2) = 0.424$.

7.2 EXERCISES

1.   Let M be a Markov chain with transition matrix

$$\mathbf{T} = \begin{pmatrix} \frac{1}{4} & \frac{3}{4} \\ \frac{1}{3} & \frac{2}{3} \end{pmatrix}$$

and initial probability vector $\mathbf{p}(0) = (\frac{1}{4}, \frac{3}{4})$. Calculate—

(a) $p_1(0)$ **(b)** $p_{12}(1)$ **(c)** $p_{12}(2)$
(d) p_{11} **(e)** $\mathbf{p}(2)$ **(f)** $p_1(2)$
(g) $p_2(1)$ **(h)** $P(X_2 = 1)$ **(i)** $P(X_1 = 2)$
(j) $P(X_2 = 2 \mid X_o = 1)$ **(k)** $P(X_3 = 2) \mid X_1 = 2, X_2 = 1)$ **(l)** $\mathbf{T}^{(2)}$

2.   Let M be a Markov chain with transition matrix

$$\mathbf{T} = \begin{pmatrix} 0.1 & 0.9 \\ 0.8 & 0.2 \end{pmatrix}$$

and initial probability vector $\mathbf{p}(0) = (0.4, 0.6)$.
(a) Use tree diagrams to calculate $\mathbf{T}^{(2)}$ and $\mathbf{p}(1)$.
(b) Use Theorem 7.1 to calculate $\mathbf{T}^{(2)}$ and $\mathbf{p}(1)$.
(c) Are $\mathbf{T}^{(2)}$ and $\mathbf{p}(1)$ stochastic?

3.   (a) Find the transition matrix $\mathbf{T}$ of Exercise 3 following Section 7.1.
(b) Write the verbal equivalents of p_{12} and $p_{12}(2)$.

4.   Redo Exercise 3, Section 7.1, using Theorem 7.1.

5.   (a) Find the transition matrix $\mathbf{T}$ of Exercise 10, Section 7.1.
(b) Calculate $\mathbf{T}^{(2)}$.
(c) Redo Exercise 10, Section 7.1, using Theorem 7.1.

6.   Consider the chimpanzee of Exercise 10, Section 7.1.
(a) If his first response is correct, what is the probability that his fourth response will also be correct? (Hint: Calculate $\mathbf{T}^3 = \mathbf{T}^{(3)}$ and then find $p_{11}(3)$.)
(b) If the chimpanzee's first response is wrong, what is the probability that his third and fourth responses will be correct? (Hint: Calculate $p_{21}(2) \cdot p_{11}$.)

7.   Let Al, Bob, and Carl by the boys of Exercise 1, Section 1. Find the probability that Carl has the ball after three throws if Al has the ball now.

8.   In order to win a game of Ping-Pong, a player must have at least twenty-one points and lead by at least two points. If two players, A and B, are tied twenty–twenty, then the sequence of scores that follow the tie generates a Markov chain with five states: E_1 (A wins the game), E_2 (A leads by one stroke), E_3 (A and B are tied), E_4 (B leads by one stroke), and E_5 (B wins the game).
(a) If A and B are evenly matched (i.e., if each one is equally likely to get the next point), find the transition matrix $\mathbf{T}$.
(b) Find $\mathbf{T}^2$.
(c) Find $\mathbf{T}^3$.
(d) If A and B are tied, what is the probability that B wins in three volleys or less?

*9.   Using the data of Example 8, find the probability of rain—
(a) in two days given that the probability of rain today is $\frac{1}{2}$
(b) in three days given that $\mathbf{p}(0) = (\frac{1}{4}, \frac{3}{4})$
(c) in five days given that $p_1(0) = \frac{1}{3}$. (Hint: First calculate $\mathbf{T}^2$, $\mathbf{T}^3$, and $\mathbf{T}^5 = \mathbf{T}^2 \cdot \mathbf{T}^3$. Then apply Theorem 7.1(b). Use a hand calculator to shorten the time spent doing calculations.)

7.3 SUBMATRICES AND SUBCHAINS

If a Markov chain has a large number of states, it may be tedious to calculate powers of the entire matrix $\mathbf{T}$. In this section we will see how we can sometimes avoid these calculations.

Example 20   Suppose that a Markov chain M has the following transition matrix:

$$
\begin{array}{c|ccccccc}
 & E_1 & E_2 & E_3 & E_4 & E_5 & E_6 & E_7 \\
\hline
E_1 & 0 & 0 & 0 & 0 & \frac{1}{2} & \frac{1}{2} & 0 \\
E_2 & \frac{1}{3} & 0 & \frac{1}{3} & 0 & 0 & 0 & \frac{1}{3} \\
E_3 & 0 & 0 & 1 & 0 & 0 & 0 & 0 \\
E_4 & 0 & 0 & 0 & \frac{1}{2} & 0 & 0 & \frac{1}{2} \\
E_5 & \frac{2}{3} & 0 & 0 & 0 & \frac{1}{3} & 0 & 0 \\
E_6 & \frac{1}{2} & 0 & 0 & 0 & \frac{1}{2} & 0 & 0 \\
E_7 & 0 & 0 & 0 & \frac{1}{3} & 0 & 0 & \frac{2}{3}
\end{array}
$$

If M is now in state E_5, calculate $\mathbf{p}(2)$.

Solution   Since the only nonzero entries in the fifth row are in the first and fifth columns, M can enter only states E_1 and E_5 from E_5 (i.e., M can only be in E_1 or E_5 next if it is now in E_5). From E_1, M can enter only E_5 or E_6; from E_6 it can enter only E_1 or E_5. These facts are conveniently illustrated by using arrows as shown in Figure 7.6.

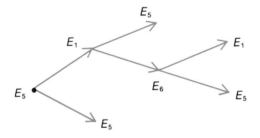

Figure 7.6   Diagram of states that can be reached from state E_5 of Example 20

It is therefore clear that, since M is initially in state E_5, it can never reach states E_2, E_3, E_4, or E_7. It follows that

$$p_{52}(2) = p_{53}(2) = p_{54}(2) = p_{57}(2) = 0 \tag{5}$$

We must still calculate $p_{51}(2)$, $p_{55}(2)$, and $p_{56}(2)$.

As Figure 7.6 shows, the states E_1, E_5, and E_6 are *accessible* (i.e., can be reached) from E_5. Let

$$R_5 = \{E_1, E_5, E_6\}$$

and consider the transition matrix $\mathbf{T}$. If we delete the rows and columns of $\mathbf{T}$

that are associated with states not in R_5 (i.e., states E_2, E_3, E_4, and E_7), we are left with the following matrix $\mathbf{T}_5$:

$$
\begin{array}{c}
 \\
E_1 \\
E_2 \\
E_3 \\
E_4 \\
E_5 \\
E_6 \\
E_7
\end{array}
\begin{array}{ccccccc}
E_1 & E_2 & E_3 & E_4 & E_5 & E_6 & E_7 \\
\left(0 \right. & 0 & 0 & 0 & \frac{1}{2} & \frac{1}{2} & 0 \\
\frac{1}{3} & 0 & \frac{1}{3} & 0 & 0 & 0 & \frac{1}{3} \\
0 & 0 & 1 & 0 & 0 & 0 & 0 \\
0 & 0 & 0 & \frac{1}{2} & 0 & 0 & \frac{1}{2} \\
\frac{2}{3} & 0 & 0 & 0 & \frac{1}{3} & 0 & 0 \\
\frac{1}{2} & 0 & 0 & 0 & \frac{1}{2} & 0 & 0 \\
0 & 0 & 0 & \frac{1}{3} & 0 & 0 & \left. \frac{2}{3} \right)
\end{array}
$$

$$
=
\begin{array}{c}
E_1 \\
E_5 \\
E_6
\end{array}
\begin{pmatrix}
0 & \frac{1}{2} & \frac{1}{2} \\
\frac{2}{3} & \frac{1}{3} & 0 \\
\frac{1}{2} & \frac{1}{2} & 0
\end{pmatrix}
= \mathbf{T}_5
$$

The matrix $\mathbf{T}_5$ is called a *submatrix* of $\mathbf{T}$ because it is formed by eliminating some of the rows and columns of $\mathbf{T}$. Notice that $\mathbf{T}_5$ is the matrix associated with R_5—its headings are the states of R_5. Since $\mathbf{T}_5$ stochastic and square, R_5 and its associated matrix $\mathbf{T}_5$ form a Markov chain; that is, there is a Markov chain having states E_1, E_5, and E_6 and transition matrix $\mathbf{T}_5$. The transition probabilities $p_{ij}(n)$, where $i,j = 1,5,6$, can be calculated by using $\mathbf{T}_5$ instead of $\mathbf{T}$. Since $\mathbf{T}_5$ is smaller than $\mathbf{T}$, the calculation of $\mathbf{T}_5{}^n$ is easier than the calculation of $\mathbf{T}^n$.

In particular,

$$
\mathbf{T}_5{}^2 =
\begin{pmatrix}
0 & \frac{1}{2} & \frac{1}{2} \\
\frac{2}{3} & \frac{1}{3} & 0 \\
\frac{1}{2} & \frac{1}{2} & 0
\end{pmatrix}
\begin{pmatrix}
0 & \frac{1}{2} & \frac{1}{2} \\
\frac{2}{3} & \frac{1}{3} & 0 \\
\frac{1}{2} & \frac{1}{2} & 0
\end{pmatrix}
$$

$$
=
\begin{array}{c}
E_1 \\
E_5 \\
E_6
\end{array}
\begin{array}{ccc}
E_1 & E_5 & E_6 \\
\left(\frac{7}{12} \right. & \frac{5}{12} & 0 \\
\frac{2}{9} & \frac{4}{9} & \frac{1}{3} \\
\frac{1}{3} & \frac{5}{12} & \left. \frac{1}{4} \right)
\end{array}
$$

It follows that

$$
p_{51}(2) = \tfrac{2}{9}, \qquad p_{55}(2) = \tfrac{4}{9}, \qquad p_{56}(2) = \tfrac{1}{3}
$$

Combining equations (5) and (6), we have $\mathbf{p}(2) = (\tfrac{2}{9},0,0,0,\tfrac{4}{9},\tfrac{1}{3},0)$.   (6)

The method used in Example 20 can be applied to any Markov chain. We will first formally define some of the concepts used in the example.

Definition 7.6   A matrix formed by deleting certain rows and columns of a matrix $\mathbf{A}$ is called a **submatrix** of $\mathbf{A}$.

Definition 7.7   Let M be a Markov chain with transition matrix $\mathbf{T}$. Then M_1 is a **subchain** of M if—

(a) M_1 is a Markov chain, and
(b) The transition matrix of M_1 is a submatrix of $\mathbf{T}$.

Example 21 Let M be a Markov chain with transition matrix

$$
\mathbf{T} = \begin{array}{c} \\ E_1 \\ E_2 \\ E_3 \end{array}
\begin{array}{ccc} E_1 & E_2 & E_3 \\ \left(\begin{array}{ccc} 0.5 & 0 & 0.5 \\ 1 & 0 & 0 \\ 0.6 & 0 & 0.4 \end{array}\right) \end{array}
$$

(a)

$$
\mathbf{U} = \begin{array}{c} \\ E_1 \\ E_2 \\ E_3 \end{array}
\begin{array}{ccc} E_1 & E_2 & E_3 \\ \left(\begin{array}{ccc} 0.5 & 0 & 0.5 \\ 1 & 0 & 0 \\ 0.6 & 0 & 0.4 \end{array}\right) \end{array}
= \begin{array}{c} \\ E_1 \\ E_2 \end{array}
\begin{array}{cc} E_1 & E_2 \\ \left(\begin{array}{cc} 0.5 & 0 \\ 1 & 0 \end{array}\right) \end{array}
$$

and

$$
\mathbf{V} = \begin{array}{c} \\ E_1 \\ E_2 \\ E_3 \end{array}
\begin{array}{ccc} E_1 & E_2 & E_3 \\ \left(\begin{array}{ccc} 0.5 & 0 & 0.5 \\ 1 & 0 & 0 \\ 0.6 & 0 & 0.4 \end{array}\right) \end{array}
= \begin{array}{c} \\ E_1 \\ E_3 \end{array}
\begin{array}{cc} E_1 & E_3 \\ \left(\begin{array}{cc} 0.5 & 0.5 \\ 0.6 & 0.4 \end{array}\right) \end{array}
$$

are submatrices of $\mathbf{T}$.

(b) The set $\{E_1, E_2\}$ and its associated matrix $\mathbf{U}$ do not form a Markov chain because $\mathbf{U}$ is not a transition matrix (the sum of the entries in the first row of $\mathbf{U}$ is 0.5). However, $\{E_1, E_3\}$ and its associated matrix $\mathbf{V}$ do form a Markov chain $\bar{M}$, because $\mathbf{V}$ is a transition matrix. Moreover, $\bar{M}$ is a subchain of M because $\mathbf{V}$ is a submatrix of $\mathbf{T}$.

Definition 7.8 A state E_j can be *reached*, or is an **accessible state**, from a state E_i (written $E_i \rightarrow E_j$) if $p_{ij}(n) > 0$ for some $n \geq 1$.

In other words, E_j can be reached from E_i if it is possible to go from E_i to E_j in n steps, for some $n \geq 1$.

In applying Definition 7.8, we need only to know whether $p_{ij}(n)$ is *positive*; we do not need to know its exact value. We can determine whether E_j can be reached from E_i by using a diagram like the one in Figure 7.6.

In Example 20 we used R_i to denote the set of all states that can be reached from E_i. In other words,

$$R_i = \{E_j \mid E_i \rightarrow E_j\} \tag{7}$$

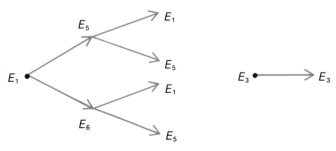

Figure 7.7 Diagrams of states that can be reached from states E_1 and E_3 of Example 20

These are the states that appear in a diagram for E_i like Figure 7.6. In Example 20, $R_5 = R_1 = R_6 = \{E_1, E_5, E_6\}$, and $R_3 = \{E_3\}$. (See Figures 7.6 and 7.7.)

The following theorem states an important consequence of Definition 7.8.

THEOREM   7.3   If $E_i \rightarrow E_j$ and $E_j \rightarrow E_k$, then $E_i \rightarrow E_k$.

Notice that Theorem 7.3 states that if E_j can be reached from E_i and E_k can be reached from E_j, then E_k can be reached from E_i.

In Example 20 we saw that R_5 and its associated transition matrix $\mathbf{T}_5$ form a Markov chain M_5, and that M_5 is a subchain of the original chain M. These facts allowed us to use $\mathbf{T}_5$ to calculate certain probabilities instead of using the larger matrix $\mathbf{T}$. A similar result is true for any R_i and any Markov chain M. This result is stated in the following theorem.

THEOREM   7.4   Let M be a Markov chain with transition matrix T and let E_i be a fixed state of M. Let R_i be defined by equation (7), and let $\mathbf{T}_i$ be the submatrix formed by deleting from $\mathbf{T}$ all rows and columns associated with states that are not in R_i. Then R_i and its associated matrix $\mathbf{T}_i$ form a subchain M_i of M.

If M is a Markov chain, the subchain M_i described in Theorem 7.4 can be used to calculate absolute and n-step probabilities, as was done in Example 20. The following examples will show how useful this can be.

Example 22   If the Markov chain of Example 20 is initially in state E_4, what are the probabilities that in three moves it will be (*a*) in state E_5, (*b*) in state E_7?

Solution

(**a**) From Figure 7.8 we see that the set of all states that can be reached from

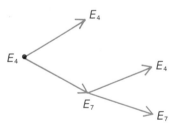

Figure 7.8   Diagram for R_4, Example 22

E_4 is $R_4 = \{E_4, E_7\}$. Hence, M will never reach E_5 and so the probability is 0.

(**b**) To find the required probability, $p_{47}(3)$, we need only consider the submatrix $\mathbf{T}_4$ formed by deleting the rows and columns associated with states that are not in R_4:

$$\begin{array}{cc} & E_4 \quad E_7 \\ \mathbf{T}_4 = \begin{array}{c} E_4 \\ E_7 \end{array} & \begin{pmatrix} \frac{1}{2} & \frac{1}{2} \\ \frac{1}{3} & \frac{2}{3} \end{pmatrix} \end{array}$$

$$\mathbf{T}_4{}^2 = \begin{pmatrix} \frac{1}{2} & \frac{1}{2} \\ \frac{1}{3} & \frac{2}{3} \end{pmatrix} \begin{pmatrix} \frac{1}{2} & \frac{1}{2} \\ \frac{1}{3} & \frac{2}{3} \end{pmatrix} = \begin{pmatrix} \frac{5}{12} & \frac{7}{12} \\ \frac{7}{18} & \frac{11}{18} \end{pmatrix}$$

and

$$\mathbf{T}_4{}^3 = \mathbf{T}_4 \cdot \mathbf{T}_4{}^2 = \begin{pmatrix} \frac{1}{2} & \frac{1}{2} \\ \frac{1}{3} & \frac{2}{3} \end{pmatrix} \begin{pmatrix} \frac{5}{12} & \frac{7}{12} \\ \frac{7}{18} & \frac{11}{18} \end{pmatrix}$$

$$\begin{array}{cc} & E_4 \quad\;\; E_7 \\ = \begin{array}{c} E_4 \\ E_7 \end{array} & \begin{pmatrix} \frac{29}{72} & \frac{43}{72} \\ \frac{43}{108} & \frac{65}{108} \end{pmatrix} \end{array}$$

Thus, $p_{47}(3) = \frac{43}{72}$.

Example 23   Suppose the initial probability vector of the Markov chain in Example 20 is $\mathbf{p}(0) = (0,0,0,\frac{1}{3},0,0,\frac{2}{3})$. Find $\mathbf{p}(3)$.

Solution   Since M is initially in either state E_4 or state E_7, we need to consider the states that can be reached from E_4 or from E_7. It is clear from Figure 7.8 that $R_4 = R_7 = \{E_4, E_7\}$. We can therefore restrict our attention to the states E_4 and E_7. Hence, let M_4 be a Markov chain with transition matrix $\mathbf{T}_4$ of Example 22(**b**) and initial probability vector

$$\begin{array}{c} E_4 \;\; E_7 \\ \mathbf{p}(0) = (\frac{1}{3}, \;\; \frac{2}{3}) \end{array}$$

Using Theorem 7.1(a) and $\mathbf{T}_4{}^3$ as calculated in Example 22(b),

$$\mathbf{p}(3) = \mathbf{p}(0)\mathbf{T}_4{}^3 = (\tfrac{1}{3},\tfrac{2}{3}) \begin{pmatrix} \frac{29}{72} & \frac{43}{72} \\ \frac{43}{108} & \frac{65}{108} \end{pmatrix} = \overset{E_4 \quad E_7}{(\tfrac{259}{648}, \tfrac{389}{648})}$$

This is, of course, the probability vector at time $n = 3$ for M_4. However, what we need to find is the probability vector at time $n = 3$ for the original Markov chain M. Since $E_1, E_2, E_3, E_5,$ and E_6 can never be reached from E_4 or E_7, $p_1(3) = p_2(3) = p_3(3) = p_5(3) = p_6(3) = 0$. Therefore,

$$\mathbf{p}(3) = (0,0,0,\tfrac{259}{648},0,0,\tfrac{389}{648})$$

7.3 EXERCISES

1.   Determine which of the following matrices are submatrices of

$$\mathbf{T} = \begin{pmatrix} 1 & 2 & 1 \\ 2 & 1 & 3 \\ 1 & -1 & 2 \end{pmatrix}$$

(a) $\begin{pmatrix} 1 & 2 \\ 2 & 1 \end{pmatrix}$ **(b)** $\begin{pmatrix} 2 & 1 & 3 \\ 1 & -1 & 2 \end{pmatrix}$ **(c)** $\begin{pmatrix} 2 & 1 & 3 \\ 1 & 2 & 1 \end{pmatrix}$

(d) $\begin{pmatrix} 1 & 1 \\ 2 & 3 \end{pmatrix}$ **(e)** $\begin{pmatrix} 1 \\ 2 \end{pmatrix}$

2. Let M be a Markov chain with transition matrix

$$\mathbf{T} = \begin{array}{c} \\ E_1 \\ E_2 \\ E_3 \end{array} \begin{array}{ccc} E_1 & E_2 & E_3 \\ \begin{pmatrix} \frac{1}{3} & 0 & \frac{2}{3} \\ \frac{1}{2} & \frac{1}{2} & 0 \\ 0 & 0 & 1 \end{pmatrix} \end{array}$$

Let R_i be defined by equation (7) and let $\mathbf{T}_i$ be the matrix associated with R_i.
Determine—
(a) R_1 **(b)** $\mathbf{T}_1$ **(c)** the matrix associated with $\{E_1, E_2\}$
(d) R_2 **(e)** $\mathbf{T}_2$

3. Let M be the Markov chain of Exercise 2.
   (a) Which of the following matrices are transition matrices?
       (*i*) $\mathbf{T}_1$
       (*ii*) the matrix associated with $\{E_1, E_2\}$
       (*iii*) the matrix associated with $\{E_1, E_3\}$
       (*iv*) the matrix associated with R_2
   (b) Which of the following are subchains of M?
       (*i*) $\{E_1, E_2\}$ and its associated matrix
       (*ii*) $\{E_1, E_3\}$ and its associated matrix
       (*iii*) $\{E_2, E_3\}$ and its associated matrix
       (*iv*) R_1 and $\mathbf{T}_1$

4. Let M be a Markov chain with transition matrix

$$\begin{array}{c} \\ E_1 \\ E_2 \\ E_3 \\ E_4 \\ E_5 \\ E_6 \\ E_7 \\ E_8 \end{array} \begin{array}{cccccccc} E_1 & E_2 & E_3 & E_4 & E_5 & E_6 & E_7 & E_8 \\ \begin{pmatrix} 0 & 0.4 & 0 & 0 & 0 & 0 & 0.6 & 0 \\ 0.1 & 0 & 0 & 0 & 0.6 & 0 & 0.3 & 0 \\ 0 & 0 & 0 & 0.2 & 0 & 0.1 & 0 & 0.7 \\ 0 & 0 & 0.5 & 0 & 0 & 0 & 0 & 0.5 \\ 0.3 & 0 & 0 & 0 & 0 & 0 & 0.7 & 0 \\ 0.1 & 0.2 & 0 & 0 & 0.7 & 0 & 0 & 0 \\ 0.1 & 0.3 & 0 & 0 & 0.6 & 0 & 0 & 0 \\ 0 & 0 & 0.4 & 0 & 0.6 & 0 & 0 & 0 \end{pmatrix} \end{array}$$

(a) Calculate—
    (*i*) R_1 (*ii*) R_2 (*iii*) R_3
(b) If M is now in state E_1, what is the probability that it will be—
    (*i*) in state E_3 after 3 moves
    (*ii*) in state E_2 after 3 moves
    (*iii*) in state E_2 at time $n = 2$ and in state E_1 at time $n = 3$?

5. Let M be the Markov chain of Exercise 4. What are the probabilities that M will be in state E_5 at time $n = 2$ if M is now (a) in state E_3, (b) in state E_1?

6. In a study of the effects of the personalities of parents upon their offspring, rats were classified as aggressive (A), normal (N), and passive (P). Two rats were mated (first pair), and then two of their offspring of opposite sex were randomly chosen and mated (second pair), and so on. The pairs were classified AA (both aggressive), AN (one aggressive and one normal), and so on. It was found that the transfer of personality from one pair of rats to the next pair is a Markov chain with the following transition matrix:

$$\mathbf{T} = \begin{array}{c} \\ AA \\ AN \\ AP \\ NN \\ NP \\ PP \end{array} \begin{array}{c} \begin{array}{cccccc} AA & AN & AP & NN & NP & PP \end{array} \\ \left(\begin{array}{cccccc} 0.6 & 0.4 & 0 & 0 & 0 & 0 \\ 0.4 & 0.5 & 0 & 0.1 & 0 & 0 \\ 0.1 & 0.3 & 0 & 0.6 & 0 & 0 \\ 0 & 0.3 & 0 & 0.7 & 0 & 0 \\ 0 & 0 & 0 & 0 & 0.8 & 0.2 \\ 0 & 0 & 0 & 0 & 0 & 1 \end{array} \right) \end{array}$$

For example, if both members of the first pair are aggressive (AA), the probability that both members of the second pair are aggressive is 0.6; the probability that one is aggressive and the other is normal (AN) is 0.4.

(a) If both rats in the first pair are normal, what is the probability that—

(i) one of the rats in the second pair is normal and the other is aggressive

(ii) both members of the second pair are aggressive

(iii) both members of the third pair are aggressive

(iv) both members of the fourth pair are passive?

(b) If one member of the first pair is aggressive and the other is normal, what is the probability that—

(i) all members of the second and fourth pair are normal

(ii) at least one member of the third pair is normal?

7. Consider the rat pairs of Exercise 6.

(a) If 20 percent of the pairs of the first generation are AN and 80 percent are AA, what percentages of the third-generation pairs will be AA, AN, AP, NN, NP, and PP?

(b) Answer the question of part **(a)** if 10 percent of the pairs of the first generation are NP and 90 percent are PP.

7.4 REGULAR MATRICES AND ERGODIC CHAINS

In the preceding sections we considered what happens at the nth step of a Markov chain. We saw that information about the nth step could be found by calculating $\mathbf{T}^n$. In this section we will consider what happens in the long run. To do this, we will see what happens to $\mathbf{T}^n$ as n gets larger and larger.

Consider a typical situation: Weekly sales in a certain store are denoted successful (S), moderate (M), or unsuccessful (U). Assume that the weekly sales form a Markov chain with transition matrix

$$\mathbf{T} = \begin{array}{c} \\ S \\ M \\ U \end{array} \begin{array}{ccc} S & M & U \\ \begin{pmatrix} 0.7 & 0.2 & 0.1 \\ 0.2 & 0.6 & 0.2 \\ 0.1 & 0.4 & 0.5 \end{pmatrix} \end{array}$$

Calculating several representative powers of $\mathbf{T}$ yields

$$\mathbf{T}^2 = \begin{pmatrix} 0.54 & 0.30 & 0.16 \\ 0.28 & 0.48 & 0.24 \\ 0.20 & 0.46 & 0.34 \end{pmatrix}$$

$$\mathbf{T}^4 = \begin{pmatrix} 0.4076 & 0.3796 & 0.2128 \\ 0.3336 & 0.4248 & 0.2416 \\ 0.3048 & 0.4372 & 0.2580 \end{pmatrix}$$

$$\mathbf{T}^8 = \begin{pmatrix} 0.3576 & 0.4090 & 0.2334 \\ 0.3513 & 0.4127 & 0.2360 \\ 0.3487 & 0.4142 & 0.2371 \end{pmatrix}$$

$$\mathbf{T}^{16} = \begin{pmatrix} 0.3529 & 0.4117 & 0.2353 \\ 0.3529 & 0.4118 & 0.2353 \\ 0.3529 & 0.4118 & 0.2354 \end{pmatrix}$$

We see that, although the rows of $\mathbf{T}$ are quite different, the rows of $\mathbf{T}^2, \mathbf{T}^3, \mathbf{T}^4, \ldots$ become more and more alike.[4] In fact, if we were to calculate higher powers of $\mathbf{T}$, we would see that the entries of $\mathbf{T}^n$ become closer and closer to those of

$$\mathbf{L} = \begin{pmatrix} 0.3529 & 0.4118 & 0.2353 \\ 0.3529 & 0.4118 & 0.2353 \\ 0.3529 & 0.4118 & 0.2353 \end{pmatrix} = \begin{pmatrix} \frac{6}{17} & \frac{7}{17} & \frac{4}{17} \\ \frac{6}{17} & \frac{7}{17} & \frac{4}{17} \\ \frac{6}{17} & \frac{7}{17} & \frac{4}{17} \end{pmatrix}$$

The matrix $\mathbf{L}$ is called the *limit matrix* of $\mathbf{T}$. Note that the rows of $\mathbf{L}$ are exactly the same. It follows that if n is large, $p_{11}(n)$, $p_{21}(n)$, and $p_{31}(n)$ are all approximately the same number, 0.353. In other words, if n is large, the probability of being in state S at time n does not depend upon whether the chain was initially in state S, M, or U. Hence, in the long run, sales will be successful approximately 35.3 percent of the time, regardless of what state sales are in now. Similarly, sales will be moderate 41.2 percent of the time and unsuccessful 23.5 percent of the time.

Not all Markov chains have a limit matrix. However, if the Markov chain M has a limit matrix $\mathbf{L}$, M is called an **ergodic chain**. In this section, we will study certain ergodic chains and a method for finding the limit matrix $\mathbf{L}$. We will begin by defining regular matrices. Later we will see that all regular matrices are ergodic.

Definition 7.9   A square stochastic matrix is a **regular matrix** if one of its powers has all positive entries.

---

4. The sum of the entries in row 1 of $\mathbf{T}^{16}$ is 0.9999 instead of 1 because these figures were rounded to four decimal places.

Example 24 A study of social mobility resulted in the following table of probabilities:

Parents' Social Class	Son's or Daughter's Social Class		
	Upper	Middle	Lower
Upper	0.448	0.484	0.068
Middle	0.054	0.699	0.247
Lower	0.011	0.503	0.486

Suppose that transitions between social classes of successive generations of a family can be regarded as transitions of a homogeneous Markov chain. Then the array of numbers in the data chart is the transition matrix $\mathbf{T}$ of the chain. The matrix $\mathbf{T}$ is regular because $\mathbf{T}^1 = \mathbf{T}$ satisfies the condition of the definition.

Example 25 The transition matrix

$$\mathbf{T} = \begin{pmatrix} \frac{1}{6} & \frac{5}{6} & 0 & 0 \\ \frac{1}{6} & 0 & \frac{5}{6} & 0 \\ \frac{1}{6} & 0 & 0 & \frac{5}{6} \\ \frac{1}{6} & 0 & 0 & \frac{5}{6} \\ \frac{1}{6} & 0 & 0 & \frac{5}{6} \end{pmatrix}$$

is regular because

$$\mathbf{T}^2 = \begin{pmatrix} \frac{6}{36} & \frac{5}{36} & \frac{25}{36} & 0 \\ \frac{6}{36} & \frac{5}{36} & 0 & \frac{25}{36} \\ \frac{6}{36} & \frac{5}{36} & 0 & \frac{25}{36} \\ \frac{6}{36} & \frac{5}{36} & 0 & \frac{25}{36} \\ \frac{6}{36} & \frac{5}{36} & 0 & \frac{25}{36} \end{pmatrix}$$

$$\mathbf{T}^4 = \begin{pmatrix} \frac{6}{36} & \frac{5}{36} & \frac{25}{216} & \frac{125}{216} \\ \frac{6}{36} & \frac{5}{36} & \frac{25}{216} & \frac{125}{216} \\ \frac{6}{36} & \frac{5}{36} & \frac{25}{216} & \frac{125}{216} \\ \frac{6}{36} & \frac{5}{36} & \frac{25}{216} & \frac{125}{216} \\ \frac{6}{36} & \frac{5}{36} & \frac{25}{216} & \frac{125}{216} \end{pmatrix}$$

and $\mathbf{T}^4$ satisfies the condition of Definition 7.9.

Example 26 The matrix

$$\mathbf{T} = \begin{pmatrix} 1 & 0 \\ 0 & 0 \end{pmatrix}$$

is not regular, because $\mathbf{T}^n = \mathbf{T}$ for all n. Therefore, all powers of $\mathbf{T}$ have zero entries.

The following theorem states that if the transition matrix of M is regular, then M is ergodic. It also tells us how to find the limit matrix $\mathbf{L}$.

THEOREM 7.5   Let **T** be a regular $m \times m$ matrix.

(a) As n increases, $\mathbf{T}^n$ approaches a stochastic matrix **L**.

(b) The rows of **L** are all equal to the same vector $\mathbf{v} = (\pi_1, \pi_2, \ldots, \pi_m)$.

(c) The vector **v** of part (b) and the matrix **T** satisfy the equation

$$\mathbf{v T} = \mathbf{v}$$

The matrix **L** of Theorem 7.6 is called the **limit matrix** of **T**. The vector $\mathbf{v} = (\pi_1, \pi_2, \ldots, \pi_m)$ is called the **limiting distribution** of **T**. The components π_j tell us what proportion of time the Markov chain M will be in state E_j in the long run. It follows that

$$\pi_1 + \pi_2 + \cdots + \pi_m = 1 \qquad (8)$$

Note that the limiting distribution $\mathbf{v} = (\pi_1, \pi_2, \ldots, \pi_m)$ does not depend upon the initial distribution $(p(0), p(1), \ldots, p(m))$. The vector **v** can be found by solving the matrix equation $\mathbf{vT} = \mathbf{T}$ and using equation (8).

Suppose that M is now in state E_j and let μ_j be the average number of steps that occur before M first returns to E_j. Then μ_j is called the **mean recurrence time**. It can be shown that

$$\mu_j = \frac{1}{\pi_j} \qquad \text{if } \pi_j \neq 0 \qquad (9)$$

If $\pi_j = 0$, M will never return to E_j.

Example 27   A financier is considering investing money in one of the two local supermarkets. A market survey indicates that store A now has 75 percent of the market. However, 30 percent of the customers who buy at store A each week switch to store B the following week, while only 20 percent of those who buy at B switch to A. Which store should she invest in?

Solution   The weekly distribution of customers forms a Markov chain with transition matrix

$$\mathbf{T} = \begin{array}{c} \\ A \\ B \end{array} \begin{array}{cc} A & B \\ \begin{pmatrix} 0.7 & 0.3 \\ 0.2 & 0.8 \end{pmatrix} \end{array}$$

and initial probability vector

$$\begin{array}{cc} A & B \end{array}$$
$$\mathbf{p}(0) = (0.75, 0.25)$$

(Notice that the entry in row 1, column 2 of **T** is 0.3 because 30 percent of the customers switch from A to B; the remaining entry in that row is $1 - 0.3 = 0.7$ because a customer must buy from either store A or store B.)

Although she knows that store A is doing better now, the prospective investor is naturally interested in which store will have the greater share of the market *in the long run*. That is, she is interested in the limiting distribution $\mathbf{v} = (\pi_1, \pi_2)$. Theorem 7.5 can be used because $\mathbf{T}$ is regular ($\mathbf{T}$ has no zero entries).

Using Theorem 7.5(c) and equation (8), the values of π_1 and π_2 can be found by simultaneously solving the matrix equation

$$(\pi_1, \pi_2)\begin{pmatrix} 0.7 & 0.3 \\ 0.2 & 0.8 \end{pmatrix} = (\pi_1, \pi_2) \tag{10}$$

and the equation

$$\pi_1 + \pi_2 = 1 \tag{11}$$

Equation (10) yields

$$(0.7)\pi_1 + (0.2)\pi_2 = \pi_1 \tag{12}$$

$$(0.3)\pi_1 + (0.8)\pi_2 = \pi_2 \tag{13}$$

Equation (12) is equivalent to

$$(0.2)\pi_2 = (0.3)\pi_1 \qquad \text{or} \qquad \pi_2 = \tfrac{3}{2}\pi_1 \tag{14}$$

Substituting equation (14) in equation (11) yields

$$\pi_1 + \tfrac{3}{2}\pi_1 = 1, \qquad \text{or} \qquad \tfrac{5}{2}\pi_1 = 1$$

It follows that

$$\pi_1 = \tfrac{2}{5} \qquad \text{and} \qquad \pi_2 = 1 - \tfrac{2}{5} = \tfrac{3}{5} \tag{15}$$

Hence, in the long run, store A will have 40 percent of the market and store B will have 60 percent of the market. The financier should therefore invest in store B.

Notice that equation (13) was not used to find π_1 and π_2 in Example 27. However, the values $\pi_1 = \tfrac{2}{5}$ and $\pi_2 = \tfrac{3}{5}$ must satisfy equation (13) also. In general, one of the equations resulting from the matrix equation $\mathbf{vT} = \mathbf{v}$ can be omitted when solving for π_i $(i = 1,2,\ldots,m)$. (It does not matter which one is omitted.) The values of π_i, however, must also satisfy the omitted equation, so this equation can be used as a check of the calculation. Notice also that the limiting distribution in Example 27 did not depend in any way upon the initial condition $\mathbf{p}(0)$.

Example 28   Suppose that a customer is now shopping in store B of Example 27. On the average, when will he or she first return to store B?

Solution   Applying equations (9) and (15), the customer will first return to store B, on the average, after

$$\mu_2 = \frac{1}{\pi_2} = \frac{1}{\tfrac{3}{5}} = \frac{5}{3} \text{ weeks}$$

The following examples show that Theorem 7.5 and equation (9) can be *indirectly* applied to matrices that are not regular.

Example 29   Let M be a Markov chain with transition matrix

$$
\mathbf{T} = \begin{array}{c} \\ E_1 \\ E_2 \\ E_3 \\ E_4 \\ E_5 \\ E_6 \end{array}
\begin{array}{c}
\begin{array}{cccccc} E_1 & E_2 & E_3 & E_4 & E_5 & E_6 \end{array} \\
\left(\begin{array}{cccccc}
\frac{7}{10} & 0 & \frac{3}{10} & 0 & 0 & 0 \\
0 & \frac{1}{4} & 0 & \frac{3}{4} & 0 & 0 \\
\frac{1}{5} & 0 & \frac{4}{5} & 0 & 0 & 0 \\
0 & \frac{1}{5} & 0 & \frac{4}{5} & 0 & 0 \\
0 & 1 & 0 & 0 & 0 & 0 \\
0 & 0 & 0 & 0 & 0 & 1
\end{array}\right)
\end{array}
$$

What percentage of the time will M spend in states $E_1, E_2, \ldots, E_6$, respectively, if M is now (a) in state E_1, (b) in state E_2, (c) in state E_3, (d) in state E_5, (e) in state E_6?

Solution   The transition matrix $\mathbf{T}$ is not a regular matrix because $\mathbf{T}^n$ has zero entries in the first five positions of row 6 for all n. Therefore, Theorem 7.5 does not apply to $\mathbf{T}$. However, the submatrices associated with $R_1 = R_3 = \{E_1, E_3\}$, $R_2 = R_4 = R_5 = \{E_2, E_4\}$, and $R_6 = \{E_6\}$ are all regular and Theorem 7.5 will apply to them. Let $\mathbf{T}_i$ be the matrix associated with R_i. Then

$$
\mathbf{T}_1 = \begin{array}{c} \\ E_1 \\ E_3 \end{array}
\begin{array}{c}
\begin{array}{cc} E_1 & E_3 \end{array} \\
\left(\begin{array}{cc} \frac{7}{10} & \frac{3}{10} \\ \frac{1}{5} & \frac{4}{5} \end{array}\right)
\end{array},
\qquad
\mathbf{T}_2 = \begin{array}{c} \\ E_2 \\ E_4 \end{array}
\begin{array}{c}
\begin{array}{cc} E_2 & E_4 \end{array} \\
\left(\begin{array}{cc} \frac{1}{4} & \frac{3}{4} \\ \frac{1}{5} & \frac{4}{5} \end{array}\right)
\end{array},
$$

$$
\mathbf{T}_6 = \begin{array}{c} \\ E_6 \end{array} \, E_6 \qquad (1)
$$

(a) If M is now in state E_1, M will always be in state E_1 or in state E_3. It follows that the percentage of time spent in states E_2, E_4, E_5, and E_6 is zero. Matrix $\mathbf{T}_1$ is equal to the matrix $\mathbf{T}$ of Example 27. The limiting distribution will therefore be the same. Therefore, M will be in state E_1 40 percent of the time and in state E_2 60 percent of the time.

(b) If M is now in state E_2, it will always be in state E_2 or in state E_4. The limiting distribution can be found by solving the equations

$$
(\pi_2, \pi_4)\begin{pmatrix} \frac{1}{4} & \frac{3}{4} \\ \frac{1}{5} & \frac{4}{5} \end{pmatrix} = (\pi_2, \pi_4) \tag{16}
$$

and

$$
\pi_2 + \pi_4 = 1 \tag{17}
$$

Equation (16) becomes two linear equations:

$$
\tfrac{1}{4}\pi_2 + \tfrac{1}{5}\pi_4 = \pi_2 \tag{18}
$$

and

$$
\tfrac{3}{4}\pi_2 + \tfrac{4}{5}\pi_4 = \pi_4 \tag{19}
$$

Simultaneously solving equations (17) and (18) yields $\pi_2 = \frac{4}{19} = 0.211$ and $\pi_4 = \frac{15}{19} = 0.789$. Substituting these values in the left side of equation (19) yields

$$\left[\frac{3}{4} \cdot \frac{4}{19} + \frac{4}{5} \cdot \frac{15}{19} = \frac{15}{19}\right]$$

Since $\pi_4 = \frac{15}{19}$, these values satisfy equation (19) as well, and so are correct. It follows that M will be in state E_2 about 21.1 percent of the time and in state E_4 about 78.9 percent of the time; M will never be in any other state.

(c) The transition matrix $\mathbf{T}_1$ is regular; hence, the limiting probabilities do not depend on the initial probabilities. It follows that (π_1, π_3) is the same as in part (a). Therefore, M will be in states $E_1, E_2, \ldots, E_6$, respectively, 40, 0, 60, 0, 0, 0 percent of the time.

(d) State E_5 is a **nonreturn state**; that is, once M leaves E_5 it will never return. If M is now in E_5, it will enter E_2 on the first trial. After that, it can only be in E_2 or E_4. In the long run, the fact that M was once in E_5 will not affect the percentage of time M spends in each state, and the limiting distribution of part (b) applies. Therefore, M will be in state $E_1, E_2, \ldots, E_6$, respectively for 0, 21.1, 0, 78.9, 0, 0 percent of the time.

(e) State E_6 is an **absorbing state**; that is, once M enters E_6, it stays there. (Mathematically, this means that $p_{66} = 1$.) If M is now in E_6, M will spend 100 percent of its time in E_6 and never enter any other state.

Example 30 Let M be the Markov chain of Example 29. Suppose M is now (a) in state E_1, (b) in state E_2, (c) in state E_5, (d) in state E_6. How long, on the average, will it take before M first returns to this state?

Solution

(a) Using the results of Example 29(a) and equation (9), the required average is

$$\mu_1 = \frac{1}{\pi_1} = \frac{1}{0.4} = 2.5 \text{ steps}$$

(b) Similarly, using the results of Example 29(b), the required average is

$$\mu_2 = \frac{1}{\frac{4}{19}} = 4.75 \text{ steps}$$

(c) M will never return to E_5.

(d) Since M is certain to return on the first step, $\mu_6 = 1$. Alternatively, equation (9) and Example 29(e) can be used to obtain the same result.

The following example reviews many of the concepts studied in this chapter. It also shows how Markov chains can be applied to the study of genetics.

Example 31 Suppose that a simple Mendelian trait is being studied (see Section 2.5). An animal of unknown genetic character is crossed with a hybrid (genotype aA). The first offspring is then crossed with a hybrid, and so on. This process is a Markov chain with three states: dominant (D) (genotype AA), hybrid (H), and recessive (R) (genotype aa).

(a) Find the transition matrix **T**.

(b) Is **T** regular?

(c) After repeated crossings, what is the probability that the offspring is a hybrid if the original animal was a hybrid? If the original animal was a recessive?

(d) If the original animal was a recessive (the initial generation), in what generation can we expect the first recessive offspring?

Solution

(a) Since one of the parents in each cross is aA, and the other parent may be aa, aA, or AA, there are three possible crosses:

	a	A
a	aa	aA
a	aa	aA

	a	A
a	aa	aA
A	aA	AA

	a	A
A	aA	AA
A	aA	AA

Suppose that the nth-generation offspring is recessive. Using the first table, the probabilities of getting recessive, hybrid, dominant offspring in the $(n + 1)$st generation are $\frac{1}{2}, \frac{1}{2}, 0$, respectively. These probabilities therefore belong in the first row of the transition matrix **T**. The remaining rows are calculated similarly.

$$
\mathbf{T} = \begin{array}{c} \\ R \\ H \\ D \end{array} \begin{array}{ccc} R & H & D \\ \left(\begin{array}{ccc} \frac{1}{2} & \frac{1}{2} & 0 \\ \frac{1}{4} & \frac{1}{2} & \frac{1}{4} \\ 0 & \frac{1}{2} & \frac{1}{2} \end{array} \right) \end{array}
$$

(b) The transition matrix **T** is regular because every entry of

$$
\mathbf{T}^2 = \begin{pmatrix} \frac{3}{8} & \frac{1}{2} & \frac{1}{8} \\ \frac{1}{4} & \frac{1}{2} & \frac{1}{4} \\ \frac{1}{8} & \frac{1}{2} & \frac{3}{8} \end{pmatrix}
$$

is positive.

(c) Since **T** is regular, in the long run the genotype of the offspring does not depend upon the genotype of the initial animal. The limiting distribution is the same regardless of whether the initial animal was hybrid or recessive. To find the limiting distribution (π_1, π_2, π_3), we solve the equations

$$
(\pi_1, \pi_2, \pi_3) \begin{pmatrix} \frac{1}{2} & \frac{1}{2} & 0 \\ \frac{1}{4} & \frac{1}{2} & \frac{1}{4} \\ 0 & \frac{1}{2} & \frac{1}{2} \end{pmatrix} = (\pi_1, \pi_2, \pi_3) \tag{20}
$$

and

$$
\pi_1 + \pi_2 + \pi_3 = 1 \tag{21}
$$

Equation (20) becomes the three linear equations

$$\tfrac{1}{2}\pi_1 + \tfrac{1}{4}\pi_2 = \pi_1, \qquad \tfrac{1}{2}\pi_1 + \tfrac{1}{2}\pi_2 + \tfrac{1}{2}\pi_3 = \pi_2, \qquad \tfrac{1}{4}\pi_2 + \tfrac{1}{2}\pi_3 = \pi_3$$

from which we find $\pi_1 = \pi_3 = \tfrac{1}{2}\pi_2$. Substituting these values into equation (21) yields

$$\tfrac{1}{2}\pi_2 + \pi_2 + \tfrac{1}{2}\pi_2 = 1, \quad \text{or} \quad \pi_2 = \tfrac{1}{2}$$

Thus, $\mathbf{v} = (\tfrac{1}{4},\tfrac{1}{2},\tfrac{1}{4})$. Hence, the probability that the offspring is hybrid is $\tfrac{1}{2}$.

(d) Using the results of part **(c)** and equation (9), the first recessive offspring will appear, on the average, after

$$\mu_1 = \frac{1}{\pi_1} = \frac{1}{\tfrac{1}{4}} = 4 \text{ steps}$$

If we call the original parents generation zero, the first recessive offspring will appear (on the average) in the fourth generation.

7.4 EXERCISES

1.   Which of the following matrices are regular?

(a) $\begin{pmatrix} 1 & 0 \\ \tfrac{1}{2} & \tfrac{1}{2} \end{pmatrix}$ **(b)** $\begin{pmatrix} \tfrac{1}{2} & \tfrac{1}{2} \\ \tfrac{1}{4} & \tfrac{3}{4} \end{pmatrix}$ **(c)** $\begin{pmatrix} 2 & -1 \\ \tfrac{1}{2} & \tfrac{1}{2} \end{pmatrix}$

(d) $\begin{pmatrix} \tfrac{1}{2} & 0 & \tfrac{1}{2} \\ \tfrac{1}{4} & \tfrac{1}{2} & \tfrac{1}{4} \\ \tfrac{1}{2} & \tfrac{1}{2} & 0 \end{pmatrix}$ **(e)** $(\tfrac{1}{2},\tfrac{1}{2})$ **(f)** $\begin{pmatrix} 1 & 0 & 0 \\ \tfrac{1}{4} & \tfrac{1}{4} & \tfrac{1}{2} \\ \tfrac{1}{3} & \tfrac{1}{3} & \tfrac{1}{3} \end{pmatrix}$

2.   Let M be a Markov chain with transition matrix

$$\mathbf{T} = \begin{pmatrix} \tfrac{1}{4} & \tfrac{1}{4} & \tfrac{1}{2} \\ \tfrac{1}{2} & \tfrac{1}{2} & 0 \\ \tfrac{1}{4} & 0 & \tfrac{3}{4} \end{pmatrix}$$

(a) Is M ergodic? Explain.

(b) Find the limiting distribution of M.

(c) Find the limit matrix of T.

3.   Let M be the Markov chain of Exercise 2. Suppose M is now in state E_1.

(a) When, on the average, will it first return to state E_1?

(b) In the long run, what percentage of the time will it be in state E_2?

4.   Let M be a Markov chain with transition matrix

	E_1	E_2	E_3	E_4	E_5
E_1	$\tfrac{1}{2}$	0	0	$\tfrac{1}{2}$	0
E_2	0	$\tfrac{2}{3}$	0	0	$\tfrac{1}{3}$
E_3	$\tfrac{1}{4}$	0	0	$\tfrac{3}{4}$	0
E_4	$\tfrac{1}{3}$	0	0	$\tfrac{2}{3}$	0
E_5	0	1	0	0	0

(a) What is the probability that M will be in state E_1 after the one-thousandth transition if M is now

  (i) in state E_1,   (ii) in state E_2,   (iii) in state E_3,
  (iv) in state E_4,   (v) in state E_5?

(b) If M is now in state E_2 when, on the average, will it first return to state E_2?

5.   Supermarket A of Example 27 is considering opening a branch in location C or D. Preliminary surveys show the following weekly customer switching:

	A	B	C			A	B	D
A	0.5	0.3	0.2		A	0.6	0.2	0.2
B	0.3	0.6	0.1		B	0.2	0.7	0.1
C	0.2	0.2	0.6		D	0.2	0.1	0.7

For example, if the store opens branch C, 50 percent of the customers now using A will return to A next week, 30 percent will switch to B, and 20 percent will switch to C; if the store opens branch D, 60 percent of the customers now using A will return to A next week, 20 percent will switch to B, and 20 percent will switch to D.

(a) If store A opens a branch at location C, what portions of the market will store A, store B, and branch C get in the long run?

(b) Answer the above question if store A opens its branch at location D.

(c) Which location should store A choose if it wishes to obtain the largest total share (from both store A and its branch) of the market in the long run? What will its total share of the market be?

6.   In a genetic study similar to that of Example 31, an animal of unknown genotype is crossed with a dominant. The first offspring is then crossed with a dominant, and so on. This process is a Markov chain with three states: dominant, hybrid, and recessive.

(a) Find the transition matrix $\mathbf{T}$.

(b) Is $\mathbf{T}$ regular? Explain your answer.

7.   Consider the genetic experiment of Exercise 6.

(a) Suppose that the initial individual is dominant.
    (i) In the long run, what percentage of the offspring are dominant?
    (ii) In what generation, on the average, can we expect the first dominant offspring?

(b) Suppose that the initial individual is hybrid.
    (i) What is the probability that the first generation offspring is hybrid, the second generation offspring is hybrid, the third, the nth?
    (ii) In the long run, what percentage of the offspring are hybrid? What percentage are dominant? What percentage are recessive?

(c) Answer the questions of part (b) assuming that the initial individual is recessive.

Key Word Review

Each key word is followed by the number of the section in which it is defined.

states   7.1
homogeneous Markov chain   7.1
transition probability   7.1
transition matrix   7.1
stochastic matrix   7.1
probability vector   7.1
absolute probability vector   7.2
initial probability vector   7.2
initial trial   7.1
initial outcome   7.1
n-step transition matrix   7.2

submatrix   7.3
subchain   7.3
accessible state   7.3
ergodic chain   7.4
regular matrix   7.4
limit matrix   7.4
limiting distribution   7.4
mean recurrence time   7.4
nonreturn state   7.4
absorbing state   7.4

CHAPTER 7   **REVIEW EXERCISES**

1. Define each word in the Key Word Review.

2. What do the following symbols mean?
    (a) p_{ij} **(b)** $p_i(n)$ **(c)** $X_n = i$
    (d) $\mathbf{p}(0)$ **(e)** $\mathbf{T}^2$ **(f)** $\mathbf{T}^{(2)}$
    (g) $\mathbf{v} = (\pi_1, \pi_2, \ldots, \pi_m)$ **(h)** R_i **(i)** $\mathbf{T}_i$
    (j) $\mathbf{p}(3)$ **(k)** π_3 **(l)** $E_i \rightarrow E_j$

3. For each matrix in parts **(a)**–**(g)**, state whether it is a regular matrix, a
    stochastic matrix, a transition matrix, or a probability vector.
    (a) $(\frac{1}{2}, 0, \frac{1}{2})$ **(b)** $\begin{pmatrix} 1 & 0 & 0 \\ \frac{1}{2} & 0 & \frac{1}{2} \\ \frac{1}{3} & 0 & \frac{2}{3} \end{pmatrix}$ **(c)** $\begin{pmatrix} \frac{1}{2} & \frac{1}{2} \\ \frac{1}{4} & \frac{3}{4} \end{pmatrix}$

    (d) $(-1, 0, 1)$ **(e)** $\begin{pmatrix} \frac{1}{2} & \frac{1}{4} & \frac{1}{4} \\ \frac{1}{3} & 0 & \frac{2}{3} \end{pmatrix}$ **(f)** $\begin{pmatrix} \frac{1}{4} & \frac{3}{4} & 0 \\ \frac{1}{5} & \frac{2}{5} & \frac{2}{5} \\ \frac{1}{2} & \frac{1}{4} & \frac{1}{4} \end{pmatrix}$

    (g) $(2, 0, -1)$

4. Let M be a Markov chain with transition matrix

    $$\mathbf{T} = \begin{pmatrix} 1 & 0 & 0 \\ \frac{1}{2} & \frac{1}{2} & 0 \\ \frac{1}{4} & \frac{1}{4} & \frac{1}{2} \end{pmatrix}$$

    and initial probability vector $(\frac{1}{2}, 0, \frac{1}{2})$. Calculate the following probabilities and
    matrices:

    (a) $P(X_1 = 2)$ **(b)** $P(X_2 = 1 \mid X_1 = 2)$
    (c) $P(X_3 = 2 \mid X_1 = 2, X_2 = 1)$ **(d)** $P(X_3 = 2, X_2 = 2 \mid X_1 = 3)$
    (e) $\mathbf{T}^2$ **(f)** $\mathbf{T}^{(2)}$ **(g)** p_{12} **(h)** $\mathbf{p}(2)$,   **(i)** $p_1(2)$

5. Draw the transition diagram for the Markov chain of Exercise 4.

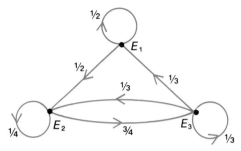

Figure 7.9

6. Let M be a Markov chain with the transition diagram shown in Figure 7.9.
    (a) Find the transition matrix **T**.
    (b) What is the probability that M is in state E_1 at time $n = 2$ if—
      (*i*) $\mathbf{p}(0) = (\frac{1}{2},\frac{1}{2},0)$
      (*ii*) $X_o = 3$
      (*iii*) M is now in state E_1.

7. A cliff in a certain area is observed upward from its base. Suppose that each
    layer of rock may be considered to be one step in a Markov chain with four
    states—sandstone (S), shale (H), lignite (L), and siltstone (I)—and with tran-
    sition matrix

    $$
    \begin{array}{c c c c c}
     & S & H & L & I \\
    \begin{array}{c} S \\ H \\ L \\ I \end{array} &
    \left(\begin{array}{cccc}
    0.75 & 0.10 & 0.10 & 0.05 \\
    0.05 & 0.75 & 0.15 & 0.05 \\
    0.20 & 0.40 & 0.30 & 0.10 \\
    0.10 & 0.30 & 0.20 & 0.40
    \end{array}\right)
    \end{array}
    $$

    If we start our observations (layer 0) with sandstone, calculate the probability
    that—
    (a) lignite will appear in the next layer (layer 1)
    (b) lignite will appear in layer 1 and sandstone will appear in layer 2
    (c) lignite will appear in layer 1 and/or layer 2
    (d) lignite will first appear in layer 2.

8. Let M be a Markov chain with transition matrix

    $$
    \mathbf{T} = \begin{pmatrix} x & y \\ 2x & \dfrac{y}{3} \end{pmatrix}
    $$

    and initial probability vector (y,x). Calculate $\mathbf{p}(0)$.

9. Let M be a Markov chain with transition matrix

    $$
    \mathbf{T} = \begin{pmatrix} \frac{1}{2} & \frac{1}{2} & 0 \\ \frac{1}{4} & \frac{1}{4} & \frac{1}{2} \\ 0 & \frac{1}{2} & \frac{1}{2} \end{pmatrix}
    $$

    and suppose that M is now in state E_1.

(a) Is **T** regular?

(b) Is M ergodic?

(c) Find the limit matrix **L** and the limiting distribution **v**.

(d) What is the probability that M will be in state E_1 at time $n = 2$? At time $n = 3$?

(e) In the long run, what percentage of the time will M spend in state E_1? In state E_2?

(f) On the average, when can we expect M to first return to state E_1?

10.   Let M be a Markov chain with transition matrix

$$
\begin{array}{c@{\quad}ccccccc}
 & E_1 & E_2 & E_3 & E_4 & E_5 & E_6 & E_7 \\
E_1 & \frac{1}{2} & 0 & 0 & \frac{1}{2} & 0 & 0 & 0 \\
E_2 & 1 & 0 & 0 & 0 & 0 & 0 & 0 \\
E_3 & 0 & 0 & \frac{1}{2} & 0 & 0 & \frac{1}{2} & 0 \\
E_4 & 0 & 0 & 0 & \frac{1}{2} & \frac{1}{2} & 0 & 0 \\
E_5 & \frac{1}{4} & 0 & 0 & 0 & \frac{3}{4} & 0 & 0 \\
E_6 & 0 & 0 & \frac{1}{4} & 0 & 0 & \frac{1}{4} & \frac{1}{2} \\
E_7 & 0 & 0 & 0 & 0 & 0 & 0 & 1
\end{array}
$$

Suppose M is now in state E_1.

(a) Find R_2 and R_6.

(b) What is the probability that—

   (*i*) M will be in state E_4 in two moves

   (*ii*) M will be in state E_2 in three moves?

(c) In the long run, what percentages of the time will M spend in states $E_1, E_2, \ldots, E_6$?

(d) On the average, when can we first expect M to return to E_1?

Suggested Readings

1. Alling, D. W. "The After-history of Pulmonary T.B., a Stochastic Model." *Biometrics* 4 (1958): 527–47.
2. Bower, G. "Application of a Model to Paired Associate Learning." *Psychometrika* 26 (1961): 255–80.
3. Bush, Robert, R., and Estes, William K., eds. *Studies in Mathematical Learning Theory.* Stanford, Calif.: Stanford University Press, 1959.
4. Cox, D. R., and Miller, H. D. *The Theory of Stochastic Processes.* New York: John Wiley & Sons, 1965.
5. Kay, Paul, ed. *Explorations in Mathematical Anthropology.* Cambridge, Mass.: M.I.T. Press, 1971.
6. Kemeny, John G., and Snell, J. Laurie. *Finite Markov Chains.* Princeton: Van Nostrand Reinhold Co., 1959.
7. Parzan, Emanuel. *Stochastic Processes.* San Francisco: Holden-Day, 1962.

FAO

CHAPTER 8 Linear Programming

INTRODUCTION

The pattern of crops and fallow lands shown in the photograph on the left is the result of a series of complex decisions. These decisions involved calculating what combination of crops would be most profitable given the farmers' resources. In this chapter we will learn to use a method, called *linear programming*, for choosing the best combination of certain items or quantities when there are restrictions on the possible choices. For example, a farmer might want to decide how much acreage to devote to each of three different crops. The restrictions on the possible choices would include the amount of land available and the amount of money and labor the farmer can invest. (See Examples 13 and 21.)

Many problems of this kind can be translated into mathematical problems called *linear programs*. In this chapter we will learn how to translate real-life problems into linear programs and how to solve these problems in order to obtain a plan for action.

Linear programming is a relatively new branch of mathematics. Much of the theory was developed during World War II to solve military problems. Today, however, linear-programming techniques are also used to solve problems in business, economics, medicine, and so on. In Chapter 9 we will see that linear-programming methods also can be used to find the best strategy for playing certain types of games.

8.1 GRAPHS OF LINEAR EQUATIONS

Before we can discuss linear programs, we must learn how to draw the graphs of linear equations and linear inequalities. The reader probably already knows how numbers can be represented on a number line and how ordered pairs of numbers and linear equations can be represented on a Cartesian plane. These concepts will be briefly reviewed in this section. In Section 8.3 we will see how the representations of linear equations can be used to solve linear programs.

We will begin by reviewing the number line. A **number line** is formed as follows: First, a line is drawn and a point 0 is arbitrarily selected. This point is called the **origin** and represents the number 0. Next, a unit of length is selected. This unit of length may be 1 centimeter or 0.5 inch, or any other measure that meets our needs. The unit of length may represent 1 inch, or 10 years, or 5000 pounds, again depending on our needs. The line is then scaled off in terms of this unit of length as shown in Figure 8.1.

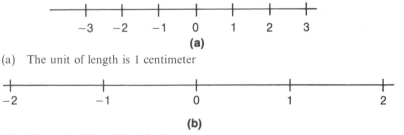

(a) The unit of length is 1 centimeter

(b) The unit of length is 1 inch

Figure 8.1 The unit of length represents 1

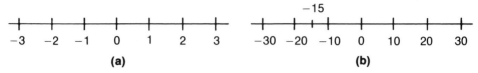

(a)   The unit of length represents 1

(b)   The unit of length represents 10

Figure 8.2

Usually the unit of length represents 1 (i.e., 1 mile, 1 year, 1 ton, etc.). In this case the scale marks to the right of the origin are labeled 1, 2, 3, ... as in Figure 8.1. The scale marks to the left of the origin are labeled -1, -2, -3, ... If the unit of length represents 10, the scale marks are labeled ..., -20, -10, 0, 10, 20, ... as in Figure 8.2(b).

Let L be any number line. It can be shown that every real number is represented by a point on L. Moreover, under this correspondence, every point on L represents some real number.

For example, consider the number line in Figure 8.2(a). The correspondence between the integers and points on the line is indicated by the labels. That is, the points lying 1, 2, 3, ... units to the right of the origin represent the positive integers 1, 2, 3, ... The points lying 1, 2, 3, ... units to the left of the origin represent the negative integers -1, -2, -3, ...

The other real numbers are also represented in a natural way by the points on this line. For example, the number $\frac{1}{2}$ is represented by the point halfway between the points labeled 0 and 1; the number $-\frac{3}{2}$ is represented by the point halfway between the points labeled -2 and -1; and so on. Figure 8.3 shows the representations of $\frac{1}{2}$, $-\frac{3}{2}$, $-\frac{5}{3}$, and 2.2.

Figure 8.3   A number line

A similar correspondence exists for every other number line. For example, let L be the number line shown in Figure 8.2(b). The points 1, 2, 3, ... units to the right of the origin represent the integers 10, 20, 30, ...; those to the left represent -10, -20, -30, ... The number -15 is represented by the point halfway between the points labeled -10 and -20.

We have seen how real numbers can be represented by points on a number line. Ordered pairs of real numbers (x,y) can be represented by points on a *Cartesian plane*.[1] A **Cartesian plane** is a plane on which two perpendicular number lines have been drawn so that they intersect at their respective origins. As Figure 8.4 shows, the two number lines may be scaled differently. These two lines, called *axes*, form the basis for a correspondence between ordered pairs of real numbers and points on the plane.

Usually, ordered pairs are denoted (x,y). The first component, x, is called the *x*-**coordinate** of (x,y). The horizontal number line on the Cartesian plane is called the

1. Named after the French mathematician René Descartes (1596–1650).

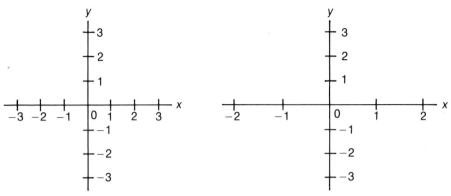

Figure 8.4 Two Cartesian planes

x-**axis** and is used to represent the *x*-coordinate. The vertical number line is called the *y*-**axis** and is used to represent the *y*-coordinate. A similar notation is used if the ordered pair is denoted by another pair of letters, for example, (*s*,*t*) or (*u*,*v*).

To Represent the Ordered Pair (a,b) on a Cartesian Plane—

(1) Notice that the *x*-coordinate of the ordered pair (*a*,*b*) is *a*. Use the scale on the *x*-axis to represent this coordinate by drawing a vertical line through *a* on the *x*-axis. (See Figure 8.5.)

(2) The *y*-coordinate, *b*, is represented by using the *y*-axis. Draw a horizontal line through *b* on the *y*-axis. (See Figure 8.5.)

(3) The point at which the two lines intersect represents the ordered pair (*a*,*b*). (See Figure 8.5.)

By following these three steps, we can represent every ordered pair (*x*,*y*) by a point in the plane. Likewise, any point in the plane represents some ordered pair of numbers. Because of this correspondence, an ordered pair of numbers is often called a point. Finding the point in the plane that represents the ordered pair is called *plotting the point.*

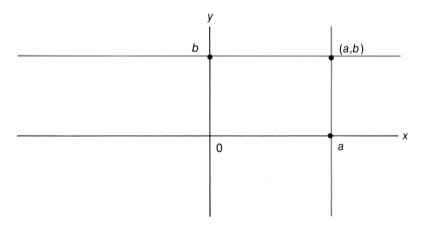

Figure 8.5 Representing (*a*,*b*) on a Cartesian plane

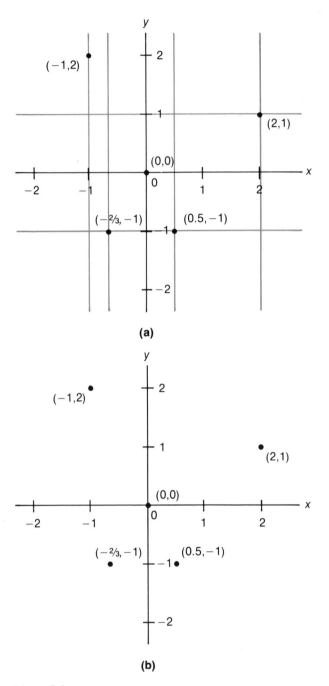

Figure 8.6

The defining lines described in steps 1 and 2 do not have to be actually drawn; they can simply be imagined. The points $(2,1)$, $(-1,2)$, $(0.5,-1)$, $(-\frac{2}{3},-1)$, and $(0,0)$ are shown in Figures 8.6(a) and (b). In Figure 8.6(a) the defining lines are drawn, but in Figure 8.6(b) they are omitted. The point $(0,0)$ is called the **origin** of the Cartesian plane.

The representation of ordered pairs as points in the plane can be used to represent linear equations as lines in the plane. Consider, for example, the linear equation

$$2x + y = 5 \qquad (1)$$

Two ordered pairs that satisfy equation (1) are (0,5) and (1,3), since

$$2(0) + 5 = 5 \quad \text{and} \quad 2(1) + 3 = 5$$

(See Definition 6.10.) The set of all ordered pairs that satisfy equation (1) is called the *graph* of equation (1). In general, we have the following definition:

Definition 8.1 The set of all ordered pairs that satisfy a linear equation of the form

$$ax + by = c$$

is called the **graph** of the linear equation.

Symbolically, the graph of $ax + by = c$ is $\{(x,y) \mid ax + by = c\}$.
A linear equation can be pictorially represented by plotting all the points in its graph. This is called *drawing the graph* of the equation. The graph of a linear equation can be drawn using the following method:

Method for Drawing the Graph of a Linear Equation

(1) Plot any two distinct points that satisfy the linear equation.
(2) Draw the straight line that passes through these two points.

The following examples illustrate the use of this method.

Example 1 Draw the graph of the linear equation

$$2x - y = 5 \qquad (2)$$

Solution To graph equation (2), we use the two steps outlined above.

(1) Although we may plot any two points satisfying equation (2), it is often convenient to use the points whose x-coordinates are 0 and 1. Setting $x = 0$ in equation (2) yields $y = 2(0) - 5 = -5$, and setting $x = 1$ in equation (2) yields $y = 2(1) - 5 = -3$. It follows that $(0, -5)$ and $(1, -3)$ are two points that satisfy equation (2). These points are plotted in Figure 8.7.
(2) Figure 8.7 also shows the straight line through $(0, -5)$ and $(1, -3)$. This line is the graph of equation (2); we will assume that it extends indefinitely in both directions. Every point on this line is a solution of equation (2).

Example 2 Draw the graph of the linear equation

$$3x + 2y = 6$$

Solution The two plotted points and the straight line through them are shown in Figure 8.8.

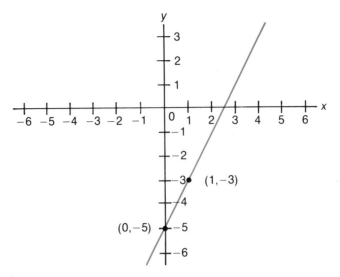

Figure 8.7 Graph of $2x - y = 5$

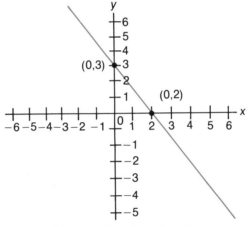

Figure 8.8 Graph of $3x + 2y = 6$

Notice that the pictorial representation of a linear equation is a line. The linear equation is therefore sometimes called a line.

In Chapter 6 we learned that a system of two linear equations in two variables can have no solution, exactly one solution, or infinitely many solutions. In the next three examples we will see what happens when pairs of equations of these three types are drawn on the same Cartesian plane.

Example 3 (A system with no solution)

   (a) Draw the graphs of

$$2x - y = -1 \tag{3}$$

and

$$2x - y = -2 \tag{4}$$

on the same Cartesian plane.

(b) Where do the lines intersect?

Solution

(a) Notice that equations (3) and (4) form an inconsistent system because subtracting equation (4) from equation (3) yields the impossible statement

$$0 = 1$$

Thus, equations (3) and (4) have no solution. Their graphs are shown in Figure 8.9.

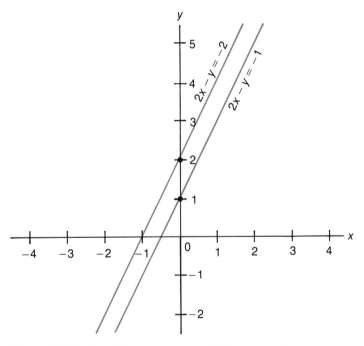

Figure 8.9 Graphs of $2x - y = -1$ and $2x - y = -2$

(b) Figure 8.9 indicates that the lines do not intersect. Recall from geometry that lines that do not intersect are called **parallel lines**.

In general, suppose that a, b, and k_1 are fixed numbers. It can be shown that a line L is parallel to the line with equation

$$ax + by = k_1$$

if and only if the equation of L can be written in the form

$$ax + by = k_2$$

for some number k_2 not equal to k_1. For example, the lines

$$2x + 3y = 12, \qquad 2x + 3y = -8$$
$$2x + 3y = 4, \qquad \text{and} \qquad 2x + 3y = 0.5$$

are all parallel to the line

$$2x + 3y = 7$$

because they are all of the form

$$2x + 3y = k$$

where $k \neq 7$. The line

$$4x + 6y = 280$$

is also parallel to $2x + 3y = 7$ because its equation can be written in the form $2x + 3y = 140$.

Example 4   (A system with exactly one solution)

   (a) Draw the graphs of

$$2x - y = 0 \tag{5}$$

   and

$$3x - 2y = -1 \tag{6}$$

   on the same Cartesian plane.
   (b) Where do the graphs intersect?

Solution

   (a) Equations (5) and (6) are equivalent to the system

$$2x - y = 0$$
$$-x \qquad = -1$$

   obtained by replacing equation (6) by the sum of equation (6) and -2 times equation (5). It is clear from this system that $x = 1$. Substituting this value into equation (5) yields

$$2(1) - y = 0$$
$$-y = -2$$
$$y = 2$$

   Thus, $(1,2)$ is the unique solution of the system consisting of equations (5) and (6).
   The graphs of equations (5) and (6) are shown in Figure 8.10.

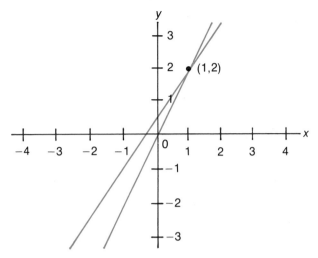

Figure 8.10 Graphs of $2x - y = 0$ and $3x - 2y = -1$

(b) Figure 8.10 indicates that the lines intersect at $(1,2)$. Notice that the point of intersection of the two lines is also the solution to the system of equations that the two lines represent.

Example 5 (A system with infinitely many solutions)

(a) Draw the graphs of

$$x + y = 1 \qquad \text{and} \qquad 2x + 2y = 2$$

on the same Cartesian plane.

(b) Where do they intersect?

Solution

(a) Notice that $x + y = 1$ and $2x + 2y = 2$ are equivalent equations because the second equation is 2 times the first. This means that any solution of

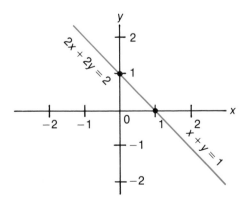

Figure 8.11 Graphs of $x + y = 1$ and $2x + 2y = 2$

the first equation is a solution of the second, and vice versa. Consequently, any two ordered pairs that we use to graph the first equation (e.g., (0,1) and (1,0)) can be used to graph the second. This means that the graphs of the two equations will be the same line. Figure 8.11 shows this graph.

(b) Because the graphs of the two equations are identical, they intersect at every point.

Because of their pictorial representation, equivalent lines are often called **identical lines**.

8.1 EXERCISES

1. Draw two different number lines and two different Cartesian planes.

2. Represent the following numbers on the same number line:

    1,   0.5,   3, -2.5, $-\frac{8}{3}$

3. Represent the following numbers on the same number line:

    10, -40, -25,   15

4. State the y-coordinates of (a) (1,2), (b) (2,1), (c) the origin.

5. Represent the following ordered pairs on the same Cartesian plane:

    (1,2), $(3,-1)$, $(-1,4)$, $(-4,1)$,   (1,4),

    $(-1,-4)$,   the origin

6. The gross monthly sales (in thousands of dollars) of the Better Buy Buggy Company are listed in the following table:

Month	Gross Sales
1 (January)	10.0
2 (February)	5.0
3 (March)	5.0
4 (April)	7.5
5 (May)	15.0

    Each row of the table can be represented by an ordered pair. For example, the first row is represented by the ordered pair (1,10), the second row by the ordered pair (2,5), and so on. Plot these ordered pairs on a Cartesian plane.

7. Length measured in inches (i) can be transformed into centimeters (c) by using the formula

    $$c = 2.54i \tag{7}$$

(a) Express 0 inches and 1 inch in centimeters.

(b) Express 2 centimeters in inches.

(c) Draw the graph of equation (7). (Hint: Remember that equation (7) is equivalent to $c - 2.54i = 0$. Label the horizontal axis c and the vertical axis i.)

8. Draw the graphs of—
(a) $y - 3x = 0$ (b) $t - 2s = 1$ (c) $2h + 3s = 2$

9. Draw the graphs of $y - 2x = 0$ and $2y - 4x = 0$ on the same Cartesian plane. Do you notice anything?

10. Draw the graphs of $y - 2x = 0$, $y - 2x = 1$, $y - 2x = 2$, and $y - 2x = 3$ on the same Cartesian plane. Do you notice anything?

11. Draw the graphs of $x = 3$, $x = 5$, $x = -4$, and $y = 3$.

12. Draw the following pairs of lines on the same Cartesian plane. In each case, state whether the lines are (*i*) parallel, (*ii*) identical, or (*iii*) intersecting. If the lines intersect, state the point of intersection.

(a) $2x - y = 3$ (b) $2x - y = 3$ (c) $2x - y = 3$

$2x - y = 1$ $x - y = 1$ $4x - 2y = 6$

(d) $2x - y = 3$ (e) $2x - y = 3$ (f) $2x - y = 3$

$4x - 2y = -1$ $2x - 2y = 1$ $4x - 2y = 3$

(g) $y = 5$ (h) $x = 5$

$x = 3$ $x = 7$

8.2 LINEAR INEQUALITIES

In Section 8.1 we learned how to draw the graphs of linear equations like

$$x + y = 3$$

These equations are also called *linear equalities*. In this section, we will discuss *linear inequalities* and see how to draw their graphs. In Section 8.4 we will use these graphs to solve linear programs.

Consider the inequality

$$x + 2y < 3 \tag{8}$$

If we replace the inequality sign ($<$) of statement (8) by the equality sign ($=$), we obtain the corresponding equality

$$x + 2y = 3$$

An inequality like inequality (8) is called a **linear inequality** because the corresponding equality is linear. Other examples of linear inequalities are

$$2x + 3y - 5 < 0, \qquad x \geq 4y, \qquad \text{and} \qquad 5y - 7 \leq x$$

Before we study linear inequalities, we will briefly review inequalities in general. Tables 8.1 and 8.2 review the four inequality symbols and three important properties of inequalities.

Table 8.1 The Four Inequality Symbols

Symbol	Meaning	Examples	
$<$	less than	$2 < 4$,	$-7 < 3$
$\leq$	less than or equal to	$2 \leq 4$,	$4 \leq 4$
$>$	greater than	$4 > 2$,	$3 > -7$
$\geq$	greater than or equal to	$4 \geq 2$,	$4 \geq 4$

Table 8.2 Three Properties of Inequalities

Property		Examples
1. If $a < b$, then $a + c < b + c$		$2 < 4$; hence, $2 + 5 < 4 + 5$
If $a > b$, then $a + c > b + c$		$4 > 2$; hence, $4 + 5 > 2 + 5$
2. *Suppose* $c > 0$	Let $c =$	$5 > 0$
If $a < b$, then $ac < bc$		$2 < 4$; hence, $2(5) < 4(5)$
If $a > b$, then $ac > bc$		$4 > 2$; hence, $4(5) > 2(5)$
3. *Suppose* $c < 0$	Let $c = -5 < 0$	
If $a < b$, then $ac > bc$		$2 < 4$; hence, $2(-5) > 4(-5)$
If $a > b$, then $ac < bc$		$4 > 2$; hence, $4(-5) < 2(-5)$

Table 8.2 gives the properties of the inequality symbols $<$ and $>$. Similar laws hold for the symbols $\leq$ and $\geq$.

Property 1 states that if any number is added to both sides of an inequality, the inequality sign remains the same.

Property 2 states that if both sides of an inequality are multiplied by any *positive* number, the inequality sign remains the same.

Property 3 states that if both sides of an inequality are multiplied by any *negative* number, the inequality sign changes from $<$ to $>$ or from $>$ to $<$.

Many of the definitions and concepts that apply to equalities can be easily extended to inequalities. For example, if a and b are numbers and a is not equal to b, we write $a \neq b$. Similarly, if a is not less than b, we write $a \not< b$; if a is not greater than b, we write $a \not> b$; and so on. Examples of this notation are: $2 \not> 5$, $3 \not< 3$, and $2 \not\leq 1$.

When we studied equalities, we learned that the point $(2,1)$ is a solution of, or satisfies, the equality

$$3x + y = 7 \tag{9}$$

because the substitution of $x = 2$ and $y = 1$ in inequality (9) yields the true statement

$$3(2) + 1 = 7$$

Similarly, the point $(1,3)$ is a *solution* of (or *satisfies*) the inequality

$$4x - y \leq 5 \tag{10}$$

because the substitution of $x = 1$ and $y = 3$ in inequality (10) yields the true statement

$$4(1) - 3 \leq 5$$

Other points that satisfy inequality (10) are (0,1), (2,4), and (2,5). However, (5,2) is not a solution of inequality (10) because

$$4(5) - 2 \nleq 5$$

The concepts of graph and equivalence can also be extended to inequalities. The *graph of an inequality* L is the set of all points that satisfy L. Two inequalities L and M are *equivalent* if every point that satisfies L also satisfies M, and vice versa. (In other words, L and M are equivalent if their graphs are identical.)

In Section 8.1 we learned how to draw the graph of a linear equality. To draw the graph G of a linear inequality I, we shade in all parts of the Cartesian plane containing points that satisfy I. (See Figure 8.12.) This picture is also called the graph of I.

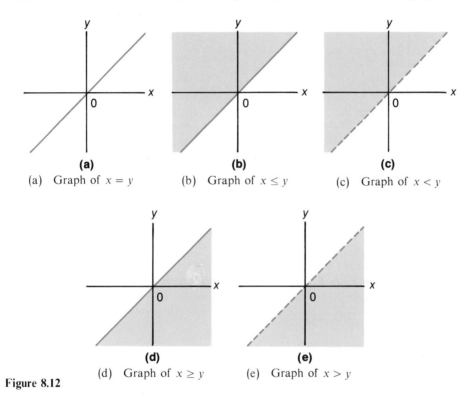

(a) Graph of $x = y$ (b) Graph of $x \leq y$ (c) Graph of $x < y$

(d) Graph of $x \geq y$ (e) Graph of $x > y$

Figure 8.12

The dotted lines in Figures 8.12(c) and (e) indicate that the line $x = y$ is not in the shaded area; that is, the points on the line $x = y$ do not satisfy the linear inequality.

Example 6 will lead us to a quick method for drawing the graph of a linear inequality.

Example 6 Draw the graph of the linear inequality

$$x \leq 5 \tag{11}$$

Solution We must shade in all parts of the Cartesian plane that represent the set

$$G = \{(x,y) \mid x \leq 5\}$$

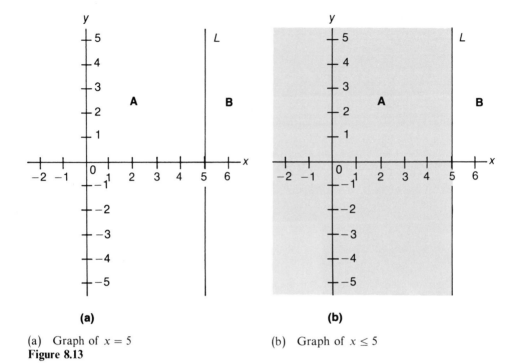

(a)   (b)

(a)   Graph of $x = 5$ (b)   Graph of $x \leq 5$
Figure 8.13

To do this efficiently, let us first consider the equality corresponding to inequality (11). This equality is

$$x = 5 \tag{12}$$

Figure 8.13(a) is the graph of equality (12).

It is clear from Figure 8.13(a) that the graph of equation (12) divides the Cartesian plane into three disjoint regions

$$A = \{(x,y) \,|\, x < 5\}$$
$$L = \{(x,y) \,|\, x = 5\}$$
$$B = \{(x,y) \,|\, x > 5\}$$

We will see that either

(1)   every point of the region A is in the graph G

or

(2)   no point of A is in G.

Moreover, we will see that similar statements are true of L and B.

Let (x,y) be any element of A. Since $x < 5$, it follows that $x \leq 5$, and so $(x,y) \in G$. Hence, every point of A is in G and the *entire* region A should be shaded.

Let (x,y) be any element of L. Since $x = 5$, it follows that $x \leq 5$, and so $(x,y) \in G$. Hence, every point of L is in G, and so these points should also be in

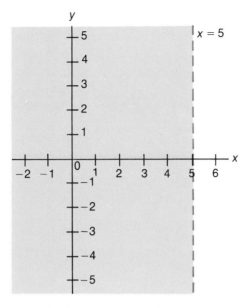

Figure 8.14   Graph of $x < 5$

the shaded region. We represent this fact by representing L by a solid line in Figure 8.13(b). If no point of L was in G, we would represent L by a dotted line, as in Figure 8.14.

We can similarly prove that no point of B is in G. This entire region should therefore remain unshaded. Figure 8.13(b) shows the completed diagram.

Figure 8.14 is the graph of $x < 5$. Note that the dotted line indicates that the line $x = 5$ does not satisfy the inequality.

In general, let M be any linear inequality and let G be its graph. The corresponding linear equality E divides the plane into three disjoint regions A, B, and L, where L is the set of all points that satisfy E. (See Figure 8.15.) As in Example 6, it can be shown that—

(1) either every point of L is in G, or no point of L is in G

and

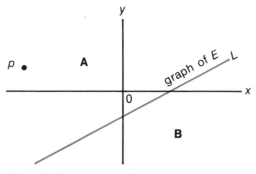

Figure 8.15

(2) either: every point of A is in G and no point of B is in G
  or: every point of B is in G and no point of A is in G.

It follows that the graph of a linear inequality can be drawn by using the following method:

Quick Method for Drawing the Graph of a Linear Inequality

(All points satisfying the inequality will be shaded.)

(1) Let L be the set of all points that satisfy the corresponding equality. If every point in L is in the graph G, represent L by a solid line. If no point of L is in G, represent L by a dotted line.

(2) Choose a point p not in L. Let A be the region containing p and let B be the other region. Test p to see whether p satisfies the inequality.

(3) If p satisfies the inequality, shade the entire region A; if p does not satisfy the inequality, shade the entire region B.

 The following example illustrates this method.

Example 7 Draw the graph of

$$2x - 3y > 4 \tag{13}$$

Solution

 (1) Let L be the set of points that satisfy the corresponding equality

$$2x - 3y = 4 \tag{14}$$

 (See Figure 8.16(a).) Since no point (x,y) that satisfies equation (14) can also satisfy inequality (13), L is represented by a dotted line. (See Figure 8.16(b).)

 (2) For this step, a point not in L must be chosen. Although there are an infinite number of points that can be used, it is convenient to choose $p = (0,0)$. Let A be the region containing p and let B be the other region.

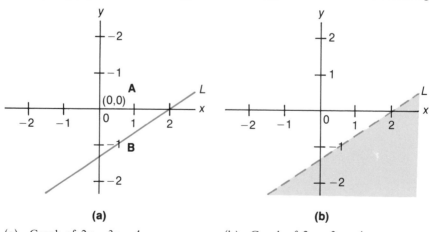

    (a)   **(b)**

(a) Graph of $2x - 3y = 4$ (b) Graph of $2x - 3y < 4$
Figure 8.16

(See Figure 8.16(a).) Since

$$2(0) - 3(0) = 0 \not> 4$$

p does not satisfy inequality (13).

(3) Since p does not satisfy the inequality, the entire region B must be shaded and the entire region A must be left unshaded. Figure 8.16(b) shows the required graph.

Until now, we have considered only one linear inequality at a time. In Section 8.4, however, we will see that many linear programs can be solved by drawing the graph of a set of one or more linear inequalities. Such a set is called a **system of linear inequalities** and its graph is the set of all points that satisfy every inequality in the system. The following examples will show how to draw the graph of a system of linear inequalities.

Example 8   Draw the graph of the system of linear inequalities

$$2x + y > 4 \tag{15}$$

$$x - y \leq 1 \tag{16}$$

Solution   We need to shade in all points (x,y) that satisfy *both* inequality (15) and inequality (16). The graph of (15) and the graph of (16) are drawn in Figures 8.17(a) and (b), respectively. The required graph is the intersection of the two regions just shaded. This region is sketched in Figure 8.17(c).

Example 9   Draw the graph of the system

$$3x + 2y > 4 \tag{17}$$

$$3x + 2y \geq 4 \tag{18}$$

Solution   Because any number that is greater than 4 is certainly greater than or equal to 4, it is clear that every point that satisfies inequality (17) also satisfies inequality (18). It follows that the system of linear inequalities (17) and (18) is equivalent to the single linear inequality (17). The graph of this inequality (and hence the graph of the system) is given in Figure 8.18.

Example 10   Draw the graph of the system

$$x + y \geq 3 \tag{19}$$

$$x - y \leq 4 \tag{20}$$

$$x \geq 1 \tag{21}$$

$$y \geq 2 \tag{22}$$

Solution   We must shade in all points that satisfy each of the inequalities (19)–(22). This can be done by first drawing the graphs shown in Figure 8.19 of the four corresponding equalities. Once these lines have been drawn, we can mentally identify the four regions representing inequalities (19)–(22) and shade the intersection of these regions. Figure 8.19 is the required graph.

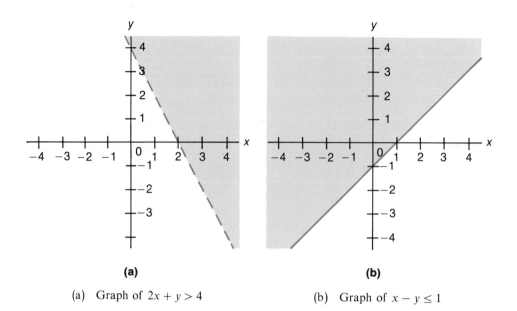

(a)

(a)   Graph of $2x + y > 4$

(b)

(b)   Graph of $x - y \leq 1$

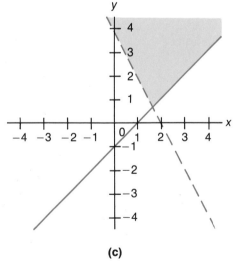

(c)

(c)   Graph of the system

Figure 8.17

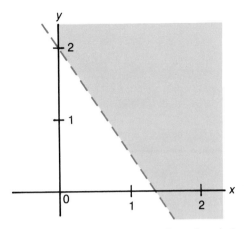

Figure 8.18   Graph of the system $3x + 2y > 4$ and $3x + 2y \geq 4$

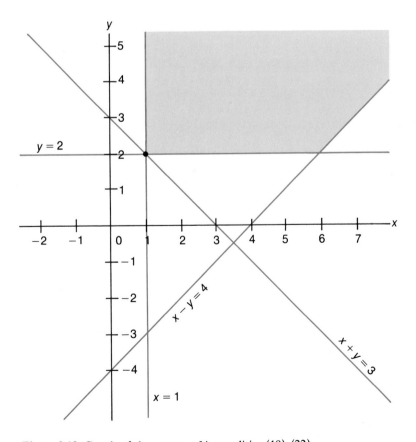

Figure 8.19 Graph of the system of inequalities (19)–(22)

8.2 EXERCISES

1. Which of the following statements are true?
    (a) $2 < 3$ **(b)** $2 \nless 3$ **(c)** $5 > 2$ **(d)** $-5 \geq 2$
    (e) $3 \leq 3$ **(f)** $3 < 3$ **(g)** $7 \nless -9$ **(h)** $0.5 > 0$

2. Determine which of the following ordered pairs satisfy

$$3x + 2y \leq 5$$

    (a) $(1,0)$ **(b)** $(0,1)$ **(c)** $(2,1)$
    (d) $(1,2)$ **(e)** $(-4,0)$ **(f)** $(0.5,0.2)$

3. Let G be the graph of

$$x + 2y < 3$$
$$2x - 3y \leq 5$$

    Which of the following points are in G?
    (a) $(1,0)$ **(b)** $(1,1)$ **(c)** $(0,0)$ **(d)** $(2,0)$ **(e)** $(0,2)$

4. Let G be the graph of

$$2x + y \geq 7 \tag{23}$$

    (a) Find the corresponding linear equality E.
    (b) Let L be the set of points that satisfy E. Are the points in L also in G?
    (c) Is the point $p = (0,0)$ in L?
    (d) Is p in G?
    (e) Draw the graph of inequality (23).

5. Draw the graphs of the following inequalities on separate axes.
    (a) $y > 0$ **(b)** $x \leq 0$ **(c)** $x > -1$

6. Draw the graphs of the following on separate axes.
    (a) $y = x$ **(b)** $y \geq x$ **(c)** $y < x$ **(d)** $y \leq x$ **(e)** $y > x$

7. Draw the graphs of the following inequalities on separate axes.
    (a) $x - 2y \leq -4$ **(b)** $3x + 6y < 9$ **(c)** $2x + y \geq 4$

8. Draw the graphs of the following systems of inequalities.
    (a) $x - 2y \geq 4$ **(b)** $x - 2y \leq 4$
           $x - 2y \leq 2$ $x - 2y \geq 2$
    (c) $x - 2y \geq 4$ **(d)** $x - 2y \leq 4$
           $x - 2y \geq 2$ $x - 2y \leq 2$

9. Draw the graphs of the following systems of inequalities.
    (a) $x - y \geq 0$ **(b)** $x - y > 1$
           $2x + y \geq 4$ $2x - 2y > 2$

(c) $x - y \geq 1$ **(d)** $x - y \geq 1$

$2x - 2y \leq 3$ $x - y > 0$

10.   Draw the graph of the system

$x - y \leq 1$

$2x - y \leq 3$

$x \geq 0, \quad y \geq 0$

8.3 LINEAR PROGRAMS

In this section we will finally learn what a linear program is and how it can be used. In Section 8.4 we will see how certain linear programs can be solved by graphing linear inequalities. Additional methods for solving linear programs will be discussed in Section 8.5.

The following example illustrates the kind of problem that linear programming can solve and will lead us to the definition of a linear program.

Example 11   The city zoo has two kinds of lion food, brand A and brand B. Brand A contains 20 grams of protein and 500 calories per pound, while brand B contains 15 grams of protein and 1000 calories per pound. Suppose that Leo, the lion, requires at least 60 grams of protein and 2000 calories per meal. Moreover, suppose that brand A costs 40 cents per pound while brand B costs 60 cents per pound. The zoo would like to determine the least expensive mixture of the two foods that will meet Leo's minimum requirements.

Solution   The information stated in this problem is conveniently summarized in Table 8.3.

Table 8.3	Brand A (per pound)	Brand B (per pound)	Minimum
Protein	20 g	15 g	60 g
Calories	500	1000	2000
Cost	40¢	60¢	

Let x and y be the numbers of pounds of brands A and B respectively, that the zoo puts in the mixture for one meal. Since each pound of brand A contains 20 grams of protein and there are x pounds of brand A in the mixture, brand A will provide

$20x$

grams of protein in the mixture. Similarly, brand B will provide

$15y$

grams of protein in the mixture. It follows that the total amount of protein in the mixture will be

$$20x + 15y$$

Since Leo must receive at least 60 grams of protein,

$$20x + 15y \geq 60 \tag{24}$$

Notice that the coefficients of inequality (24) are found in row 1 of Table 8.3. (The *coefficients* of the linear inequality $ax + by \geq c$ are the numbers a, b and c.)

Using the calorie data of row 2, the reader can similarly show that

$$500x + 1000y \geq 2000 \tag{25}$$

Notice that the coefficients of inequality (25) are found in row 2 of the table.

We now know that the zoo's apportionment of brands A and B is subject to certain constraints. The zoo cannot, for example, use a mixture of $x = 1$ pound of brand A and $y = 2$ pounds of brand B because of inequality (24). Nor can the zoo use $x = -3$ pounds of brand A, because the number of pounds of either brand cannot be negative. In other words,

$$x \geq 0 \quad \text{and} \quad y \geq 0$$

Summarizing the preceding remarks, the zoo can feed Leo any mixture of x pounds of brand A and y pounds of brand B as long as the following constraints are satisfied:

$$20x + \quad 15y \geq 60$$
$$500x + 1000y \geq 2000 \tag{26}$$
$$x \geq 0, \quad y \geq 0$$

There are many mixtures that satisfy each of the inequalities (26) (e.g., $x = 2$, $y = 4$; $x = 4$, $y = 2$; etc.) Since the zoo wishes to minimize its cost, some of these mixtures are better than others.

In Section 5.1 we discussed functions of one variable, x, such as

$$f(x) = 2x + 1 \quad \text{and} \quad F(x) = 3x^2$$

The definitions and notation of that section can be easily extended to functions of two variables (x and y), functions of three variables (x, y, and z), and so on. For example, the cost of feeding Leo is a function of two variables, because the cost depends on the number x of pounds of brand A and the number y of pounds of brand B that Leo is fed. Using a symbol similar to the symbols used to denote functions of one variable, let us denote the cost of feeding Leo by

$$F(x,y)$$

Using row 3 of Table 8.3, the reader can show that

$$F(x,y) = 40x + 60y \tag{27}$$

For example, if Leo is fed $x = 2$ pounds of brand A and $y = 4$ pounds of brand B, the cost of feeding him is calculated by replacing x by 2 and y by 4 in equation (27); that is, the cost of feeding him is

$$F(2,4) = 40(2) + 60(4) = 320 \text{ cents}$$

On the other hand, if he is fed $x = 4$ pounds of brand A and $y = 2$ pounds of brand B, the cost of feeding Leo is only

$$F(4,2) = 40(4) + 60(2) = 280 \text{ cents}$$

Since we are interested in minimizing costs, this method of feeding Leo is better than the first method.

The function $F(x,y)$ defined in equation (27) is called a **linear function in two variables** because it has the form

$$F(x,y) = ax + by + c$$

where x and y are variables, and a, b, and c are fixed numbers. This definition is similar to the definitions of a linear equation and a linear inequality. Naturally, other letters may be used instead of x and y to denote the variables. Other examples of linear functions are

$$F(x,y) = 2x - 3y, \qquad G(x,y) = 3x + 10y - 5, \qquad \text{and}$$
$$H(r,s) = 4r - 2s$$

Returning to Example 11, we can combine inequalities (26) and equation (27) and restate the problem of finding the proper diet for Leo as follows:

Find the minimum value of the linear function

$$F(x,y) = 40x + 60y \tag{28}$$

Subject to the linear constraints

$$20x + 15y \geq 60 \tag{29}$$
$$500x + 1000y \geq 2000 \tag{30}$$
$$x \geq 0, \qquad y \geq 0 \tag{31}$$

A problem like the one defined by statements (28)–(31) is called a **linear program** in two variables. Other linear programs may have a different linear function in place of (28) and/or a different set of inequalities or equalities in place of (29)–(31). Still other linear programs ask for the *maximum* value of a linear function. The following is the general form of a linear program in two variables:

Find the maximum (or minimum) value of the linear function

$$F(x,y) = ax + by + c \tag{32}$$

Subject to the constraints

$$a_1 x + b_1 y \geq c_1$$
$$a_2 x + b_2 y \geq c_2 \qquad\qquad (33)$$
$$\vdots \qquad \vdots \qquad \vdots$$
$$a_n x + b_n y \geq c_n$$

where x and y are variables and the remaining letters represent fixed numbers.

A linear program may also be written with $\leq$ or $=$ signs, instead of $\geq$, in certain of the constraints. It may also be written with $<$ or $>$ signs, instead of $\geq$, in certain of the constraints. However, in order to simplify our discussion, we will not consider these last cases.

The function $F(x,y)$ of equation (32) is called the **objective function**; the inequalities (33) are called the **constraints**. The set C of all points that satisfy all of the constraints is called the **constraint set**. (Notice that the constraint set is the graph of the system of inequalities (33).)

For example, in the linear program defined by statements (28)–(31), the objective function is $F(x,y) = 40x + 60y$ and the constraints are the inequalities (29)–(31). Two elements of the constraint set C are (0,4) and (4,0); these ordered pairs satisfy each of the constraints (29)–(31). However, the ordered pair (1,2) is not in C, because (1,2) does not satisfy constraint (29).

The linear function (32) and the linear inequalities (33) can be easily altered to define a linear program in three, four, or more variables. For example, the general form of a linear program in three variables is:

Find the maximum (or minimum) value of the linear function

$$F(x,y,z) = ax + by + cz + d \qquad\qquad (34)$$

Subject to the constraints

$$a_1 x + b_1 y + c_1 z \geq d_1$$
$$a_2 x + b_2 y + c_2 z \geq d_2 \qquad\qquad (35)$$
$$\vdots \qquad \vdots \qquad \vdots \qquad \vdots$$
$$a_n x + b_n y + c_n z \geq d_n$$

where x, y, and z are variables and the remaining letters represent fixed numbers.

As in the two-variable situation, a linear program in three or more variables may be written with $\leq$ or $=$ signs, instead of $\geq$, in certain of the constraints. The words *objective function*, *constraints*, and *constraint set* which were defined for linear programs in two variables are similarly defined for the three-or-more variable case.

The following examples illustrate how certain real-life situations can be translated into linear programs. The basic method that we will use is the following:

Method for Translating a Real-Life Problem into a Linear Program

(1)   Make a table similar to Table 8.3.

(2)   Using the information given in the last row of the table, find the objective function $F(x,y)$ (or $F(x,y,z)$, or $F(x,y,z,w)$, etc.), and decide whether it is to be minimized or maximized.

(3)   Using the information given in the remaining rows of the table, find some of the linear constraints. Additional constraints such as $x \geq 0$ may also be needed.

(4)   Place the objective function and the linear constraints in the form indicated in (32) and (33), or in a similar form if there are more than two variables.

Example 12   A firm that sells magazine subscriptions has forty employees and ten cars. The manager is responsible for assigning the cars and people to full teams (one car and five people) or half teams (one car and three people). A full team averages fifty subscriptions per day, while a half team averages forty subscriptions per day. The manager wishes to divide her employees and cars into full and half teams in a way that will maximize the number of subscriptions they sell. Write a linear program whose solution will solve the manager's problem.

Solution   This problem can be translated into a linear program by using the four steps listed before this example.

(1) The information stated in this problem is summarized in Table 8.4.

Table 8.4	Full Team	Half Team	Number Available
Cars	1	1	10
People	5	3	40
Subscriptions	50	40	

(2) Let x, y be the number of full, half teams that the manager forms, and let $F(x,y)$ be the number of subscriptions that the firm can expect to sell. Using the information in the last row of the table, the objective function is

$$F(x,y) = 50x + 40y$$

This function is to be maximized.

(3) Using the information in row 1 of Table 8.4, the total number of cars that are used is $x + y$. Since there are only 10 cars available,

$$x + y \leq 10 \tag{36}$$

Similarly, using row 2,

$$5x + 3y \leq 40 \tag{37}$$

Moreover, since the number of full teams and the number of half teams cannot be negative, we also have

$$x \geq 0 \quad \text{and} \quad y \geq 0 \tag{38}$$

The linear constraints are inequalities (36)–(38).

(4) Placing the objective function and the linear constraints in the form of (32) and (33) yields the required linear program:

Find the maximum value of the linear function

$$F(x,y) = 50x + 40y$$

Subject to the constraints

$$x + y \leq 10$$

$$5x + 3y \leq 40$$

$$x \geq 0, \qquad y \geq 0$$

Example 13 A farmer has a large number of acres of land lying fallow. He wants to plant alfalfa, barley, and corn. The cost, labor, and profit per acre for each of these crops is given in the following table:

Crop	Alfalfa	Barley	Corn
Cost/acre	$350	$300	$200
Man-hours/acre	20	20	30
Profit/acre	$500	$400	$300

Suppose that the farmer wishes to invest at most 2500 dollars and 150 man-hours. Write a linear program whose solution will maximize the farmer's profit.

Solution Following the four steps for formulating a linear program, we have:

(1) The information stated in this problem is summarized in Table 8.5.

Table 8.5

	Alfalfa	Barley	Corn	Amount Available
Cost	350	300	200	2500
Man-hours	20	20	30	150
Profit	500	400	300	

(2) Let x, y, and z be the numbers of acres allotted to alfalfa, barley, and corn, respectively, and let $F(x,y,z)$ be the farmer's profit. Using the information in the last row of the table, the objective function is

$$F(x,y,z) = 500x + 400y + 300z$$

This function is to be maximized because the farmer desires maximum profits.

(3) Using the information in rows 1 and 2 of Table 8.5,

$$350x + 300y + 200z \leq 2500 \tag{39}$$

$$20x + 20y + 30z \leq 150 \tag{40}$$

Since the number of acres allotted to each crop cannot be negative,

$$x \geq 0, \quad y \geq 0, \quad \text{and} \quad z \geq 0 \tag{41}$$

The linear constraints are (39)–(41).

(4) Since this is a linear program in *three* variables, x, y, and z, it will have the same form as (34)–(35). The required linear program is:

Find the maximum value of the linear function

$$F(x,y,z) = 500x + 400y + 300z$$

Subject to the constraints

$$350x + 300y + 200z \leq 2500$$
$$20x + 20y + 30z \leq 150$$
$$x \geq 0, \quad y \geq 0, \quad z \geq 0$$

8.3 EXERCISES

1. The sanitation department of Sun City divides its trucks and employees into half teams (one truck and two employees) and full teams (one truck and three employees). A half team can collect 0.5 tons of garbage while a whole team can collect 1.0 ton of garbage. Suppose that the department has twenty trucks and forty-five employees.

   (a) Summarize the information in this problem in a table similar to that of Table 8.3.

   (b) Write a linear program whose solution will maximize the amount of garbage collected.

   (c) How much garbage will be collected if the department forms (*i*) five half teams and three full teams, (*ii*) three half teams and five full teams?

2. A health-food store has 75 pounds of raisins and 100 pounds of walnuts. These are mixed in 1-pound packages as follows:

	Mixture 1	Mixture 2
Raisins	12 oz	8 oz
Walnuts	4 oz	8 oz
Profit	25¢	40¢

   The store wishes to ascertain how many packages of each mixture it should make if it wishes to maximize its profits. Formulate the store's problem as a linear program.

3. A certain refinery produces three grades of gasoline: high test, low test, and unleaded. The profit on high test is $2.00 per unit, the profit on low test is $1.50 per unit, and the profit on unleaded is $1.00 per unit. Suppose that a certain

customer has been promised a minimum of 400, 600, and 500 units of high-test, low-test, and unleaded gasolines, respectively, and that the maximum daily production of the refinery is 2000 units. Moreover, assume that government regulations require that the production of unleaded gasoline must be at least twice the production of high-test gasoline. Write a linear program whose solution will maximize the refinery's profits.

4.   A dietician in a certain pediatric hospital can feed her patients a mixture of two kinds of prepared baby foods, brand A and brand B. The following table gives the cost and nutritional makeup of one unit of each brand.

	Brand A	Brand B
Calories	300	400
Protein	8 g	6 g
Fat	3 g	4 g
Cost	40¢	30¢

Suppose that a certain baby requires between 2000 and 3600 calories (including the endpoints). Write linear programs whose solutions will—

(a) maximize the amount of protein in the baby's diet
(b) minimize the amount of fat in the diet
(c) minimize the cost of the diet if the baby also requires at least 36 grams of protein.

8.4 LINEAR PROGRAMS AND GRAPHS

In Section 8.3 we learned how certain practical problems can be stated as linear programs. Our problem now, of course, is to actually solve these linear programs. In this section we will learn how to solve linear programs in two variables by using graphs. In Section 8.5 we will see how to solve linear programs in any number of variables.

Let us now try to solve the linear program of Example 11. This linear program, which minimizes the cost of feeding Leo the lion is restated here for convenience:

Find the minimum value of the linear function

$$F(x,y) = 40x + 60y \tag{42}$$

Subject to the linear constraints

$$20x + 15y \geq 60 \tag{43}$$

$$500x + 1000y \geq 2000 \tag{44}$$

$$x \geq 0, \quad y \geq 0 \tag{45}$$

The constraint set C of this linear program is the set of all points that satisfy inequalities (43)–(45). Figure 8.20 is the graph of C.

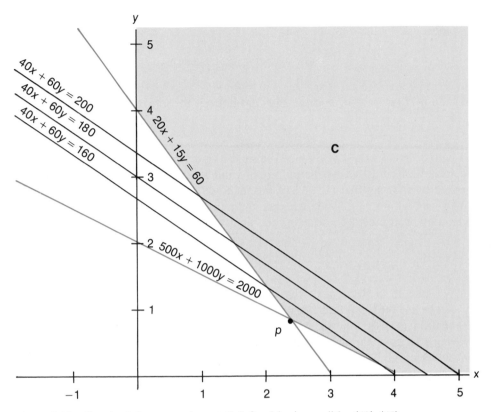

Figure 8.20   Graph of the constraint set C defined by inequalities (43)–(45)

Every point (x,y) in the shaded area of Figure 8.20 is called a *feasible solution* of the problem because its coordinates satisfy each of the inequalities (43)–(45). These points represent the only possible ways that brands A and B may be mixed. Obviously, however, some of these ways will be better (i.e., yield a lower cost) than others. Our problem, therefore, is to find which of these points yields the lowest possible cost (i.e., minimizes $F(x,y)$).

From the graph, we see that one feasible solution is (0,4). This means that Leo can be fed 0 pounds of brand A and 4 pounds of brand B. If this option is taken, the cost will be

$$F(0,4) = 40(0) + 60(4) = 240 \text{ cents}$$

We now know that Leo can be fed for 240 cents, or $2.40. We may wonder if he can be fed for $2.00. Clearly, there are many points (x,y) such that

$$40x + 60y = 200 \tag{46}$$

for example, $(-1,4)$. However, $(-1,4)$ is not in C, because it does not satisfy inequality (45). In order to show that the solution is at most $2.00, we must find a point (x,y) *in* C that satisfies equation (46). The line with equation (46) has been drawn in Figure 8.20. It is clear from this figure that there are many points that are both in C and on the line (46)—for example, (5,0). This can be checked by showing that (5,0) satisfies (43)–(46). It follows that the minimum cost is at most $2.00.

Similarly, we can decide if the minimum value of $F(x,y)$ subject to the constraints is at most 180 by drawing the line

$$40x + 60y = 180 \tag{47}$$

and seeing if this line intersects the constraint set. (See Figure 8.20.) Notice that lines (46) and (47) are parallel because they are both in the form

$$40x + 60y = k$$

but with different values of k. Since Figure 8.20 indicates that the line (47) intersects the constraint set at (4.5,0), the minimum value is at most 180. (This should also be checked by showing that (4.5,0) satisfies inequalities (43)–(45) and equation (47).)

In general, we can decide whether the minimum value of $F(x,y)$ subject to the constraints is at most k by drawing the line

$$40x + 60y = k \tag{48}$$

and seeing if it intersects C.

We learned in Section 8.1 that every line having the form of equation (48) is parallel to line (46), and that every line parallel to line (46) can be written in the form of equation (48). It follows that the minimum value of $F(x,y)$ can be found by drawing a series of parallel lines such as those in Figure 8.20 until we find the last one that intersects C. Notice that k gets smaller as the lines move to the left.

Let k_1 be the minimum value of $F(x,y)$ subject to the constraints. Figure 8.20 indicates that k_1 occurs at the point labeled $p = (x,y)$. The value of k_1 can be found by first finding p.

The point p is the intersection of the lines

$$20x + \quad 15y = \quad 60 \tag{49}$$

and

$$500x + 1000y = 2000 \tag{50}$$

To find p, therefore, we must find the point that satisfies both of these equations. Dividing equation (49) by 5 and equation (50) by 500 yields the equivalent equations

$$4x + 3y = 12 \tag{51}$$

$$x + 2y = \quad 4 \tag{52}$$

Multiplying equation (52) by 4 and subtracting (51) yields

$$5y = 4$$

It follows that $y = \frac{4}{5} = 0.8$. Substituting this value in equation (52) yields $x = 2.4$. Hence, $p = (2.4, 0.8)$. Substituting $x = 2.4$ and $y = 0.8$ in equation (42) yields

$$k_1 = 40(2.4) + 60(0.8) = 144$$

In other words, the minimum value of $F(x,y)$ subject to the constraints is 144. This occurs at the point (2.4,0.8). It follows that the zoo should feed Leo a mixture of 2.4 pounds of brand A and 0.8 pounds of brand B. The cost of the food will be $1.44 per meal.

The preceding linear program was solved by first restricting our attention to the constraint set C. The points of C are the only points at which $F(x,y)$ may be evaluated. Therefore, each of these points is called a **feasible solution**. A point of C (if it exists) that gives the minimum (or maximum) value of $F(x,y)$ is called an **optimal solution**. Notice that we spoke of *an* optimal solution rather than *the* optimal solution. The reader will see in Exercise 4 at the end of this section that a linear program may have many optimal solutions. However, the only optimal solution of the lion-feeding problem is (2.4,0.8). If no optimal solution exists, the linear program has no minimum (or maximum) value subject to the constraints.

The method used to solve the preceding linear program is summarized in the following four steps. These four steps give directions for finding the minimum value of $F(x,y)$ subject to the constraints; directions for finding the maximum value of $F(x,y)$ are given in parentheses.

Graphing Method for Solving a Linear Program with Objective Function F(x,y)

(1) Draw the constraint set C.
(2) Draw the lines $F(x,y) = k$ for several suitable values of k. These lines are all parallel. Notice that the value of k increases as the lines move in one direction, and decreases as the lines move in the other direction.
(3) From the diagram it should be evident that either—
    (a) a line L_o exists which is the last parallel line that intersects C in the direction of decreasing (increasing) k, or
    (b) no such line exists.
(4) (c) If L_o exists, let $p = (x_o, y_o)$ be a point of intersection of L_o with C. The point p is an optimal solution. The minimum (maximum) value of $F(x,y)$ subject to the constraints is $F(x_o, y_o)$.
    (d) If L_o does not exist, the program has no optimal solution and the objective function has no minimum (maximum) value subject to the constraints.

Notice that the solution of a linear program L may be that L has no optimal solution and hence the objective function has no minimum (or maximum) value subject to the constraints.

The following examples show how to use the graphing method for solving a linear program.

Example 14 Find the maximum value of the function

$$F(x,y) = 2x + y$$

subject to the constraints

$$x + y \le 1, \qquad x \ge 0, \qquad y \ge 0$$

Solution This problem may be solved by following the four steps of the graphing method.

(1) The constraint set C is drawn in Figure 8.21.

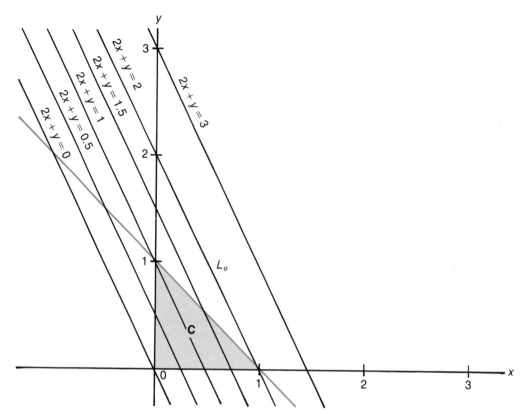

Figure 8.21   Graph of the constraint set for Example 14

(2) The parallel lines $2x + y = 1$, $2x + y = 2$, and $2x + y = 3$ are also drawn in Figure 8.21. Notice that the value of k increases as the lines move to the right.

(3) Several additional parallel lines are drawn in Figure 8.21. It is clear from the diagram that the last line which intersects the constraint set C is the line labeled L_o.

(4) The line L_o intersects the constraint set at $p = (1,0)$. It follows that p is an optimal solution and the maximum value of $F(x,y)$ is $F(1,0) = 2(1) + 0 = 2$.

Example 15   Find the maximum value of the function

$$F(x,y) = 2x + y$$

subject to the constraints

$$x + y \geq 1, \qquad x \geq 0, \qquad y \geq 0$$

Solution   Applying the graphing method for solving a linear program yields the following:

(1)–(2) The constraint set C and the lines $2x + y = 1$, $2x + y = 2$, and $2x + y = 3$ are drawn in Figure 8.22. The value of k increases as k moves to the right.

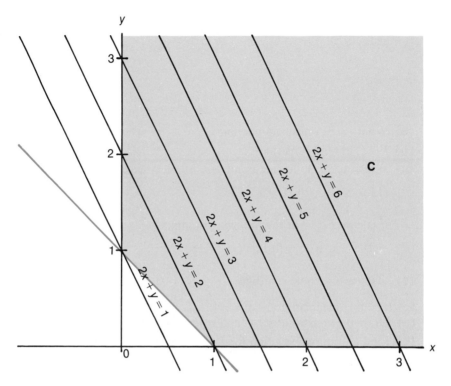

Figure 8.22 Graph of the constraint set for Example 15

(3)–(4) Several additional parallel lines are drawn in Figure 8.22 in the direction of increase. It is clear that there is no *last* line which intersects the constraint set. It follows that this linear program has no optimal solution, and therefore $F(x,y)$ has no maximum value subject to the constraints.

Example 16 Find the minimum value of the function of Example 15 subject to the same constraints.

Solution Figure 8.22 indicates that the minimum value of $F(x,y)$ subject to the constraints occurs at the point $(0,1)$. In other words, the optimal value of the linear program is $(0,1)$. Since $F(0,1) = 1$, the minimum value of $F(x,y)$ is 1.

8.4 EXERCISES

1. Consider the following linear program:

Find the maximum value of the linear function

$$F(x,y) = 2x + y$$

Subject to the constraints

$$x + y \leq 3$$
$$x - y \leq 2$$
$$x \geq 0, \qquad y \geq 0$$

(a) What is the objective function?

(b) What are the constraints?

(c) Which of the following points are in the constraint set C?

$$(1,2), \quad (3,0), \quad (0,3), \quad (1,-1)$$

(d) Evaluate $F(x,y)$ at the following points:

$$(1,2), \qquad (3,0), \qquad (0,3), \qquad (1,-1)$$

(e) State two feasible solutions.

2.   Let L be the linear program of Exercise 1.
   (a) Draw the constraint set C and several parallel lines $2x + y = k$.
   (b) Find an optimal solution for L (if one exists).
   (c) If possible, solve L.

3.   Let $F(x,y)$ and C be the linear function and constraint set of Exercise 1. Find the minimum value of $F(x,y)$ over C and the corresponding optimal solution (if one exists).

4.   (a) *Find the maximum and minimum values of*

   $$F(x,y) = 2x - y$$

   Subject to the constraint

   $$2x - y \leq 5$$

   (b) How many optimal solutions does each of the two linear programs in part (a) have? (The first linear program is finding the maximum value; the second is finding the minimum value.)
   (c) Give three optimal solutions of the first program.

5.   Let $F(x,y) = x + 2y$. Find the minimum and maximum values of $F(x,y)$ subject to the constraints—

   (a) $x + y \leq 4$ (b) $x + y \leq 4$

   $\quad x - y \leq 2$ $\quad x - y \geq 2$

   $\quad x \geq 0, \qquad y \geq 0$ $\quad x \geq 0, \qquad y \geq 0$

   (c) $\ x + y \geq 4$ (d) $x + y \geq 4$

   $\quad x - y \leq 2$ $\quad x - y \geq 2$

   $\quad x \geq 0, \qquad y \geq 0$ $\quad x \geq 0, \qquad y \geq 0$

6. (a) Into how many full teams and half teams should the manager of the magazine-selling firm described in Example 12 divide her workforce?[3]

(b) If the manager forms the optimal number of full teams and half teams, approximately how many subscriptions will the firm sell?

7. Into how many full and half teams should the sanitation department of Exercise 1, Section 8.3, divide its resources if it wishes to collect the most garbage?

8. Consider the health-food store of Exercise 2, Section 8.3.
(a) How much profit will be made if—

(i) fifty packages of mixture 1 and seventy packages of mixture 2 are produced

(ii) seventy packages of mixture 1 and fifty packages of mixture 2 are produced

(iii) sixty packages of mixture 1 and sixty packages of mixture 2 are produced?

(b) Which of the combinations of mixtures 1 and 2 described in part (a) are feasible?

(c) Calculate the number of packages of each mixture that should be made in order to maximize profits.

(d) If profits are maximized, how many pounds of walnuts and how many pounds of raisins will be used?

8.5 CORNER POINTS

In the preceding section we saw that linear programs in two variables can be solved by graphing the constraint set. For linear programs in three variables, this method is impractical if the constraint set is complicated and therefore difficult to graph. Moreover, the graphing method cannot be used for linear programs having more than three variables because their constraint sets cannot be graphed. In this section we will discuss a second method for solving linear programs. This method does not require drawing a graph and can therefore be used to solve linear programs with more than two variables.

In order to find a new method for solving linear programs, let us first consider linear programs in two variables. Every linear program L requires finding the maximum or minimum value of a linear function subject to certain linear constraints. If the constraint set of L is drawn, the region of the Cartesian plane that is shaded may be *bounded* or *unbounded*. (See Figure 8.23 for two examples.) If the region is bounded, then the constraint set is also said to be bounded. A similar definition holds if the region is unbounded.

---

3. It is fortunate that when the graphing method for solving linear programs is applied to this problem, the answer is x full teams and y half teams, where x and y are *integers*. Obviously, this is a requirement of the problem, because the manager cannot form 2.7 full teams or 4.8 half teams. If the answers provided by the graphing method for solving linear programs are not integers when physical considerations require integers, then other mathematical techniques beyond the level of this book must be used. These techniques are known as *integer programming*.

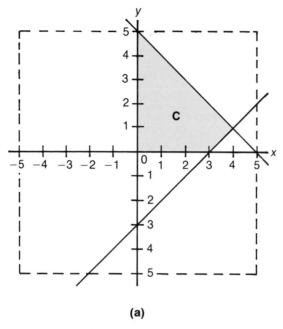

(a)

(a)   The constraint set is bounded
Constraints: $x + y \le 5,$ $x - y \le 3$
$x \ge 0,$ $y \ge 0$

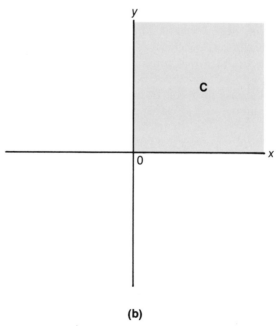

(b)

(b)   The constraint set is unbounded
Constraints: $x \ge 0,$ $y \ge 0$

Figure 8.23

The preceding "definition" is intuitive because we have not defined a bounded region but only looked at pictures. Moreover, this "definition" cannot be directly extended to linear programs with four, five, or more variables, because we cannot draw the graph of a set having more than three variables. To overcome these limitations, we need a mathematical definition that does not depend on pictures.

Definition 8.2   A constraint set C in two variables, x and y, is a **bounded constraint set** if there exists a positive number M such that

$$|x| \leq M \qquad \text{and} \qquad |y| \leq M \tag{53}$$

for all points (x,y) in C.

Geometrically, this means that the constraint set C lies within a square whose center is the origin, and whose sides are of length $2M$. Figure 8.23(a) shows a bounded constraint set C where M is equal to 5.

The reader can easily extend Definition 8.2 to three, four, or more variables. Example 17 shows how it can be applied to two-dimensional constraints sets.

Example 17   Prove (a) that the constraint set of Figure 8.23(a) is bounded, and (b) that the the constraint set of Figure 8.23(b) is unbounded.

Solution

(a) To show that the constraint set C of Figure 8.23(a) is bounded, we must find a positive number M for which condition (53) is satisfied. Let (x,y) be any element of C. Since the constraints state that

$$x + y \leq 5, \qquad x \geq 0, \qquad \text{and} \qquad y \geq 0$$

it follows that

$$|x| = x \leq x + y \leq 5, \qquad \text{or} \qquad |x| \leq 5$$

Similarly,

$$|y| \leq 5$$

The inequalities of condition (53) are therefore satisfied with $M = 5$. It follows from the definition that C is bounded.

Notice that we could have used any number larger than 5 instead of 5 in this proof; for example, we could have let M be 6 or 7.3.

(b) To show that the constraint set C_1 of Figure 8.23(b) is unbounded, we must show that there is no positive number M for which condition (53) is satisfied. To show this, let M be any positive number. The point $(M + 1, 0)$ is in C_1 because $M + 1 \geq 0$ and $0 \geq 0$, but

$$|M + 1| \not\leq M$$

Hence, it is not the case that condition (53) is satisfied for all points (x,y) in C_1. It follows that no number M satisfying condition (53) can be found, and so C_1 is unbounded.

If the constraint set is unbounded, as in Examples 15 and 16, the linear program may or may not have an optimal solution. It is often easy to determine whether an optimal solution exists by reading the problem or looking at the graph of the constraint set. (See Example 15.) More sophisticated techniques for determining whether or not an optimal solution exists when the constraint set is unbounded are beyond the level of this book. We will therefore consider only linear programs with bounded constraint sets in the remainder of this section.[4]

The following discussion will show that every linear program in two variables with a bounded constraint set has an optimal solution. A similar statement can be made for linear programs in more than two variables. Our discussion will also lead us to another method for finding this optimal solution. Unlike the graphing method, this method will also apply to linear programs in any number of variables.

Suppose that L is any linear program in two variables that has a bounded constraint set C. We know that L can be solved by applying the graphing method discussed in Section 8.5. Remember, however, that the solution of a linear program may be that the linear program has no optimal solution, and so the objective function has no minimum (or maximum) value subject to the constraints. We will now see that this solution cannot happen when C is bounded.

Applying steps (1)–(3) of the graphing method will result in a diagram like the one shown in Figure 8.24(a) or the one shown in Figure 8.24(b). In Figure 8.24(a) the

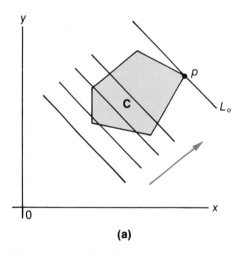

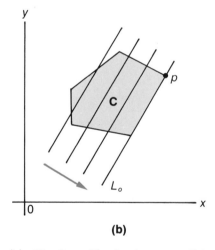

(a)   The lines $F(x,y) = k$ are *not* parallel to any of the lines defining the boundary of the constraint set

(b)   The lines $F(x,y) = k$ are parallel to one of the lines defining the boundary of the constraint set

Figure 8.24

---

4. In more specialized texts, a bounded constraint set C is called a bounded *linear polygonal convex set*. The set C is called *linear polygonal* because its boundary is composed of line segments (see Fig. 8.23(a)). It is called *convex* because the following property is satisfied: If p_1 and p_2 are any two points in C, then the line segment joining these points is a subset of C.

lines $F(x,y) = k$ are not parallel to any of the boundary lines of the constraint set. However, in Figure 8.24(b) the lines $F(x,y) = k$ *are* parallel to one of the boundary lines of the constraint set. The direction of increasing k is indicated by the arrow in each diagram.

To find the maximum value of $F(x,y)$, we must continue drawing parallel lines in the direction of increasing k until we find the last line L_o that intersects C, or until it is clear that no such line exists. However, it is clear from Figure 8.24 that L_o must exist because C is bounded. In other words, the linear program must have an optimal solution, and hence the objective function must have a maximum value. Similarly, the objective function must also have a minimum value.

To find the optimal solution, let us first consider Figure 8.24(a). The line L_o of Figure 8.24(a) intersects the constraint set C at p. This point p is therefore an optimal solution. Moreover, p is the unique intersection of two boundary lines of C. These lines correspond to two of the inequalities that define C.

Now consider Figure 8.24(b). The line L_o of Figure 8.24(b) intersects the constraint set C at an infinite number of points. Each of these points is an optimal solution. Two of these optimal solutions (in particular, the point p) are the unique intersections of pairs of boundary lines of C.

Hence, in both cases the line L_o obtained in the graphing method intersects C at a point which is the unique intersection of two boundary lines of C. A moment's reflection should convince the reader that this will always happen, because L_o is the last parallel line that intersects C. (Also see Figure 8.21.)

Summarizing the above results, we see that, if L is a linear program in two variables with a bounded constraint set, then L has an optimal solution p which is the unique intersection of two of the boundary lines of C. Looking at it from an algebraic rather than a geometric point of view, p is the solution of two of the linear equations whose graphs are boundary lines of C. A similar result is true for linear programs with three, four, or more variables.

Suppose L is a linear program in two variables. If point (x,y) is the unique solution of the linear equations defining the boundary of the constraint set C, it is called a *corner point* of C.

A corner point of a linear program in *three* variables is the unique solution of *three* of the linear equations that define the boundary of the constraint set. Definition 8.3 extends this concept to linear programs in any number of variables.

Definition 8.3 Let L be a linear program in m variables, and let C be its constraint set. The unique solution (if it exists) of m of the linear equations defining the boundary of C is called a **corner point**, or *extreme point*, of C.

Example 18 shows how to find the corner points of a constraint set C. It also shows that it is possible that a corner point will not be a feasible point.

Example 18 Find all the corner points of the constraint sets defined by the following systems of linear inequalities:

(a) $x \geq 0,$ $\quad y \geq 0,$ $\quad y \leq 1$ **(b)** $x \geq 0,$ $\quad y \geq 0,$ $\quad y \geq 1$

$\quad x - y \leq 1$ $\qquad\qquad\qquad\quad x + y + z \leq 2$

Solution

(a) The linear equations defining the boundary of the constraint set C are

$$x = 0, \qquad y = 0, \qquad y = 1, \qquad \text{and} \qquad x - y = 1$$

Since this is a constraint set in 2 variables (x and y), each corner point p is the unique solution of 2 of the 4 linear equations defining the boundary of C. The 2 linear equations defining p can be chosen from the 4 linear equations defining the boundary of C in $_4C_2 = 6$ ways. It follows that there are at most 6 corner points. There may be less than 6 corner points, either because one or more of the pairs of linear equations have no solution, or because some point p is the solution of two pairs of linear equations, or because the solution of a pair of equations is not unique.

The corner points and corresponding pairs of defining equations are:

Equations	*Corner point*
$x = 0$ $y = 0$	$(0,0)$
$x = 0$ $y = 1$	$(0,1)$
$x = 0$ $x - y = 1$	$(0,-1)$
$y = 0$ $y = 1$	Undefined
$y = 0$ $x - y = 1$	$(1,0)$
$y = 1$ $x - y = 1$	$(2,1)$

The constraint set and the corner points are shown in Figure 8.25. Notice that the corner point $(0,-1)$ does not lie in C; that is, it is not a feasible point.

(b) The linear equations defining the boundary of the constraint set are

$$x = 0, \qquad y = 0, \qquad y = 1, \qquad \text{and} \qquad x + y + z = 2$$

Since this is a constraint set in 3 variables (x, y, and z), each corner point is the solution of 3 of the 4 boundary lines of C. It follows that there are at most $_4C_3 = 4$ corner points. The corner points and their corresponding pairs of defining equations are:

Equations	**Corner point**
$x = 0$ $y = 0$ $y = 1$	Undefined
$x = 0$ $y = 0$ $x + y + z = 2$	$(0,0,2)$
$x = 0$ $y = 1$ $x + y + z = 2$	$(0,1,1)$
$y = 0$ $y = 1$ $x + y + z = 2$	Undefined

Notice that this constraint set has only 2 corner points, although there are 4 groups of defining equations.

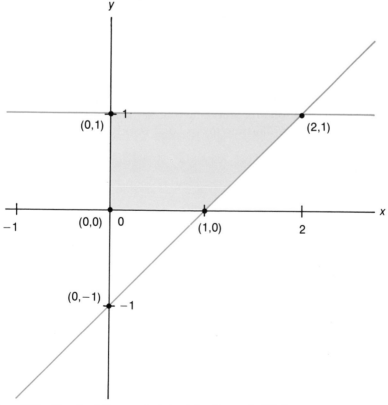

Figure 8.25 Graph of the constraint set for Example 18**(a)**

The following theorem summarizes the preceding discussion.

THEOREM 8.1   Let L be a linear program with a bounded constraint set C. Then L has an optimal solution which is a corner point of C.

Suppose L is a linear program with a bounded constraint set. Then, applying Theorem 8.1, L has an optimal solution. This optimal solution must be a feasible point. Therefore, the optimal solution can be found by first listing all corner points and then finding the feasible corner point that yields the minimum (or maximum) value of the objective function. This process is described in greater detail in the following four steps:

Algebraic Method for Solving a Linear Program in m Variables with a bounded Constraint Set

(1)   Find all corner points as follows:
    (a)   Select m of the linear equations that define the boundary of the constraint set C. If the linear program has n constraints, this can be done in $_nC_m$ ways.[5]
    (b)   Find the solution of the selected system of linear equations, if it exists.
    (c)   Repeat steps a and b for all possible combinations of m linear equations.

(2)   Determine which corner points are feasible.
(3)   Evaluate the objective function F at each feasible corner point.
(4)   Using step 3, find a feasible corner point that yields the minimum (or maximum) value of the objective function. This point is an optimal solution. The minimum (or maximum) value of the objective function F is its value at the optimal solution point.

Since there are only a finite number of corner points (at most $_nC_m$, it is clear that this method will give the solution in a finite (but perhaps large) number of steps.

The following examples illustrate the use of this method.

Example 19

Find the maximum value of

$$F(x,y) = 2x + 3y \tag{54}$$

Subject to the constraints

$$x + 2y \le 3 \tag{55}$$

$$2x - y \le 2 \tag{56}$$

$$x \ge 0 \tag{57}$$

$$y \ge 0 \tag{58}$$

---

5. It can be proved that, since C is bounded, $n \ge m$.

Solution It follows from (57), (58), and (55) that

$$|x| = x \le x + 2y \le 3$$

Similarly,

$$|y| \le 3$$

so the constraint set C is bounded. This problem can therefore be solved by applying the algebraic method.

(1) We must find all corner points.

(a) The equations defining the boundary of the constraint set C are:

$$x + 2y = 3 \tag{59}$$

$$2x - y = 2 \tag{60}$$

$$x = 0 \tag{61}$$

$$y = 0 \tag{62}$$

Since this is a linear program in 2 variables (x and y), we must select 2 of the 4 equations defining the boundary of C. This can be done in $_4C_2 = 6$ ways. For example, we may select equations (60) and (61) or equations (59) and (60).

(b) We can solve equations (60) and (61) by setting $x = 0$ in equation (60). This yields $y = -2$. It follows that $(0, -2)$ is a corner point.

To solve equations (59) and (60), we first multiply equation (59) by 2. This yields

$$2x + 4y = 6 \tag{63}$$

Subtracting equation (60) from equation (63) yields $5y = 4$, so $y = 0.8$. Substituting $y = 0.8$ into equation (59) yields $x = 3 - 1.6 = 1.4$. It follows that $(1.4, 0.8)$ is a second corner point.

We can find the remaining corner points similarly. They are listed in Table 8.6. "Constraints 12" means that the point is the

Table 8.6	Constraints	Corner Point	Feasible?	$F(x, y) = 2x + 3y$
	12	(1.4, 0.8)	Yes	5.2
	13	(0, 1.5)	Yes	4.5
	14	(3, 0)	No	—
	23	(0, −2)	No	—
	24	(1, 0)	Yes	2
	34	(0, 0)	Yes	0

solution of the linear equations corresponding to the first and second constraints (i.e., constraints (55) and (56)); "constraints 24" means that the point is the solution of the equations corresponding to the second and fourth constraints (i.e., constraints (56) and (58)); and so on.

(2) The point $(0, -2)$ is not feasible because it does not satisfy constraint (58). However, $(1.4, 0.8)$ is feasible because it satisfies each of the inequalities (55)–(58). The remaining corner points are tested similarly. The feasibility of each corner point is indicated in the third column of Table 8.6.

(3) The objective function (54) must now be evaluated at each feasible corner point. For example,

$$F(1.4, 0.8) = 2(1.4) + 3(0.8) = 5.2$$

and

$$F(1,0) = 2(1) + 3(0) = 2$$

The value of $F(x,y)$ at each feasible corner point is listed in the last column of Table 8.6.

(4) It is clear from Table 8.6 that the maximum value of $F(x,y)$ subject to the constraints is 5.2, and that the optimal solution is $(1.4, 0.8)$.

Example 20   Find the minimum value of the objective function (54) subject to the same constraints as in Example 19. What is the optimal solution?

Solution   It is clear from Table 8.6 that the minimum value of $F(x,y)$ subject to the constraints is 0, and that the optimal solution is $(0,0)$.

Example 21   How many acres of each crop should the farmer of Example 13 (page 326) plant in order to maximize his profits? If he plants these crops, how much profit will he make?

Solution   In Example 13, we saw that the following linear program will maximize the farmer's profits:

Find the maximum value of the linear function

$$F(x,y,z) = 500x + 400y + 300z \tag{64}$$

Subject to the constraints

$$350x + 300y + 200z \leq 2500 \tag{65}$$

$$20x + 20y + 30z \leq 150 \tag{66}$$

$$x \geq 0, \qquad y \geq 0, \qquad z \geq 0 \tag{67}$$

where x, y, and z are the numbers of acres allotted to alfalfa, barley, and corn, respectively.

The reader can easily show that the above constraint set is bounded, and hence that this linear program can be solved using the algebraic method.

(1) We must find all corner points.

(a) The equations defining the boundary of the constraint set C are:

$$350x + 300y + 200z = 2500 \tag{68}$$

$$20x + 20y + 30z = 150 \tag{69}$$

$$x = 0 \tag{70}$$

$$y = 0 \tag{71}$$

$$z = 0 \tag{72}$$

Since this is a linear program in 3 variables (x, y, and z), we must select 3 of the 5 equations defining the boundary of C. This can be done in $_5C_3 = 10$ ways. For example, we may select equations (68)–(70), or equations (70)–(72).

(b) We can solve equations (68)–(70) by first setting $x = 0$ in equations (68) and (69). (This can be done because of equation (70).). This yields

$$300y + 200z = 2500 \tag{73}$$

$$20y + 30z = 150 \tag{74}$$

Multiplying both sides of equation (73) by 2 and both sides of equation (74) by 30 yields

$$600y + 400z = 5000 \tag{75}$$

$$600y + 900z = 4500 \tag{76}$$

Subtracting equation (76) from equation (75) yields $-500z = 500$. It follows that $z = -1$. We do not have to solve for y, because no point $(x, y, -1)$ can be a feasible solution, because any such point violates constraint (67).

The remaining corner points can be found similarly. They are listed in Table 8.7.

Table 8.7

Constraints	Corner Points	Feasible ?	Profit: $F(x, y, z) =$ $500x + 400y + 300z$
123	$(0, 9, -1)$	No	——
124	$(\frac{90}{13}, 0, \frac{5}{13})$	Yes	3577
125	$(5, \frac{5}{2}, 0)$	Yes	3500
134	$(0, 0, \frac{25}{2})$	No	——
135	$(0, \frac{25}{3}, 0)$	No	——
145	$(\frac{50}{7}, 0, 0)$	Yes	3571
234	$(0, 0, 5)$	Yes	1500
235	$(0, \frac{15}{2}, 0)$	Yes	3000
245	$(\frac{15}{2}, 0, 0)$	No	——
345	$(0, 0, 0)$	Yes	0

(2) We have already seen that the corner point that is the solution of equations (68)–(70) is not feasible. However, the corner point $(0,0,0)$, which is the solution of equations (70)–(72), is feasible, because it satisfies each of the constraints (65)–(67). The remaining corner points can be tested similarly. The feasibility of each corner point is indicated in Table 8.7.

(3) We must now evaluate the objective function (64) at each feasible corner point. For example,

$$F(0,0,0) = 500(0) + 400(0) + 300(0) = 0$$

We can evaluate the objective function at the other feasible corner points in a similar manner. These values are listed in Table 8.7.

(4) Using Table 8.7, we see that the maximum value of $F(x,y,z)$ is 3577, and that the corner point $(\frac{90}{13},0,\frac{5}{13})$ gives this value. It follows that this point is the optimal solution.

In order to maximize his profits, the farmer should therefore plant $\frac{90}{13}$, or 6.9, acres of alfalfa and $\frac{5}{13}$, or 0.4, acres of corn. His profit will be 3577 dollars.

The method used in this section has an obvious difficulty. The number of corner points, although finite, can be extremely large. A linear program with 7 variables and 11 constraints, for example, requires the solution of $_{11}C_7 = 330$ systems of 7 equations in 7 variables. There is another method, called the *simplex algorithm*, which eliminates certain of the corner points from consideration. This method, an important application of matrix theory, is presented in Section 8.6.

8.5 EXERCISES

1. **(a)** Prove that the constraint set defined by the following inequalities is bounded.

$$3x + 5y \leq 6$$
$$2x - 2y \leq 8$$
$$x \geq 0, \qquad y \geq 0$$

(b) List all corner points.

(c) Which corner points are feasible?

2. *Find the maximum and minimum values of the function*

$$F(x,y) = 2x + 3y + 1$$

Subject to the constraints

$$3x + 4y \leq 2$$
$$x - y \leq 1$$
$$x \geq 0, \qquad y \geq 0$$

(a) by listing all corner points

(b) by graphing the constraint set.

3. **(a)** List all the corner points of the constraint set defined by the following inequalities.

$$2x + y + 3z \leq 12$$

$$x - y + 2z \leq 6$$

$$x \geq 0, \qquad y \geq 0, \qquad z \geq 0$$

(b) Which corner points are feasible?

(c) Find the maximum and minimum values of

$$F(x,y,z) = 5x + y + 8z$$

subject to the constraints listed in part **(a)**.

(d) At what point or points do the maximum and minimum values of F occur?

4. **(a)** To maximize profits, how many units of high-test, low-test, and unleaded gasolines should the oil refinery of Exercise 3, Section 8.3, produce?

(b) How much profit will the refinery earn if it produces the optimal amount of these three gasolines?

5. **(a)** Referring to Exercise 4, Section 8.3, find an optimal mixture of brands A and B baby food that maximizes the protein content of the baby's diet. What is the amount of protein in the optimal diet?

(b) Find an optimal mixture that minimizes the fat content of the baby's diet. What is the amount of fat in this diet?

(c) Find an optimal mixture that minimizes the cost of the diet and supplies at least 36 grams of protein. What is the cost of this diet?

6. A coffee company has 16,000 pounds of Brazilian coffee and 4000 pounds of Columbian coffee. The contents (in ounces per pound) and profits (in cents per pound) of the three blends of coffee that the company sells are given in the following table:

Blend	Brazilian	Columbian	Profit
A	4	12	40
B	8	8	30
C	12	4	20

For example, a pound of blend A coffee contains 4 ounces of Brazilian coffee and 12 ounces of Columbian coffee, and earns a profit of 40 cents. Write a linear program whose solution will answer the following question: How many pounds of each blend of coffee should be made in order to maximize profits? (Hint: Let x, y, and z be the numbers of pounds of blends A, B, and C, respectively, and let $F(x, y, z)$ be the profit. Show that $4x + 8y + 12z \leq 16$ (16,000).)

7. How many pounds of each blend of coffee should the company of Exercise 6 produce? What is their maximum profit?

8. A dress manufacturer produces garments for children and women. The hours of labor required to make 100 garments of each type and the hours available each week are given in the following table:

Worker	Children's Garments	Women's Garments	Maximum Available Hours
Patternmaker	4	4	36
Cutter	15	20	120
Seamer	110	160	880
Finisher	110	120	880

Profits are 120 dollars per 100 children's garments and 160 dollars per 100 women's garments.

(a) Determine the number of garments of each type that should be produced in order to maximize profits.

(b) What is the maximum profit?

(Hint: Although this problem can be solved using corner points, you may find that the graphing method is easier.)

8.6 THE SIMPLEX METHOD

In previous sections, linear programs were solved by drawing graphs or listing all corner points. Both of these methods have obvious difficulties if there are a large number of variables or many constraints. In this section, we shall discuss a third method, called the *simplex algorithm*, which is better adapted to handling large-scale linear programs. To apply this algorithm, we must first master an operation called *pivoting*.

Let's solve a simple linear programming problem using the method of listing corner points and then use that same problem as a model for our discussion of pivoting.

Example 22

Find the maximum value of the linear function

$$F = F(x,y) = 3x + 4y$$

Subject to the constraints

$$4x + 3y \leq 12 \tag{77}$$

$$x + 2y \leq 6 \tag{78}$$

$$x \geq 0, \quad y \geq 0 \tag{79}$$

Solution   Let C be the constraint set defined by inequalities (77)–(79) and let $(x, y) \in C$. Since (78) and (79) imply that

$$|x| = x \leq x + 2y \leq 6$$

and, similarly,

$$|y| \leq 6$$

the linear program has a solution. The constraint set is graphed in Figure 8.26; A, B, C, and D are the feasible corner points. (The variables r and s in Figure 8.26 will be explained later.)

   Table 8.8 lists the corner points of C and the values of F for the feasible corner points. It is clear from the table that the maximum value of F is 13.2, and that the optimal solution is $(1.2, 2.4)$.

Table 8.8	Corner Point	Feasible?	F
	$A = (0,0)$	Yes	0
	$B = (0,3)$	Yes	12
	$C = (1.2, 2.4)$	Yes	13.2
	$D = (3,0)$	Yes	9
	$E = (0,4)$	No	—
	$F = (0,6)$	No	—

   Although Example 22 was easily solved by the corner-point method, we shall now re-solve it using the simplex algorithm, or *simplex method*. This method requires writing the linear program in a matrix form that allows us to evaluate the objective function F at a feasible corner point easily. Then a series of matrix transformations, called *pivots*, are used to transform the first matrix into matrices corresponding to other corner points, increasing the value of F each time.

   As our first step in applying this method to Example 22, we will find a feasible corner point—say $A = (0,0)$—and write the linear program in a matrix form which displays the value of F at this point.

   Inequality (77) implies that there is a nonnegative number r satisfying

$$4x + 3y + r = 12$$

Similarly, inequality (78) implies that there exists a nonnegative number s satisfying

$$x + 2y + s = 6$$

The variables r and s are called **slack variables**, and the original variables, x and y, are called **basic variables**. The linear program of Example 22 is therefore equivalent to the following linear program:

Find the maximum value of

$$F = F(x, y) = 3x + 4y \tag{80}$$

Subject to the constraints

$$4x + 3y + r = 12 \tag{81}$$

$$x + 2y + s = 6 \tag{82}$$

$$x \geq 0, \qquad y \geq 0, \qquad r \geq 0, \qquad s \geq 0 \tag{83}$$

Rewriting equation (80) as

$$-3x - 4y + F = 0 \tag{84}$$

our linear programming problem can be formulated as a system of three linear equations in five unknowns:

$$4x + 3y + r = 12 \tag{81}$$

$$x + 2y + s = 6 \tag{82}$$

$$-3x - 4y + F = 0 \tag{84}$$

There are two ways to display these equations in matrix form. One way is to use the corresponding matrix

$$\mathbf{I'} = \begin{matrix} & x & y & r & s & F & \\ & 4 & 3 & 1 & 0 & 0 & \bigm| & 12 \\ & 1 & 2 & 0 & 1 & 0 & \bigm| & 6 \\ & -3 & -4 & 0 & 0 & 1 & \bigm| & 0 \end{matrix}$$

A second, more compact form is the following:

$$\mathbf{I} = \begin{matrix} & & x & y & & \text{Corresponding equations} \\ r & (& 4 & 3 & \bigm| & 12 &) & \to & 4x + 3y + r = 12 \\ s & (& 1 & 2 & \bigm| & 6 &) & \to & x + 2y + s = 6 \\ F & (& -3 & -4 & \bigm| & 0 &) & \to & -3x - 4y + F = 0 \end{matrix}$$

The matrix $\mathbf{I}$ is called the initial matrix; it corresponds to the feasible corner point $x = 0$, $y = 0$. Notice that the last column of either matrix gives the values of r, s, and F at the feasible corner point $A = (0,0)$. At this point,

$$x = 0, \qquad y = 0, \qquad r = 12, \quad \text{and} \quad s = 6$$

The value of $F(0,0)$, 0, appears in the intersection of the last row and the last column of the matrix. This value is clearly not the maximum value we seek.

We can increase the value of F by—

(1) letting $x = 0$ and increasing y

or

(2) letting $y = 0$ and increasing x.

Of course, we must be careful not to increase x or y too much, or r and s will become negative, contradicting inequalities (83).

Suppose we arbitrarily decide to use option (1). We must find the value y_1 of y which will make $r = 0$ (any larger value of y will make r negative) and the value y_2 of y which will make $s = 0$. We cannot increase y more than $\min(y_1, y_2)$.

Looking at the problem geometrically (referring to Figure 8.26), options (1) and (2) can be accomplished by—

(1) starting at the feasible corner point A and moving along the line $x = 0$ to the first feasible corner point, B

or

(2) starting at A and moving along the line $y = 0$ to the first feasible corner point, D.

We must decide which option is best. The entry in the y column of the last row of **I**, -4, is more negative than the corresponding entry in the x column, -3. This means that the coefficient of y in equation (80) is larger than the coefficient of x. It follows that increasing y will increase F faster than increasing x; that is, option (1) is better. We should evaluate F at B.

The corner point $A = (0,0)$ is the intersection of the lines $x = 0$ and $y = 0$. The value of F at this point was easily calculated from equation (80) because F is written in terms of x and y. The corner point B is the intersection of the lines $x = 0$ and $s = 0$. (See Figure 8.26.) Therefore, we can evaluate F at B more easily if equation (80) is rewritten in terms of x and s instead of x and y. Let us therefore use elementary row operations (see Section 6.6) on **I'** to solve for y, r, and F in terms of x and s.

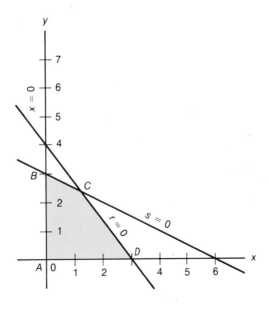

Figure 8.26 Graph of the constraint set for Example 21

Dividing row 2 of $\mathbf{I'}$ by 2, we have:

$$\begin{array}{ccccc} x & y & r & s & F \end{array}$$
$$\left(\begin{array}{ccccc|c} 4 & 3 & 1 & 0 & 0 & 12 \\ \frac{1}{2} & 1 & 0 & \frac{1}{2} & 0 & 3 \\ -3 & -4 & 0 & 0 & 1 & 0 \end{array}\right)$$

Adding 4 times row 2 of this matrix to row 3, we have:

$$\begin{array}{ccccc} x & y & r & s & F \end{array}$$
$$\left(\begin{array}{ccccc|c} 4 & 3 & 1 & 0 & 0 & 12 \\ \frac{1}{2} & 1 & 0 & \frac{1}{2} & 0 & 3 \\ -1 & 0 & 0 & 2 & 1 & 12 \end{array}\right)$$

Finally, adding -3 times row 2 to row 1, we have:

$$\begin{array}{ccccc} x & y & r & s & F \end{array}$$
$$\mathbf{J'} = \left(\begin{array}{ccccc|c} \frac{5}{2} & 0 & 1 & -\frac{3}{2} & 0 & 3 \\ \frac{1}{2} & 1 & 0 & \frac{1}{2} & 0 & 3 \\ -1 & 0 & 0 & 2 & 1 & 12 \end{array}\right)$$

The matrix $\mathbf{J'}$ corresponds to the equations

$$\tfrac{5}{2}x + r - \tfrac{3}{2}s = 3 \tag{85}$$
$$\tfrac{1}{2}x + y + \tfrac{1}{2}s = 3 \tag{86}$$
$$-x + 2s + F = 12 \tag{87}$$

and the compact matrix corresponding to equations (85)–(87) is

$$\begin{array}{cc} x & s \end{array}$$
$$\mathbf{J} = \begin{array}{c} r \\ y \\ F \end{array}\left(\begin{array}{cc|c} \frac{5}{2} & -\frac{3}{2} & 3 \\ \frac{1}{2} & \frac{1}{2} & 3 \\ -1 & 2 & 12 \end{array}\right)$$

The matrix $\mathbf{J}$ corresponds to the feasible corner point B, where

$$x = 0, \qquad s = 0$$

The value of F at this point, 12, appears in the last row and last column of $\mathbf{J}$. This value is a decided improvement over 0, the value of F at A.

Equation (87) implies that

$$F = 12 + x - 2s \tag{88}$$

so that we have indeed expressed F as a function of x and s. Comparing $\mathbf{J}$ to $\mathbf{I}$, we see that the roles of y and s have been interchanged; y is now acting as a slack variable, and s as a basic variable.

We must now decide whether the new value for F is a maximum, or whether the process should be repeated. Using Figure 8.26, we see that, from corner point B, our only option is to go along the line $s = 0$ to corner point C. Since this would increase x, equation (88) shows that the value of F would be increased. We must therefore continue the process if we wish to maximize F.

The preceding process is what we call the simplex algorithm. However, to make the algorithm practical, we must find a simple method for interchanging the roles of two variables (one a slack variable and one a basic variable) in a system of simultaneous equations. This process of exchange is called **pivoting**.

In the preceding discussion, we pivoted from equations (80)–(82) to equations (85)–(87) by using elementary row operations on the matrix $\mathbf{I}'$ to obtain the matrix $\mathbf{J}'$ corresponding to (85)–(87). However, it would be much more convenient and efficient if we could use an equivalent process that would take us directly from the compact initial matrix $\mathbf{I}$ to the compact matrix $\mathbf{J}$. This is especially true if we consider the fact that in real-life applications the simplex method is applied to programs with large numbers of variables and large numbers of constraints. (The corresponding matrix for a linear programming problem with 10 variables and 10 constraints is an 11×20 matrix, whereas the compact initial matrix is only an 11×10 matrix.)

Suppose that $w_1, w_2, \ldots, w_m$ are slack variables, $x_1, x_2, \ldots, x_n$ are basic variables, and $c_1, c_2, \ldots, c_m$ are constants. Then the system of linear equations

$$w_1 + a_{11}x_1 + a_{12}x_2 + \cdots + a_{1n}x_n = c_1$$

$$w_1 + a_{21}x_1 + a_{22}x_2 + \cdots + a_{2n}x_n = c_2$$

$$\vdots$$

$$w_m + a_{m1}x_1 + a_{m2}x_2 + \cdots + a_{mn}x_n = c_m$$

can be represented by the compact matrix

$$
\mathbf{K} =
\begin{array}{c}
\begin{array}{cccccc}
\quad\ x_1 & \ x_2 & \cdots & x_j & \cdots & x_n
\end{array} \\
\left(
\begin{array}{c|c}
\begin{array}{c}
w_1 \\
w_2 \\
\vdots \\
w_i \\
\vdots \\
w_m
\end{array}
\begin{array}{cccccc}
a_{11} & a_{12} & \cdots & a_{1j} & \cdots & a_{1n} \\
a_{21} & a_{22} & \cdots & a_{2j} & \cdots & a_{2n} \\
\vdots & \vdots & & \vdots & & \vdots \\
a_{i1} & a_{i2} & & \boxed{a_{ij}} & & a_{in} \\
\vdots & \vdots & & \vdots & & \vdots \\
a_{m1} & a_{m2} & & a_{mj} & & a_{mn}
\end{array}
&
\begin{array}{c}
c_1 \\
c_2 \\
\vdots \\
c_i \\
\vdots \\
c_m
\end{array}
\end{array}
\right)
\end{array}
$$

If the roles of w_i and x_j are interchanged in the above set of linear equations, the new set of linear equations will be represented by a new compact matrix $\mathbf{L}$. The following method for obtaining $\mathbf{L}$ from $\mathbf{K}$, called the **pivot transformation**, can be proved by the method used to change $\mathbf{I}$ to $\mathbf{J}$ in the preceding discussion.

Pivot Transformation   To interchange the roles of w_i and x_j—

(1)   Let $p = a_{ij}$ be called the **pivot**, and let the row labeled w_i and the column labeled x_j be called the **pivot row, pivot column** respectively. (The pivot p is circled in $\mathbf{K}$.)

(2)   Replace p by $\dfrac{1}{p}$.

(3)   Let x be any other entry in the pivot row. Replace x by $\dfrac{x}{p}$.

(4)   Let y be any other entry in the pivot column. Replace y by $-\dfrac{y}{p}$.

(5)   Let z be any other entry in the matrix and let z row and z column denote the row and column in which z occurs. Let u be the entry in the intersection of the z row and pivot column and let v be the entry in the intersection of the z column and pivot row. Replace z by $z - \dfrac{uv}{p}$.

The pivot transformation can be summarized as follows:

$$-\begin{pmatrix} p & x \\ y & z \end{pmatrix} \rightarrow \begin{pmatrix} \dfrac{1}{p} & \dfrac{x}{p} \\ -\dfrac{y}{p} & z - \dfrac{uv}{p} \end{pmatrix}$$

Example 23   Use the pivot transformation to interchange the roles of s and y in the system of equations (81), (82), and (84).

Solution   The matrix representation of the system is matrix **I**.

$$\mathbf{I} = \begin{array}{c} \\ r \\ s \\ F \end{array}\begin{pmatrix} \begin{array}{cc} x & y \end{array} & \\ \begin{array}{cc} 4 & 3 \end{array} & \left| \begin{array}{c} 12 \end{array} \right. \\ \begin{array}{cc} 1 & \boxed{2} \end{array} & \left| \begin{array}{c} 6 \end{array} \right. \\ \begin{array}{cc} -3 & -4 \end{array} & \left| \begin{array}{c} 0 \end{array} \right. \end{pmatrix}$$

Applying the pivot transformation, we use the following steps:

(1) The pivot row is the row labeled s, and the pivot column is the column labeled y, so the pivot p is 2, circled above.

(2) Replacing $p = 2$ by $\dfrac{1}{p} = \dfrac{1}{2}$ yields

$$\begin{array}{c} \\ r \\ y \\ F \end{array}\begin{pmatrix} \begin{array}{cc} x & s \end{array} & \\ \begin{array}{cc} - & - \end{array} & \left| \begin{array}{c} - \end{array} \right. \\ \begin{array}{cc} - & \frac{1}{2} \end{array} & \left| \begin{array}{c} - \end{array} \right. \\ \begin{array}{cc} - & - \end{array} & \left| \begin{array}{c} - \end{array} \right. \end{pmatrix}$$

(3) Replacing each entry of the pivot row (row 2) other than 2 by that entry divided by $p = 2$ results in

replacing 6 by $\frac{6}{2} = 3$

replacing 1 by $\frac{1}{2}$

This yields

$$
\begin{array}{c} r \\ y \\ F \end{array}
\left(
\begin{array}{cc|c}
\overset{x}{-} & \overset{s}{-} & - \\
\frac{1}{2} & \frac{1}{2} & 3 \\
- & - & -
\end{array}
\right)
$$

(4) Replacing each entry in the pivot column (column 2) other than 2 by the negative of that entry divided by $p = 2$ yields

$$
\begin{array}{c} r \\ y \\ F \end{array}
\left(
\begin{array}{cc|c}
\overset{x}{-} & \overset{s}{-\frac{3}{2}} & - \\
\frac{1}{2} & \frac{1}{2} & 3 \\
- & 2 & -
\end{array}
\right)
$$

(5) Let a be any other entry in the matrix—that is, $a = 0, -3, 12$, or 4. For example, let $a = 0$. The entry in the intersection of the a row (row 3) and the pivot column (column 2) of **I** is $u = -4$. The entry in the intersection of the a column (column 3) and the pivot row (row 2) is $v = 6$. We must replace $a = 0$ by

$$
a - \frac{uv}{p} = 0 - \frac{(-4)(6)}{2} = 12
$$

Similarly, we must

replace -3 by $-3 - \dfrac{(-4)(1)}{2} = -1$

replace 12 by $12 - \dfrac{(3)(6)}{2} = 3$

replace 4 by $4 - \dfrac{(3)(1)}{2} = \dfrac{5}{2}$

This yields the matrix

$$
\mathbf{L} =
\begin{array}{c} r \\ y \\ F \end{array}
\left(
\begin{array}{cc|c}
\overset{x}{\frac{5}{2}} & \overset{s}{-\frac{3}{2}} & 3 \\
\frac{1}{2} & \frac{1}{2} & 3 \\
-1 & 2 & 12
\end{array}
\right)
$$

This matrix **L** is identical to the compact matrix **J** corresponding to the system of equations (85)–(87). Thus, the pivot transformation yields the same results as the more time-consuming process of using elementary row operations to solve equations (80)–(82) for F, r, and y in terms of x and s.

As we saw in our discussion of Example 22, when the simplex method is used, the value of the objective function is improved at each step.[5] Each of these steps required using the pivot transformation or its equivalent. Now that we know how to apply the pivot transformation, we can discuss the simplex method for *maximizing* a linear program in more detail. We will first state the simplex algorithm and then solve some more complicated linear programs. For simplicity, we will only discuss linear programs in which the constraints are inequalities, and which have the additional constraints $x_i \geq 0$, $i = 1, 2, \ldots, n$.

Let us use the following simple linear program as a model:

Find the maximum value of

$$F(x, y) = 2x + 3y \tag{89}$$

Subject to the constraints

$$x + 2y \leq 3 \tag{90}$$
$$2x - y \leq 2 \tag{91}$$
$$x \geq 0, \quad y \geq 0 \tag{92}$$

This linear equation was solved by the corner-point method in Example 19, page 342. We will now solve it by using the simplex method. This method will then be applied to more complicated problems.

If we review the procedure used to solve the linear program given in Example 22, we see that the **simplex method** for maximizing a function has three basic steps:

Outline of the Simplex Method

Step 1   Form the initial matrix.

Step 2   Decide whether the initial matrix provides an optimal solution.

Step 3   If the solution is not optimal, use the pivot transformation to find another matrix and another solution that will increase the value of the objective function.

Example 24   Find an initial matrix for the linear program described by (89)–(92).

Solution   Introducing slack variables r and s and rewriting equation (89), we have

$$x + 2y + r = 3$$
$$2x - y + s = 2$$
$$-2x - 3y + F = 0$$
$$x \geq 0, \quad y \geq 0, \quad r \geq 0, \quad s \geq 0$$

---

5. If there is a zero in the last entry of the pivot row or pivot column, the value of F will remain the same. In this case, the simplex algorithm may continue indefinitely. Such cases are called *degenerate* and rarely occur. A careful choice of pivot will avoid this.

The initial matrix is

$$
\mathbf{I} = \begin{array}{c} \\ \\ \\ \\ \end{array}
\begin{array}{c}
 \\ r \\ s \\ F
\end{array}
\begin{pmatrix}
\begin{array}{cc} x & y \end{array} & \\
\begin{array}{cc} 1 & 2 \end{array} & 3 \\
\begin{array}{cc} 2 & -1 \end{array} & 2 \\
\begin{array}{cc} -2 & -3 \end{array} & 0
\end{pmatrix}
$$

This matrix corresponds to the solution

$$x = 0, \qquad y = 0, \qquad r = 3, \qquad s = 2$$

This is a feasible corner point because all of the inequalities (90)–(92) are satisfied. The value of F at this point, 0, appears in the last row and last column of the initial matrix.

Example 25   Determine whether the initial matrix $\mathbf{I}$ of Example 24 gives an optimal solution to the linear program.

Solution   The initial matrix corresponds to the feasible corner point $x = 0$, $y = 0$. Since $F = 2x + 3y$, it is obvious that the value of F can be increased by increasing x or y.

Notice that the entries in the last row of the initial matrix are negative if and only if F is a linear function of x and y with positive coefficients. In such a case, F can be increased by increasing one of these variables, so the solution represented by the matrix is not optimal. In general, the matrix representing a linear program corresponds to an optimal solution if and only if all of the entries in the last row (the row that represents the objective function) but the one in the last column are positive.

Example 26   Use the pivot transformation to increase the value of F.

Solution   There are three steps to this process: (a) choosing the pivot column, (b) choosing the pivot row, and (c) performing the pivot transformation.

(a) *Choosing the pivot column.* The entries in the last row of $\mathbf{I}$ tell us that $-2x - 3y + F = 0$, or that

$$F = 2x + 3y$$

To increase F, we can either—

(1) increase x and let $y = 0$

or

(2) increase y and let $x = 0$.

Since the coefficient of y in the equation for F is larger than the coefficient of x, increasing y will increase F faster. Notice that in the initial matrix the coefficient of y, -3, is the most negative coefficient in the last row. Thus, the y column (column 2) is our pivot column.

(b) *Choosing the pivot row.* We have decided to let $x = 0$ and increase y. However, we cannot increase y too much, or either r or s will become negative. For example, we cannot increase y to more than $\frac{3}{2}$, or r will become negative, because

$$x + 2y + r = 3 \tag{93}$$

Thus, when $x = 0$,

$$r + 2y = 3, \quad \text{or} \quad r = 3 - 2y$$

Notice that $\frac{3}{2}$ is the smallest positive ratio obtained by dividing each entry in the last column (except the last entry) by the corresponding entry in the pivot column:

$$
\begin{array}{ccc}
& x & y & & \text{ratios} \\
\mathbf{I} = \begin{array}{c} r \\ s \\ F \end{array} \left(\begin{array}{ccc} 1 & 2 & 3 \\ 2 & -1 & 2 \\ -2 & -3 & 0 \end{array}\right) & \begin{array}{c} \frac{3}{2} \\ -2 \\ \end{array}
\end{array}
$$

Thus, y will be at its maximum possible value when $y = \frac{3}{2}$. When $x = 0$ and $y = \frac{3}{2}$, then, by equation (93), $r = 0$. Thus, the new feasible corner point to which we wish to pivot is determined by $x = 0$ and $r = 0$. Therefore, the pivot row is the r row, and the pivot element is the 3 circled above.

(c) *Performing the pivot transformation.* This yields the new matrix

$$
\mathbf{L} = \begin{array}{c} y \\ x \\ F \end{array} \begin{array}{cc} s & r \\ \left(\begin{array}{cc|c} -\frac{1}{5} & \frac{2}{5} & \frac{4}{5} \\ \frac{2}{5} & \frac{1}{5} & \frac{7}{5} \\ \frac{1}{5} & \frac{8}{5} & \frac{26}{5} \end{array}\right) \end{array}
$$

This matrix represents an optimal solution because all of the entries in the bottom row are positive. The optimal solution is

$$x = \tfrac{7}{5}, \quad \text{or} \quad 1.4; \qquad y = \tfrac{4}{5}, \quad \text{or} \quad 0.8$$

The maximum value of F is $\frac{26}{5}$, or 5.2. These results are the same as those we obtained in Example 19 (see page 344).

Before we try applying the simplex method to some more complex examples, let's summarize the process for choosing a pivot element.

To Choose the Pivot Element—

(1)   The **pivot column** is the column that has the most negative (i.e., the smallest) entry in the last row (the F row). The last column is never used as the pivot column.

(2)   The **pivot row** is determined by dividing each entry in the last column, except the entry in the last row, by the positive entry in the same row of the pivot column. The pivot row is the row for which the smallest quotient is obtained.

(3)   The **pivot element** is the element that appears in the intersection of the pivot row and the pivot column.

We can now formally state the simplex method. Before applying this method to any linear program, we should determine whether the constraint set C is non-empty and bounded. This will be relatively simple for the linear programs of this book. If C is empty or not bounded, an optimal solution may not exist. On the other hand, if C is nonempty and bounded, an optimal solution exists and can be found by using the simplex method. In general, however, deciding on the bounded-ness of a constraint set may be quite difficult. In such cases the simplex method itself will show that no solution exists.

Simplex Method for Maximizing a Linear Function

$$F(x_1, x_2, \ldots, x_n) = c_0 + c_1 x_1 + c_2 x_2 + \cdots + c_n x_n$$

Subject to the Constraints

$$a_{11} x_1 + a_{12} x_2 + \cdots + a_{1n} x_n \le d_1$$

$$a_{21} x_1 + a_{22} x_2 + \cdots + a_{1n} x_n \le d_2$$

$$\vdots \qquad \vdots \qquad \qquad \vdots \qquad \vdots$$

$$a_{m1} x_1 + a_{m2} x_2 + \cdots + a_{mn} x_n \le d_m$$

$$x_i \ge 0, \qquad i = 1, 2, \ldots, n$$

where $d_j \ge 0, \qquad j = 1, 2, \ldots, n.$

(1)   Rewrite the m constraints as m equations by introducing m nonnegative slack variables $s_1, s_2, \ldots, s_n$.

(2)   Construct the compact initial matrix $\mathbf{I}$ from the equations formed in step (1):

$$\mathbf{I} = \begin{array}{c} \\ s_1 \\ s_2 \\ \vdots \\ s_m \\ F \end{array} \begin{pmatrix} \begin{array}{ccccc} x_1 & x_2 & \cdots & x_n & \\ a_{11} & a_{12} & \cdots & a_{1n} & d_1 \\ a_{21} & a_{22} & \cdots & a_{2n} & d_2 \\ \vdots & \vdots & & \vdots & \vdots \\ a_{m1} & a_{m2} & \cdots & a_{mn} & d_m \\ -c_1 & -c_2 & \cdots -c_n & c_0 \end{array} \end{pmatrix}$$

(3)   If the first n entries in the bottom row of $\mathbf{I}$ $(-c_1, -c_2, \ldots, -c_n)$ are all non-negative, the optimal solution is

$$x_i = 0 \qquad (i = 1, 2, \ldots, n)$$

and the maximum value of F is c_0.

If any one of the $-c_i$ $(i = 1, 2, \ldots, n)$ is negative, proceed to step (4).

(4)   Choose a pivot, p.

(5)   Use the pivot transformation with pivot p to obtain a new compact matrix $\mathbf{J}$ from $\mathbf{I}$. The new matrix will have basic variables $u_1, u_2, \ldots, u_n$, where each u_k is either an x_i or an s_j.

(6)   If the first n entries in the bottom row of $\mathbf{J}$ are all nonnegative, the optimal solution is

$$u_k = 0 \qquad (k = 1, 2, \ldots, n)$$

The values of all other variables can be read from the last column in $\mathbf{J}$, and the maximum value of F is the entry in the last row and last column of $\mathbf{J}$.

If any of the first n entries in the bottom row of $\mathbf{J}$ is negative, proceed to step (4), using matrix $\mathbf{J}$ in place of $\mathbf{I}$.

Notice that there are a number of restrictions on the kind of linear program to which the above method can be applied: All of the constraints must use $\leq$, all of the d_j must be positive, the problem must be one of *maximizing* the linear function, and so on. If a particular linear program fails to meet any of these restrictions, then modifications to the simplex method are necessary.

Let's now consider two examples that will show how to modify the simplex method when certain of these restrictions are violated. The first example illustrates what happens when one or more of the constraints uses $\geq$.

Example 27   Use the simplex method to solve the following linear program:

Find the maximum value of

$$F(x, y) = x + 2y \tag{94}$$

Subject to the constraints

$$x + y \leq 3 \tag{95}$$

$$x - y \geq 2 \tag{96}$$

$$x \geq 0, y \geq 0 \tag{97}$$

Solution   Since all of the linear programs we have considered so far have had only constraints using $\leq$, let us reformulate (96):

$$-x + y \leq -2$$

Notice, however, that we are now in violation of the restriction that all of the d_j be nonnegative.

Rewriting equation (94) and introducing slack variables r and s, we have:

$$x + \ y + r = \ \ 3 \tag{98}$$

$$-x + \ y + s = -2 \tag{99}$$

$$-x - 2y + F = \ \ 0 \tag{100}$$

$$x \geq 0, \qquad y \geq 0, \qquad r \geq 0, \qquad s \geq 0 \tag{101}$$

The initial matrix for this linear program is

$$
\mathbf{I} = \begin{array}{c} \\ r \\ s \\ F \end{array}
\begin{array}{cc} x & y \\ \left(\begin{array}{cc|c} 1 & 1 & 3 \\ -1 & 1 & -2 \\ -1 & -2 & 0 \end{array} \right) \end{array}
$$

This matrix corresponds to the corner point

$$x = 0, \qquad y = 0, \qquad r = 3, \qquad s = -2$$

Unfortunately, this corner point is not feasible, since $s = -2$ contradicts the inequalities (101).

In each of the linear programs we have discussed in this section, our initial matrix corresponded to the feasible corner point $x = 0$, $y = 0$. However, as we now see, this corner point is not always a feasible one. In order for the simplex method to work, we must be able to start with an initial matrix corresponding to a feasible corner point and then pivot to adjacent feasible corner points. Thus, our first step in solving the problem described by (94)–(97) must be to find a feasible corner point and formulate an initial matrix that corresponds to that point.

How can we find such a point without graphing the constraint set? (This is an important thing to know, since constraint sets for linear programs with more than two variables cannot be graphed.) We know that each feasible corner point corresponds to setting two of the four variables x, y, r, and s equal to 0. We already know that $x = 0$, $y = 0$ is not feasible, so let us try some of the five remaining possibilities.

We can make our work easier by examining equations (98) and (99). Looking at equation (99), it is obvious that if $x = 0$, then

$$s + y = -2$$

which means that one or both of s and y would have to be negative, contradicting (101). This eliminates the corner points $x = 0$, $r = 0$ and $x = 0$, $s = 0$ from consideration.

Let's now consider those corner points for which $y = 0$. If $y = 0$ and $r = 0$, then by equation (98), $x = 3$, and by equation (99), $s = 1$. Since all of these values satisfy constraints (95)–(97),

$$y = 0, \qquad r = 0$$

is a feasible corner point. Similarly, we can show that

$$y = 0, \quad s = 0$$

is a feasible corner point.

We can arbitrarily select the second of these two feasible corner points and apply the pivot transformation to **I** to obtain a matrix corresponding to this point. Since $y = 0$ and $s = 0$, y and s will be our new basic variables, and x and r will be the new slack variables. Thus, we will be interchanging the roles of x and s, and so the pivot element is the -1 that appears in the intersection of the x column and s row in **I**. Applying the pivot transformation, we have

$$
\mathbf{J} = \begin{array}{c} \\ r \\ x \\ F \end{array}
\begin{array}{cc} s & y \\ \end{array}
\left(\begin{array}{cc|c} 1 & ② & 1 \\ -1 & -1 & 2 \\ -1 & -3 & 2 \end{array}\right)
\begin{array}{c} \text{ratios} \\ 2 \\ -2 \\ \end{array}
$$

Because the last row contains negative entries in the first two columns, we know that the solution corresponding to the matrix **J** is not optimal. Pivoting on column 2 and row 1 (the ratios for picking the pivot row are shown next to **J**), we have:

$$
\mathbf{K} = \begin{array}{c} \\ y \\ x \\ F \end{array}
\begin{array}{cc} s & r \\ \end{array}
\left(\begin{array}{cc|c} \frac{1}{2} & \frac{1}{2} & \frac{1}{2} \\ -\frac{1}{2} & \frac{1}{2} & \frac{5}{2} \\ \frac{1}{2} & \frac{3}{2} & \frac{7}{2} \end{array}\right)
$$

Since all of the entries in the last row of **K** are positive, the solution corresponding to this matrix,

$$x = 5/2, \quad y = 1/2$$

is optimal, and the maximum value of F is $7/2$.

We can summarize the method we applied to solve Example 27 as follows:

To Apply the Simplex Method When the Initial Matrix Does Not Correspond to a Feasible Corner Point—

(1)   Find a set of variables $u_1, u_2, \ldots, u_n$, where each u_k is either an x_i or an s_j, for which $u_i = 0$ ($i = 1, 2, \ldots, n$) is a feasible corner point.

(2)   Find the compact matrix that corresponds to the feasible corner point found in step (1). (If $u_1, u_2, \ldots, u_n$ differ from $x_1, x_2, \ldots, x_n$ by only one variable, this can be done by applying an appropriate pivoting transformation to the initial matrix; otherwise, the constraint equations must be solved for the other variables in terms of $u_1, u_2, \ldots, u_n$.)

(3)   Apply the usual steps of the simplex method, using the matrix found in step (2) as the initial matrix.

The next example shows how the simplex method can be applied to a linear program in which we wish to *minimize* the objective function.

Example 28   Use the simplex method to solve the following linear program:

Find the minimum value of

$$F(x,y) = 3x + 2y + z$$

Subject to the constraints

$$x + y + z \geq 1$$
$$x + y + z \leq 2$$
$$x \geq 0, \qquad y \geq 0, \qquad z \geq 0$$

Solution   If m is the minimum value of F, then $-m$ is the maximum value of $G = -F$. Similarly, if M is the maximum value of $G = -F$, $-M$ is the minimum value of F. Therefore, minimizing F is the same as maximizing G. Thus, the given linear program is equivalent to the following linear program:

Find the maximum value of

$$G = -F = -3x - 2y - z$$

Subject to

$$-x - y - z \leq -1 \tag{102}$$
$$x + y + z \leq 2 \tag{103}$$
$$x \geq 0, \qquad y \geq 0, \qquad z \geq 0 \tag{104}$$

Introducing slack variables r and s, inequalities (102)–(104) become

$$-x - y - z + r = -1 \tag{105}$$
$$x + y + z + s = 2 \tag{106}$$
$$x \geq 0, \qquad y \geq 0, \qquad z \geq 0, \qquad r \geq 0, \qquad s \geq 0 \tag{107}$$

Using (105)–(107) and

$$3x + 2y + z + G = 0$$

we can construct the compact matrix **I** corresponding to the corner point $x = 0, y = 0, z = 0$:

$$\mathbf{I} = \begin{array}{c} \\ r \\ s \\ G \end{array} \begin{array}{ccc} x & y & z \end{array} \\ \left(\begin{array}{ccc|c} -1 & -1 & \boxed{-1} & -1 \\ 1 & 1 & 1 & 2 \\ 3 & 2 & 1 & 0 \end{array} \right)$$

At first glance, we might be tempted to say that this matrix corresponds to an optimal solution, because all of the entries in the bottom row are positive. However, the corner point $x = 0$, $y = 0$, $z = 0$ is not feasible—this is obvious because the values $r = -1$ and $s = 2$ can be read from the last column of $\mathbf{I}$, and the value $r = -1$ violates inequalities (107).

As in Example 27, we must begin by finding a feasible corner point and the matrix corresponding to that point. It is obvious from equation (105) that at least one of x, y, and z must be nonzero. In fact, it is easy to verify that

$$x = 0, \qquad y = 0, \qquad r = 0, \qquad z = 1, \qquad s = 2$$

is a feasible corner point. We can obtain an initial matrix corresponding to this point if we apply the pivot operation to $\mathbf{I}$ to interchange z and r. Thus, column 3 (the z column) is the pivot column, and row 1 (the r row) is the pivot row. The pivot element, $p = -1$, is circled in $\mathbf{I}$.

Applying the pivot transformation, we obtain

$$\mathbf{J} = \begin{array}{c} \\ z \\ s \\ G \end{array} \begin{array}{ccc} x & y & r \end{array} \\ \left(\begin{array}{ccc|c} 1 & 1 & 1 & 1 \\ 0 & 0 & 1 & 1 \\ 2 & 1 & 1 & -1 \end{array} \right)$$

This matrix represents an optimal solution because it corresponds to a feasible corner point and the first three entries in the bottom row are positive. Therefore, the maximum value of G is -1, so the minimum value of F is $-(-1) = 1$.

We can summarize the method used in Example 28 as follows:

To Minimize a Linear Function F Using the Simplex Method—

(1)   Let $G = -F$.
(2)   Use the simplex method to maximize G over the constraint set given for F.
(3)   If an optimal solution for G exists, it is an optimal solution for F, and the minimum value of F is the negative of the maximum value of G.

We will conclude this section by re-solving one of the linear programs we previously solved using the corner-point method.

Example 29   Use the simplex method to solve the problem of Example 13.

Solution   In Example 21, we solved this problem by applying the corner-point method to the following linear program:

Find the maximum value of

$$F(x,y,z) = 500x + 400y + 300z$$

Subject to the constraints

$$350x + 300y + 200z \le 2500$$
$$20x + 20y + 30z \le 150$$
$$x \ge 0, \quad y \ge 0, \quad z \ge 0$$

(1) Introducing slack variables r and s, we have

$$350x + 300y + 200z + r = 2500$$
$$20x + 20y + 30z + s = 150$$
$$x \ge 0, \quad y \ge 0, \quad z \ge 0, \quad r \ge 0, \quad s \ge 0$$

(2) The initial matrix is

$$\mathbf{I} = \begin{array}{c} \\ r \\ s \\ F \end{array} \begin{array}{ccc} x & y & z \\ \left(\begin{array}{ccc|c} 350 & 300 & 200 & 2500 \\ 20 & 20 & 30 & 150 \\ -500 & -400 & -300 & 0 \end{array} \right) \end{array}$$

This matrix corresponds to the feasible corner point $x = 0, y = 0, z = 0$.

(3) The first three entries of the last row of $\mathbf{I}$ are negative, so we know that $\mathbf{I}$ does not correspond to an optimal solution.

(4) Since -500 is the most negative entry in the bottom row, column 1 is the pivot column. The ratios obtained by dividing the entries in the last column by those in column 1 are as follows:

row 1: $\frac{2500}{350} = \frac{50}{7} = 7\frac{1}{7}$

row 2: $\frac{150}{20} = \frac{15}{2} = 7\frac{1}{2}$

The minimum is $7\frac{1}{7}$, so row 1 is the pivot row and $p = 350$ is the pivot element.

(5) Applying the pivot transformation, we obtain

$$\mathbf{J} = \begin{array}{c} \\ x \\ s \\ F \end{array} \begin{array}{ccc} r & y & z \\ \left(\begin{array}{ccc|c} 1 & \frac{6}{7} & \frac{4}{7} & \frac{50}{7} \\ -\frac{2}{35} & \frac{20}{7} & \frac{130}{7} & \frac{50}{7} \\ \frac{10}{7} & \frac{200}{7} & -\frac{100}{7} & \frac{25{,}000}{7} \end{array} \right) \end{array}$$

(6) Since one of the first three entries in the bottom row of $\mathbf{J}$ is negative, $\mathbf{J}$ does not give us an optimal solution. Therefore, we must return to step (4), using $\mathbf{J}$ in place of $\mathbf{I}$.

(4) Since the only negative entry in the bottom row of **J** is $-\frac{100}{7}$, the pivot column is column 3. The ratios obtained by dividing the entries in column 4 by those in column 3 are as follows:

row 1: $\frac{50}{7} \div \frac{4}{7} = \frac{50}{4} = \frac{25}{2} = 12\frac{1}{2}$

row 2: $\frac{50}{7} \div \frac{130}{7} = \frac{50}{130} = \frac{5}{13}$

Row 2 is the pivot row, and $p = \frac{130}{7}$ is the pivot element.

(5) Applying the pivot transformation, we obtain

$$
\mathbf{K} = \begin{array}{c} x \\ z \\ F \end{array}
\begin{array}{ccc} r & y & s \end{array}
\left(
\begin{array}{ccc|c}
\frac{2279}{2275} & \frac{10}{13} & -\frac{2}{65} & \frac{90}{13} \\
-\frac{1}{325} & \frac{2}{13} & 1 & \frac{5}{13} \\
\frac{18}{13} & \frac{400}{13} & \frac{10}{13} & \frac{46,500}{13}
\end{array}
\right)
$$

(6) All of the entries in the bottom row of **K** are positive, so **K** gives an optimal solution,

$$x = \frac{90}{13}, \qquad y = 0, \qquad z = \frac{5}{13}$$

The maximum value of F appears in the last column and last row of **K**; the maximum value is $F = \frac{46,500}{13}$, or approximately 3577.

Notice that the optimal solution found for Example 29 agrees completely with the one found in Example 21. (See page 346.) However, the simplex method required applying the pivot transformation only twice, which is equivalent to solving only two of the ten sets of equations that were solved to find the corner points in Example 21. Thus, the simplex method is clearly more efficient than the corner-point method, and this advantage increases sharply with the size of the linear program. For practical applications, linear programs involving hundreds of variables and constraints can be solved by high-speed computers using computer programs based on the simplex method.

8.6 EXERCISES

1. Consider the following linear program:
Find the maximum value of

$$F(x,y,z) = 2x + 3y + z$$

Subject to the constraints

$$x + y + 2z \leq 1$$
$$x - y + 3z \leq 2$$
$$2x + 3y - z \leq 3$$
$$x \geq 0, \qquad y \geq 0, \qquad z \geq 0$$

(a) Show that the constraint set C is bounded.

(b) Rewrite the first three linear inequalities as linear equalities by introducing three slack variables, r, s, and t.

(c) State which of the following corner points are feasible.

(i) $x = 0, y = 0, z = 0$

(ii) $y = 0, z = 0, t = 0$

(iii) $x = 0, z = 0, r = 0$

(iv) $r = 0, s = 0, z = 0$

2. (a) Write the initial matrix $\mathbf{I}$ for the linear program in Exercise 1.

(b) What is the value of F at $(0,0,0)$?

(c) Is this value a maximum? Explain.

3. (a) Find the pivot p for the initial matrix of Exercise 2.

(b) Use the pivot transformation to find a second matrix $\mathbf{J}$.

(c) What is the corner point corresponding to $\mathbf{J}$?

(d) What is the value of F at this corner point?

(e) Is this value a maximum? Explain.

4. Use the simplex algorithm to maximize the linear function

$$F(x,y) = 2x + 3y$$

subject to the constraints

$$x - 2y \leq 4$$
$$x + 3y \leq 6$$
$$x \geq 0, \qquad y \geq 0$$

5. Solve Exercise 5(b), Section 8.4, using the simplex method.

6. Use the simplex method to solve the linear program of Exercise 6, Section 8.4.

7. An investor wishes to invest $10,000 for one year. The following table gives the expected return (in percentages) for the next year, assuming three different world political climates. Each of these climates appears equally likely.

	Climate 1	Climate 2	Climate 3
Government Bonds	6	6	6
Armament Stocks	−2	6	12
Industrial Stocks	5	7	10
Airline Stocks	8	6	4

(a) How much should the investor invest in each type of investment in order to maximize her gain?

(b) What will her maximum gain be?

Key Word Review

Each key word is followed by the number of the section in which it is defined.

number line 8.1
origin 8.1
Cartesian plane 8.1
x-coordinate 8.1
y-coordinate 8.1
x-axis 8.1
y-axis 8.1
graph (of a linear equation) 8.1
parallel lines 8.1
identical line 8.1
linear inequality 8.2
system of linear inequalities 8.2
linear function in two variables 8.3
linear program 8.3
objective function 8.3

constraints 8.3
constraint set 8.3
feasible solution 8.4
optimal solution 8.4
bounded constraint set 8.5
corner point 8.5
slack variables 8.6
basic variables 8.6
pivoting 8.6
pivot transformation 8.6
pivot 8.6
pivot row 8.6
pivot column 8.6
simplex method 8.6
pivot element 8.6

CHAPTER 8   REVIEW EXERCISES

1.   Define each word in the Key Word Review.

2.   Represent the following numbers on the same number line: 2, 4, -1, and 1.5.

3.   Find the x-coordinates of (a) (1,7), (b) (7,1).

4.   Represent the following ordered pairs on the same Cartesian plane:

(10,20),   ($-13,30$),   (15,-2.5)

5.   Draw the graphs of—
(a) $2x + y = 7$ **(b)** $2x + y < 7$ **(c)** $2x + y \geq 7$

6.   Draw the graphs of the following pairs of lines and state whether the lines are parallel, intersecting, or identical.

(a) $2x + 3y = 9$ **(b)** $2x + 3y = 9$
       $2x + 3y = 4$ $4x + 6y = 18$
(c) $2x + 3y = 9$ **(d)** $2x + 3y = 5$
       $4x + 6y = 5$ $x + 6y = 8$

7.   Draw the graph of each system of inequalities.

(a) $x - y \leq 7$ **(b)** $x - y \leq 7$ **(c)** $x - y \leq 7$
       $2x - 2y \leq 7$ $2x - 2y \leq 14$ $x + y \leq 7$

8.   Which of the following statements are true?
     (a) $4 < 4$ **(b)** $4 \le 4$ **(c)** $4 \not> 4$
     (d) $4 \ge 4$ **(e)** $3 \not< 4$ **(f)** $3 > 4$

9.   Let G be the graph of

$$2x + y < 7$$

Which of the following points are in G?
     (a) $(3,1)$ **(b)** $(1,3)$ **(c)** $(-3,-2)$

10. Consider the following linear program:

Find the maximum value of

$$F(x,y) = 2x + 4y + 5$$

Subject to the constraints

$$2x + 3y \le 12$$

$$x \ge 0$$

$$y \ge 0$$

(a) What is the objective function?
(b) What are the constraints?
(c) Show that the constraint set is bounded
(d) List all corner points.
(e) Which of the following points are feasible?
     (*i*) $(1,4)$ (*ii*) $(4,1)$ (*iii*) $(4,-1)$

11. **(a)** Solve the linear program of Exercise 10 by—
     (*i*) drawing the constraint set C
     (*ii*) listing all corner points.
     (b) Find an optimal solution.

12. Answer true or false.
     (a) All corner points are feasible.
     (b) All feasible points are corner points.
     (c) A linear program can have more than one optimal solution.
     (d) A linear program in four variables that has five constraints can have six corner points.

13. Draw (*a*) a bounded constraint set, (*b*) an unbounded constraint set.

14. **(a)** *Find the maximum value of*

$$F(x,y,z) = 2x - y + 3z$$

Subject to the constraints

$$x + y + 3z \le 5$$

$$x \ge 0, \qquad y \ge 0, \qquad z \ge 0$$

(b) What is the optimal solution?

15.   The Oriental Rice Bowl Company manufactures two kinds of bowls. Each standard bowl requires 2.00 hours of grinding and 1.25 hours of finishing. Each deluxe bowl requires 2.00 hours of grinding and 3.00 hours of finishing. The company has two grinders and three finishers, each of whom works 40 hours per week. Each standard bowl produces a profit of 3 dollars and each deluxe model produces a profit of 4 dollars. Assume that every bowl produced will be sold.

(a) Write a linear program that will answer the following question: How many bowls of each model should be produced in order to maximize profits?

(b) Graph the constraint set.

(c) Calculate the number of bowls of each model that should be produced.

(d) If profits are maximized, how many hours will
   (*i*) the two grinders,
   (*ii*) the three finishers work?

(e) What is the maximum profit?

16.   Consider the following linear program.

Find the maximum value of

$$F(x,y,z,w) = 3 + 2x + 3y + 4z + w$$

Subject to the constraints

$$x + y + 3z + w \leq 1$$

$$x - 2y + 3z + w \geq 0$$

$$x - y + 3z + 2w \leq 4$$

$$x \geq 0, \quad y \geq 0, \quad z \geq 0, \quad w \geq 0$$

(a) Prove that the constraint set C is bounded.

(b) Show that $C \neq \emptyset$.

(c) Rewrite the first three linear inequalities as equations by introducing three nonnegative slack variables r, s, and t.

(d) Write the initial matrix **I**.

(e) What is the value of F at $(0,0,0,0)$? Is this value a maximum? Explain.

17.   **(a)** Find the pivot element p for the initial matrix of Exercise 16.

(b) Use the pivot transformation to obtain a second matrix **J**.

(c) What is the corner point corresponding to **J**?

(d) What is the value of F at this point?

(e) Solve the linear program.

Suggested Readings

1. Campbell, Hugh G. *Introduction to Matrices, Vectors, and Linear Programming.* New York: Appleton-Century-Crofts, 1965.
2. Dantzig, George B. *Linear Programming and Extensions.* Princeton, N.J.: Princeton University Press, 1963.
3. Hu, T. C. *Integer Programming and Network Flows.* Reading, Mass.: Addison-Wesley Publishing Co., 1969.
4. Ijiri, Y.; Levy, F. K.; and Lyon, R. C. "A Linear Programming Model for Budgeting and Financial Planning." *Journal of Accounting Research* 1 (1963): 198–212.
5. Painter, R. J., and Yantis, R. P. *Elementary Matrix Algebra with Linear Programming.* Boston: Prindle, Weber & Schmidt, 1971.

Hollywood Presbyterian Medical Center

CHAPTER 9 Game Theory

INTRODUCTION

For most of us, the word "game" refers to a pleasant leisure activity like the chess game played by the two young men at left. Some games, however, are of a more serious nature. Farmers play against the weather; businesses play against their competitors. If a garment manufacturer brings out her fall fashions early, will she lose sales to competitors who copy her designs? If she brings her designs out late, will she lose important early sales? Choosing the best strategy in this situation can mean the difference between financial failure and success.

The reader may be aware of the war games that the military plays. In greatest secrecy, top officers discuss the various strategies they might use. If a particular strategy is chosen, what will the reaction of the enemy be? What should our next move then be? Games of this kind are extremely serious.

In this chapter we will discuss certain kinds of games and several methods for choosing the best strategies for winning them. In particular, we will see that many games can be represented by matrices, and that linear programming can sometimes be used to find the best strategies for winning them.

9.1 TREE GAMES

Many recreational games, such as ticktacktoe, chess, and checkers, have the following properties:

(1) There are two players, usually denoted Red and Black, who alternate turns. We will assume that Red always goes first.

(2) Each player has a finite choice of moves for each turn.

(3) The game lasts at most n turns, where n is a positive integer. After n turns have been played, the outcome of the game is determined. The possible outcomes are

      Red wins,   *Black wins*, and *draw*

(4) Each player knows the moves of the other player.

Any game having these four characteristics is called a **tree game** and can be represented by a tree diagram. In this section we will see that if T is any tree game, then either—

    (a) it is possible for Red to win every time no matter what Black does
      or
    (b) it is possible for Black to win every time no matter what Red does
      or
    (c) Red can prevent Black from winning and Black can prevent Red from winning.

If statement **(a)** is true, we say that the **natural outcome** of the game T is "Red wins."

If statement **(b)** is true, we say that the natural outcome is "Black wins," and if statement **(c)** is true, the natural outcome is "draw."

Throughout this chapter, we will assume that *each player plays rationally and wants to win*. We will not consider games in which parents let their children win, or in which employees play badly to avoid humiliating their bosses.

The following example illustrates these concepts.

Example 1 The following game, called tic-tac, is a simplified version of ticktacktoe.

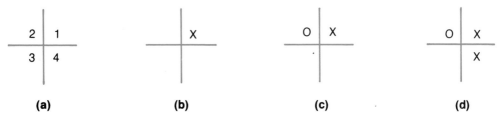

(a) The game   (b) Red chooses space 1   (c) Black chooses space 2   (d) Red wins

Figure 9.1 A game of tic-tac

The first player (Red) marks an X in any one of the four spaces of Figure 9.1(a). For example, he might choose space 1 as indicated in Figure 9.1(b). The second player (Black) then marks an O in any one of the remaining spaces—for example, she might choose space 2 as indicated in Figure 9.1(c). This procedure is repeated until one player gets two of his symbols in a row horizontally, vertically, or diagonally.

Tic-tac is a tree game because it satisfies criteria 1–4. Notice that, no matter what Black does, Red will always win. It therefore follows that the natural outcome of this game is "Red wins."

The natural outcome *N* of most tree games cannot be determined as easily as the natural outcome of tic-tac. However, *N* can often be found by using a tree diagram. Figure 9.2 shows the tree diagram for tic-tac.

In the tree diagram of a tree game—

(a) all the vertices except those in the top row represent the player who is making the move

(b) the edges represent the moves that the player can make

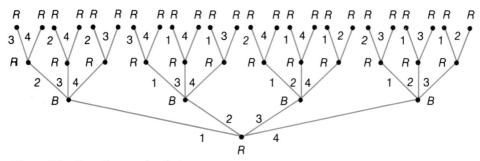

Figure 9.2 Tree diagram for tic-tac

(c) the vertices in the top row (called the *endpoints*) represent the final outcome.

For example, the bottom vertex of Figure 9.2 is labeled *R* because Red goes first. The edges connected to this vertex are labeled 1, 2, 3, 4 because Red can initially choose any one of these positions. The first endpoint on the left is labeled *R* (denoting Red wins) because this endpoint represents the outcome of the following sequence of moves: Red chooses space 1, Black chooses space 2, Red chooses space 3. According to the rules of the game, this sequence of moves results in Red's winning because two of Red's symbols are in a row diagonally.

The tree diagram clearly shows that the natural outcome of this game is " Red wins," because Red is the winner of every possible sequence of moves. (Notice that all the endpoints are labeled *R*.) In other words, every strategy is Red's "best" strategy because Red will win no matter what moves are made. Most games, of course, are not that simple. The following example shows how a tree diagram can help us find the best strategy for each player and the natural outcome of a more complicated tree game.

Example 2   Red and Black each have three numbers: 1, 2, and 3. Red first plays one of his numbers, Black then plays one of her numbers, and then Red plays one of his *remaining two* numbers. The outcome of the game is determined by the sum of the three numbers that were played in the following manner:

Sum	*Outcome*
4 or 5	Red wins (*R*)
6 or 7	Black wins (*B*)
8 or 9	Draw (*D*)

The tree diagram for this game is given in Figure 9.3.

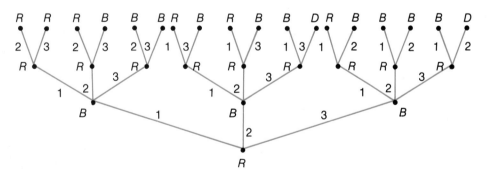

Figure 9.3   Tree diagram for Example 2

Red must decide which number to play first. If he plays number 1, then Black can play either 1, 2, 3.

(a) If Black plays 1, the tree diagram shows that *Red will win.*

(b) If Black plays 2, Red can play 2 and win, or Red can play 3 and let Black win. Since we are assuming that Red plays rationally and wishes to win, Red will play 2. Hence, if Black plays 2, *Red will win.*

(c) If Black plays 3, *Black will win,* no matter what Red does.

It follows that, if Red plays 1, then Black will play 3, and *Black will win.*

By using similar arguments, the reader can show that, if Red plays 2, Black will play 3, and the game will result in a *draw.* Moreover, if Red plays 3, Black will play 2, and *Black will win.*

Red's three initial choices and their final outcomes are summarized in the following table:

Choice	*Outcome*
1	Black will win
2	Draw
3	Black will win

It follows that Red's best choice is to play number 2. Black will then play 3 and the game will be a draw. Therefore, the natural outcome of this game is " draw."

The tree games of Examples 1 and 2 both have natural outcomes. The following theorem will help us show that every tree game has a natural outcome.

THEOREM 9.1 Let n be any positive integer. If every tree game having n moves has a natural outcome, then every tree game having $n + 1$ moves has a natural outcome.

Proof Let T be a tree game having $n + 1$ moves. Red, the first player, has a finite number of choices $r_1, r_2, \ldots, r_m$ for his first move. Whatever his choice, he then presents Black with a tree game having n moves. (See Figure 9.4; in this diagram we have arbitrarily let m be 3.)

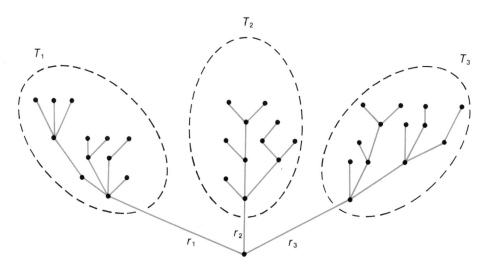

Figure 9.4

By assumption, each of these tree games $T_1, T_2, \ldots, T_m$ has a natural outcome. If one of these tree games (say T_1) has the natural outcome "Red wins," then Red would of course begin the game with the corresponding choice r_1. The natural outcome of T would then be "Red wins." If all of the games that Red could present to Black have "Black wins" as natural outcome, then the natural outcome of the original game is "Black wins," regardless of the choice Red makes. The only other possibility is that none of the natural outcomes of $T_1, T_2, \ldots, T_m$ is "Red wins," but at least one is "draw." In this case, Red will choose a first move that leads to one of these games, and the natural outcome of T will be "draw."

THEOREM 9.2   Every tree game has a natural outcome.

Proof   It is clear that every tree game having one move has a natural outcome. Red merely looks at his possible choices and picks the one that is best for him. The outcome of this choice is the natural outcome of the game.

It then follows from Theorem 9.1 (letting $n = 1$), that every tree game with two moves has a natural outcome. Applying Theorem 9.1 again, with $n = 2$, it follows that every tree game with three moves has a natural outcome. Continuing in this way, we see that every tree game has a natural outcome.

It follows from Theorem 9.2 that games such as ticktacktoe, checkers, and chess have natural outcomes. However, Theorem 9.2 does not tell us how to determine the natural outcome. If the tree diagram is relatively small, the natural outcome can be found by inspection, as in Example 2. However, the tree diagram may be quite large. The tree diagram for ticktacktoe, for example, has $9! = 362,880$ endpoints. In such a case, more sophisticated techniques are needed. Using these techniques, it can be proved that the natural outcome of ticktacktoe is "draw," and that Red's best initial move is to choose one of the four corner spaces.

The definition of a tree game can be easily extended to games that satisfy criteria 1–4 on page 373, but whose possible outcomes are numbers or other symbols instead of "Red wins," "Black wins," and "draw." If the outcomes are numbers, then a *positive* number x denotes the fact that *Red wins* x units and *Black loses* x units. These units may be dollars, pounds of sugar, percentage of votes, or whatever. On the other hand, a *negative* number y denotes the fact that *Red loses* y units and *Black wins* y units. The number 0 denotes the fact that neither Red nor Black win anything. Numerical outcomes are often called **payoffs**. Examples of tree games with numerical outcomes are given in Exercises 2 and 3.

9.1 EXERCISES

1.   In the following tree games, R, B, and D denote the outcomes "Red wins," "Black wins," and "draw," respectively. The vertices are unlabeled because, as usual, Red goes first, Black goes next, and so on. Find Red's best initial move and the natural outcome for each of these tree games.

(a)

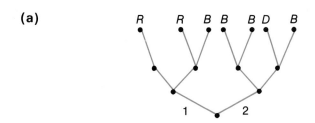

(b)

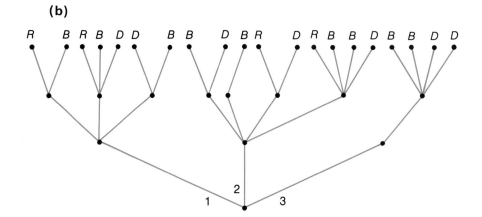

2.   The outcomes of the following tree games are numerical. Assume that the payoffs are in dollars. For example, the number 4, -7, and 0 represent the events "Red wins 4 dollars," "Red loses 7 dollars," and "Red wins nothing," respectively. Remember that Red wishes to win as much as possible. Find Red's best initial move and the natural outcome of each game.

(a)

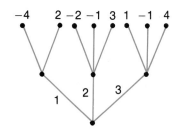

(b)

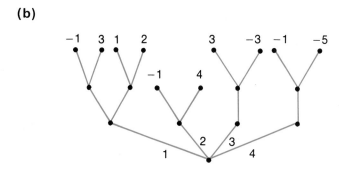

3.   A dress manufacturer (R) has decided to show his fall line of dresses early, even though he knows that his chief competitor (B) will then be able to copy him. Both the manufacturer and his competitor have three choices as to dress length: mini (M), knee length (K), and maxi (A). The manufacturer's market research department has informed him that, if both he and his competitor use a mini hemline, he will receive 70 percent of the market; if both he and his competitor use the knee-length hemline, he will receive 55 percent of the market. These outcomes have been denoted 70 and 55 on the tree diagram in Figure 9.5. The other numbers used to label the endpoints also represent percentage shares of the market.

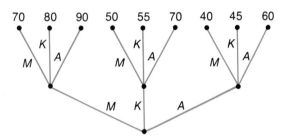

Figure 9.5

(a) What hemline should the manufacturer use?

(b) What share of the market can he then expect to get?

4.   A penny is placed by Red on one of the squares bordering one edge of a checkerboard. It is then alternatively moved by Red and Black. Each move is either forward toward the opposite edge or sideways. The object of the game is to reach the last row; the winner is the one who pushes the penny into that row.

(a) Is this a tree game? Explain.

(b) Suppose that the players are not permitted to return the penny to a position it occupied previously. Is this modified game a tree game? Explain.

5.   Red and Black each toss a coin simultaneously. If both coins are heads or both are tails, Red pays Black 1 dollar; otherwise, Black pays Red 1 dollar. Is this a tree game? Explain.

9.2 MATRIX GAMES

According to criterion 4 for a tree game (page 373), both players in such a game have **perfect information**; that is, each player knows every move made by his opponent. This, of course, is not true of many practical competitive situations. For example, department store A, does not know which items store B plans to put on sale for Washington's Birthday until it is too late to use this information. The remainder of this chapter will be devoted to games in which the players may *not* have perfect information.

The games that we will now study have the following characteristics:

(1) There are two players who are usually denoted Red and Black.

(2) Each player has a finite choice of moves.

(3) The outcome of the game is determined after each player has made one move.

(4) The outcomes, called *payoffs*, are numerical. A positive payoff x means that Red wins x units and Black loses x units; a negative payoff y means that Red loses y units and Black wins y units. A zero payoff means that neither Red nor Black wins or loses anything.

In this section we will see that any game having these four characteristics can be represented by a matrix. Because of this representation, such a game is called a **matrix game**.

Let M be a matrix game. Since Black loses what Red wins, and vice versa, there are no payoff units left over. Therefore, M is also called a **two-player zero-sum game**.

When a person bets against a race track, the situation is not a two-player zero-sum game because the race track does not win everything that the bettor loses. This is because a percentage of every bet goes to the government.

The following example describes a simple matrix game and shows how to find its matrix representation.

Example 3   Red places his coin either heads up (h) or tails up (t). Then Black, not knowing Red's move, calls either heads or tails. If Black guesses correctly, Red pays her 1 dollar; otherwise, Black pays Red 1 dollar.

This game is not a tree game because the players do not have perfect information. It is however, a matrix game because the four criteria for a matrix game are satisfied. We shall now show how this game can be represented by a matrix. In Section 9.5 we will show how this matrix representation can be used to find the best strategy for each player.

Red has two possible choices: to play heads (h), or to play tails (t). Black can also choose to play heads or tails. These moves and their corresponding payoffs are conveniently displayed in the following matrix.

$$
\begin{array}{c}
\text{Black} \\
\begin{array}{cc}
h & t
\end{array} \\
\text{Red} \quad
\begin{array}{c}
h \\
t
\end{array}
\begin{pmatrix}
-1 & 1 \\
1 & -1
\end{pmatrix}
\end{array}
$$

For example, the entry -1 in row 1, column 1 indicates that, if Red and Black both play heads, the payoff to Red is -1 (i.e., Red pays Black 1 dollar). Similarly, the entry 1 in row 2, column 1 indicates that if Red plays tails and Black plays heads, the payoff to Red is 1 (i.e., Black pays Red 1 dollar).

The following definition shows how every matrix game can be represented by a matrix.

Definition 9.1   Let G be a matrix game in which Red has m choices $r_1, r_2, \ldots, r_m$ and Black has n choices $b_1, b_2, \ldots, b_n$. Let a_{ij} be the payoff that occurs if Red plays r_i and Black plays b_j. Then G can be represented by the $m \times n$ matrix.

Black

$$
\mathbf{A} = \quad \text{Red} \quad
\begin{array}{c}
r_1 \\ r_2 \\ \vdots \\ r_m
\end{array}
\begin{pmatrix}
a_{11} & a_{12} & \cdots & a_{1n} \\
a_{21} & a_{22} & \cdots & a_{2n} \\
\vdots & \vdots & & \vdots \\
a_{m1} & a_{m2} & \cdots & a_{mn}
\end{pmatrix}
\begin{array}{cccc}
b_1 & b_2 & \cdots & b_n
\end{array}
$$

The matrix $\mathbf{A}$ is called the **payoff matrix** of G.

Notice that the *rows* of the payoff matrix $\mathbf{A}$ represent the strategies of Red. Red is therefore called the **row player**. If Red plays r_i, we also say that *Red plays the ith row*. For similar reasons, Black is called the **column player** and is said to *play column i* if she plays b_i.

Recall, from criterion 4 for a matrix game, that the payoffs a_{ij} are given from the point of view of the row player. For example, a *positive* payoff a_{ij} means that the row player Red *wins a_{ij}* units and the column player Black *loses a_{ij}* units; a *negative* payoff a_{ij} means that the row player *loses a_{ij}* units and the column player *wins a_{ij}* units. Red therefore wants the payoff to be as *large* as possible. Black, however, wants the payoff to be as *small* as possible.

The following example shows how the matrix representation of a matrix game can be used to determine the best strategy for each player. Most games, however, are not as simple to solve as this one and require the techniques that we will study in the remainder of this chapter.

Example 4   Candidates Smith and Jones are vying for the electoral votes of two states, X and Y. These states have forty and sixty votes, respectively. According to a public-opinion poll, these votes would all go to Jones if the election were held tomorrow. Each candidate has the time to make personal appearances in exactly one of these states. Smith feels that if she and Jones both visit the same state, Smith will change the voter's minds and win the votes of that state. On the other hand, if Smith visits one state and Jones visits the other state, each will receive the votes of the state in which he or she campaigns. Which state should Smith visit?

Solution   This competitive situation is a matrix game because it satisfies the four criteria for such games. Because we are interested in Smith's problem, let Smith take the place of Red (i.e., let Smith be the row player) and let Jones take the place of Black (i.e., let Jones be the column player).

Smith has two possible moves: visit state X or visit state Y. Her opponent, Jones, can also visit either X or Y. These moves and their corresponding payoffs (in electoral votes for Smith) are conveniently displayed in the following payoff matrix.

Jones

$$
\text{Smith} \quad
\begin{array}{c}
X \\ Y
\end{array}
\begin{pmatrix}
40 & 40 \\
60 & 60
\end{pmatrix}
\begin{array}{cc}
X & Y
\end{array}
$$

For example, if both Smith and Jones visit state X, Smith will win the 40 electoral votes of that state. (Notice that a win of 40 votes for Smith is a loss of 40 votes for Jones.) Hence, the number 40 was placed in the intersection of the row labeled X and the column labeled X. Similarly, 60 was placed in the intersection of the row labeled Y and the column labeled X because Smith will win the 60 votes of state Y if she visits Y and Jones visits X.

It is clear from this matrix that, if Smith visits X, she will win 40 electoral votes no matter what Jones does. On the other hand, if Smith visits Y, she will win 60 votes no matter what Jones does. It follows that Smith should visit Y.

In the following example we will see that the payoffs of a matrix game can represent the intensity of feelings such as satisfaction. Such payoffs are called **utilities**.

Example 5   Two gasoline stations, A and B, are contemplating a price war. Each station has three alternatives: keep prices normal, reduce prices 2 cents per gallon, or reduce prices 5 cents per gallon. The entries in the following matrix are utilities.

$$
\begin{array}{c}
\\
\\
\text{Station A}
\end{array}
\begin{array}{c}
\\
\text{Normal} \\
\text{Reduce } 2\cent \\
\text{Reduce } 5\cent
\end{array}
\overset{\displaystyle \text{Station B}}{
\begin{array}{ccc}
\text{Normal} & \text{Reduce } 2\cent & \text{Reduce } 5\cent \\
\left(\begin{array}{ccc}
0 & -4 & -3 \\
3 & 2 & -2 \\
2 & 3 & 2
\end{array}\right)
\end{array}}
$$

Payoff a_{ij} represents the satisfaction to gas station A if row i and column j are played.

Some of the factors that influence the value of a_{ij} in a situation such as this one are profit, customer relations, the desire to drive one's opponent out of business, and so on. Since the method of representing the intensity of feelings by numbers is complicated and controversial, we will not discuss it further. The reader need only remember that *the higher the utility, the better the row player likes it, and the lower the utility, the better the column player likes it.* In the above matrix, the best utility from station A's viewpoint is 3; the best utility from station B's viewpoint is -4.

In each game that we have studied so far, the two players have been people, or human institutions such as businesses. In many practical situations, however, one of the players may be the weather, the economy, the policial situation, or some other impersonal form. A player who is pitted against one of these disinterested opponents is said to be playing against *nature*. Nature, of course, unlike a human opponent, is not at all interested in winning. Nor can we say that a loss of 400 dollars worth of fruit for the farmer is a gain of 400 dollars for the weather. The safest approach, however, is to *assume* that the worst thing that can happen will happen; that is, to assume that nature will *act* like an intelligent opponent who wishes to minimize our winnings. Of course, if the worst does not happen, we will be pleasantly surprised. We make a similar assumption of intelligence when playing against a human opponent. We are also pleasantly surprised when a human opponent plays badly and we win more than we expected.

Therefore, in a game that pits a human being (or a human organization such as a business or a government) against nature, we will waive the requirement that a win for Red is a loss for Black, and vice versa. We will, however, assume that nature wishes to minimize our winnings. Many practical situations involving nature then become matrix games. One such situation is given in the next example.

Example 6   A farmer wishes to plant either wheat or corn. The following matrix represents the profit (in thousands of dollars) that he will make from each crop under the following weather conditions: severe, average, and excellent.

Weather

		Severe	Average	Excellent
Farmer	Wheat	10	8	6
	Corn	7	9	11

This is a matrix game. The two players are the farmer and the weather (i.e., nature). Since one of the players is nature, part of criterion 4 for a matrix game is waived: the fact that the farmer wins 10,000 dollars does not mean that the weather loses 10,000 dollars.

As in Example 6, we will use the convention that *nature is always the column player.*

9.2 EXERCISES

1. Red and Black each have three cards numbered 1, 2, and 3. The two players each simultaneously put down one card. If the two numbers add up to an even number, Black pays Red 1 dollar; otherwise, Red pays Black 1 dollar.
   (a) Is this a tree game? Explain.
   (b) Is this a matrix game? Explain.
   (c) Represent this game by a matrix.

2. Red has two letters, A and O; Black has three letters, F, M, and N. The players simultaneously put down one letter each. If the two letters can form one two-letter word, Red pays Black 1 dollar; if they can form two two-letter words, Red pays Black 2 dollars; otherwise, Black plays Red 3 dollars. (Foreign words, interjections, and abbreviations are not counted.) Show that this game is a matrix game and represent it by a matrix.

3. Let

Black

$$A = \begin{array}{c} \\ \text{Red} \end{array} \begin{array}{c} r_1 \\ r_2 \\ r_3 \end{array} \begin{pmatrix} 0 & 1 & 1 \\ 2 & -1 & 0 \\ 3 & 1 & 2 \end{pmatrix}$$

be the matrix of a matrix game. Assume that the payoff is in dollars.

(a) Calculate the payoff to Red if—

   (i) Red plays r_1 and Black plays b_2
   (ii) Red plays r_3 and Black plays b_1.

(b) Let (r_i, b_j) mean that Red plays r_i and Black plays b_j. Who pays whom how much under each of the following conditions:

   (i) (r_1, b_1) (ii) (r_2, b_1) (iii) (r_3, b_2) (iv) (r_2, b_3)

4. In two-fingered Morra each player holds up either one or two fingers. If the total number of fingers is even, then Red gets that amount of money from Black. (For example, if there are four fingers, Red gets 4 dollars; if there are two fingers, Red gets 2 dollars.) If the total numbers of fingers is odd, Black collects that amount of money from Red. Construct a payoff matrix for this game.

5. Attacking bombers are trying to sink a fleet of enemy ships that is protected by an antiaircraft carrier. The bombers can either attack from high altitudes or from low altitudes. The carrier can either send its planes high or low to search for the attackers. If the bombers avoid the defending planes, they will sink four or seven ships if they are flying high or low, respectively. If they meet the defenders, they will sink one ship if flying high and three ships if flying low. Give the payoff matrix for this problem.

6. A woman is considering investing her money in either a savings certificate or a certain stock. A savings certificate will yield an 8-percent return regardless of the economic situation. The woman feels that the stock will yield a 6-percent, 8-percent, or 10-percent return, respectively, depending upon whether the economy is recessionary, normal, or inflationary. Show that this is a matrix game with nature as one of the opponents and give the payoff matrix.

9.3 PURE STRATEGIES AND SADDLE POINTS

In Section 9.2 we saw how to represent certain games as matrices. Our problem now is to decide which strategy is best for each player. By a **strategy** we mean a course of action such as " Play row 1 " or " Play column 4," or even " Play row 1 30 percent of the time, row 2 60 percent of the time, and row 3 10 percent of the time." (This last strategy assumes that the game is played many times.) The last strategy is called a **mixed strategy** because it involves more than one row. Mixed strategies will be discussed in the next section. A strategy like the first two mentioned, which do not involve more than one row or column, is called a **pure strategy**. Pure strategies will be discussed in this section.

In order to simplify the notation for matrix games, we will frequently omit the matrix headings. It will then be understood that Red plays the rows (moves $r_1, r_2, \ldots,$ r_m) and Black plays the columns (moves $b_1, b_2, \ldots, b_n$). We will also assume that the payoffs are in dollars unless otherwise stated. In addition, since the payoff matrix **A** of a matrix game tells us everything that we need to know about the game mathematically, we will refer to **A** itself as a matrix game.

In order to see how the best strategies for Red and Black can be determined, let us consider the following matrix game:

$$
\begin{array}{c}
\\
\text{Red}
\end{array}
\begin{array}{cc}
 & \text{Black} \\
 & \begin{array}{ccc} b_1 & b_2 & b_3 \end{array} \\
\begin{array}{c} r_1 \\ r_2 \\ r_3 \end{array}
&
\left(\begin{array}{ccc}
3 & 4 & 6 \\
-1 & 10 & -8 \\
-2 & 3 & -1
\end{array}\right)
\end{array}
$$

Red has three choices for his move: r_1, r_2, and r_3. He might be tempted to play r_2 in the hope of winning 10 dollars. Similarly, Black might be tempted to play b_3 in the hope of winning 8 dollars. This pair of moves that Red and Black plan to make is conveniently denoted by

$$(r_2, b_3)$$

Now suppose that Red reconsiders his move. He might reason as follows: "Since Black wishes to make as much money as possible, she will play b_3. If I play r_2 as I had planned, I will lose 8 dollars instead of winning 10 dollars. If, on the other hand, I play r_1, I will win 6 dollars and if I play r_3 I will lose 1 dollar. Therefore, I should play r_1." The pair of moves that Red and Black now plan to make is

$$(r_1, b_3)$$

Black may also reconsider her move and reason as follows: "Red has probably decided that I intend to play b_3. He will therefore play r_1. Since he is playing r_1, my best move is to play b_1. I will then lose only 3 dollars instead of losing 6 dollars." The pair of moves that Red and Black now plan to make is

$$(r_1, b_1)$$

Suppose that Red, using similar logic, now decides that Black will play b_1. Red knows that the payoffs will be 3, -1, or -2, respectively, according to whether he plays r_1, r_2, or r_3. Red's best move is therefore still r_1. Similarly, even if Black knows that Red intends to play r_1, Black will not change her mind about playing b_1. It appears that r_1 is the best move for Red and b_1 is the best move for Black. The pair (r_1, b_1) is called an *equilibrium pair*, and the payoff 3 that occurs if Red plays r_1 and Black plays b_1 is called a *saddle point* of the game.

In general, we have the following definition:

Definition 9.2   Let

$$
\mathbf{A} =
\begin{array}{c}
\begin{array}{c} r_1 \\ r_2 \\ \vdots \\ r_i \\ \vdots \\ r_m \end{array}
\end{array}
\begin{array}{c}
\begin{array}{cccccc}
b_1 & b_2 & \cdots & b_j & \cdots & b_n
\end{array} \\
\left(\begin{array}{cccccc}
a_{11} & a_{12} & \cdots & a_{1j} & \cdots & a_{1n} \\
a_{21} & a_{22} & \cdots & a_{2j} & \cdots & a_{2n} \\
\vdots & \vdots & & \vdots & & \vdots \\
a_{i1} & a_{i2} & \cdots & a_{ij} & \cdots & a_{in} \\
\vdots & \vdots & & \vdots & & \vdots \\
a_{m1} & a_{m2} & \cdots & a_{mj} & \cdots & a_{mn}
\end{array}\right)
\end{array}
$$

be any matrix game. The pair (r_i, b_j) is called an **equilibrium pair** if the following statements are true:

(a)   If Red plays r_i, then Black's best move is b_j.

(b)   If Black plays b_j, then Red's best move is r_i.

Definition 9.3   Suppose that (r_i, b_j) is an equilibrium pair. The payoff a_{ij} that occurs if Red plays r_i and Black plays b_j is called a **saddle point**.

If a game has an equilibrium pair (r_i, b_j), then Red's best strategy is to play r_i and Black's best strategy is to play b_j. The payoff will then be the saddle point a_{ij}.

Let G be any game that has a saddle point. (We will soon see that there are many games that do not have saddle points.) For convenience, let us arbitrarily assume that the payoff matrix **A** of this game has three rows and four columns, and that the saddle point is the circled entry v in row 2, column 3.

$$\mathbf{A} = \begin{array}{c} \\ r_1 \\ r_2 \\ r_3 \end{array} \begin{array}{cccc} b_1 & b_2 & b_3 & b_4 \\ \left(\begin{array}{cccc} p & q & r & s \\ t & u & \textcircled{v} & w \\ x & y & z & a \end{array} \right) \end{array}$$

(The letters $p, q, r, \ldots$ can stand for any numbers; some of these numbers may even be the same.) The remarks we make about **A** will be equally true for payoff matrices of any size having any entry as saddle point. In the following discussion, we will discover certain characteristics of the saddle point v that will enable us to quickly determine the saddle point of any game that has one.

Since the saddle point v is found in the intersection of the row labeled r_2 and the column labeled b_3, (r_2, b_3) is an equilibrium pair. This means that if Red plays r_2, then Black's best move is b_3. Suppose that Red plays r_2. The payoff will be $t, u, v,$ or w, according to whether Black plays $b_1, b_2, b_3,$ or b_4, respectively. Since b_3 is best for Black, v must be the *smallest* of the numbers $t, u, v,$ and w. In other words, *the saddle point v is the smallest entry in its row*.

The fact that (r_2, b_3) is an equilibrium pair also indicates that if Black plays b_3, then Red should play r_2. Using logic similar to that of the preceding paragraph, the reader can show that this implies that *the saddle point v is the largest entry in its column*.

We have now proved that if a game has a saddle point v, then v is the smallest number in its row and the largest number in its column. We can also easily prove that if v is any number that satisfies these two criteria, then v is a saddle point. We therefore have the following theorem:

THEOREM 9.3   Let **A** be a matrix games as defined in Definition 9.1. Then the payoff a_{ij} is a saddle point of **A** if and only if—

(a) a_{ij} is the smallest number in its row
and

(b) a_{ij} is the largest number in its column.

Suppose that a_{ij} is a saddle point of **A**. If Red plays r_i, then Red will win $a_{i1}, a_{i2},$

..., or a_{in} dollars, depending upon whether Black plays $b_1, b_2, \ldots,$ or b_n. (See matrix **A** of Definition 9.2.) Since a_{ij} is the smallest entry in its row,

$$a_{ij} \leq a_{i1}, \qquad a_{ij} \leq a_{i2}, \ldots, \qquad a_{ij} \leq a_{in}$$

It follows that, if Red plays r_i, then Red will win *at least* a_{ij} dollars, regardless of what Black does. Similarly, if Black plays b_j, Black will lose *at most* a_{ij} dollars, regardless of what Red does.

The following theorem, coupled with Theorem 9.3, summarizes all the important information that we have learned about saddle points.

THEOREM 9.4 If a_{ij} is a saddle point of **A**, then—

- (a) (r_i, b_j) is an equilibrium pair
- (b) Red's best strategy is to play r_i
- (c) Black's best strategy is to play b_j
- (d) if Red and Black both use their best strategies, the payoff is a_{ij}
- (e) if Red plays r_i, Red will win at least a_{ij} dollars, regardless of what Black does
- (f) if Black plays b_j, Black will lose at most a_{ij} dollars, regardless of what Red does.

In any matrix game G a best strategy for either player is called an **optimal strategy**. Thus, Theorem 9.4 implies that, if a_{ij} is a saddle point, then r_i and b_j are optimal strategies for Red and Black, respectively.

Suppose that game G is played many times. If G has saddle point a_{ij}, then Red should always choose row i and Black should always choose column j. Red will gain nothing by sometimes playing a different row, even though Black will soon know Red's strategy. This type of strategy is called a *pure strategy*. We shall see in Section 9.4 that—under certain conditions—it may be advantageous to choose row i sometimes and a different row at other times. This type of strategy is called a *mixed strategy*.

Example 7 illustrates each of the concepts studied in this section.

Example 7 Consider the following matrix game:

$$
\begin{array}{c}
 \\
r_1 \\
r_2 \\
r_3
\end{array}
\begin{array}{ccc}
b_1 & b_2 & b_3 \\
\left(\begin{array}{ccc}
1 & 2 & 3 \\
5 & 4 & 6 \\
-3 & -1 & -2
\end{array} \right)
\end{array}
$$

(a) Find a saddle point if one exists.
(b) Find an equilibrium pair if one exists.
(c) Find optimal strategies for Red and Black.
(d) If Red and Black use their optimal strategies, what will the payoff be?

Solution

(a) Applying Theorem 9.3, a saddle point is the smallest entry in its row and the largest entry in its column. The smallest entries in rows 1, 2, and 3 are $a_{11} = 1$, $a_{22} = 4$, and $a_{31} = -3$, respectively. These numbers are circled in the following matrix:

$$\begin{pmatrix} ① & 2 & 3 \\ 5 & ④ & 6 \\ ③ & -1 & -2 \end{pmatrix}$$

Of these numbers, only $a_{22} = 4$ is the largest number in its column. It follows that $a_{22} = 4$ is the only saddle point. This number is circled twice in the matrix.

(b) The equilibrium pair is the pair of moves associated with the saddle point $a_{22} = 4$. Since a_{22} appears in the intersection of the row labeled r_2 and the column labeled b_2, the equilibrium pair is (r_2, b_2).

(c) Since the equilibrium pair is (r_2, b_2), the optimal strategy for Red is "play r_2" and the optimal strategy for Black is "play b_2."

(d) If Red plays r_2 and Black plays b_2, the payoff is 4.

The following examples show how Theorem 9.3 can be used to solve practical problems. They also show that a matrix game may have no saddle point, exactly one saddle point, or more than one saddle point.

Example 8   A certain patient's symptoms indicate that he is suffering from disease a, b, or c. Two different medicines, m and n, can be prescribed. The probability that disease X is cured by medicine Y is given in the following matrix:

$$\begin{array}{c} \\ m \\ n \end{array} \begin{array}{ccc} a & b & c \\ \begin{pmatrix} 0.5 & 0.4 & 0.6 \\ 0.7 & 0.1 & 0.8 \end{pmatrix} \end{array}$$

For example, the probability that disease b is cured by medicine m is the entry in the intersection of the row labeled m and the column labeled b. This probability is 0.4.

Which medicine should the doctor prescribe? If this medicine is prescribed, what is the probability that the patient will be cured?

Solution   This conflict situation is a matrix game with two players, the doctor (the row player) and nature (the column player). In order to find the best strategy for the doctor, we must look for a saddle point. Applying Theorem 9.3, a saddle point is the smallest entry in its row and the largest entry in its column. The smallest entry in each row is circled in this matrix:

$$\begin{pmatrix} 0.5 & ⦿0.4 & 0.6 \\ 0.7 & ⦿0.1 & 0.8 \end{pmatrix}$$

Of the circled entries, the only one that is the largest entry in its column is 0.4. This is the entry that is circled twice. Since the saddle point, 0.4, is in the row labeled m, the doctor should use medicine m.

If the doctor uses medicine m, the probability that his patient will be cured is *at least* 0.4. Notice that if the doctor uses medicine n, he risks having a probability of only 0.1 of curing his patient.

The matrix games of Examples 7 and 8 have exactly one saddle point. In the next example we will see that a matrix game may have no saddle point, or more than one saddle point.

Example 9 Find all saddle points (if any exist) of the following matrix games.

(a) $\begin{pmatrix} 1 & 2 & -1 \\ -1 & 2 & 3 \\ -1 & 3 & -2 \end{pmatrix}$ (b) $\begin{pmatrix} 0 & 2 & 3 \\ 0 & 3 & 4 \\ -2 & 0 & 3 \end{pmatrix}$

In addition, find optimal strategies for Red and Black.

Solution

(a) The smallest number in each row is circled in the following matrix:

$$\begin{pmatrix} 1 & 2 & \boxed{-1} \\ \boxed{-1} & 2 & 3 \\ -1 & 3 & \boxed{-2} \end{pmatrix}$$

However, none of these circled numbers is the largest number in its column. It follows from Theorem 9.3 that this game has no saddle point, and so the methods of this section do not apply. In the remaining sections of this chapter you will learn how to find optimal strategies for games that do not have saddle points.

(b) The smallest number in each row is circled in the following matrix:

$$\begin{pmatrix} \boxed{0} & 2 & 3 \\ \boxed{0} & 3 & 4 \\ \boxed{-2} & 0 & 3 \end{pmatrix}$$

Of these circled numbers, both $a_{11} = 0$ and $a_{21} = 0$ are the largest numbers in their columns. These numbers, which are both saddle points, are circled twice. This matrix game therefore has two saddle points.

Since both saddle points are in the first column, Black's best strategy is to play column 1. Since one of the saddle points is in row 1 and the other is in row 2, Red has two optimal strategies, "play row 1" and "play row 2." Notice that, if Black uses his optimal strategy and plays column 1, it does not matter whether Red plays row 1 or row 2. In either case, the payoff is 0. Of course, Red will probably choose row 2 hoping that Black will make a mistake and let him win 4 dollars.

Example 9(b) shows that a matrix game may have more than one saddle point and that Red and Black may have more than one optimal strategy. Suppose that v and w are both saddle points of a game G. It can be shown that $v = w$ and that the payoff will be the same no matter which optimal strategy is chosen by Red and Black. (In fact, the payoff will be v.)

In examples 7 and 9(b), optimal strategies were found for Red and Black. If Red and Black both play their optimal strategies in these two games, Red will win 4 dollars and 0 dollars, respectively. Each of these numbers, 4 and 0, is called the *value of the game*. Clearly, the game of Example 7 is not *fair*; Black should insist that Red pay her 4 dollars for the privilege of playing. However, the game of Example 9(b) is fair: neither player wins or loses anything if each plays his best strategy.

Definition 9.4　If Red and Black both play their optimal strategies, the corresponding payoff is called the **value of the game**, denoted v. The game is called a **fair game** if and only if $v = 0$.

If a_{ij} is a saddle point of **A**, then Red's optimal strategy is to play row i and Black's optimal strategy is to play column j. The corresponding payoff will then be $v = a_{ij}$. In other words, *the value of the game is its saddle point* (if it has one).

9.3 EXERCISES

1.　Circle all saddle points of the following matrix games:

$$A = \begin{pmatrix} 1 & 2 & 3 \\ -4 & 1 & 5 \\ -2 & -1 & -3 \end{pmatrix} \qquad B = \begin{pmatrix} 2 & 1 & -4 \\ 0 & -2 & -3 \\ 1 & 3 & -3 \end{pmatrix}$$

$$C = \begin{pmatrix} 4 & 2 & 3 \\ -3 & -1 & 5 \end{pmatrix} \qquad D = \begin{pmatrix} 2 & 3 & 6 & 6 \\ 4 & 1 & -3 & 4 \\ 2 & 5 & -7 & 1 \end{pmatrix}$$

2.　Considering the following matrix game:

$$A = \begin{pmatrix} 0 & 2 & 1 \\ -1 & 4 & 3 \\ 1 & 2 & 1 \end{pmatrix}$$

(a) Find all saddle points of **A**.
(b) What is the value of the game?
(c) Is this game fair? Explain.
(d) What are the best strategies for Red and Black?
(e) If Red uses his optimal strategy, how much will be win?
(f) Find all equilibrium pairs.

3.　Answer the questions of Exercise 2 for the following matrix game:

$$\begin{pmatrix} -1 & 7 & -5 & -10 \\ 0 & 2 & 0 & 3 \\ 0 & 3 & 0 & 6 \\ -3 & 4 & -7 & 10 \end{pmatrix}$$

4. Consider the following matrix game:

$$
\begin{array}{c}
 & \begin{array}{cccc} b_1 & b_2 & b_3 & b_4 \end{array} \\
\begin{array}{c} r_1 \\ r_2 \\ r_3 \\ r_4 \end{array} &
\left(\begin{array}{cccc}
1 & 2 & 1 & 4 \\
-1 & 2 & -6 & -1 \\
1 & 3 & 1 & 5 \\
-4 & -1 & -3 & -7
\end{array}\right)
\end{array}
$$

Let (r_i, b_j) denote the fact that Red plays r_i and Black plays b_j.
   (a) Circle all saddle points.
   (b) Show that r_1 and r_3 are optimal strategies for Red, and that b_1 and b_3 are optimal strategies for Black.
   (c) Find the payoff if Red and Black choose the following strategies:
      (*i*) (r_1, b_1) (*ii*) (r_1, b_3) (*iii*) (r_3, b_3) (*iv*) (r_3, b_1).
   (d) What is the value of this game?

5. A manufacturer can buy an inexpensive component that costs 4 dollars, or a guaranteed component that costs 6 dollars. If the component does not go bad within one year, the manufacturer's only expense is the cost of the component. If, however, the component does go bad within the year, the manufacturer must pay 6 dollars more to replace the inexpensive part and 2 dollars more (for service) to replace the guaranteed part. This conflict situation is a matrix game with two players, the manufacturer and nature.
   (a) Give the matrix of this game.
   (b) Which component should the manufacturer buy?
   (c) If the manufacturer buys this component, what will be his cost?

6. Suppose that the manufacturer of Exercise 5 has a third option: he can purchase a third type of component for 9 dollars. If this component goes bad during the year, he will have no additional expense to replace the part, and his 9 dollars will be refunded. Adding this third option to the two described in Exercise 5, we have a new conflict situation between the manufacturer and nature. This conflict situation is a matrix game.
   (a) State the matrix for this game.
   (b) Does this game have a saddle point?

7. **(a)** Find optimal strategies for the gas station owners of Example 5.
   (b) Which investment should the woman of Exercise 6, Section 9.2, make?

9.4 MIXED STRATEGIES AND EXPECTED PAYOFFS

Many matrix games, such as those in Exercises 5–7 of the preceding section, can be played many times. In Section 9.3, we saw that, if a game has a saddle point, an optimal strategy for each player is to play a particular row or a particular column repeatedly. As mentioned before, this plan of action is called a *pure strategy*. Neither player will gain any advantage by using a *mixed strategy*; that is, by sometimes playing one row (or column) and sometimes playing another row (or column). In this section we will discuss games that do not have a saddle point. We will see that these games require a mixed strategy.

Consider the following matrix game:

$$
\begin{array}{c}
 \quad b_1 \quad b_2 \quad b_3 \\
\begin{array}{c} r_1 \\ r_2 \\ r_3 \end{array}
\left(\begin{array}{rrr}
1 & 2 & 4 \\
-1 & 3 & -2 \\
4 & 0 & -5
\end{array} \right)
\end{array}
\qquad (1)
$$

The reader can easily show that this game does not have a saddle point.

Suppose that Red wishes to use a pure strategy. He then has three choices: play r_1 repeatedly, play r_2 repeatedly, or play r_3 repeatedly.

(1) If he plays r_1 repeatedly, Black will soon notice what Red is doing and will then play b_1. Red will then notice what Black is doing, and Red will change his move to r_3. It follows that playing r_1 repeatedly is not Red's best strategy.

(2) Similarly, if Red decides to play r_2 repeatedly, Black will play b_3, and Red will change to r_1. It follows that playing r_1 repeatedly is not Red's best strategy either.

(3) The reader can similarly show that playing r_3 repeatedly is also not Red's best strategy.

It follows that, if Red wishes to win, he should not use a pure strategy. Instead, he must use a mixed strategy. In a similar way, it can be shown that Black must also use a mixed strategy.

This result is true in general.

THEOREM 9.5   If **A** is any matrix game that does not have a saddle point, then both Red and Black must use a mixed strategy.

Remember that we are assuming in this theorem—and in the entire chapter—that both Red and Black wish to win the most of whatever is at stake—money, happiness, sales of their product, or whatever.

Returning to the particular matrix game (1) that we have been discussing, Red now realizes that he should not play any one of his moves continuously. He might therefore arbitrarily decide to randomly play r_1 50 percent of the time, r_2 30 percent of the time, and r_3 20 percent of the time. This can be accomplished by using a card deck consisting of five hearts, three clubs, and two spades. If after the deck is shuffled a heart is the top card, r_1 is played; if a club is the top card, r_2 is played; and if a spade is the top card, r_3 is played. The associated probabilities, 0.5, 0.3, and 0.2, are conveniently displayed in a 1×3 row vector. This vector,

$$(0.5, 0.3, 0.2)$$

is called a *mixed strategy* for Red. For reasons that will become apparent later, Black's strategies are displayed in a 3×1 column vector. For example, if Black chooses to play b_1 40 percent of the time, b_2 10 percent of the time, and b_3 50 percent of the time, the mixed strategy for Black is

$$\begin{pmatrix} 0.4 \\ 0.1 \\ 0.5 \end{pmatrix}$$

Notice that

$$\begin{pmatrix} 0.4 \\ 0.1 \\ 0.5 \end{pmatrix} \neq \begin{pmatrix} 0.1 \\ 0.4 \\ 0.5 \end{pmatrix}$$

The second mixed strategy means that Black chooses b_1 10 percent of the time, b_2 40 percent of the time, and b_3 50 percent of the time.

The following is a general definition of mixed strategies:

Definition 9.5 Let **A** be a matrix game with m rows and n columns.

(a) A $1 \times m$ probability vector

$$\mathbf{R} = (p_1, p_2, \ldots, p_m)$$

is called a **strategy** *for Red*. If none of the components p_i is 1, then **R** is a **mixed strategy** *for Red*; if exactly one of $p_1, p_2, \ldots,$ or p_m is 1, then **R** is a **pure strategy** *for Red*.

(b) A $n \times 1$ probability vector

$$\mathbf{B} = \begin{pmatrix} p_1 \\ p_2 \\ \vdots \\ p_n \end{pmatrix}$$

is called a **strategy** *for Black*. If none of the components p_i is 1, then **B** is a **mixed strategy** *for Black*; if exactly one of $p_1, p_2, \ldots,$ or p_n is 1, then **B** is a **pure strategy** *for Black*.

Notice that, if **A** has m rows and n columns, then strategies for Red have m components and strategies for Black have n components. Notice also that Red's strategies are row vectors while Black's strategies are column vectors.

Example 10 illustrates mixed strategies.

Example 10 Let

$$\mathbf{A} = \begin{pmatrix} 4 & -1 & 2 \\ 0 & 1 & -1 \end{pmatrix}$$

be a matrix game. Which of the following vectors are (*a*) mixed strategies for Red, (*b*) mixed strategies for Black, (*c*) pure strategies for Red?

$$\mathbf{C} = (0.1, 0.9), \qquad \mathbf{D} = (0.7, 0.4), \qquad \mathbf{E} = (0.1, 0.3, 0.6)$$

$$\mathbf{F} = \begin{pmatrix} 0.2 \\ 0.7 \\ 0.1 \end{pmatrix}, \qquad \mathbf{G} = (0, 1)$$

Solution

(a) Applying Definition 9.5, a vector is a mixed strategy for Red if (1) it is a row vector with two components, (2) it is a probability vector, and (3) none of the components is 1. The vector **C** is the only mixed strategy for Red because it is the only vector that satisfies these three criteria. The vector **D** is not a mixed strategy for Red because it is not a probability vector; **E** is not a mixed strategy for Red because it has three components instead of two. The vector **F** is not a mixed strategy for Red because it is not a row vector; **G** is not a mixed strategy for Red because one of its components is 1.

(b) Applying Definition 9.5, a vector is a mixed strategy for Black if (1) it is a column vector with three components, (2) it is a probability vector, and (3) none of its components is 1. The vector **F** is the only mixed strategy for Black because it is the only vector that satisfies these three criteria. None of the other vectors are column vectors.

(c) A vector is a pure strategy for Red if (1) it is a probability vector, (2) it is a row vector with two components, and (3) one of the components is 1. Since the vector is a two-component probability vector, the last requirement implies that the component that is not 1 must be 0. Therefore, the only pure strategy for Red is **G**, because it is the only vector that satisfies these criteria. The other vectors are not pure strategies because they do not have 1 as one of their components.

Consider the matrix game discussed in the beginning of this section:

$$
\begin{array}{c}
\quad b_1 \quad b_2 \quad b_3 \\
\begin{array}{c} r_1 \\ r_2 \\ r_3 \end{array}
\left(
\begin{array}{ccc}
1 & 2 & 4 \\
-1 & 3 & -2 \\
4 & 0 & -5
\end{array}
\right)
\end{array}
$$

The preceding discussion has shown that Red cannot use a pure strategy, but must use a mixed strategy instead. We must still decide which mixed strategy is best for Red. Before we can make this decision, we must see what happens when Red uses a particular strategy **R** and Black uses a particular strategy **B**.

First let us consider what happens when Red and Black each use a pure strategy. For example, suppose that Red plays r_1 repreatedly and Black plays b_2 repeatedly. In this case, the payoff is the entry in the intersection of the row labeled r_1 and the column labeled b_2. The payoff will *always* be 2.

Now suppose that Red and Black each use a mixed strategy. For example, suppose that Red plays r_1 10 percent of the time and r_2 90 percent of the time (the mixed strategy $(0.1,0.9,0)$), and suppose that Black plays b_1 40 percent of the time and b_3 60 percent of the time (the mixed strategy $\begin{pmatrix} 0.4 \\ 0 \\ 0.6 \end{pmatrix}$). In this case the payoff will sometimes be 1 (when Red plays r_1 and Black plays b_1), sometimes be 4 (when Red plays r_1 and Black plays b_3), and sometimes be -1 or -2. Because we do not know what the payoff will be any particular time that the game is played, we must find what the *average payoff*, or **expected payoff**, will be when the game is played many times.

The following example shows how the expected payoff can be determined.

Example 11   Let

$$b_1 \ b_2 \quad b_3$$

$$A = \begin{array}{c} r_1 \\ r_2 \\ r_3 \end{array} \begin{pmatrix} 1 & 2 & 4 \\ -1 & 3 & -2 \\ 4 & 0 & -5 \end{pmatrix}$$

be the matrix game discussed previously, and suppose that Red and Black choose to play the mixed strategies

$$R = (0.1, 0.9, 0) \quad \text{and} \quad B = \begin{pmatrix} 0.4 \\ 0 \\ 0.6 \end{pmatrix}$$

What is the expected payoff of this game?

Solution   Let X be the payoff and let R_i and B_j be the events "Red plays r_i" and "Black plays b_j," respectively. We wish to calculate $E(X)$.

Since Red plays r_1 and r_2 and Black plays b_1 and b_3, X can be 1, 4, -1, or -2. Now, $X = 1$, if and only if $R_1 B_1$ occurs. Since we naturally assume that the players choose their strategies independently,

$$P(X = 1) = P(R_1 B_1) = P(R_1)P(B_1) = (0.1)(0.4) = 0.04$$

Similarly,

$$P(X = 4) = P(R_1 B_3) + P(R_3 B_1) = P(R_1)P(B_3) + P(R_3)(B_1)$$
$$= (0.1)(0.6) + 0(0.4) = 0.06$$

The remaining values in the following table are calculated similarly.

x	1	4	-1	-2
$P(X = x)$	0.04	0.06	0.36	0.54

Multiplying each entry in row 1 of the above table by the corresponding entry in row 2 and then adding the products yields

$$E(X) = 1(0.04) + 4(0.06) + (-1)(0.36) + (-2)(0.54) = -1.16$$

The expected payoff is therefore -1.16. This means that, if Red uses strategy **R** and Black uses strategy **B**, then in the long run Red will lose about $1.16 per game.

Let **R**, **A**, and **B** be the matrices of Example 11 and let X be the payoff if strategies **R** and **B** are used. If we calculate the matrix product

$$\mathbf{RAB} = (\mathbf{RA})\mathbf{B} = (0.1,0.9,0) \begin{pmatrix} 1 & 2 & 4 \\ -1 & 3 & -2 \\ 4 & 0 & -5 \end{pmatrix} \begin{pmatrix} 0.4 \\ 0 \\ 0.6 \end{pmatrix}$$

$$= (-0.8, 2.9, -1.4) \begin{pmatrix} 0.4 \\ 0 \\ 0.6 \end{pmatrix} = -1.16$$

we see that $\mathbf{RAB} = E(X)$. This result holds for any matrix game **A** and any strategies **R** and **B**.

THEOREM 9.6   Let **A** be a matrix game and let X be the payoff if Red and Black use strategies **R** and **B**, respectively. Then

$$E(X) = \mathbf{RAB}$$

Notice that the matrix multiplication of Theorem 9.6 is not defined if Black's strategy **B** is written as a row vector instead of a column vector.

In particular, suppose that **R*** and **B*** are optimal strategies for Red and Black. Then, extending Definition 9.4 to mixed strategies, the value of the game, v, is the corresponding expected payoff. It then follows from Theorem 9.6 that

$$v = \mathbf{R^*AB^*}$$

(2)

The following example is an application of Theorem 9.6.

Example 12   Two stores, Y and Z, are the only supermarkets in a certain town. Each store is considering sharply reducing its prices on one of the following items: eggs, coffee, ice cream. The following matrix shows the percentage of shoppers who will shop in store Y if store Y reduces the price of item y and store Z reduces the price of item z. The remaining shoppers will shop in store Z.

Store Z

		Eggs	Coffee	Ice cream
	Eggs	40	30	20
Store Y	Coffee	60	40	50
	Ice cream	70	50	40

For example, if stores Y and Z reduce the prices of ice cream and eggs, respectively, then 70 percent of the shoppers will shop in store Y and the remaining 30 percent of the shoppers will shop in store Z. Find the average percentage of shoppers that will shop in store Y if—

(a) store Y reduces the prices of eggs, coffee, and ice cream 40 percent, 30 percent, and 30 percent of the time, respectively, and store Z reduces the price of eggs, coffee, and ice cream 20 percent, 50 percent, and 30 percent of the time, respectively.

(b) store Y reduces the price of eggs, coffee, and ice cream 10 percent, 60 percent, and 30 percent of the time, respectively, and store Z reduces the price of eggs, coffee, and ice cream 40 percent, 10 percent, and 50 percent of the time, respectively.

Solution   First notice that the matrix for this game does not have a saddle point, and so neither store Y nor store Z should use a pure strategy.

(a) In this case stores Y and Z wish to use strategies

$$\mathbf{R} = (0.4,0.3,0.3) \quad \text{and} \quad \mathbf{B} = \begin{pmatrix} 0.2 \\ 0.5 \\ 0.3 \end{pmatrix}$$

respectively. Let X be the payoff if store Y uses $\mathbf{R}$ and store Z uses $\mathbf{B}$. We must find $E(X)$. Applying Theorem 9.6,

$$E(X) = \mathbf{RAB} = (\mathbf{RA})\mathbf{B} = (0.4,0.3,0.3) \begin{pmatrix} 40 & 30 & 20 \\ 60 & 40 & 50 \\ 70 & 50 & 40 \end{pmatrix} \mathbf{B}$$

$$= (55,39,35) \begin{pmatrix} 0.2 \\ 0.5 \\ 0.3 \end{pmatrix} = 41$$

It follows that, on the average, 41 percent of the shoppers will shop in store Y.

(b) In this case stores Y and Z wish to use strategies

$$\mathbf{R} = (0.1,0.6,0.3) \quad \text{and} \quad \mathbf{B} = \begin{pmatrix} 0.4 \\ 0.1 \\ 0.5 \end{pmatrix}$$

respectively. Let X be the payoff if store Y uses $\mathbf{R}$ and store Z uses $\mathbf{B}$. Applying Theorem 9.6,

$$E(X) = \mathbf{RAB} = (\mathbf{RA})\mathbf{B} = (0.1,0.6,0.3) \begin{pmatrix} 40 & 30 & 20 \\ 60 & 40 & 50 \\ 70 & 50 & 40 \end{pmatrix} \mathbf{B}$$

$$= (61,42,44) \begin{pmatrix} 0.4 \\ 0.1 \\ 0.5 \end{pmatrix} = 50.6$$

Hence, on the average, 50.6 percent of the shoppers will shop in store Y.

Of the two pairs of strategies discussed in Example 12, the first pair is better for store Z and the second pair is better for store Y. However, we still do not know how to find the *best* strategies for stores Y and Z. Methods for finding optimal strategies for games that do not have saddle points will be developed in the remainder of this chapter.

9.4 EXERCISES

1.   Let

$$\begin{array}{ccc} & b_1 & b_2 & b_3 \end{array}$$
$$A = \begin{array}{c} r_1 \\ r_2 \end{array} \begin{pmatrix} 0.4 & 0.1 & 0.7 \\ 0.6 & 0.5 & 0.4 \end{pmatrix}$$

be a matrix game. Write the following strategies in vector form.
(a) Red will use r_1 10 percent of the time and r_2 90 percent of the time.
(b) Red will only use r_1.
(c) Black will use b_1 10 percent of the time, b_2 40 percent of the time, and b_3 50 percent of the time.

2.   Let A be the matrix game of Exercise 1. Give a verbal description of the following strategies:

(a) $(0.3, 0.7)$ (b) $(0,1)$ (c) $\begin{pmatrix} 1 \\ 0 \\ 0 \end{pmatrix}$ (d) $\begin{pmatrix} 0.4 \\ 0.5 \\ 0.1 \end{pmatrix}$

3.   Let

$$\begin{array}{ccc} & b_1 & b_2 & b_3 \end{array}$$
$$\begin{array}{c} r_1 \\ r_2 \end{array} \begin{pmatrix} 11 & 20 & -1 \\ 3 & 4 & -9 \end{pmatrix}$$

be a matrix game. Which of the following vectors are mixed strategies for Red?

$A = (0.1, 0.9),$ $B = (0.2, 0.4, 0.4),$ $C = (0.3, 0.8),$ $D = (-1, 2)$

$E = (0.4, 0.3, 0.4),$ $F = (1, 0),$ $G = \begin{pmatrix} 0.1 \\ 0.9 \end{pmatrix}$

4.   Let

$$A = \begin{pmatrix} 1 & 2 & -1 \\ 0 & -1 & -4 \end{pmatrix}$$

(a) What is the expected payoff if the following pairs of "strategies" are used?

(i) $R = (0.1, 0.9),$ $B = \begin{pmatrix} 0.4 \\ 0 \\ 0.6 \end{pmatrix}$

(ii) $R = (0.5, 0.5),$ $B = \begin{pmatrix} 1 \\ 0 \\ 0 \end{pmatrix}$

(iii) $R = (0.4, 0.3, 0.3),$ $B = \begin{pmatrix} 0.1 \\ 0.4 \\ 0.5 \end{pmatrix}$

(b) Of the preceding pairs of strategies, which is best for Red and which is best for Black?

5. Let **A** be the matrix game of Exercise 4. If Red uses the following mixed strategies, what percentage of the time will he be using row 1?
   (a) $(0.4, 0.6)$ **(b)** $(0.3, 0.7)$ **(c)** $(0, 1)$

6. A farmer wishes to plant lettuce, tomatoes, or melons. His profit (in thousands of dollars) for each crop if the rainfall is light, average, or heavy is given in the following matrix:

$$
\begin{array}{c c c c}
 & \text{Light} & \text{Average} & \text{Heavy} \\
\text{Lettuce} & 10 & 5 & 2 \\
\text{Tomatoes} & 8 & 9 & 11 \\
\text{Melons} & 4 & 7 & 12 \\
\end{array}
$$

It is known that the rainfall in the farmer's area is light 30 percent of the time, average 50 percent of the time, and heavy 20 percent of the time.

(a) What will be the farmer's average profit if he—

   (*i*) plants lettuce every year

   (*ii*) plants tomatoes every year

   (*iii*) plants lettuce 60 percent of the time and tomatoes 40 percent of the time

   (*iv*) plants lettuce, tomatoes, and melons 10 percent, 40 percent, and 50 percent of the time, respectively.

   (Hint: Let **A** be the payoff matrix and let **B** be nature's strategy. Show that $\mathbf{B} = \begin{pmatrix} 0.3 \\ 0.5 \\ 0.2 \end{pmatrix}$ and then calculate **AB**.)

(b) Which of these four strategies is best for the farmer?

7. Find the best strategy for the farmer of Exercise 6. (Hint: Show that the farmer's best strategy is a pure strategy because (1) nature's strategy is known, and (2) nature will not change its strategy when it finds out what the farmer is doing.)

9.5 SOLUTION OF 2 × 2 GAMES

In Section 9.4 we decided that games with no saddle point require mixed strategies. Although we were able to use Theorem 9.6 to calculate the expected payoff of a particular game given a particular pair of mixed strategies, we were not able to find the best strategy for each player. In this section we will develop a method for finding the optimal strategies for any 2 × 2 game (i.e., any game with 2 rows and 2 columns). In Section 9.6 we will show that many larger games can be reduced in size to a 2 × 2 game and then solved by the methods of this section.

We will begin our study of 2 × 2 games by stating a theorem that tells how to solve every 2 × 2 game that does not have a saddle point. Although the proof of this theorem is beyond the scope of this text, we will show by example that the results it gives agree with those we might arrive at through trial-and-error calculations.

THEOREM 9.7 Let

$$A = \begin{pmatrix} a & b \\ c & d \end{pmatrix}$$

be any 2×2 matrix game that has no saddle point; and let

$$e = a + d - b - c$$

Furthermore, let

$$p_1 = \frac{d - c}{e}, \qquad p_2 = \frac{a - b}{e}, \qquad q_1 = \frac{d - b}{e}, \qquad q_2 = \frac{a - c}{e}$$

and

$$v = \frac{ad - bc}{e}$$

Then v is the value of the game, and

$$\mathbf{R^*} = (p_1, p_2) \qquad \text{and} \qquad \mathbf{B^*} = \begin{pmatrix} q_1 \\ q_2 \end{pmatrix}$$

are optimal strategies for Red and Black, respectively.

It can be shown that, if a 2×2 matrix $\mathbf{A}$ has no saddle point, then the number e defined in Theorem 9.7 is not zero. It then follows that $\mathbf{R^*}$, $\mathbf{B^*}$, and v of Theorem 9.7 are all well-defined. (Remember that division by zero is not defined.)

The following example shows how to apply Theorem 9.7.

Example 13 Consider the coin-tossing game of Example 3, page 380. The matrix representation of this game is

$$\begin{array}{cc} & \begin{array}{cc} h & t \end{array} \\ \begin{array}{c} h \\ t \end{array} & \begin{pmatrix} -1 & 1 \\ 1 & -1 \end{pmatrix} \end{array}$$

Find optimal strategies for Red and Black.

Solution The reader can easily show that this is a 2×2 matrix game with no saddle point. It follows that Theorem 9.7 can be applied, with $a = -1$, $b = 1$, $c = 1$, and $d = -1$. Hence,

$$e = a + d - b - c = -1 + (-1) - 1 - 1 = -4$$

and so

$$p_1 = \frac{d - c}{e} = \frac{-1 - 1}{-4} = \frac{1}{2}, \qquad p_2 = \frac{a - b}{e} = \frac{-1 - 1}{-4} = \frac{1}{2}$$

$$q_1 = \frac{d - b}{e} = \frac{-1 - 1}{-4} = \frac{1}{2}, \qquad q_2 = \frac{a - c}{e} = \frac{-1 - 1}{-4} = \frac{1}{2}$$

and

$$v = \frac{ad - bc}{e} = \frac{(-1)(-1) - 1(1)}{-4} = 0$$

It follows that optimal strategies for Red and Black are

$$\mathbf{R}^* = (p_1, p_2) = (\tfrac{1}{2}, \tfrac{1}{2}) \quad \text{and} \quad \mathbf{B}^* = \begin{pmatrix} \tfrac{1}{2} \\ \tfrac{1}{2} \end{pmatrix}$$

and the value of the game is $v = 0$. (This game is fair.)

Suppose that the player Red in Example 13 does not know about Theorem 9.7. How might he go about finding an optimal strategy? He might try a trial-and-error approach: he would calculate the expected payoffs of a number of different strategies and then try to decide which strategy is best. Let's follow through on this approach and see if the results obtained agree with those calculated using Theorem 9.7.

Suppose, for example that Red decides to use the pure strategy (1,0). Black will soon notice that Red is playing heads continuously. She will therefore also play heads, and Red will lose 1 dollar on each game.

Since Red does not wish to lose a dollar a game if he can avoid it, he might decide to use the mixed strategy (0.9,0.1). Black will respond by using some strategy $\begin{pmatrix} x \\ y \end{pmatrix}$. (Since Red does not know Black's strategy, he only knows that x and y can be any two numbers, as long as $x + y = 1$, $0 \le x \le 1$, and $0 \le y \le 1$.) Using Theorem 9.6, we can calculate the expected payoff:

$$E(X) = (0.9, 0.1) \begin{pmatrix} -1 & 1 \\ 1 & -1 \end{pmatrix} \begin{pmatrix} x \\ y \end{pmatrix} = (-0.8, 0.8) \begin{pmatrix} x \\ y \end{pmatrix} = -(0.8)x + (0.8)y$$

The worst possible expected payoff will occur when $y = 0$, because this value of y will make $E(X)$ as negative as possible. Since $x + y = 1$, x will then be $1 - 0 = 1$. Therefore, the worst possible payoff for Red will occur if Black uses the pure strategy $\begin{pmatrix} 1 \\ 0 \end{pmatrix}$. In that case, the payoff will be -0.8.

Therefore, if Red plays (0.9,0.1), he can lose as much as approximately 80 cents per game. (If Red is lucky and Black does not play $\begin{pmatrix} 1 \\ 0 \end{pmatrix}$, Red may lose less than 80 cents per game.)

Red does not wish to risk losing 80 cents per game if he can avoid it. Therefore, he might consider using the strategy (0.8,0.2) instead of (0.9,0.1). Using the same method as before, the reader can show that, if Red uses the strategy (0.8,0.2), he risks losing approximately 60 cents per game. This worst expected payoff will occur if Black uses the pure strategy $\begin{pmatrix} 1 \\ 0 \end{pmatrix}$.

In general, suppose that Red considers using the strategy (a,b), where a and b are any two numbers satisfying $0 \le a \le 1$, $0 \le b \le 1$, and $a + b = 1$. One can use reasoning similar to that of the preceding paragraphs to show that, if $a \ge b$, Red risks losing $a - b$ dollars per game. This loss will occur if Black uses the pure strategy $\begin{pmatrix} 1 \\ 0 \end{pmatrix}$. If

$a \leq b$, Red risks losing $b - a$ dollars per game. This loss will occur if Black uses the pure strategy $\binom{0}{1}$.

For example, if Red uses the strategy $(0.4, 0.6)$, he risks losing $0.6 - 0.4 = 0.2$ dollars; if he uses the strategy $(0.7, 0.3)$, he risks losing $0.7 - 0.3 = 0.4$ dollars. It therefore follows that, if $a \neq b$, Red risks losing some money. However, if $a = b$, Red risks losing $a - b = a - a = 0$ dollars. Therefore, Red's best strategy is (a,b) where $a = b$. Since $a + b = 1$, and both a and b are numbers between 0 and 1, it follows that $a = b = 0.5$. That is, Red's best strategy is

$$\mathbf{R}^* = (0.5, 0.5)$$

If Red uses this strategy, the worst expected payoff that can occur is 0.

Using similar logic, we can show that Black's best strategy is

$$\mathbf{B}^* = \binom{0.5}{0.5}$$

Notice that these results for $\mathbf{R}^*$ and $\mathbf{B}^*$ agree with the results we calculated in Example 13.

We must still check the value v of the game. Using equation (2), we have

$$v = \mathbf{R}^*\mathbf{AB}^* = (0.5, 0.5) \begin{pmatrix} -1 & 1 \\ 1 & -1 \end{pmatrix} \binom{0.5}{0.5} = (0,0) \binom{0.5}{0.5} = 0$$

This agrees with the result we calculated using Theorem 9.7.

Thus, in this example the results we obtained by using Theorem 9.7 agree with those we obtained using a trial-and-error approach. It is also clear that Theorem 9.7 is a much quicker and more efficient way to calculate optimal strategies than is trial and error. In the following examples we will show how Theorems 9.3, 9.4, and 9.7 can be used to find optimal strategies for any 2×2 game.

Example 14   Ralph suspects that his French teacher may give a quiz tomorrow. He must decide whether he should go to the movies or stay home and study. The following matrix gives his payoffs in satisfaction.[1]

$$
\begin{array}{cc}
& \text{Quiz} \quad \text{No quiz} \\
\begin{matrix} \text{Movies} \\ \text{Study} \end{matrix} & \begin{pmatrix} -3 & 4 \\ 3 & 1 \end{pmatrix}
\end{array}
$$

What should Ralph do?

Solution   Applying Theorem 9.3, this matrix game has no saddle point. It follows that Theorem 9.7 can be applied with $a = -3$, $b = 4$, $c = 3$, and $d = 1$. Then

$$e = a + d - b - c = -3 + 1 - 4 - 3 = -9$$

---

1. Payoffs that represent intensity of feelings such as satisfaction are called *utilities*. This concept was discussed in Example 5, page 382.

and so

$$p_1 = \frac{d-c}{e} = \frac{1-3}{-9} = \frac{2}{9}, \qquad p_2 = \frac{a-b}{e} = \frac{-3-4}{-9} = \frac{7}{9}$$

$$q_1 = \frac{d-b}{e} = \frac{1-4}{-9} = \frac{3}{9}, \qquad q_2 = \frac{a-c}{e} = \frac{-3-3}{-9} = \frac{6}{9}$$

and

$$v = \frac{ad-bc}{e} = \frac{(-3)(1)-4(3)}{-9} = \frac{15}{9}$$

Hence,

$$\mathbf{R}^* = (\tfrac{2}{9}, \tfrac{7}{9}), \qquad \mathbf{B}^* = \begin{pmatrix} \tfrac{3}{9} \\ \tfrac{6}{9} \end{pmatrix}$$

and $v = \frac{15}{9}$.

Ralph's best strategy, therefore, is to go to the moves two-ninths of the times that this situation occurs and study the remaining times.

Example 15   Mr. Red and Dr. Black are the leading contenders for school board president. They have both been invited to a series of neighborhood discussions about the major issues. Both candidates can tailor their speeches so as to appeal either to the younger voters or to the older voters. The following matrix gives the payoffs in the percentage of voters who will vote for Mr. Red. The remaining voters will vote for Dr. Black. What strategies should Mr. Red and Dr. Black use?

Dr. Black

Younger   Older

$$\text{Mr. Red} \quad \begin{array}{c} \text{Younger} \\ \text{Older} \end{array} \begin{pmatrix} 40 & 60 \\ 20 & 70 \end{pmatrix}$$

Solution   It follows from Theorem 9.3 that 40 is a saddle point of this game. Hence, Theorem 9.7 does not apply. However, since 40 is a saddle point, parts (b) and (c) of Theorem 9.4 imply that Mr. Red should always play the row containing 40, and Dr. Black should always play the column containing 40. In other words, both Mr. Red and Dr. Black should tailor their speeches to appeal to the younger voters. If they do this, Mr. Red will receive 40 percent of the votes and Dr. Black will receive 60 percent of the votes.

9.5 EXERCISES

Answer the following questions for each of the matrix games in Exercises 1–3.
  (a) Find optimal strategies for Red and Black.
  (b) Find the value of the game.

(c) How should Red play? (Use words to describe Red's strategy.)

(d) If Red and Black use their optimal strategies, what will happen in the long run?

1. $\begin{pmatrix} 1 & -5 \\ 0 & 8 \end{pmatrix}$ **2.** $\begin{pmatrix} -1 & 3 \\ 4 & 7 \end{pmatrix}$ **3.** $\begin{pmatrix} 0 & 3 \\ 11 & 2 \end{pmatrix}$

4. Red and Black simultaneously show either three or four fingers. If the sum of the fingers shown is odd, Red pays this sum to Black. If the sum is even, Black pays this sum to Red.

(a) Represent this game by a matrix.

(b) Find optimal strategies for Red and Black.

(c) Is this game fair? Explain.

5. Find optimal strategies for the bombers and the carriers of Exercise 5, Section 9.2. If both sides use their optimal strategies, what is the average number of ships that will be sunk?

6. A police officer has just lost sight of the suspect he is chasing. The officer knows that the suspect turned either left or right at the corner. The probabilities that the officer will catch the suspect are given in the following matrix:

$$\begin{array}{cc} & \text{Suspect} \\ & \begin{array}{cc} \text{Left} & \text{Right} \end{array} \\ \text{Officer} \begin{array}{c} \text{Left} \\ \text{Right} \end{array} & \begin{pmatrix} 0.5 & 0.6 \\ 0.4 & 0.7 \end{pmatrix} \end{array}$$

(a) What should the police officer do?

(b) If the officer uses his optimal strategy, what is the probability that he will catch the suspect?

9.6 RECESSIVE ROWS AND COLUMNS

In this section we will see that many larger games can be reduced to 2×2 games. Optimal strategies can then be determined by applying Theorems 9.3, 9.4, and 9.7. The following examples illustrate the method used to reduce a matrix game to a smaller one.

Example 16 Let

$$\mathbf{A} = \begin{array}{c} \\ r_1 \\ r_2 \\ r_3 \end{array} \begin{array}{c} \begin{array}{cc} b_1 & b_2 \end{array} \\ \begin{pmatrix} 3 & -2 \\ 2 & 3 \\ 1 & -10 \end{pmatrix} \end{array}$$

be a matrix game. Find optimal strategies for Red and Black and the value of the game.

Solution   This matrix game cannot be solved by previous methods because it is not 2 × 2 and it does not have a saddle point. However, consider the entries in rows 1 and 3 of **A**. *Each entry in row 1 is greater than or equal to the corresponding entry in row 3*; that is, $3 \geq 1$ and $-2 \geq -10$. Clearly, each payoff associated with row 1 is better (from Red's point of view) than the corresponding payoff associated with row 3. Therefore, row 1 is said to *dominate* row 3. Red would be foolish to even consider using row 3. It follows that row 3 can be eliminated from the game. Crossing out row 3 yields a new matrix $\mathbf{A}_1$:

$$\mathbf{A} = \begin{array}{c} \\ r_1 \\ r_2 \\ r_3 \end{array} \begin{array}{cc} b_1 & b_2 \\ \left(\begin{array}{cc} 3 & -2 \\ 2 & 3 \\ 1 & -10 \end{array} \right) \end{array}, \qquad \mathbf{A}_1 = \begin{array}{c} \\ r_1 \\ r_2 \end{array} \begin{array}{cc} b_1 & b_2 \\ \left(\begin{array}{cc} 3 & -2 \\ 2 & 3 \end{array} \right) \end{array}$$

Theorem 9.7 can be applied to $\mathbf{A}_1$ because $\mathbf{A}_1$ is a 2 × 2 matrix game with no saddle point. The value of $\mathbf{A}_1$ will be exactly the same as the value of **A**. The optimal strategies for **A** can be quickly found from the optimal strategies for $\mathbf{A}_1$ by using the fact that Red never uses r_3.

Applying Theorem 9.7 with $a = 3$, $b = -2$, $c = 2$, and $d = 3$ yields

$$e = 6, \qquad p_1 = \tfrac{1}{6}, \qquad p_2 = \tfrac{5}{6}$$

$$q_1 = \tfrac{5}{6}, \qquad q_2 = \tfrac{1}{6}, \qquad \text{and} \qquad v = \tfrac{13}{6}$$

It follows that the value of the game $\mathbf{A}_1$ is $\tfrac{13}{6}$ and optimal strategies for $\mathbf{A}_1$ are

$$\mathbf{R}_1{}^* = (\tfrac{1}{6}, \tfrac{5}{6}) \qquad \text{and} \qquad \mathbf{B}_1{}^* = \begin{pmatrix} \tfrac{5}{6} \\ \tfrac{1}{6} \end{pmatrix}$$

Since Red never uses r_3, the optimal strategies for **A** are

$$\mathbf{R}^* = (\tfrac{1}{6}, \tfrac{5}{6}, 0) \qquad \text{and} \qquad \mathbf{B}^* = \begin{pmatrix} \tfrac{5}{6} \\ \tfrac{1}{6} \end{pmatrix}$$

The value of the game **A** is $\tfrac{13}{6}$.

Example 17   The matrix representation of the word game of Exercise 2, Section 9.2, is given below.

$$\mathbf{W} = \begin{array}{c} \\ A \\ O \end{array} \begin{array}{ccc} F & M & N \\ \left(\begin{array}{ccc} -1 & -2 & -1 \\ -1 & 3 & -2 \end{array} \right) \end{array}$$

Find the optimal strategies for Red and Black and the value of the game.

Solution   This game, like the game in Example 16, cannot be solved by previous methods because it does not have a saddle point and it is not a 2 × 2 game. However, consider columns 1 and 3 of **W**. *Each entry in column 3 is less than or equal to the corresponding entry in column 1*; that is, $-1 \leq -1$ and $-2 \leq -1$.

Since Black wishes to make the payoffs as small as possible, each entry in column 3 is better (from Black's point of view) than the corresponding entry in column 1. Column 3 is therefore said to *dominate* column 1. Since Black would be foolish to even consider using column 1, this column can be eliminated from the game. Crossing out column 1 yields a new matrix $\mathbf{W}_1$:

$$\mathbf{W} = \begin{matrix} & F & M & N \\ A & \\ O \end{matrix} \left(\begin{matrix} 1 & -2 & -1 \\ 1 & 3 & -2 \end{matrix} \right), \qquad \mathbf{W}_1 = \begin{matrix} & M & N \\ A \\ O \end{matrix} \left(\begin{matrix} -2 & -1 \\ 3 & -2 \end{matrix} \right)$$

Since $\mathbf{W}_1$ is a 2×2 matrix with no saddle point, Theorem 9.7 applies. Using this theorem with $a = -2$, $b = -1$, $c = 3$, and $d = -2$ yields

$$e = -6, \qquad p_1 = \tfrac{5}{6}, \qquad p_2 = \tfrac{1}{6}$$

$$q_1 = \tfrac{1}{6}, \qquad q_2 = \tfrac{5}{6}, \qquad v = \tfrac{7}{6}$$

It follows that the optimal strategies for $\mathbf{W}_1$ are

$$\mathbf{R}_1{}^* = (\tfrac{5}{6}, \tfrac{1}{6}) \qquad \text{and} \qquad \mathbf{B}_1{}^* = \begin{pmatrix} \tfrac{1}{6} \\ \tfrac{5}{6} \end{pmatrix}$$

The value of $\mathbf{W}_1$ is $v = \tfrac{7}{6}$.

Since Black will never play F, optimal strategies for $\mathbf{W}$ are

$$\mathbf{R}^* = (\tfrac{5}{6}, \tfrac{1}{6}) \qquad \text{and} \qquad \mathbf{B}^* = \begin{pmatrix} 0 \\ \tfrac{1}{6} \\ \tfrac{5}{6} \end{pmatrix}$$

The value of $\mathbf{W}$ is $v = \tfrac{7}{6}$. Red should therefore play A five-sixths of the time and O one-sixth of the time. Black should play M one-sixth of the time, N five-sixths of the time, and F never.

In general, we have the following definition:

Definition 9.6   Let $\mathbf{A} = (a_{ij})$ be an $m \times n$ matrix.

(a)   Row i *dominates* row j if $a_{ik} \geq a_{jk}$ for all k. If row i dominates row j, row j is called a **recessive row**.

(b)   Column i *dominates* column j if $a_{ki} \leq a_{kj}$ for all k. If column i dominates column j, column j is called a **recessive column**.

In other words, row i dominates row j if each entry of row i is *greater than or equal to* the corresponding entry of row j. However, column i dominates j if each entry of column i is *less than or equal to* the corresponding entry of column j.

Examples 16 and 17 show that a recessive row or a recessive column can be deleted from a payoff matrix. Example 18 shows that this can be done several times. However, *before deleting any rows or columns from a matrix* $\mathbf{A}$, *one should always be certain that* $\mathbf{A}$ *does not have a saddle point*. If $\mathbf{A}$ has a saddle point point a_{ij}, then Red should always play row i, Black should always play column j, and the value of the game is a_{ij}. Therefore, deleting recessive rows and columns from games that have saddle points is unnecessary.

Example 18 In the game Master Mind, the code maker, Red, must insert colored pegs into four hidden holes. The code breaker, Black, must then guess the color of the peg that is in each hole. If Black does not guess all the pegs correctly, Red scores a point. Red then tells Black how many pegs are correct (but not which pegs are correct), and Black must guess again. Black keeps guessing and Red keeps scoring points until Black guesses all the pegs correctly.

Red may use one, two, three, or four colors in his code. Red feels that the following matrix gives his average score if he uses x colors in his code and Black initially guesses that Red is using y colors.

Black

	1 color	2 colors	3 colors	4 colors
1 color	3	4	4	4
2 colors	4	5	5	6
3 colors	7	6	6	7
4 colors	7	8	7	6

$$A = \text{Red}$$

For example, if Red uses four colors, his average score will be 7 if Black guesses that Red is using one color, 8 if Black guesses that Red is using two colors, and 7 if Black guesses that he is using three colors.

Find optimal strategies for Red and Black. If Red and Black use their optimal strategies, what will Red's average score be?

Solution Since A does not have a saddle point and is not 2×2, we must try to remove recessive rows and columns. The first row is recessive because it is dominated by row 2. Removing this row yields A_1.

$$A = \begin{pmatrix} 3 & 4 & 4 & 4 \\ 4 & 5 & 5 & 6 \\ 7 & 6 & 6 & 7 \\ 7 & 8 & 7 & 6 \end{pmatrix}, \qquad A_1 = \begin{array}{c} 2 \\ 3 \\ 4 \end{array}\begin{pmatrix} 4 & 5 & 5 & 6 \\ 7 & 6 & 6 & 7 \\ 7 & 8 & 7 & 6 \end{pmatrix}$$

Row 1 of A_1 can now be deleted because it is dominated by row 3 of A_1. Removing this row yields A_2.

$$A_1 = \begin{pmatrix} 4 & 5 & 5 & 6 \\ 7 & 6 & 6 & 7 \\ 7 & 8 & 7 & 6 \end{pmatrix}, \qquad A_2 = \begin{array}{c} 3 \\ 4 \end{array}\begin{pmatrix} 7 & 6 & 6 & 7 \\ 7 & 8 & 7 & 6 \end{pmatrix}$$

Column 1 of A_2 can be deleted because it is dominated by column 3 of A_2. Notice that column 1 is not recessive in the original matrix A. Deleting column 1 from A_2 yields

$$A_3 = \begin{array}{c} 3 \\ 4 \end{array}\begin{pmatrix} 6 & 6 & 7 \\ 8 & 7 & 6 \end{pmatrix}$$

Finally, deleting column 1 (2 colors) from A_3 yields

$$A_4 = \begin{array}{c} \\ 3 \\ 4 \end{array} \begin{array}{cc} 3 & 4 \\ \begin{pmatrix} 6 & 7 \\ 7 & 6 \end{pmatrix} \end{array}$$

The matrix game A_4 can now be solved using Theorem 9.7. Applying this theorem with $a = 6$, $b = 7$, $c = 7$, and $d = 6$ yields

$$e = -2, \quad p_1 = 0.5, \quad p_2 = 0.5$$

$$q_1 = 0.5, \quad q_2 = 0.5, \quad \text{and} \quad v = 6.5$$

Hence, optimal strategies for Red and Black in game A_4 are

$$R_4{}^* = (0.5, 0.5) \quad \text{and} \quad B_4{}^* = \begin{pmatrix} 0.5 \\ 0.5 \end{pmatrix}$$

and $v = 6.5$.

We still must find optimal strategies and the value of the game for A. Since Red will never use rows 1 or 2, Red's optimal strategy is

$$R^* = (0, 0, 0.5, 0.5)$$

Similarly, since Black will never use columns 1 or 2, Black's optimal strategy is

$$B^* = \begin{pmatrix} 0 \\ 0 \\ 0.5 \\ 0.5 \end{pmatrix}$$

The value of the game is $v = 6.5$.

Red should therefore use three colors for half of his codes and four colors for the remaining half. Black should assume that Red is using three colors for half of his codes and four colors for the rest. If Red and Black use their optimal strategies, Red will average 6.5 points per game.

9.6 EXERCISES

Answer the following questions for each of the matrix games in Exercises 1–3.

(a) Find optimal strategies for Red and Black.

(b) Find the value of the game.

(c) How should Red play? (Use words to describe Red's strategy.)

(d) If Red and Black use their optimal strategies, what will happen in the long run?

1. $\begin{pmatrix} 4 & 5 \\ -3 & 0 \\ 12 & 4 \end{pmatrix}$ 2. $\begin{pmatrix} 4 & 0 & 2 \\ 0 & -2 & -2 \end{pmatrix}$ 3. $\begin{pmatrix} 1 & 3 & 2 & 4 \\ 10 & 11 & -3 & 12 \\ 2 & 3 & 4 & 3 \end{pmatrix}$

4.   Consider the following matrix game:

$$\begin{pmatrix} 0 & 2 & -6 & 0 \\ 2 & 8 & 2 & 10 \\ 4 & -6 & 4 & 8 \\ 2 & 4 & -2 & 6 \end{pmatrix}$$

(a) Find an optimal strategy for Red and two optimal strategies for Black.

(b) Find the value of the game.

5.   Lewis cannot decide whether today is Andrea's birthday. In an attempt to decide mathematically whether he should bring her a present, he decides that his opponent, fate, has two choices: f_1, it is Andrea's birthday; and f_2, it is not her birthday. He, in turn, has three choices: r_1, to bring her an expensive present; r_2, to bring her an inexpensive present; and r_3, to bring her no present. After great deliberation, he decides that the following matrix represents his payoffs (in happiness).

$$\begin{array}{c} \\ r_1 \\ r_2 \\ r_3 \end{array} \begin{array}{cc} f_1 & f_2 \\ \begin{pmatrix} 4 & -2 \\ 1 & 1 \\ -3 & 0 \end{pmatrix} \end{array}$$

Find Lewis's optimal strategy by—
(a) deleting recessive rows and columns
(b) finding a saddle point.

6.   A manufacturer is considering the four following suggestions for his advertising campaign: sponsoring a contest, broadcasting an endorsement by a famous personality, distributing free T-shirts with the name of the company on them, and offering customers discount tickets to sports events. The following matrix represents the percent increases in the manufacturer's sales if he uses suggestion x and his chief competitor uses suggestion y.

	Competitor			
	Contest	Endorse-ment	T-shirts	Tickets
Contest	5	7	6	6
Endorsement	6	4	8	6
T-shirts	8	7	6	7
Tickets	9	5	3	2

(a) Find optimal strategies for the manufacturer and his chief competitor.

(b) Find the percent increase in sales if the manufacturer uses his optimal strategy.

9.7 SOLUTION OF $m \times n$ GAMES

In the previous sections we learned how to find optimal strategies for all 2×2 matrix games and all games that have a saddle point. We have also seen that many larger games can be reduced to 2×2 games by deleting recessive rows and columns. In this section we will see that any game that cannot be handled by these methods can be translated into a pair of linear programs. The optimal solutions of these linear programs determine the optimal strategies for the corresponding game.

To see how this can be done, we will translate the following matrix game into two linear programs whose solutions will give the optimal strategies for the game.

$$
C = \begin{array}{c} \\ r_1 \\ r_2 \\ r_3 \end{array}
\begin{array}{ccc}
b_1 & b_2 & b_3 \\
\left(\begin{array}{ccc}
1 & 3 & -1 \\
2 & 3 & -5 \\
-1 & -4 & 7
\end{array}\right)
\end{array}
$$

This game has no saddle point, has no recessive rows or columns, and is not 2×2. It follows that the optimal strategies for this game cannot be determined by any of the methods studied previously.

Since C does not have a saddle point, Red knows that he must use a mixed strategy. Suppose Red arbitrarily uses the strategy $R_1 = (0.4, 0.3, 0.3)$.

Consider a single play of the game. On a single play, Black must use either b_1, b_2, or b_3. Suppose that Black uses b_1 and let X be the payoff. The random variable X may be 1, 2, or -1, according to whether Red uses r_1, r_2, or r_3. These values of X and the corresponding probabilities are given in the following table:

x	1	2	-1
$P(X=x)$	0.4	0.3	0.3

Multiplying each entry in row 1 of the table by the corresponding entry in row 2 and adding the products yields

$$E(X) = 1(0.4) + 2(0.3) + (-1)(0.3) = 0.7$$

Similarly, if Black uses b_2,

$$E(X) = 3(0.4) + 3(0.3) + (-4)(0.3) = 0.9$$

and if Black uses b_3,

$$E(X) = (-1)(0.4) + (-5)(0.3) + 7(0.3) = 0.2$$

Hence, if Red uses the strategy $R_1 = (0.4, 0.3, 0.3)$, the expected payoff $E(X)$ will be 0.7, 0.9, or 0.2, according to whether Black uses b_1, b_2, or b_3.

Remember that the expected payoff estimates the average payoff after the game has been played a large number of times. For example, if the expected payoff is 0.7, Red will average about 0.7 dollars per game after playing many times. If the expected payoff is 0.9, Red will average about 0.9 dollars per game.

The worst possible expected payoff from Red's point of view is the *smallest* of the numbers 0.7, 0.9, and 0.2. The worst possible expected payoff is therefore 0.2. Red now knows that, if he uses strategy (0.4,0.3,0.3), his expected winnings will be at least 0.2 dollars (i.e., 20 cents) per game, no matter what Black does. If Black sometimes uses b_1 or b_2, Red's expected winnings will be more than 20 cents per game.

Red may now wonder if he can find a different mixed strategy that will increase his minimum expected winnings. Suppose, therefore, that Red decides to use the mixed strategy $\mathbf{R} = (x_1, x_2, x_3)$, where x_1, x_2, and x_3 are any three numbers that satisfy

$$x_1 + x_2 + x_3 = 1 \tag{3}$$

and

$$0 \le x_1 \le 1, \qquad 0 \le x_2 \le 1, \qquad 0 \le x_3 \le 1 \tag{4}$$

Let X be the payoff and let a_1, a_2, and a_3 be the expected payoffs if Black uses b_1, b_2, and b_3, respectively. Then—

(a) if Black uses b_1, the expected payoff will be

$$a_1 = E(X) = 1x_1 + 2x_2 - x_3 \tag{5}$$

(b) if Black uses b_2, the expected payoff will be

$$a_2 = E(X) = 3x_1 + 3x_2 - 4x_3 \tag{6}$$

(c) if Black uses b_3, the expected payoff will be

$$a_3 = E(X) = -x_1 - 5x_2 + 7x_3 \tag{7}$$

The worst possible payoff w from Red's points of view is the *smallest* of the three numbers a_1, a_2, and a_3. Hence, $w \le a_1, w \le a_2$, and $w \le a_3$, and so, using equations (5)–(7),

$$w \le \quad 1x_1 + 2x_2 - 1x_3 \tag{8}$$

$$w \le \quad 3x_1 + 3x_2 - 4x_3 \tag{9}$$

$$w \le -x_1 - 5x_2 + 7x_3 \tag{10}$$

We now see that each strategy $\mathbf{R} = (x_1, x_2, x_3)$ that Red might use has a corresponding worst possible payoff w. This payoff will occur if Black chooses a particular strategy b_i. Black will definitely use b_i if she knows that Red plans to use $\mathbf{R}$, because Black wishes to win as much as possible (or lose little as possible). However, since Red is playing a mixed strategy, Black may not be able to discover which strategy Red is using. Red may therefore hope that Black will choose a different strategy, b_j. However, in this book we will assume that the worst thing that can happen will happen. This is the assumption that we have used in games in which one of the players is nature. In the present case, this assumption means that, if Red chooses $\mathbf{R}$, Black will choose b_i and the payoff will be w. This conservative approach guarantees that Red will win *at least* w.

Red naturally wishes to find the strategy that will guarantee him the best payoff. Since the best payoff from Red's point of view is the biggest payoff, Red wishes to find the strategy $\mathbf{R}^* = (p_1, p_2, p_3)$ that yields the maximum value of w.

Using statements (3)–(4) and inequalities (8)–(10), we can now restate Red's problem as follows:

Find the maximum value v of w

Subject to the constraints

$$w \leq \ 1x_1 + 2x_2 - 1x_3 \tag{11}$$

$$w \leq \ 3x_1 + 3x_2 - 4x_3 \tag{12}$$

$$w \leq -x_1 - 5x_2 + 7x_3 \tag{13}$$

$$x_1 + x_2 + x_3 = 1 \tag{14}$$

$$x_1 \geq 0, \qquad x_2 \geq 0, \qquad x_3 \geq 0 \tag{15}$$

Red's problem has now been reduced to a linear program L. Using the methods of Chapter 8, we can find the values p_1, p_2, p_3, and v of x_1, x_2, x_3, and w, respectively, that will maximize w and satisfy the constraints. Then (p_1,p_2,p_3,v) is an optimal solution of L, and $\mathbf{R}^* = (p_1,p_2,p_3)$ is an optimal strategy for Red and v is the value of the game.

A similar result is true in general. Let $\mathbf{A} = (a_{ij})$ be any $m \times n$ matrix game and let $\mathbf{R} = (x_1,x_2,\ldots x_m)$ be any mixed strategy for Red. To find Red's optimal strategy $\mathbf{R}^* = (p_1,p_2,\ldots,p_m)$, we must find the maximum value of w subject to certain constraints. This maximum value of w will be denoted v.

The constraints that are needed are similar to constraints (11)–(15). Constraints (14) and (15) can be generalized to

$$x_1 + x_2 + \cdots + x_m = 1 \tag{16}$$

and

$$x_1 \geq 0, \qquad x_2 \geq 0, \ldots, \qquad x_m \geq 0 \tag{17}$$

The general form of the remaining constraints will be the same for every game. However, the coefficients on the right side of each inequality will differ from game to game. To see how these coefficients can be quickly determined, let us study matrix $\mathbf{C}$ again. The linear program for finding Red's optimal strategy for this game is defined by constraints (11)–(15).

Consider the first constraint, inequality (11):

$$w \leq 1x_1 + 2x_2 + (-1)x_3$$

The numerical coefficients of the terms on the right side of this inequality are the entries in the first *column* of $\mathbf{C}$:

$$\mathbf{C} = \begin{pmatrix} \begin{array}{|c|} 1 \\ 2 \\ -1 \end{array} & \begin{array}{cc} 3 & -1 \\ 3 & -5 \\ -4 & 7 \end{array} \end{pmatrix}$$

Similarly, the coefficients on the right side of the second constraint,

$$w \leq 3x_1 + 3x_2 + (-4)x_3$$

are the entries in the second column of $\mathbf{C}$:

$$\mathbf{C} = \begin{pmatrix} 1 & \boxed{\begin{matrix} 3 \\ 3 \\ -4 \end{matrix}} & -1 \\ 2 & & -5 \\ -1 & & 7 \end{pmatrix}$$

The coefficients on the right side of the third constraint,

$$w \le (-1)x_1 + (-5)x_2 + 7x_3$$

are the entries in the third column of $\mathbf{C}$.

Notice that $\mathbf{C}$ has three columns. We know that the coefficients on the right sides of the first, second, and third constraints are the entries in the first, second, and third columns of $\mathbf{C}$, respectively.

Let us now return to our general matrix game

$$\mathbf{A} = \begin{pmatrix} a_{11} & a_{12} & \cdots & a_{1n} \\ a_{21} & a_{22} & \cdots & a_{2n} \\ \vdots & \vdots & & \vdots \\ a_{m1} & a_{m2} & \cdots & a_{mn} \end{pmatrix} \tag{18}$$

Notice that $\mathbf{A}$ has n columns because $\mathbf{A}$ is $m \times n$. Therefore, the coefficients to be used on the right sides of the first, second, ..., nth constraints corresponding to this game are the entries in the first, second, ..., nth columns of $\mathbf{A}$.

We therefore have the following theorem:

THEOREM   9.8   Let $\mathbf{A}$ of equation (18) be any $m \times n$ matrix game, and let L be the following linear program:

Find the maximum value of w

Subject to the constraints

$$w \le a_{11}x_1 + a_{21}x_2 + \cdots + a_{m1}x_m$$

$$w \le a_{12}x_1 + a_{22}x_2 + \cdots + a_{m2}x_m$$

$$\vdots \qquad \vdots \qquad \vdots \qquad \vdots$$

$$w \le a_{1n}x_1 + a_{2n}x_2 + \cdots + a_{mn}x_m$$

$$x_1 + x_2 + \cdots + x_m = 1$$

$$x_1 \ge 0, \qquad x_2 \ge 0, \ldots, \qquad x_m \ge 0$$

Furthermore, assume that the feasible solutions of L are written in the form $(x_1, x_2, \ldots, x_m, w)$. Then $(p_1, p_2, \ldots, p_m, v)$ is an optimal solution of L if and only if—

(a) $\mathbf{R^*} = (p_1, p_2, \ldots, p_m)$ is an optimal strategy for Red

and

(b) v is the value of the game.

Notice that the coefficients on the right side of the first inequality of Theorem 9.8 are the entries in the first column of matrix $\mathbf{A}$; the coefficients on the right side of the second inequality are the entries in the second column of $\mathbf{A}$; and so on.

Fortunately, it can be proved that the linear program of Theorem 9.8, unlike some linear programs, has at least one optimal solution. Therefore, optimal strategies for Red can be determined by solving the corresponding linear program.

A similar method can be used to find optimal strategies for Black. Remember, however, that the best payoff for Black is the *smallest* payoff. Black, therefore, must find the *minimum* value of her worst possible payoff, w. Moreover, Black's worst possible payoff w is the *largest* of the possible expected payoffs. Therefore, w is *greater than or equal to* the expected payoffs. In the case of the row player, Red, we saw that the coefficients of these expected payoffs were the entries in the columns of the matrix $\mathbf{A}$. In the case of the column player, Black, the coefficients of these expected payoffs are the entries in the *rows* of $\mathbf{A}$.

We now have the following theorem:

THEOREM 9.9 Let

$$\mathbf{A} = \begin{pmatrix} a_{11} & a_{12} & \cdots & a_{1n} \\ a_{21} & a_{22} & \cdots & a_{2n} \\ \vdots & \vdots & & \vdots \\ a_{m1} & a_{m2} & \cdots & a_{mn} \end{pmatrix}$$

be any $m \times n$ matrix game, and let L be the following linear program:

Find the minimum value of w
Subject to the constraints

$$w \geq a_{11}x_1 + a_{12}x_2 + \cdots + a_{1n}x_n$$

$$w \geq a_{21}x_1 + a_{22}x_2 + \cdots + a_{2n}x_n$$

$$\vdots \qquad \vdots \qquad \vdots \qquad \qquad \vdots$$

$$w \geq a_{m1}x_1 + a_{m2}x_2 + \cdots + a_{mn}x_n$$

$$x_1 + x_2 + \cdots + x_n = 1$$

$$x_1 \geq 0, \qquad x_2 \geq 0, \ldots, \qquad x_n \geq 0$$

Furthermore, assume that the feasible solutions of L are written in the form $(x_1, x_2, \ldots, x_n, w)$. Then $(q_1, q_2, \ldots, q_n, v)$ is an optimal solution of L if and only if—

(a) $\mathbf{B}^* = \begin{pmatrix} q_1 \\ q_2 \\ \vdots \\ q_n \end{pmatrix}$ is an optimal strategy for Black

and

(b) v is the value of the game.

Notice that the coefficients on the right side of the first inequality of Theorem 9.9 are the entries in the first row of matrix A; the coefficients on the right side of the second inequality are the entries in the second row of A; and so on.

As in the case of the linear program of Theorem 9.8, it can be proved that the linear program of Theorem 9.9 has at least one optimal solution. It follows that optimal strategies for Red and Black can be determined by solving two linear programs. Of course, this method should only be used if simpler methods—such as finding a saddle point—do not apply. When the two linear programs are solved, the maximum value of w found in the first linear program and the minimum value of w found in the second linear program will both equal the same number v. This number v is the value of the game.

Example 19 shows how to apply Theorems 9.8 and 9.9.

Example 19 Two countries, Red and Black, are engaged in a battle. Suppose that Red has three different antiaircraft systems, r_1, r_2, and r_3, and its enemy, Black, has three different airplanes, b_1, b_2, and b_3. The following matrix gives the probability that Red's gunners will shoot down airplane b_j if they use antiaircraft system r_i.

$$\mathbf{A} = \begin{array}{c} \\ r_1 \\ r_2 \\ r_3 \end{array} \begin{array}{ccc} b_1 & b_2 & b_3 \\ \left(\begin{array}{ccc} 0.1 & 0.4 & 0.3 \\ 0.4 & 0.1 & 0.6 \\ 0.3 & 0.2 & 0.5 \end{array} \right) \end{array}$$

For example, if Red's gunners use antiaircraft system r_1, the probability that they shoot down plane b_1 is 0.1, the probability thay they shoot down b_2 is 0.4, and the probability that they shoot down b_3 is 0.3.

Suppose that Red's gunners are able to use only one antiaircraft system at a time and that Black's air force is able to use only one airplane at a time. Write linear programs that will find (a) Red's best strategy, (b) Black's best strategy, and (c) the value of the game.

Solution We should first simplify this problem by deleting column 3, which is dominated by column 1. This will eliminate a great deal of unnecessary calculation. Deleting column 3 yields

$$\mathbf{A}_1 = \begin{array}{c} \\ r_1 \\ r_2 \\ r_3 \end{array} \begin{array}{cc} b_1 & b_2 \\ \left(\begin{array}{cc} 0.1 & 0.4 \\ 0.4 & 0.1 \\ 0.3 & 0.2 \end{array} \right) \end{array}$$

Since $\mathbf{A}_1$ does not have a saddle point, does not have any recessive rows or columns, and is not 2×2, linear programs must be used to find optimal strategies for Red and Black.

(a) Let $\mathbf{R}^* = (p_1, p_2, p_3)$ be an optimal strategy for Red, and let v be the value of the game. Applying Theorem 9.8, (p_1, p_2, p_3, v) is an optimal solution of a certain linear program. According to the theorem, the coefficients on the right side of the first constraint of this program are the entries in the first column of A_1.

$$\begin{pmatrix} \boxed{\begin{matrix} 0.1 \\ 0.4 \\ 0.3 \end{matrix}} & \begin{matrix} 0.4 \\ 0.1 \\ 0.2 \end{matrix} \end{pmatrix}$$

It follows that the first constraint is

$$w \leq (0.1)x_1 + (0.4)x_2 + (0.3)x_3$$

Similarly, the coefficients on the right side of the second constraint are the entries in the second column of A_1.

$$\begin{pmatrix} \begin{matrix} 0.1 \\ 0.4 \\ 0.3 \end{matrix} & \boxed{\begin{matrix} 0.4 \\ 0.1 \\ 0.2 \end{matrix}} \end{pmatrix}$$

Hence, the second constraint is

$$w \leq (0.4)x_1 + (0.1)x_2 + (0.2)x_3$$

The required linear program is:

Find the maximum value of w
Subject to the constraints

$$w \leq (0.1)x_1 + (0.4)x_2 + (0.3)x_3$$
$$w \leq (0.4)x_1 + (0.1)x_2 + (0.2)x_3$$
$$x_1 + x_2 + x_3 = 1$$
$$x_1 \geq 0, \qquad x_2 \geq 0, \qquad x_3 \geq 0$$

(b) Similarly, let $B^* = \begin{pmatrix} q_1 \\ q_2 \end{pmatrix}$ be an optimal strategy for Black. Applying Theorem 9.9, (q_1, q_2, v) is an optimal solution of a certain linear program. According to the theorem, the coefficients on the right sides of the first, second, and third constraints are the entries in the first, second, and third rows of A_1, respectively.

$$\begin{pmatrix} \boxed{0.1 \quad 0.4} \\ \boxed{0.4 \quad 0.1} \\ \boxed{0.3 \quad 0.2} \end{pmatrix}$$

It follows that these constraints are

$$w \geq (0.1)x_1 + (0.4)x_2$$
$$w \geq (0.4)x_1 + (0.1)x_2$$
$$w \geq (0.3)x_1 + (0.2)x_2$$

The required linear program is:

Find the minimum value of w
Subject to the constraints

$$w \geq (0.1)x_1 + (0.4)x_2 \tag{19}$$

$$w \geq (0.4)x_1 + (0.1)x_2 \tag{20}$$

$$w \geq (0.3)x_1 + (0.2)x_2 \tag{21}$$

$$x_1 + x_2 = 1 \tag{22}$$

$$x_1 \geq 0, \qquad x_2 \geq 0 \tag{23}$$

(c) Applying Theorems 9.8 and 9.9, either the linear program defined in part **(a)** or the one defined in part **(b)** may be used to find the value of v. In the first linear program, v is the maximum value of w subject to the constraints; in the second linear program, v is the minimum value of w subject to the constraints.

As we saw in Chapter 8, solving linear programs often requires a great deal of calculation. Therefore, these programs should be made as simple as possible. We have already seen that a matrix game (and hence its corresponding linear program) can sometimes be simplified by deleting recessive rows and columns. In the next example, we will see how the linear programs themselves can be further simplified by using the constraint $x_1 + x_2 + \cdots + x_m = 1$ or the constraint $x_1 + x_2 + \cdots + x_n = 1$.

Example 20   Calculate the best strategy for Black in Example 19. If Black's air force uses this strategy, what is the probability that its airplane will be shot down?

Solution   We saw in Example 19 that Black's optimal strategy can be determined by solving the linear program defined by statements (19)–(23). This linear program can be simplified by using constraint (22). Solving for x_2 in equation (22) yields

$$x_2 = 1 - x_1 \tag{24}$$

Substituting equation (24) into constraint (19) yields

$$w \geq (0.1)x_1 + (0.4)(1 - x_1) = 0.4 - (0.3)x_1$$

Hence, constraint (19) becomes

$$w \geq 0.4 - (0.3)x_1 \tag{25}$$

Similarly, substituting equation (24) into constraints (20) and (21) yields

$$w \geq 0.1 + (0.3)x_1 \quad \text{and} \quad w \geq 0.2 + (0.1)x_1 \tag{26}$$

Using constraints (22), (23), (25), and (26), we now know that Black's optimal strategy can be determined by solving the following linear program:

Find the minimum value of w
Subject to the constraints

$$w \geq 0.4 - (0.3)x_1$$

$$w \geq 0.1 + (0.3)x_1$$

$$w \geq 0.2 + (0.1)x_1$$

$$x_1 + x_2 = 1$$

$$x_1 \geq 0, \qquad x_2 \geq 0$$

Notice that this linear program is simpler than the first, because the first 3 constraints only have 2 variables, x_1 and w, instead of 3 variables, x_1, x_2, and w.

The algebraic method for solving a linear program (Section 8.5) can now be applied. The equations defining the boundary of the constraint set C are

(a) $w = 0.4 - (0.3)x_1$

(b) $w = 0.1 + (0.3)x_1$

(c) $w = 0.2 + (0.1)x_1$

(d) $x_1 + x_2 = 1$

(e) $x_1 = 0$

(f) $x_2 = 0$

Since there are 6 equations in 3 variables (x_1, x_2, and w), there will be at most $_6C_3 = 20$ corner points. These corner points are listed in Table 9.1. "Constraints *ijk*" refers to equations (*i*), (*j*), and (*k*). Notice that there is no corner point associated with equations (a)–(c). This is because the solution to the system consisting of these equations is not unique.

From Table 9.1 it is clear that the minimum value of w (subject to the constraints) is 0.25 and that this minimum occurs at the corner point (0.5,0.5,0.25). It follows that (0.5,0.5,0.25) is an optimal solution for the linear program. Hence, applying Theorem 9.9,

$$\begin{pmatrix} 0.5 \\ 0.5 \end{pmatrix}$$

is an optimal strategy for Black in A_1, and the value of the game is $v = 0.25$.

Since Black never uses b_3, it follows that Black's optimal strategy in A is

$$\begin{pmatrix} 0.5 \\ 0.5 \\ 0 \end{pmatrix}$$

The value of the game A is $v = 0.25$. Black's best strategy, therefore, is to use airplane b_1 50 percent of the time, airplane b_2 50 percent of the time, and airplane b_3 never. If Black uses this strategy, the average probability that a plane will be shot down is at most 0.25. If Red's gunners do not use their optimal strategy, this average probability may be less than 0.25.

Table 9.1

Constraints	Corner Point	Feasible ?	w
abc	——	——	——
abd	(0.5, 0.5, 0.25)	Yes	0.25
abe	Inconsistent	——	——
abf	(0.5, 0, 0.25)	No	——
acd	(0.5, 0.5, 0.25)	Yes	0.25
ace	Inconsistent	——	——
acf	(0.5, 0, 0.25)	No	——
ade	(0, 1, 0.4)	Yes	0.40
adf	(1, 0, 0.1)	No	——
aef	(0,0, 0.4)	No	——
bcd	(0.5, 0.5, 0.25)	Yes	0.25
bce	Inconsistent	——	——
bcf	(0.5, 0, 0.25)	No	——
bde	(0, 1, 0.1)	No	——
bdf	(1, 0, 0.4)	Yes	0.40
bef	(0, 0, 0.1)	No	——
cde	(0, 1, 0.2)	No	——
cdf	(1, 0, 0.3)	No	——
cef	(0, 0, 0.2)	No	——
def	Inconsistent	——	——

The following outline organizes all of the methods that we have discussed for finding the optimal strategies for a matrix game.

Steps for Finding the Optimal Strategies for any Matrix Game A

(1) Determine whether **A** has a saddle point. If a_{ij} is a saddle point of **A**, then Red should always play row i and Black should always play column j. The value of the game is a_{ij}.

(2) If **A** has no saddle point, remove all recessive rows and columns. This forms a new matrix A_1. The value of the game **A** is the same as the value of the game A_1. Optimal strategies for Red and Black in **A** can be determined from the optimal strategies for Red and Black in A_1 by remembering that Red will never use any recessive rows and Black will never use any recessive columns.

(3) If A_1 is 2×2, apply Theorem 9.7 of Section 9.5.

(4) If A_1 is not 2×2, apply Theorems 9.8 and 9.9 of this section.

In addition, packaged computer programs for finding the optimal strategies for matrix games are available in most large computing centers.

9.7 EXERCISES

Apply the steps for finding the optimal strategies for any matrix game to each of the games in Exercises 1–5. If linear programs must be used to find optimal strategies for

Red and Black, *write the linear programs without actually solving them.* If linear programs are not necessary, find optimal strategies for Red and Black and find the value of the game.

1. $\begin{pmatrix} 1 & 2 & 4 \\ 2 & -1 & -3 \\ -1 & -4 & 10 \end{pmatrix}$ **2.** $\begin{pmatrix} 2 & 4 & -1 \\ 9 & 5 & 7 \\ -1 & 0 & 3 \end{pmatrix}$

3. $\begin{pmatrix} 7 & 1 & -5 \\ 2 & 4 & -1 \\ 6 & 3 & 5 \end{pmatrix}$ **4.** $\begin{pmatrix} 1 & -1 & 4 & 5 & -2 \\ 2 & 3 & -1 & 0 & 3 \end{pmatrix}$

5. $\begin{pmatrix} 2 & 1 & 3 \\ -1 & 5 & 5 \\ 0 & -1 & 1 \\ 13 & -4 & 5 \end{pmatrix}$

6. At a certain firm, the union is negotiating with the management for a new contract. Both the union and management have three options: act aggressively, act normally, or act cooperatively. The union feels that the following matrix gives their payoffs in percent wage increases.

Management

		Aggressive	Normal	Cooperative
	Aggressive	3	5	7
Union	Normal	4	4	5
	Cooperative	5	3	6

(a) Show that this matrix has no saddle point.

(b) Simplify this matrix by removing all recessive rows and columns.

(c) Write two linear programs whose solution will yield optimal strategies for union and management.

(d) Simplify these linear programs.

(e) Solve the linear programs of part **(d)** and then find the optimal strategies for union and management.

(f) If the union and management use their optimal strategies, what will the average percent wage increase be?

Key Word Review

Each key word is followed by the number of the section in which it is defined.

tree game   9.1
natural outcome   9.1
payoffs   9.1
perfect information   9.2
matrix game   9.2
two-player zero-sum game   9.2
payoff matrix   9.2
row player   9.2
column player   9.2
utilities   9.2
strategy   9.3, 9.4

mixed strategy   9.3, 9.4
pure strategy   9.3, 9.4
equilibrium pair   9.3
saddle point   9.3
optimal strategy   9.3
value of the game   9.3
fair game   9.3
expected payoff   9.4
recessive row   9.6
recessive column   9.6

CHAPTER 9   REVIEW EXERCISES

1.   Define each word in the Key Word Review.

2.   (a) What is the natural outcome of the tree game in Figure 9.6?

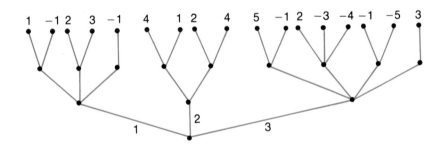

Figure 9.6

(b) What should Red's first move be?

3.   Let

$$A = \begin{pmatrix} 1 & 2 & -10 \\ -1 & -3 & -2 \end{pmatrix}$$

be a matrix game.
(a) What is the expected payoff if Red uses the mixed strategy (0.5, 0.5) and

Black uses $\begin{pmatrix} 0.50 \\ 0.25 \\ 0.25 \end{pmatrix}$?

(b) Find optimal strategies for Red and Black and the value of the game.

(c) (*i*) Which of the following vectors are mixed strategies for Red?

(*ii*) Which are pure strategies for Red?

$$\mathbf{B} = (\tfrac{1}{3}, \tfrac{2}{3}), \qquad \mathbf{C} = (\tfrac{1}{4}, \tfrac{2}{4}, \tfrac{1}{4}), \qquad \mathbf{D} = (1,0), \qquad \mathbf{E} = (-\tfrac{1}{2}, \tfrac{3}{2})$$

$$\mathbf{F} = (\tfrac{1}{8}, \tfrac{7}{8}), \qquad \mathbf{G} = (0,1), \qquad \mathbf{H} = \begin{pmatrix} 0 \\ 1 \end{pmatrix}$$

In Exercises 4–8, find optimal strategies for Red and Black and the value of the game. If linear programs must be used, *write the linear programs without actually solving them.* (Be sure to simplify the matrix before writing the linear program.)

4. $\begin{pmatrix} -4 & -1 & 13 \\ 6 & 4 & 4 \\ 2 & -3 & 5 \end{pmatrix}$ **5.** $\begin{pmatrix} 1 & 4 & -1 \\ 3 & 2 & 3 \\ 0 & 2 & -5 \end{pmatrix}$

6. $\begin{pmatrix} 1 & -1 & 2 \\ 3 & 2 & 0 \\ -1 & 2 & 2 \\ 0 & -2 & 1 \end{pmatrix}$ **7.** $\begin{pmatrix} 1 & 2 & -1 & 3 \\ 2 & -3 & 2 & 4 \\ 0 & 1 & -3 & 0 \\ 1 & 4 & 1 & 5 \end{pmatrix}$

8. $\begin{pmatrix} 2 & -1 & 3 & -2 \\ 7 & 2 & 4 & 5 \\ 1 & 0 & -4 & 4 \end{pmatrix}$

9. Two stores, Red and Black, are planning to build branches in either town A or town C. Suppose that 70 percent of the population of the area lives in town A and its suburbs and the remaining 30 percent live in town C and its suburbs. Suppose further that, if both stores locate in the same town, they will share the business (i.e., 50 percent of the population will be Red's customers.) However, if each store locates in a different town, each will get the business of that town (i.e., if Red locates in A and Black in C, 70 percent of the population will be Red's cusomers.)

(a) Write a payoff matrix for this matrix game.

(b) Find optimal strategies for Red and Black.

(c) If both stores use their optimal strategies, what percentage of the population will be Red's customers?

10. Two baseball teams, the Red Rangers and the Black Knights, each have three choices for pitcher. The choices for the Red Rangers are pitchers r_1, r_2, and r_3, while the choices for the Black Knights are pitchers b_1, b_2, and b_3. The Red Rangers have decided that the following matrix gives the probabilities that they will win the game if they use pitcher x and the Black Knights use pitcher y.

Black Knights

		b_1	b_2	b_3
	r_1	0.4	0.7	0.3
Red Rangers	r_2	0.6	0.9	0.4
	r_3	0.2	0.4	0.7

(a) Which of the following vectors are strategies for the Red Rangers?

   (*i*) (0.1,0.3,0.4),   (*ii*) (1,0.5,−0.5),   (*iii*) (0.5,0.5)

   (*iv*) $\begin{pmatrix} 0.3 \\ 0.1 \\ 0.6 \end{pmatrix}$,   (*v*) (0,1,0)

(b) What is the expected payoff if the Red Rangers use the strategy (0.1,0.3,0.6) and the Black Knights use the strategy $\begin{pmatrix} 0.9 \\ 0.1 \\ 0 \end{pmatrix}$?

(c) Write the following strategies in vector form.

   (*i*) The Red Rangers use r_1 40 percent of the time, r_2 20 percent of the time, and r_3 40 percent of the time.

   (*ii*) The Black Knights always use b_1.

(d) State the following strategies verbally.

   (*i*) (0.5,0.3,0.2)   (*ii*) $\begin{pmatrix} 0 \\ 1 \\ 0 \end{pmatrix}$

(e) Find optimal strategies for the Red Rangers and the Black Knights.

(f) If the Red Rangers use their optimal strategy, what is the average probability that they will win the game?

(g) If the Black Knights use their optimal strategy, what is the average probability that they will win the game? (Hint: Let v be the value of the game. If Red and Black both use their optimal strategies, show that the average probability that Red wins a game is v and the average probability that Black wins is $1 - v$. What happens if Red plays badly?)

11.   Consider the following matrix game:

$$\begin{pmatrix} 1 & 2 & -1 \\ 3 & -4 & 5 \end{pmatrix}$$

(a) Why must linear programs be used to find optimal strategies for Red and Black?

(b) Write these linear programs.

(c) Simplify the linear programs written in part (b).

(d) Calculate Red's optimal strategy and the value of the game.

Suggested Readings

1. Bellman, R., and Blackwell, D. "Red Dog, Blackjack, Poker." *Scientific American*, October 1951, pp. 44–47.
2. Farquharson, Robin. *Theory of Voting*. New Haven, Conn.: Yale University Press, 1969.
3. Kuhn, H. W., and Tucker, A. W. "Theory of Games." *Encyclopaedia Britannica*, 1973.
4. Luce, Robert D., and Raiffa, Howard. *Games and Decisions*. New York: John Wiley & Sons, 1957.
5. MacDonald, John. *Strategy in Poker, Business and War*. New York: W. W. Norton & Co., 1950.

6. McKinsey, John C., *Introduction to the Theory of Games*. New York: McGraw-Hill Book Co., 1952.
7. *Mathematical Thinking in Behavioral Sciences, Readings from Scientific American*. Introduction by D. Messick. San Francisco: W. H. Freeman & Co., 1968.
8. Morgenstern, Oskar. "The Theory of Games." *Scientific American* 180 (1950): 294–308.
9. Rapoport, Anatol. *Fights, Games and Debates*. Ann Arbor: University of Michigan Press, 1960.
10. Shubik, Martin. *Strategy and Market Structure*. New York: John Wiley & Sons, 1959.
11. Thompson, G. L. "Game Theory." *McGraw-Hill Encyclopedia of Science and Technology*. New York: McGraw-Hill Book Co., 1971.
12. Von Neumann, John, and Morgenstern, Oskar. *Theory of Games and Economic Behavior*. Rev. ed. Princeton, N.J.: Princeton University Press, 1953. (See Chapter 1.)
13. Williams, John D. *The Compleat Strategyst*. Rev. ed. New York: McGraw-Hill Book Co., 1965.

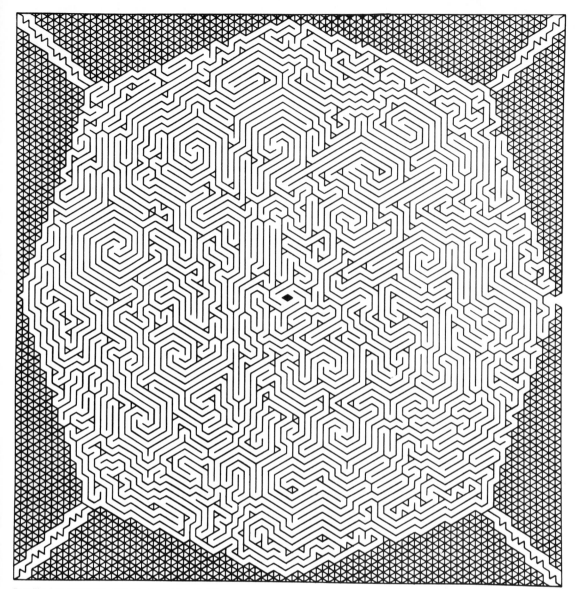

From The Second Great Maze Book by Greg Bright, © 1975 by Greg Bright.   Reprinted by permission of Pantheon Books, a Division of Random House, Inc.

CHAPTER 10 Graphs
and Their
Applications

INTRODUCTION

The drawing on the facing page is a *maze*. To solve this maze, we must find a continuous path from the opening on the right edge of the maze to the black diamond in the center. (For the solution to this maze, see page A34.)

How do we go about finding the solution to a maze? The usual approach is to use trial and error: we try various paths until we find one that works. However, there are better methods. In this chapter we will learn how to analyze mazes to eliminate some of the guesswork in solving them. In the process, we will learn that mazes, family trees, electrical circuits, and highway systems all have certain common characteristics, and that these common characteristics are described by *graph theory*. The techniques that we develop for analyzing mazes will later prove useful in analyzing each of these configurations and in solving many interesting problems; for example, graph theory can be used to help plan a manned-rocket flight to Mars.

10.1 MAZES AND GRAPHS

The most familiar mazes are those found in puzzle books and amusement parks. They may be drawn on paper, like the maze that begins this chapter, or they may be made out of bricks. Psychologists use mazes to study the learning behavior of men and animals. Other researchers have constructed mechanical mice, operated by computers, that choose a path through a maze on the basis of experience, rather than randomly. These mechanical mice may be the forerunners of the thinking robots envisioned by science-fiction writers.

In the following example we will learn how to find a path through any maze drawn on paper.

Example 1 We can quickly find a path through the maze of Figure 10.1(a)—or any other maze drawn on paper—by picking a route at random. If a blind alley is reached, we shade it in, and retrace our path until the first alternate path is found. (See Figure 10.1(b).) Continuing in this manner will eventually lead us to the goal. (See Figure 10.1(c).)

A maze—such as the one in Example 1—that is drawn on paper may represent a three-dimensional maze (e.g., a maze made out of bricks, hedges, or stones). Suppose you are faced with the task of threading your way through a three-dimensional maze. If the maze has only one entrance and only one exit (as in Figure 10.1), the way out can be found by placing your hand against the right (or left) wall and keeping it there as you walk. You are then certain to reach the exit, although your route will probably not be the shortest one possible.

In more traditional mazes, like those diagrammed in Figures 10.2 and 10.3, the goal is within the labyrinth. The hand-on-wall method will also work in this type of maze, provided—as in Figure 10.2—that there is no route by which you can walk around the goal and return to where you started. If, however, the goal is surrounded by one or more "closed circuits," the hand-on-wall method simply takes you around

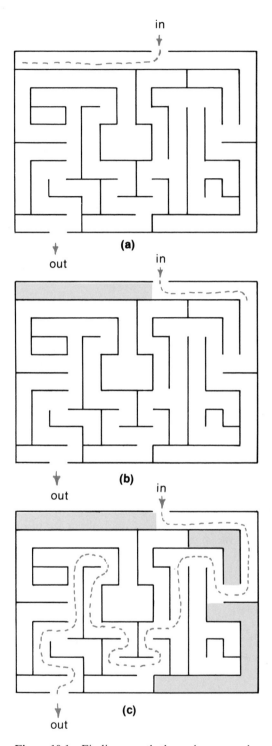

Figure 10.1 Finding a path through a maze drawn on paper

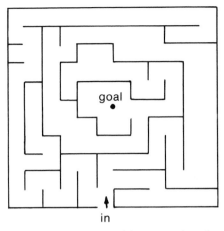

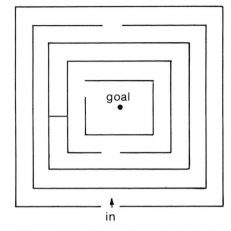

Figure 10.2   Maze with a central goal   **Figure 10.3**   Maze with a closed circuit

the largest circuit and back out of the maze. It can never lead you to the goal inside the circuit. The maze in Figure 10.3 has such a closed circuit.

The paper mazes in Figures 10.1–10.3 can represent many three-dimensional mazes: some made of stone, others of cardboard; some large enough to hide a person, others small enough to be traversed by a mouse.

Since we are only interested in the *path* through the maze, the size and composition of the maze are extraneous features. To find our way through a maze, we need a model that will show only those features of the maze that interest us. We will see that such a model can also be applied to many other situations. Example 2 shows how to find such a model.

Example 2   A path through the maze of Figure 10.4(a) can be easily found by trial and error. However, we will use this simple maze to illustrate a technique for finding all paths through any maze. Not only can this technique be applied to more difficult mazes, but it can also be applied to many practical situations that we will study later.

Every path through the maze of Figure 10.4(a) must start at some point a outside the maze. The exact location of a is unimportant as long as it satisfies this condition. For convenience, we put it in front of the entrance. (See Figure 10.4(b).) After entering the maze, each path must pass through a point b. The path then has two choices: it can either go left or right to c. The reader should verify the significance of the remaining letters. All possible paths, including those that do not lead to the goal, have been denoted by dotted lines in Figure 10.4(b).

In Figure 10.4(c) the paths are shown without the maze, because it is the paths which interest us, not the maze. This diagram shows how the points a,b, $c,\ldots,h$ are connected. The same information is conveyed in simpler form by Figure 10.4(d).

Figure 10.4(d) clearly indicates all possible paths through the maze. Denoting the path from d to e to h by deh and the top and bottom paths from b to c by $b_T c$ and $b_B c$, respectively, it is clear that the shortest possible paths through the maze are $ab_T cdeh$ and $ab_B cdeh$. Other possible paths are $ab_T cdedfdeh$, $ab_B cdfdegedeh$, and so on.

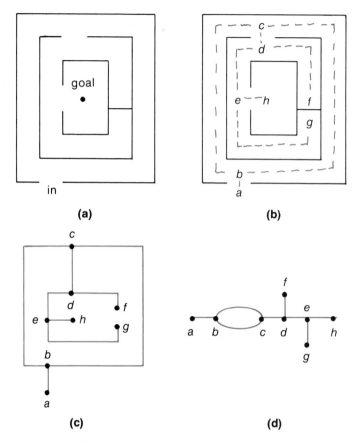

Figure 10.4   Steps in analyzing a maze

Figure 10.4(d) does not show as much information as Figure 10.4(b): it does not tell how long or how wide the passageways are, or whether they twist or turn or wrap themselves around each other. However, if we are only interested in the way the passageways lead into one another and to the goal, then Figure 10.4(d) is sufficient, in spite of the fact that it is not the exact geometric counterpart of the maze.

Figure 10.4(d) represents the paths through a certain maze. It could also represent the telephone communications between several cities or the sewer system of a neighborhood. By removing all extraneous features, we are left with a diagram that applies to many different situations. Diagrams like Figure 10.4(d) that consist of points connected by lines or curves are called *graphs*.[1] Example 3 describes two more graphs.

Example 3   The graph of Figure 10.5(a) shows the street system of a certain town. However, it is not as useful as the graph of Figure 10.5(b). In this graph, the lines with arrows represent one-ways streets; the arrows indicate the direction of traffic. Lines (or curves) with arrows are called *directed lines* (or *directed*

---

1. Do not confuse this new concept with the definition of the word "graph" given in Chapter 9. The word "graph" has two entirely different definitions, just as the word "bill" refers to both a part of a bird and a notice of money owed.

curves). Directed lines and curves are also called *arcs*. The undirected lines of Figure 10.5(b) (i.e., the lines without arrows) represent two-way streets. They indicate that traffic can flow in either direction. Undirected lines or curves are also called edges.

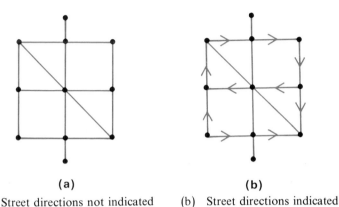

(a)   **(b)**

(a)   Street directions not indicated   (b)   Street directions indicated

Figure 10.5   Graphs showing the streets of a town

Definition 10.1 summarizes the new concepts we have discussed so far.

Definition 10.1   A **graph** is a diagram consisting of a nonempty set of points, called **vertices**, connected by—

(a) directed lines or curves called **arcs**

or

(b) undirected lines or curves called **edges**.

A graph that contains at least one arc is called a **directed graph**. An **undirected graph** contains only edges. For example, the graph of Figure 10.5(a) is undirected; the graph of Figure 10.5(b) is directed.

Arcs are used to represent energies or objects, such as traffic, electric current, nerve impulses, or blood, that flow in one direction only. The direction of the flow is indicated by the arrow. Edges are used to represent objects or energies that can move in either direction.

Although a graph can have an infinite number of vertices or an infinite number of arcs and edges, the graphs that we will study will all be *finite*—that is, they will contain a finite number of vertices, arcs, and edges. The adjective "finite" will be omitted in the remainder of our discussion.

If there is only one arc from vertex *a* to vertex *b*, this arc will be denoted by (*a,b*). The vertex *a* is called the **initial vertex**; the vertex *b* is called the **terminal vertex**. Notice that the arcs (*a,b*) and (*b,a*) have opposite direction.

The *edge* joining *c* and *d* (if there is only one) is denoted by either (*c,d*) or (*d,c*). If we wish to show that a certain object is moving from *c* to *d*, we use the notation (*c,d*); if we wish to show that it is moving in the opposite direction, we use (*d,c*), Notice that we can think of the edge (*c,d*) as consisting of two arcs—the arc (*c,d*) and the arc (*d,c*).

If two vertices *x* and *y* are connected by an arc or an edge (i.e., if an arc or an edge (*x,y*) or (*y,x*) exists), then the vertices *x* and *y* are said to be **adjacent vertices**.

The following examples illustrate these concepts and show how they can be applied to practical situations.

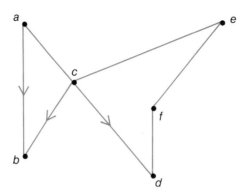

Figure 10.6   Flow of electric current in a machine

Example 4   The graph in Figure 10.6 shows the flow of electric current through the circuits of a certain machine. This graph has three arcs: (a,b), (c,b), and (c,d). It also has four edges: (a,c), (c,e), (e,f), and (d,f). Notice that the direction of each arc (and hence the direction of the current) is denoted by an arrow. Current can flow from a to b, for example, but not vice versa. The edges are not directed; this means that current can flow in either direction. In particular, current can flow from c to e or from e to c.

Notice that the initial vertex of the arc (a,b) is a; the terminal vertex of this arc is b. The vertices c and d are adjacent because they are connected by the arc (c,d). The vertices c and f, however, are not adjacent.

Example 5   Construct a graph representing the friendship relation among x, y, z, and w if the following pairs are friends:

x and y, z and w, x and w, y and z

All other pairs are not friends.

Solution   Let the vertices of the graph represent the various people and let the edge (m,n) indicate that m and n are friends. We may begin our graph by drawing dots for the four vertices x, y, z, and w. The pattern that these vertices form is arbitrary, as shown in Figures 10.7(a), (b), and (c). Vertices m and n are then connected by an edge if and only if m and n are friends. The three graphs of Figure 10.7 show three possible results; each of these graphs is correct.

Alternatively, we can begin by arbitrarily choosing one vertex, say x, and drawing it, as in Figure 10.8(a). We then connect x by edges to vertices representing the friends of x (i.e., with vertices y and w). This yields Figure 10.8(b). The new vertices y and w are then connected by edges to vertices representing their friends. This process yields the graph of Figure 10.8(c), which is also correct.

Example 5 illustrates the following two methods for drawing a graph.

Method 1
(1)   Draw all vertices.
(2)   Connect appropriate vertices by edges or arcs.

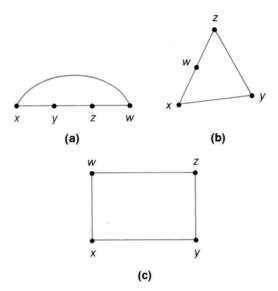

Figure 10.7   Three graphs showing the friendships among x, y, z, and w

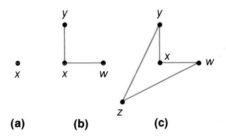

Figure 10.8   Three steps for drawing an alternative graph of the friendships among x, y, z, and w

Method 2

(1)   Arbitrarily choose one vertex and draw it.

(2)   Draw arcs or edges to connect this vertex to all adjacent vertices.

(3)   Draw arcs or edges to connect the vertices just drawn to all additional adjacent vertices.

(4)   Repeat step 3 until all vertices, edges, and arcs in the graph have been drawn.

The four graphs of Figures 10.7 and 10.8(c) all represent the same situation. Although each of these graphs *appears* different from the others, they all consist of exactly four vertices and four edges connecting the corresponding vertices. Graphs such as these, which are identical in a graph-theoretic sense, are called *isomorphic*.

Definition 10.2   Two graphs G and H are **isomorphic graphs** if and only if—

(a) G and H have the same number of vertices
       and

(b)   a correspondence between the vertices of G and H can be made so that, if two vertices of G are connected by an arc or edge, then the corresponding vertices of H are connected by a similarly directed arc or edge, and vice versa.

The following examples show how to prove whether two graphs are isomorphic or not isomorphic.

Example 6   Are the graphs in Figure 10.9 isomorphic?

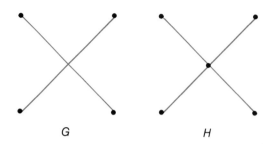

Figure 10.9

Solution   Graphs G and H are not isomorphic because G has four vertices while H has five.

Notice that the two edges in the graph G of Figure 10.9 cross. This does not mean, however, that there is a vertex at the intersection: a vertex at the intersection would be indicated by a dot, as in graph H of Figure 10.9. Therefore, the geometric intersection in G has no significance for the graph.

Example 7   Are graphs in Figure 10.10 isomorphic?

Figure 10.10

Solution   Yes, these graphs are isomorphic. Both G and H have two vertices connected by an arc. The required correspondence of vertices is $a \leftrightarrow d$, $b \leftrightarrow c$. This is the only correct correspondence.

Example 8   Which of the graphs in Figure 10.11 are isomorphic?

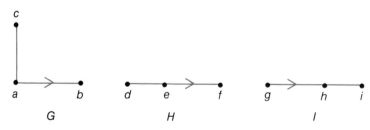

Figure 10.11

Solution   Graphs G and H are isomorphic. Both graphs contain three vertices, one arc, and one edge. The required correspondence of vertices is $a \leftrightarrow e$, $b \leftrightarrow f$, and $c \leftrightarrow d$. This is the only correct correspondence.

Graphs H and I have three vertices, one arc, and one edge. If H and I are isomorphic, there should be a correspondence of vertices satisfying Definition 10.2(b). Notice that the vertex e of H is the only vertex that is the initial vertex of an arc. It follows that e must correspond to the vertex g of I which has the same property. Since the arcs that begin at e and g end at f and h, respectively, f must correspond with h. The remaining vertex d must correspond with i. This correspondence, $e \leftrightarrow g$, $f \leftrightarrow h$, $d \leftrightarrow i$, is the only possible correspondence that might satisfy Definition 10.2. However, even this correspondence fails to satisfy the definition, because h and i in I are connected by an edge, but the corresponding vertices f and d in H are not connected by an edge.

Example 9 shows how graphs can be used to solve puzzles. In this example, two isomorphic graphs apply to the problem; however, one of these graphs is much easier to use than the other.

Example 9   Two white and two black knights have been placed on the abbreviated chessboard shown in Figure 10.12. A knight can be moved two squares vertically and then one square horizontally, or two squares horizontally and then one vertically. For example, the white knight in square 1 can be moved to square 6 or square 8. If white and black alternate turns, what is the least number of moves required for the white and black knights to change places?

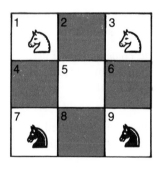

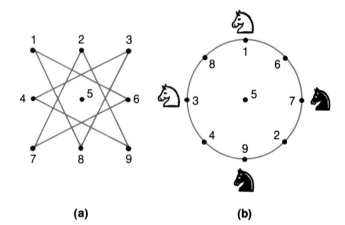

(a) (b)

Figure 10.12 Abbreviated chess game

Figure 10.13 Graphs for analyzing the abbreviated chess game

Solution   The possible moves of the knights are shown in the graphs in Figures 10.13(a) and (b). Each square is represented by a similarly numbered vertex. Two vertices are connected by an edge (i,j) if and only if it is possible for a knight to be moved from square i to square j. For example, vertex 1 is connected by edges to vertices 6 and 8 because it is possible to move a knight from square 1 to squares 6 and 8. Notice that vertex 5 is not connected to any other vertex.

The graphs of Figures 10.13(a) and (b) appear different, but they are isomorphic. Both graphs are useful because they help us to organize our thoughts about the problem. We can immediately see, for example, that no knight can ever be moved to square 5. However it is easy to see from Figure 10.13(b)—but not so easily from Figure 10.13(a)—that *each* knight must be moved at least four times before the black and white knights change places.

To see why each knight must be moved at least four times, let us mentally label the knights as follows: Let the knight that is initially on square 1 be w_1 (white 1), the knight that is initially on square 3 be w_3, the knight initially on square 9, b_9 (black 9), and the knight initially on square 7, b_7. It is clear from a careful examination of Figure 10.13(b) that, no matter how the knights are moved, their relative positions will always be the same—that is, b_7 will always be to the right of w_1, b_9 will always be to the right of b_7, and so on. Hence, the black and white knights can only interchange positions if w_1 is moved to square 9, w_3 to square 7, b_9 to square 1, and b_7 to square 3. Clearly, the quickest way of achieving this is to move each knight from its initial position to its final position in exactly four moves.

Since each knight must be moved at least four times before black knight and white knight interchange positions, and since there are four knights, at least $4 \cdot 4 = 16$ moves are required.

10.1 EXERCISES

1. Suppose that Lewis has two children, Carole and Robert; Carole has three children, Ruth, Deborah, and Naomi; and Robert has two children, Peter and Valerie. Make a graph (family tree) showing Lewis's direct descendants.

2. In a business organization, two people a and b are said to *communicate* if it is not unusual for a to telephone b. Suppose the following pairs communicate: a and b, c and e, a and c, d and b. Draw two graphs showing the foregoing communication system.

3. The assembly of a certain type of toy car is performed in the stages listed in Table 10.1. Activity x is an *immediate predecessor* of activity y if x must be done before y and if there is no other operation that must be done between x and y.

Table 10.1	Activity		Immediate Predecessor
	(a)	Start	None
	(b)	Put wheels on axle	a
	(c)	Attach assembled wheels to base	b
	(d)	Bolt motor to base	c
	(e)	Attach rear axle to motor	d
	(f)	Put doors on body	a
	(g)	Fasten body to base	e, f
	(h)	Finish	g

Draw a graph showing the assembly of such a car. Let an arc (i, j) denote the fact that activity i is an immediate predecessor of activity j.

4. In a basketball tournament each of four teams, a, b, c, and d, played every other team exactly once. Represent the following outcomes by a graph: a won

its games with b, c, and d; b won its games with c and d; and c and d tied. (Hint: Let the fact that x won its game with y be represented by the arc (x,y).)

5.   (a) Are the following pairs of graphs isomorphic? State your reasons.

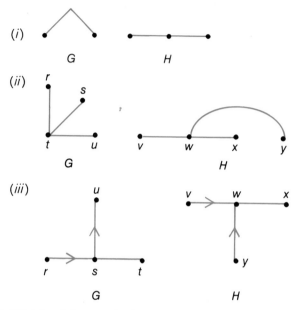

(b) Which of the graphs in part (a) are directed graphs?

10.2 PUZZLES AND SIMPLE PATHS

In this section and the next we will apply our knowledge of graphs to the solution of some well-known puzzles. We will see that many problems can be solved by finding a *path* from one vertex of a graph to another vertex. To be mathematically precise, we state the following definition:

Definition 10.3   Let a and b be vertices of a graph. A **path** from a to b is a sequence of arcs and edges

$$(a,a_1),(a_1,a_2),\ldots,(a_{n-1},a_n),(a_n,b)$$

The vertex a is called the **initial vertex** of the path; the vertex b is called the **terminal vertex**. If no ambiguity arises, the path may be denoted

$$a,a_1,a_2,\ldots,a_n,b$$

If $a = b$, the path is called a **circuit**.

Note that, in a path, the terminal vertex of each arc or edge (except the last) is the initial vertex of the next arc or edge.

The following examples illustrate these new concepts.

Example 10   Consider the graph in Figure 10.14. Three paths from a to d are: $a,b,$ d; a,b,c,d; and a,b,c,b,d. The initial vertex of each of these paths is a; the terminal vertex of each of these paths is d. There is no path from a to e. The path b,d,c,b is a circuit; it begins and ends at b. There is no circuit beginning at e.

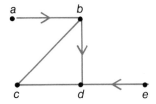

Figure 10.14

Example 11   In the graph in Figure 10.15, we cannot denote a path from a to c as a,b,c because there are two edges connecting a and b. To be precise, the two paths from a to c must be written E_1,E_3 and E_2,E_3.

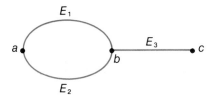

Figure 10.15

Example 12   In the graph in Figure 10.16, the three paths a,b,c; a,b,d,c; and a,b,d,b,c all connect a with c. The first path, a,b,c, is, of course, the most efficient, because it is the shortest—that is, it contains the fewest arcs. The third path, a,b,d,b,c, is particularly inefficient because it passes through the same vertex, b, twice. Paths that do not pass through any vertex more than once are called *simple paths*.

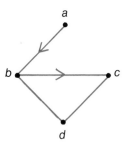

Figure 10.16

Definition 10.4

(a) A path that does not pass through any vertex more than once is called a **simple path**.

(b) A circuit beginning at a that does not pass through any vertex other than a more than once is called an **elementary circuit**.

The following example shows how certain puzzles can be solved by finding a simple path.

Example 13   A farmer (f) wishes to bring a dog (d), a goat (g), and a bag of cabbage (c) across the river. His little rowboat can only carry himself and one of these items at a time. However, he cannot leave the dog alone with the goat nor the goat alone with the cabbage. How should the farmer take them across?

Solution   This well-known puzzle can be solved without too much difficulty just by using common sense, since there are only three objects to be carried across. However, it is easy to see how common sense would flounder if there were more objects to be carried across, or if there were more restrictions upon the objects that can be left together. The following graph-theoretic approach also works in these more complicated cases.

In order to solve the farmer, goat, and cabbage puzzle using a graph, we will let the vertices of the graph represent the items left on the initial bank of the river. For example, the first vertex is labeled $fgdc$, which means that the farmer and all of his possessions are on the initial side of the river. We will draw an edge between two vertices if and only if the corresponding states can be changed into one another when the farmer transports one item across the river.

Method 2 for drawing a graph (page 433) can be used. Although we may begin drawing our graph by choosing any vertex, it is most efficient to begin by drawing the vertex that represents the initial state of the problem. This vertex, $fgdc$, indicates that the farmer (f) and all his possessions—the goat (g), dog (d), and cabbage (c)—are on the initial side of the river. We must now connect $fgdc$ with all adjacent vertices. According to the statement of the puzzle, the vertex $fgdc$ can only be connected with vertex dc; this is because the farmer must do the rowing and gdc, gc, and gd are not allowed. The vertex dc must now be connected with the vertex dcf. Using similar logic, the reader can show that the two graphs in Figure 10.17 are both correct. The problem then reduces to finding a path from $fgdc$ to $\varnothing$.

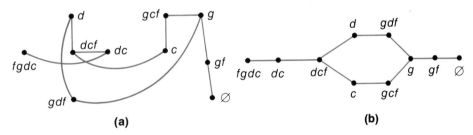

Figure 10.17   Two graphs for the farmer, goat, and cabbage puzzle

Using the second graph, Figure 10.17(b), it is easy to see that there are two simple paths from $fgdc$ to $\varnothing$, and hence there are two efficient solutions to the farmer's problem. The two simple paths are $fgdc$, dc, dcf, d, gdf, g, gf, $\varnothing$ and $fgdc$, dc, dcf, c, gcf, g, gf, $\varnothing$.

The solution corresponding to the first path is the following: The farmer should—

(1) bring the goat to the other side (this leaves the dog and the cabbage on the initial side)

(2) row himself back to the initial side (this puts the dog, cabbage, and farmer on the initial side)

(3) row the cabbage to the other side

(4) row the goat to the initial side

(5) row the dog to the other side

(6) row himself to the initial side

and, finally,

(7) row the goat to the other side.

The reader can easily describe the solution that corresponds to the second simple path.

The two graphs of Figure 10.17 are isomorphic. If the second graph is used, the two simple paths from *fgde* to ∅ are easy to find. However, it is not as easy to find the solution from the first graph because of the confusion caused by the crossing edges.

From experience, one can learn to draw graphs with the fewest possible crossing edges or arcs. Usually, it is easy to find a simple path from such graphs. Sometimes, however, one may have trouble finding a solution from the graph that has been drawn. In the next section, therefore, we will study two systematic methods for finding simple paths.

10.2 EXERCISES

1.   A graph *G* has six distinct vertices: *a*, *b*, *c*, *d*, *e*, and *f*.

   (a) What are the initial vertex and terminal vertex of the path *a,c,e,f,c,e*?

   (b) Is this path simple? Explain.

   (c) Is this path a circuit? Explain.

2.   Consider the graph in Figure 10.18.

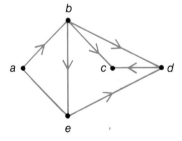

Figure 10.18

   (a) Find a path from *a* to *d*.

   (b) Find a circuit beginning at *b*.

  (c) Find an elementary circuit beginning at *a*.

  (d) Find a circuit beginning at *a* that is not elementary.

  (e) Find all simple paths from *b* to *c*.

  (f) Find a path from *b* to *c* that is not simple.

  (g) Find a path containing exactly two arcs.

3.   State in words the solution to Example 13 that corresponds to the path *fgdc*, *dc*, *dcf*, *c*, *gcf*, *g*, *gf*, $\varnothing$.

4.   How can a man cross a river with his three sons under the following conditions?

  (a) The man does all the rowing and can take no more than one son with him in the boat.

  (b) The eldest and the youngest son are the only peaceful pair if the father is absent.

5.   Three missionaries and three cannibals wish to cross a river using a boat that can only take two people at a time. If the cannibals ever outnumber the missionaries in a group, the missionaries have no chance of survival. How can the trips be arranged so that all six people cross the river safely? (Hint: Let *xy* denote the fact that *x* missionaries and *y* cannibals are on the first riverbank.)

6.   It is desired to invert an entire set of *n* upright cups by a series of moves where each move consists of turning over $n - 1$ cups. Show by drawing suitable graphs that this can be done if $n = 2$, but not if $n = 3$. (Hint: The graph for $n = 3$ has eight vertices. The vertex *abc* represents the fact that cups *a*, *b*, *c* are upright while the vertex *a'bc'* represents the fact that cups *a* and *c* are inverted. The edge (x,y) shows that situation *x* can be changed to situation *y* in one move.)

10.3 METHODS FOR FINDING SIMPLE PATHS

We saw in Section 10.2 that many problems can be solved by finding a simple path from one vertex of a graph to another. We also saw that our ability to find this path depends, in great measure, upon the way that the graph is drawn. A graph that has many crossing edges and arcs is not as easy to use as a less complicated graph. In this section we will describe a method that can be used if a simple path is not readily apparent from the graph that has been drawn. This method will not necessarily provide the easiest solution to the problem, but it will always work.

  We will illustrate the method as we state it by finding a simple path from a_1 to a_5 in Figure 10.19.

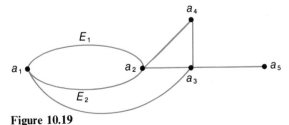

Figure 10.19

Steps for Finding One Simple Path from a_1 ***to*** a_n

(The illustrations refer to the problem of finding one simple path from a_1 to a_5 in Figure 10.19.)

If a graph G with vertices $a_1, a_2, \ldots, a_n$ fails to readily show a simple path from a_1 to a_n, draw a second graph H as follows:

(1) Draw the vertex a_1.

$$a_1 \bullet$$

(2) Suppose x_1 is adjacent to a_1, x_2 is adjacent to x_1, and so on. By adding adjacent vertices one at a time, draw a simple path

$$a_1, x_1, x_2, \ldots, x_m$$

Do not draw any vertex twice. Keep adding vertices to this path until either (a) a_n is reached, or (b) there are no previously unused vertices that can be added to the path. If x_m is the last vertex that can be drawn and $x_m \neq a_n$, then x_m is called a *dead end.* Put a cross through x_m to show that it is a dead end.

Notice that the edges (a_2, a_3) and (a_2, a_1) are not drawn because a_3 and a_1 already appear on the graph. It follows that a_2 is a dead end.

(3) When a dead end is reached, choose a vertex y that is on the simple path drawn in step (2), but which is not a dead end. Repeat step 2 forming the path

$$y, z_1, z_2, \ldots, z_k$$

For Figure 10.19, we can use the graph drawn in step 2 to choose vertex a_3 as our new starting point. Repeating step 2 yields

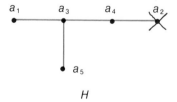

H

(4) Repeat this process using previously unused vertices until a_n is reached, or until you have shown that all paths lead to dead ends.

The graph of step 3 clearly indicates that a_1, a_3, a_5 is a simple path from a_1 to a_5 for the graph in Figure 10.19.

As our illustration shows, this method can be used to find a simple path connecting two vertices of a previously drawn graph. It can also be used to find a simple path connecting two vertices of a graph that has not yet been drawn. The following example shows how this can be done and how it can be useful.

Example 14   Three jugs, A, B, and C, hold eight, five, and three quarts, respectively. Jug A is filled with wine. Show how to divide the wine into two equal parts by repeatedly pouring it from container to container until the desired division is achieved. No measuring devices other than the jugs may be used.

Solution   The fact that jugs A, B, and C contain a, b, and c quarts of wine, respectively, can be represented by the three-digit number abc. The initial state of the jugs, for example, is represented by the number 800; the final state is represented by the number 440.

We can construct a graph that describes this problem as follows: Let $x = abc$ and $y = def$ be two distributions of wine. We will let the edge (x,y) represent the fact that distribution x can be changed into distribution y and distribution y can also be changed into distribution x. The arc (x,y) will show that distribution x can be changed into distribution y, but that the reverse cannot be done. Initially, for example, 5 quarts may be poured from A into B, or 3 quarts may be poured from A into C. Moreover, both of these operations are reversible. Hence, 800 should be joined to 350 and to 503 by edges. The pouring process, however, is not always reversible. For example, 530 should be joined to 350 by an arc but not an edge, because 2 quarts of liquid can be poured from jug A to fill jug B but 2 quarts cannot be poured from jug B because there is no way to measure them.

The problem will be solved when we have found one simple path from 800 to 440. It would therefore be extremely inefficient to draw the entire graph and then search for the required simple path. Let us instead use the four-step method and stop when we have found one simple path. Doing this yields the graph in Figure 10.20. This graph clearly indicates that one simple path from 800 to 440 is 800,350,323,620,602,152,143,440. This path can be easily translated into a solution to the problem.

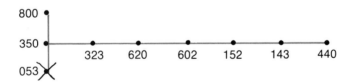

Figure 10.20   Simple path for the three-jug problem

The preceding example requires finding *one* simple path from a vertex a to a vertex b. Many problems, however, require finding *all* simple paths from a to b. Suppose, for example, that we wish to find the *best* solution to the problem of Example 14—that is, the solution that requires the fewest steps. The graph in Figure 10.20 indicates only that the problem can be solved in seven steps. There may be another solution that is shorter. To find the best solution, we must draw a graph that shows all simple paths from 800 to 440.

We will now describe a method for finding all simple paths from a to b in a graph G. For the sake of simplicity, this method will not apply to graphs—such as the one in Figure 10.19—that have more than one arc or edge connecting a given pair of vertices. However, the method can be easily extended to handle such graphs. We will illustrate this method by finding all simple paths from a_1 to a_5 in Figure 10.21.

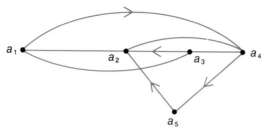

Figure 10.21

Steps for Finding All Simple Paths from a_1 to a_n

Let G be a graph with vertices $a_1, a_2, \ldots, a_n$ and at most one arc or edge from each vertex a_i to each vertex a_j. If all simple paths from a_1 to a_n cannot be easily found by examining G, draw a second graph H as follows:

(1)   Draw the vertex a_1.

a_1

(2)   Draw all arcs and edges (a_1, x).

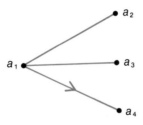

(3)   Let x be any vertex drawn in step 2. Draw all arcs and edges (x, z), unless (x, z) is already on the path from a_1 through x.

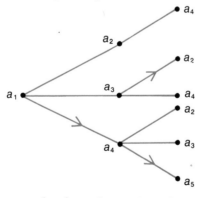

Notice that the vertex a_2 has been drawn three times because it is on the three paths a_1, a_2, a_4; a_1, a_3, a_2; and a_1, a_4, a_2.

(4)   Continue in this manner until all vertices have been used. Add an edge (y, z) only if z is not already on the path from a_1 through y. If no arc (y, z) can be drawn for some vertex $y \neq a_n$, put a cross through y to show that it is a dead end. The resulting graph H will clearly indicate all simple paths from a_1 to a_n.

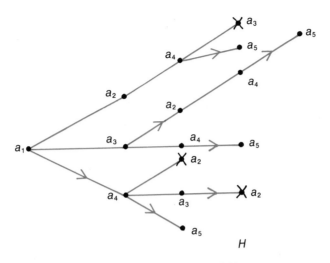

Figure 10.22   Simple paths for Figure 10.21

Figure 10.22 shows the finished graph H corresponding to G of Figure 10.21. Notice that a_2 is a dead end for the paths a_1,a_4,a_2 and a_1,a_4,a_3,a_2. (The edge (a_2,a_1) cannot be added to either one of these paths because a_1 is already on both of them.) Using this graph H, we see that there are four simple paths from a_1 to a_5 in the graph of Figure 10.21. These paths are:

$$a_1,a_2,a_4,a_5\,, \qquad a_1,a_3,a_4,a_5\,,$$

$$a_1,a_4,a_5\,, \quad \text{and} \quad a_1,a_3,a_2,a_4,a_5$$

10.3 EXERCISES

1.   Translate the path $800,350,323,620,602,152,143,440$ of Example 14 into a solution to the problem.

2.   Find a second solution to the problem of Example 14.

3.   Suppose that three jugs hold 12, 7, and 4 quarts, respectively, and that the first jug is filled with wine. Use the techniques of Example 14 to show how to divide the wine into two equal parts by pouring it from container to container.

4.   Use the first method described in this section to find one simple path from a to h in the graph in Figure 10.23.

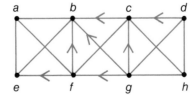

Figure 10.23

5.   Consider the graph in Figure 10.24.

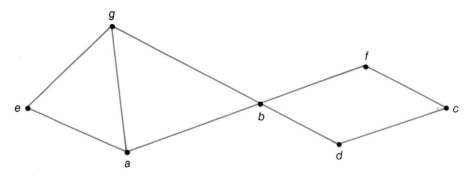

Figure 10.24

(a) Use the second method of this section to find all simple paths from a to e.

(b) Which simple paths (starting and ending at any two vertices) have the shortest length?

(c) Find all simple paths that start at vertex b and pass through four vertices.

10.4 EULER PATHS

We have seen that many problems can be solved by finding a simple path in a graph. In this section we will discuss problems that can be solved by finding a path that traverses every arc or edge of the graph exactly once. Such paths are called *Euler paths*. We will begin by formally defining Euler paths for undirected graphs. Euler paths for directed graphs will be discussed later.

Definition 10.5   An **Euler path** (**Euler circuit**) in an undirected graph is a path (circuit) that traverses each edge exactly once.

The following examples illustrate this concept and show how it applies to the solution of practical problems.

Example 15

(a) An Euler circuit for

is

(b) The graph

has two Euler paths,

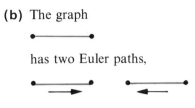

but it has no Euler circuit.

(c) The graph

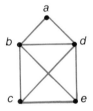

has no Euler path.

Each of the graphs in Example 15 is undirected because each contains only edges.

Example 16 Draw the picture in Figure 10.25 without retracing any edges and without lifting your pencil from the paper.

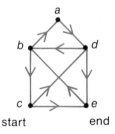

Figure 10.25

Solution Stated mathematically, this problem requires that we find an Euler circuit for the above graph. Later, we will learn several systematic techniques for doing this. However, in this case, it is possible to find a solution by trial and error, as illustrated in Figure 10.26. Start at c, go to e, and continue tracing the path c,e,b,c,d,b,a,d,e.

a
b   d
c   e
start   end

Figure 10.26

The edges of the graph in Figure 10.25 might represent highways, and the vertices might represent the cities that the highways connect. In this case, the Euler circuit might represent a route by which a highway inspector could inspect each highway exactly once without retracing any of them.

The next example will help us develop an easy technique for determining whether a graph has an Euler path.

Example 17   The city of Königsberg (now Kaliningrad, Soviet Russia) is located on two islands (*c* and *d* in Figure 10.27) and on the banks (*a* and *b*) of the river Pregel. The banks and the islands are connected by seven bridges.

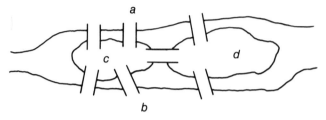

Figure 10.27   The seven bridges of Königsberg

The following famous problem was proposed by the mathematician Leonhard Euler (1707–83) for whom Euler paths and Euler circuits are named: Is it possible to plan a walk so that, starting from one of the banks or from one of the islands, one can return to the starting point after having crossed each bridge exactly once?

Solution   Since bridge crossings are our only concern, we can let *a*, *b*, *c*, and *d* be the vertices of a graph whose connecting edges correspond to the seven bridges. (See Figure 10.28.)

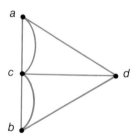

Figure 10.28   Graph for the seven bridges problem

To solve the problem, we must find an Euler circuit for this graph. Suppose the circuit starts at a vertex *y* where *y* ≠ *a*. Three edges (bridges) connect *a* to the other vertices. If an Euler circuit starting at vertex *y* exists, we will eventually have to enter vertex *a* by one of the bridges connected to *a*. We will then have to leave *a* by a second bridge. In due course, we will then have to enter *a* again by the third bridge. Since all of the bridges connected to *a* have been used, our journey must end in the vertex *a*. This is contrary to the premises of the problem. (The problem states that our journey must begin and end at the same place.) It follows that no route is possible which starts at a vertex *y* ≠ *a*.

Therefore, if there is a circuit, it must begin at vertex *a*. However, reasoning similar to that we have just used shows that a route starting at *a* must also end elsewhere. It follows that no Euler circuit exists, and so no solution to the bridge-crossing problem is possible.

It is clear that the reason we could not find a solution to Example 17 is because the number of edges having the vertex a as an endpoint is odd. The following definition will enable us to generalize this result to other graphs.

Definition 10.6   Let x be a vertex of an undirected graph G, let m be the number of edges (x,y) in G where $y \neq x$, and let n be the number of edges (x,x) in G. The **degree** $p(x)$ of x is given by the following formula:

$$p(x) = m + 2n \tag{1}$$

Notice that each edge (x,x) is counted twice in equation (1). The reason for this is that both endpoints of (x,x) are x. Edges (x,x) are called *loops*.

Example 18 illustrates this new concept.

Example 18   Find the degree of each vertex of the graph in Figure 10.29.

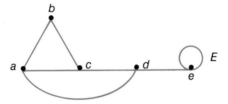

Figure 10.29

Solution   By counting the number of edges having each vertex as an endpoint, we see that $p(a) = p(c) = p(d) = 3$, while $p(b) = 2$. Note that $p(e) = 3$ because the loop E is counted twice.

The following theorem gives us one way of determining when a graph has no Euler circuit. The theorem can be proved using reasoning similar to that used in Example 17.

THEOREM   10.1   Let G be an undirected graph having at least one vertex a of odd degree. Then—

(a) G has no Euler circuit

(b) if G has an Euler path, it must either begin or end at a.

In particular, let G be the undirected graph of Figure 10.28. Since $p(a) = 3$, it follows from Theorem 10.1(a) that G has no Euler circuit. This agrees with our conclusion in Example 17.

Additional applications of Theorem 10.1 follow.

Example 19   The diagram in Figure 10.30 represents the highway system of a certain city. A highway inspector wishes to plan a route that will allow him to start from his office, inspect each of the above highways exactly once, and end up at his office. He is willing to locate his office in any one of the five cities on the route.

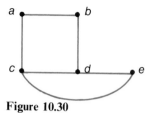

Figure 10.30

Solution   Stated in mathematical terms, the inspector must find an Euler circuit for the above graph. Applying Theorem 10.1, this cannot be done, because the degree of *c* is 3, which is odd.

Example 20   Consider the bridge of Königsberg described in Example 17 (see Figure 10.27). Determine whether one can start at a certain point *x*, traverse each bridge exactly once, and end up at a point $y \neq x$.

Solution   We already know that the graph *G* of Figure 10.28 has no Euler circuit. We must now determine whether or not this graph has an Euler path that is not a circuit.

Suppose an Euler path exists. According to Theorem 10.1(b), the two end-points of this path must be *a* and *b*, because these vertices are of odd degree ($p(a) = p(b) = 3$). However, the endpoints would also have to be *c* and *d*, because these vertices are also of odd degree ($p(c) = 5$; $p(d) = 3$). It follows that no Euler path is possible. Hence, even if we give up the luxury of returning home after our trip, we still cannot plan a walk that takes us across each of the bridges of Königsberg exactly once.

The following theorem can be proved using reasoning similar to the reasoning used in Example 20.

THEOREM   10.2   If an undirected graph has more than two vertices of odd degree, it has no Euler path.

Every graph discussed in this chapter so far has been a **connected graph**—that is, a graph in which there is a path connecting any two vertices of the graph. The graph in Figure 10.31 is clearly not connected, because there is no path connecting *b* and *c*. Since an Euler path must traverse every edge of the graph, it is clear that no Euler

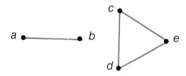

Figure 10.31   An unconnected graph

path is possible in this or in any other unconnected graph. For connected graphs, however, the following theorem is true.

THEOREM   10.3   Let *G* be an undirected, connected graph. If all of the vertices of *G* have even degree, then *G* has an Euler circuit.

To see why this theorem is true, let us imagine that the vertices of G are cities and that the edges are highways. Suppose that we start a vacation trip at some vertex a, choosing our route at random, but agreeing to use each highway exactly once. (For example, in Figure 10.32(a) our route might be a,b,c,a.) Since there are only a finite number of highways and we must use each highway exactly once, we will eventually be stuck in some city x with no way out.

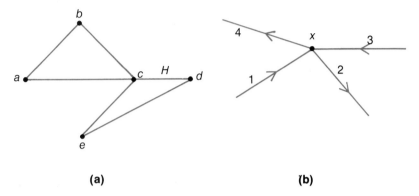

(a) **(b)**

Figure 10.32

We will now show that this city x must be our initial city a. The hypotheses of the theorem state that all the vertices of G have even degree; in particular, $p(x)$ is even. Suppose, for example, that $p(x) = 4$, as in Figure 10.32(b). (A similar argument holds if $p(x)$ is 6 or 8 or any other even number.) Since $p(x)$ is 4, there are four highways connecting x with other cities. If x is not a, our trip would start at a and eventually enter x by one of these highways. (This highway is labeled 1 in Figure 10.32(b).) We would then immediately leave x by one of the remaining three highways (e.g., highway 2 of Figure 10.32(b)). After traveling around, we would eventually reenter city x by one of the remaining two highways (e.g., highway 3) and then immediately leave x by the last highway (highway 4). Notice that we are no longer able to reenter x because there are no more untraveled highways connecting x to other cities. This means that, if x is *not* a, it is impossible for us to be stuck in x with no way out. It follows that x must be our initial city a.

If the trip that we have just taken covers all the highways, our task is finished, because the trip began and ended at the same vertex a. Suppose, however, that the trip did not cover all the highways. Since G is connected, one of the highways H not covered by our first trip must be connected to one of the cities c visited by the first trip. We may make a side trip from c over highway H and then travel randomly over unvisited highways as before. This trip will continue until we once more enter c without an escape route. We may now combine the two trips into one larger trip by starting at a, taking the first trip until c is reached, taking the second trip back to c, and then taking the remainder of the first trip back to a. This combined trip will start and end at a and will cover more highways than the first trip. (Referring to the example in Figure 10.32(a), the untraveled highway H is connected to the city c which was previously visited. Our side trip might be c,d,e,c. The combined trip would then be a,b,c,d,e,c,a, which is an Euler circuit.)

If the combined trip still does not cover all of the highways, a third side trip can be planned. Since the number of highways is finite, this method will eventually provide an Euler circuit.

The statement of Theorem 10.3 tells us that an Euler circuit exists if certain conditions are satisfied. The discussion following this theorem shows us how to find such a circuit. This process is illustrated in the next example.

Example 21   If possible, find an Euler circuit for the graph in Figure 10.33.

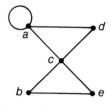

Figure 10.33

Solution   Since $p(a) = p(c) = 4$ and $p(b) = p(d) = p(e) = 2$, it follows that all the vertices are of even degree. An Euler circuit therefore exists. Using the method developed in the discussion following Theorem 10.3, we can find an Euler circuit in three steps, as illustrated in Figure 10.34. The numbers on the edges in the figure refer to the order in which the arcs in the circuit are traversed.

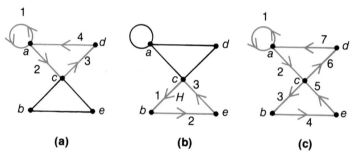

(a)   First trip, starting at a (b)   Side trip, starting at c;   (c)   Combined trip—an Euler
                                             H is the previously   circuit starting at a
                                             untraveled highway

Figure 10.34   Finding an Euler circuit for Figure 10.33

A proof similar to the discussion following Theorem 10.3 establishes the following theorem:

THEOREM   10.4   Let G be an undirected, connected graph. If G has exactly two vertices of odd degree, a and b, then G has an Euler path beginning at a and ending at b.

The next theorem can be proved by applying Theorems 10.1–10.4 and Exercise 7 following this section.

THEOREM   10.5   Let G be an undirected, connected graph.

(a)   The graph G will have an Euler circuit if and only if all of the vertices of G have even degree.

(b) The graph G will have an Euler path that is not an Euler circuit if and only if G has exactly two vertices of odd degree.

Theorems 10.3 and 10.5 completely solve the problem of finding Euler paths and Euler circuits in undirected graphs. By quickly determining the degree of each vertex, one can easily decide whether or not an Euler path or circuit exists. If such a path or circuit exists, the discussion following Theorem 10.3 provides a method for finding it.

Example 22 If possible, find an Euler circuit for each of the graphs in Figure 10.35. If the graph has no Euler circuit, find an Euler path if one exists.

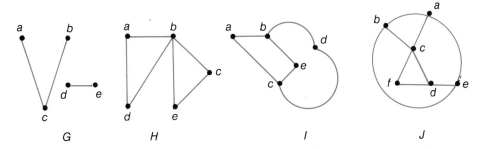

Figure 10.35

Solution

(a) Graph G has no Euler path (and therefore no Euler circuit) because the edge (d,e) is not connected to any of the other edges.

(b) None of the vertices of H has odd degree. Therefore, Theorem 10.5(a) implies that H has an Euler circuit. Using the method developed in the discussion following Theorem 10.3, we find that one Euler circuit is $a,b,c,$ e,b,d,a.

(c) Graph I has no Euler circuit because $p(b) = 3$. However, I has an Euler path beginning at b and ending at c because b and c are the only vertices of odd degree. One such path is b,a,c,d,b,e,c.

(d) Graph J has no Euler path or Euler circuit because it has more than two vertices of odd degree: $p(a) = p(b) = p(d) = p(e) = 3$.

In our discussion of Euler circuits and Euler paths in undirected graphs, we have used the example of the highway inspector who must inspect each highway in an area. We assumed that the inspector could traverse a given highway in either direction, and that he need only traverse each highway once in order to inspect it.

Suppose, however, that some of the highways are one-way roads, and that those that are two-way roads need to be inspected twice—once in each direction—in order to check the condition of both lanes. To determine how to carry out such an inspection in the most efficient manner, we must consider directed graphs. Figure 10.36 shows two graphs of this type.

In order to solve this new highway-inspection problem, we must extend the concepts of Euler path, Euler circuit, and the degree of a vertex to directed graphs. Remember that the inspector must traverse each two-way street (i.e., each edge) *twice*, once in each direction. Hence, in a directed graph, each edge (x,y) is considered to be *two* arcs, (x,y) and (y,x).

Definition 10.7 An **Euler path** (**Euler circuit**) in a directed graph is a path (circuit) that traverses each arc in the direction of the arc exactly once.

Since an edge of a directed graph G is considered to be two arcs, an Euler path (circuit) of G traverses each edge twice, once in each direction.

Definition 10.8 The **incoming degree** $p_i(x)$ of a vertex x is the number of arcs (y,x) that end at x. The **outgoing degree** $p_o(x)$ of a vertex x is the number of arcs (x,y) that begin at x.

Example 23 illustrates these concepts.

Example 23 Consider the graph of Figure 10.36(a). Find—

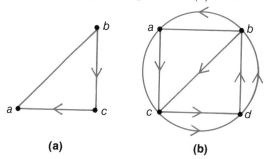

(a)   **(b)**

Figure 10.36 Two directed graphs

(a) all incoming degrees
(b) all outgoing degrees
(c) an Euler path (if possible).

Solution

(a) By counting the number of arcs ending at each vertex, and remembering that an edge is composed of two arcs, we see that the incoming degrees are

$$p_i(a) = 2 \quad \text{and} \quad p_i(b) = p_i(c) = 1$$

(b) Similarly, the outgoing degrees are

$$p_o(a) = p_o(c) = 1 \quad \text{and} \quad p_o(b) = 2$$

(c) An Euler path must traverse each arc exactly once in the direction of the arc, and each edge exactly twice, once in each direction. By trial and error, we can determine that one Euler path is b,c,a,b,a. We will soon learn a better technique for finding an Euler path.

Notice that in Example 23 the sum of the incoming degrees is the same as the sum of the outgoing degrees; that is,

$$p_i(a) + p_i(b) + p_i(c) = 4 = p_o(a) + p_o(b) + p_o(c)$$

This is true, in general, because each arc that ends at some vertex must also begin at some vertex.

Arguments similar to those that can be used to prove Theorem 10.5 will also prove the following theorem:

THEOREM   10.6   Let G be a connected, directed graph.

(a) The graph G has an Euler circuit if and only if $p_i(x) = p_o(x)$ for all vertices x.

(b) The graph G has an Euler path that is not an Euler circuit if and only if $p_i(x) = p_o(x)$ for every vertex x except exactly two vertices a and b. If an Euler path exists, the following equations will hold:

$$p_i(b) = p_o(b) + 1$$
$$p_i(a) = p_o(a) - 1$$

The Euler path will start at a and end at b.

Theorem 10.6, coupled with the method described in the discussion following Theorem 10.3, completely solves the problem of finding Euler paths and Euler circuits in directed graphs. By reviewing the incoming and outgoing degrees of each vertex and applying Theorem 10.6, one can first determine whether an Euler circuit or Euler path exists. If such a path or circuit exists, the method discussed following Theorem 10.3 can be used to find it. The following example illustrates this process.

Example 24   If possible, find an Euler circuit for the graphs of Figure 10.36. If the graph has no Euler circuit, determine whether it has an Euler path.

Solution

(a) The graph of Figure 10.36(a) has no Euler circuit because $p_i(a) = 2$, while $p_o(a) = 1$. However, it follows from Theorem 10.6(b) that the graph does have an Euler path beginning at b and ending at a. By inspection, this Euler path is b,c,a,b,a. Another Euler path is b,a,b,c,a. Note that this second path also begins at b and ends at a.

(b) The graph of Figure 10.36(b) has an Euler circuit because $p_i(x) = p_o(x)$ for all vertices x. The Euler circuit illustrated in Figure 10.37 can be found by patching together trips as discussed following Theorem 10.3. The numbers in the figure refer to the order of the arcs in the circuit. This graph also has other Euler circuits.

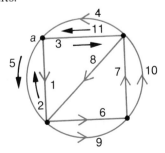

Figure 10.37   An Euler circuit for Figure 10.36(b)

10.4 EXERCISES

1. Consider the graph in Figure 10.38.

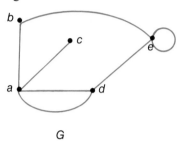

G

Figure 10.38

(a) Calculate $p(x)$ for all vertices x.

(b) Does G have an Euler circuit? If it does, give one such circuit; if it does not, state the reason why.

(c) Does G have an Euler path? If it does, give one such path; if it does not, explain why not.

2. Which of the graphs in Figure 10.39 have (a) Euler circuits, (b) Euler paths that are not circuits? Explain.

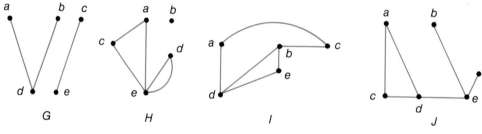

Figure 10.39

3. Consider the graph in Figure 10.40.

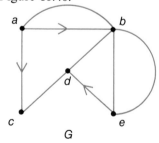

G

Figure 10.40

(a) Calculate $p_i(x)$ and $p_o(x)$ for all vertices x.

(b) Calculate—

(i) $p_i(a) + p_i(b) + \cdots + p_i(e)$

(ii) $p_o(a) + p_o(b) + \cdots + p_o(e)$

(c) Does *G* have an Euler circuit? If it does, give the circuit. If it does not, state the reason.

(d) If possible, find an Euler path that is not an Euler circuit. Explain your answer.

4.   Which of the graphs in Figure 10.41 have (*a*) Euler circuits, (*b*) Euler paths that are not circuits? Explain.

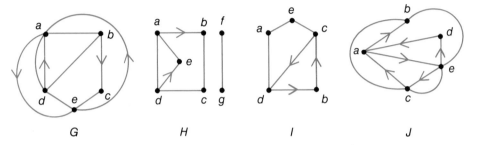

Figure 10.41

5.   The graph in Figure 10.42 represents the street system of a certain area. A mail carrier must deliver mail to all houses on both sides of the streets. If possible, design an efficient route that will begin and end at the same place.

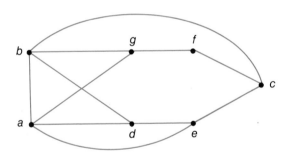

Figure 10.42

6.   The graph in Figure 10.43 represents the hallways of a museum. Pictures are to be hung on one side of each hall. If possible, design a route that will enable a person to see each exhibit exactly once. Indicate where the entrance and exit should be built.

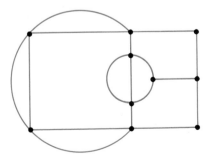

Figure 10.43

7.   Prove that a graph cannot have exactly one vertex of odd degree. (Hint: Use reasoning similar to the discussion following Theorem 10.3.)

8.   Prove Theorem 10.5.

9.   Can the hopscotch game pictured in Figure 10.44 be drawn without retracing any edges and without lifting your pencil from the paper? If so, draw it. If not, explain why not.

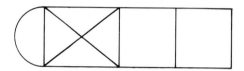

Figure 10.44

10.   Each of the diagrams in Figure 10.45 shows the floor plan of a five-room house. All doors are indicated on the drawing. If possible, design a walk that will allow a person to pass through each doorway once and only once. Show the relevant graphs. (Hint: The graph of floor plan **A** contains fifteen edges, one for each door. See Example 17.)

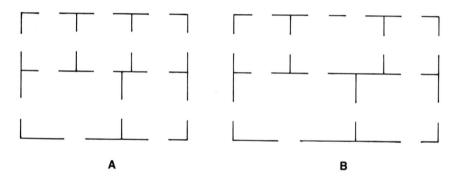

A   **B**

Figure 10.45   Two floor plans

10.5 CRITICAL PATHS

In business and in everyday life, scheduling the steps of a complicated project in the most efficient order can result in great savings of time and money. Two practical questions that naturally arise in such situations are "What is the earliest possible completion time?" and "How can the project be scheduled so that it will be completed at this earliest possible time?"

In this section we will show how these questions can be answered by using directed graphs. The following example shows one way that this can be done. It will also help us develop a simpler technique for accomplishing the same result.

Example 25   Table 10.2 lists the activities that must be completed before an airplane can leave the airport. Activity x is an **immediate predecessor** of activity y if x must precede y and there is no activity that must be done between x and y.

Table 10.2	Activity	Completion Time (Minutes)	Immediate Predecessors
	Start (a)	0	none
	Disembark passengers (b)	10	a
	Unload luggage (c)	15	a
	Check and fuel plane (d)	20	a
	Board new passengers (e)	10	b
	Load new luggage (f)	15	c
	Stop (g)	0	d, e, f

Schedule these activities so that the airplane can leave at the earliest possible time.

Solution   If these activities are done consecutively, it will take

$$0 + 10 + 15 + 20 + 10 + 15 + 0 = 70 \text{ minutes}$$

before the plane can leave the airport. It is clear, however, that some of these activities can be done simultaneously—for example, disembarking passengers, unloading luggage, and checking and fueling the plane.

The total project is represented by the directed graph in Figure 10.46. The arc (x,y) indicates that x is an immediate predecessor of y. The number L_x assigned to (x,y) is the time required to complete activity x. This number is called the *length of the arc*.

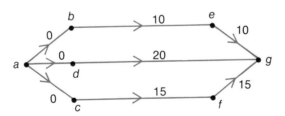

Figure 10.46   Graph for the airport schedule

We can now consider how to schedule these activities so that the airplane will be ready to leave as quickly as possible. Clearly, a (start) must be scheduled first; a will start at time 0. Figure 10.46 indicates that activities b, c, and d can then be started simultaneously. Since activity a takes 0 minutes to complete, activities b, c, and d will also start at time 0. The figure indicates that activity e can be started as soon as b is finished. Therefore, activity e can be started at time $0 + 10 = 10$ minutes because a takes 0 minutes and b takes 10 minutes. Similarly, f can be started as soon as c is finished, at time $0 + 15 = 15$ minutes. These starting times have been listed in Table 10.3.

We still must decide when to schedule g (stop). Activity g cannot begin before activities d, e, and f are *all* completed. Although e will be completed in $0 + 10 + 10 = 20$ minutes and d will be completed in $0 + 20 = 20$ minutes, f will not be completed before $0 + 15 + 15 = 30$ minutes. It follows that g cannot

Table 10.3

Activity	Earliest Starting Time
a	0
b	0
c	0
d	0
e	10
f	15
g	30

be started before 30 minutes. Since g takes 0 minutes to complete, the earliest possible completion time for the project is the earliest starting time for g, 30 minutes. This completion time will be achieved if the schedule indicated in Table 10.3 is followed.

Example 25 gives one method for finding the earliest possible completion time and best schedule for a project. A simpler method, which we will discuss soon, requires the following definition. We assume in this definition that a project consists of a list of activities, completion times, and immediate predecessors like the one given in Table 10.2. We further assume that the arc (x,y) indicates that activity x is an immediate predecessor of activity y.

Definition 10.9 Let G be the graph of a project.

(a) The *length of the arc* (x,y) is the time required to complete activity x. This number is denoted L_x and is placed over the arc (x,y) in G.

(b) The **length of the path** $a_1, a_2, \ldots, a_n$ is denoted L and is the time required to consecutively complete the activities $a_1, a_2, \ldots, a_n$. The number L satisfies the following equation:

$$L = L_{a_1} + L_{a_2} + \cdots + L_{a_n}$$

For example, Table 10.4 lists all simple paths from a to g in the graph in Figure 10.46, together with their lengths.

Table 10.4

Simple Path	Length L
a,b,e,g	$0 + 10 + 10 = 20$
a,d,g	$0 + 20 = 20$
a,c,f,g	$0 + 15 + 15 = 30$

Notice that the earliest completion time of g in Example 25 is $0 + 15 + 15 = 30$ minutes. This time depends upon the completion of activities a, c, f, and g and is the length of the path a,c,f,g. This path, a,c,f,g, is called a **critical path**; that is, it is a path whose length gives the earliest possible completion time for the project. The activities on the critical path (i.e., a,c,f, and g) are called **critical activities**. Activities that are not on the critical path (i.e., b,d, and e) can be scheduled simultaneously with the critical activities.

In Example 25 the critical path is the simple path from a to g that has the greatest length. (See Table 10.4.) With a moment's reflection, we can convince ourselves that *the critical path is always the simple path from start to stop that has greatest length.* A shorter simple path cannot be critical. This is because the length of such a path does not allow time enough to finish all of the activities on paths of greater length. The path a,b,e,g in Example 25, for example, has length 20 minutes. This is not enough time to finish activities c and f each of which requires 15 minutes. (See Figure 10.46 and Table 10.4.) Paths which are not simple are clearly not critical because it is inefficient to perform the same activity twice.

We can now outline a method for finding critical paths:

Method for Finding a Critical Path and the Shortest Completion Time for a Project

(1)   List all simple paths from a (start) to z (stop) together with their lengths.

(2)   Find a path $a, a_1, a_2, \ldots, a_n, z$ of greatest length; this path is called a *critical path*. Its length is the shortest completion time.

(3)   The critical activities should be scheduled as follows: a is begun at once; a_1 is begun as soon as a is completed; a_2 is begun as soon as a_1 is completed; and so on. There is usually some leeway in scheduling the noncritical activities.

Notice that this method finds *a* critical path rather than *the* critical path. There are many graphs that have two or more critical paths. (For example, Exercise 2 at the end of this section.)

The following example illustrates the use of the method for finding a critical path.

Example 26   Table 10.5 lists the activities necessary to send a manned flight to Mars. (The times given for these activities are fictitious.)

Table 10.5

Activity	Time (Years)	Immediate Predecessors
Start (a)	0	none
Build rocket (b)	5	a
Train crew (c)	3	a
Construct equipment (d)	4	a
Trip to Mars (e)	1	b, c, d
Explore Mars (f)	0.25	e
Return to Earth (g)	1	f
Stop (h)	0	g

(a)   Find a critical path.

(b)   What is the earliest possible completion time for this project?

(c)   Schedule the activities so that the project can be finished as quickly as possible.

Solution   The graph in Figure 10.47 represents the Mars project.

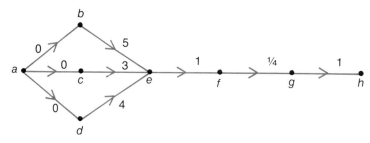

Figure 10.47   Graph for the Mars project schedule

(a) The simple paths from a to h and their lengths are:

Simple Path	Length
a,b,e,f,g,h	$0 + 5 + 1 + 0.25 + 1 = 7.25$
a,c,e,f,g,h	$0 + 3 + 1 + 0.25 + 1 = 5.25$
a,d,e,f,g,h	$0 + 4 + 1 + 0.25 + 1 = 6.25$

A critical path is a path of greatest length. It follows that the only critical path is a,b,e,f,g,h.

(b) The earliest possible completion time for the project is the length of the critical path. Therefore, the earliest possible completion time is 7.25 years.

(c) The critical path is a,b,e,f,g,h. This means that, if the project is to be completed as quickly as possible, it should begin at time 0 with building the rocket. The trip to Mars should take place as soon as the rocket is complete (at time 5 years), and exploration of the planet should start as soon as the trip is completed (at time 6 years), followed by the return trip at time 6.25 years. The entire project will be complete at time 7.25 years. The noncritical activities—training the crew and constructing the equipment—can be scheduled at time 0 because they have no predecessors except a (start). However, they do not have to be scheduled at time 0, as the remainder of this section will show.

Table 10.6 summarizes the schedule for this project.

Common sense must be applied to the solution of any scheduling problem. It is *theoretically* possible to carry out two activities simultaneously if neither activity

Table 10.6

Activity	Starting Time
a	0
b	0
c	0
d	0
e	$0 + 5 = 5$
f	$0 + 5 + 1 = 6$
g	$0 + 5 + 1 + 0.25 = 6.25$
h	$0 + 5 + 1 + 0.25 + 1 = 7.25$

depends upon the completion of the other. However, the actual performance of two activities at the same time may be physically impossible, or it may be practically impossible due to limitations of equipment and personnel. For example, in the manned-flight project, building the rocket and constructing the equipment may both require the same technical skills. If there are not enough personnel with these skills to go around, these two activities cannot be performed simultaneously.

Suppose that the manned flight of Example 26 must be sent to Mars in *less* than 7.25 years. This can be accomplished if some of the steps are carried out on a crash basis. Since this procedure is usually costly, it is important to apply it to as few activities as possible. Clearly, it does not help to carry out activity *c*, a noncritical activity, on a crash basis. Shortening the time spent on this activity cannot possibly shorten the length of the critical path, because *c* does not lie on this path. *If the completion time for the project is to be shortened, it is necessary to decrease the time required by one or more of the critical activities.*

Sometimes money can be saved by delaying certain noncritical operations. It is therefore often important to determine the maximum time that a noncritical activity can be delayed without increasing the total time spent on the project.

Activity *c* of Example 26, for example, does not have to be completed until activity *e* is scheduled to begin (i.e., at 5 years). Since *c* takes 3 years to complete, *c* does not have to be started until 2 years after the project is begun. A complete schedule for the rocket project is given in Table 10.7.

Table 10.7	Activity	Starting Time
	a	0
	b	0
	c	0–2
	d	0–1
	e	5
	f	6
	g	6.25
	h	7.25

10.5   EXERCISES

1.   The graph in Fig. 10.48 represents the scheduling (in hours) of a certain project.

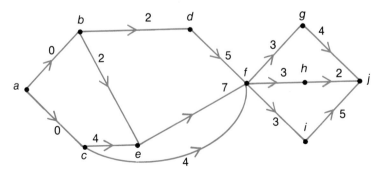

Figure 10.48

(a) What is the length of path a,c,f,i,j?

(b) Find the critical path.

(c) What are the critical activities?

(d) What is the earliest possible completion time?

(e) Schedule the project so that it can be completed as soon as possible. Show the earliest and latest possible starting times for each activity.

2.   Draw a graph with two critical paths.

Do the following for each project in Exercises 3–5.

(a) Draw the representative graph.

(b) Find the critical path.

(c) Find the earliest possible completion time.

(d) Schedule the activities so that the project can be completed at the earliest possible time. Show the earliest and latest possible starting times for each activity.

3.   Assembling a toy wagon requires the following steps:

Activity	Time (minutes)	Immediate Predecessors
Start (a)	0	None
Unpack parts and read instructions (b)	15	a
Bolt axles to body (c)	5	b
Attach wheels to axles (d)	5	c
Bolt handle to wagon (e)	2	b
Finish (f)	0	d, e

4.   The steps in installing a new door and new locks in a house are given in the following table:

Activity	Time (hours)	Immediate Predecessors
Start (a)	0	None
Remove door and widen opening (b)	3	a
Order parts (c)	48	a
Paint door and let dry (d)	24	b
Cut holes for locks (e)	1	c, d
Replace door (f)	1	e
Install locks (g)	2	f
Finish (h)	0	g

5. A contractor's plans for building a house involve the following activities:

Activity	Time (weeks)	Immediate Predecessors
Start (*a*)	0	None
Lay foundation (*b*)	3	*a*
Build basement (*c*)	2	*b*
Construct main floor (*d*)	2	*c*
Install roof(*e*)	2	*d*
Install wiring (*f*)	1	*e*
Install plumbing (*g*)	1.5	*e*
Install walls and floors (*h*)	3	*f, g*
Install electric fixtures (*i*)	2	*h*
Install plumbing fixtures (*j*)	2	*h*
Install cabinets (*k*)	1	*i, j*
Install doors and windows (*l*)	2	*h*
Install shingles and gutters (*m*)	2	*k, l*
Paint house interior (*n*)	2.5	*m*
Paint house exterior (*o*)	2	*m*
Do landscaping (*p*)	1	*o*
Stop (*q*)	0	*n, p*

10.6 MINIMAL SPANNING TREES

In this section we will study a type of graph called a *tree*. We will see that trees can be used to solve many practical problems, including finding the cheapest way to construct television or railway networks between cities.

In order to understand this new concept of tree, consider the following situation: A network of television relays is to be constructed for a certain region in which there are eight cities, *a, b, c, ..., h*. Graph *G* in Figure 10.49 shows the possible connections and their relative cost.

In constructing a television network, one certainly does not have to construct all possible connections between cities. One needs only to make certain that each city is included in the network. It is, of course, also important to minimize the cost.

To accomplish these goals, we will construct a graph *T* that contains all the vertices of the graph *G* of Figure 10.49, but only some of the edges. Each vertex of *G* will be an endpoint of some edge of *T*, and the sum of the numbers associated with these edges will be as small as possible. The graph *T* will clearly be connected, have at least two vertices, and contain no circuits. (If *T* contained a circuit, the cost would not be minimal.) Graphs of this kind are called *trees*.

To make our discussion easier, we will use the following definitions:

Definition 10.10 Let *G* and *H* be undirected graphs. Then *H* is a **partial graph** of *G* if and only if—

(a) *H* contains the same vertices as *G*

(b) *H* contains some (possibly all) of the edges of *G*

(c) Every edge of *H* is an edge of *G*.

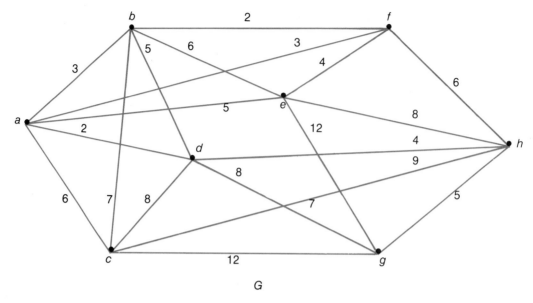

Figure 10.49 Graph for the television relay network

Definition 10.11

(a) A **tree** is an undirected, connected graph that has at least two vertices and that contains no circuit.

(b) If a graph G contains a partial graph T which is a tree, then T is called a **spanning tree** of G.

It can be shown that a graph G will contains a spanning tree if and only if G is connected.

The following examples illustrate these new concepts.

Example 27 Which of the graphs in Figure 10.50 are partial graphs of G?

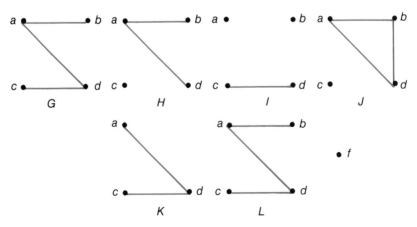

Figure 10.50

Solution   Only G, H, and I are partial graphs of G. The graph J is not a partial graph of G because (b,d) is an edge of J but not an edge of G; K is not a partial graph of G because b is a vertex of G but not a vertex of K. The graph L is not a partial graph of G because f is a vertex of L but not a vertex of G.

Example 28   Which of the graphs in Figure 10.51 are trees?

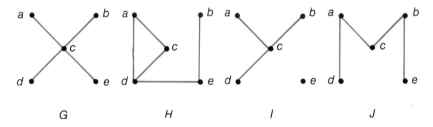

G   H   I   J

Figure 10.51

Solution   Only G and J are trees. The graph H is not a tree because it contains a circuit, a,c,d. The graph I is not a tree because it is not connected.

Example 29   Find two spanning trees for the graph in Figure 10.52.

Figure 10.52

Solution   If T is a spanning tree, it must contain all the vertices of G, but only some of the edges of G. (Since G is not a tree, T cannot contain all of the edges of G.) The reader can check that the two graphs in Figure 10.53 are spanning trees for G. The graph G also has other spanning trees.

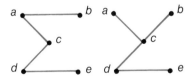

Figure 10.53   Two spanning trees for Figure 10.52

The problem of building a network of television relays at minimal cost can now be stated in general terms: Let G be a connected, undirected graph with edges $E_1, E_2, \ldots, E_n$. Assign a positive number $L(E_i)$, called the *length of E_i*, to every edge E_i. In the network problem, $L(E_i)$ represents the cost of the relays along route E_i. Find a spanning tree H whose total length

$$L = L(E_1) + L(E_2) + \cdots + L(E_n)$$

is as short as possible. Then H is called a **minimal spanning tree**.

This problem can be solved by a process known as *Kruskal's algorithm* (after mathematician J. B. Kruskal, Jr.).

Method for Finding a Minimal Spanning Tree

Let G be a connected graph. A minimal spanning tree T for G can be constructed as follows:

(1) Select the shortest edge of G. This is the first edge of T. (If there is more than one shortest edge, arbitrarily choose one of these edges.)

(2) Select the shortest remaining edge of G that does not form a circuit with the edges already included in T.

(3) Repeat step 2 until T contains all the vertices of G.

The following example illustrates the use of Kruskal's algorithm.

Example 30 Determine a network of television relays of minimum cost for the cities of Figure 10.49.

Solution
  (a) The shortest edges of G have length 2; we may therefore begin by selecting either edge (b,f) or edge (a,d). We will arbitrarily choose (b,f).

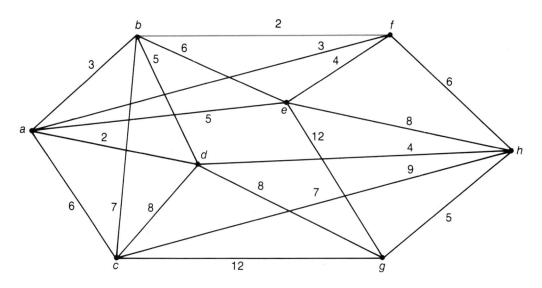

(b) The shortest remaining edge has length 2. Since (a,d) does not form a circuit when added to the graph drawn in part **(a)**, we may add (a,d). This yields the following graph:

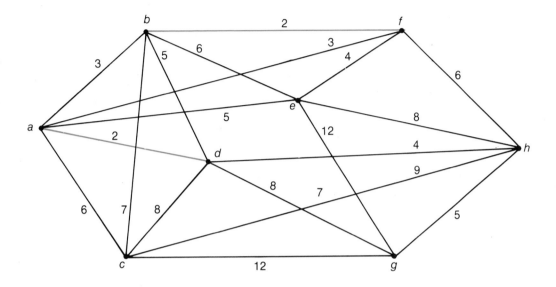

(c) The shortest remaining edge has length 3. We may choose either edge of length 3, (a,b) or (a,f), because neither of these edges forms a circuit when added to the graph of part **(b)**. Suppose we choose (a,f). Our tree now looks like this:

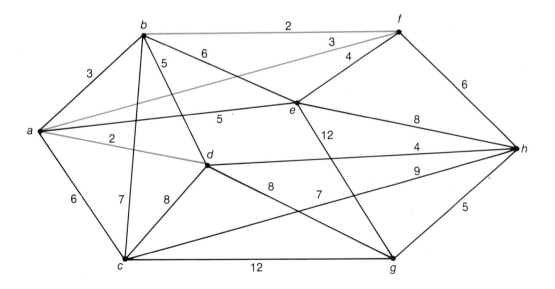

Although (a,b) is now the shortest remaining edge of G, we cannot choose (a,b) because (a,b), (b,f), and (a,f) form a circuit:

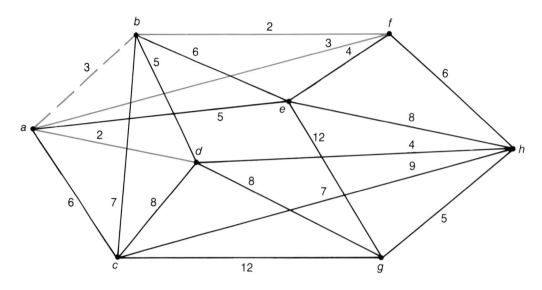

(d) Our tree now has three edges: (b,f), (a,d), and (a,f). However, we are not yet finished, because vertices c, e, g, and h are not in the tree. The two edges of length 4 (d,h), and (e,f), can both be added to the tree because no circuit will be formed.

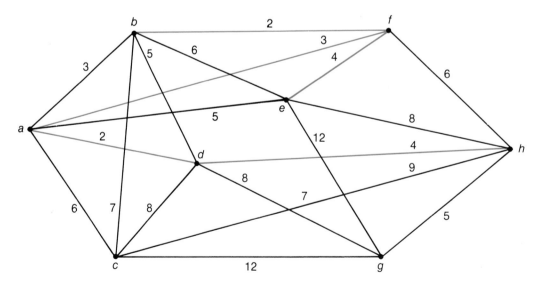

(e) The only edge of length 5 that can be added to the tree is (g,h); the others would form circuits with edges already chosen. The only edge of length 6 that can be added is (a,c). The minimal spanning tree is therefore:

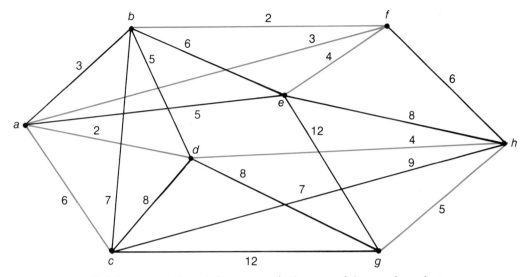

It follows that the minimum cost is the sum of the numbers that appear on the edges of this graph; that is, the minimum cost is

$$3 + 2 + 4 + 2 + 6 + 4 + 5 = 26$$

The minimum spanning tree of Example 30 is not unique: by choosing (a,b) instead of (a,f) in step **(c)**, we can obtain a minimal spanning tree with edges (a,d), (b,f), (a,b), (e,f), (d,h), (g,h), and (a,c). The minimum cost associated with this tree is

$$2 + 2 + 3 + 4 + 4 + 5 + 6 = 26$$

which agrees with our previous result.

Although the graph of Figure 10.49 represents a network of television relays between cities, it could also represent a network of railway lines, airplane or bus routes, oil pipelines, or whatever. The results of this section can therefore be applied to the problems of finding networks of minimal cost (or minimal distance) for each of these applications.

10.6 EXERCISES

1.   Which of the graphs in Figure 10.54 are trees? Explain.

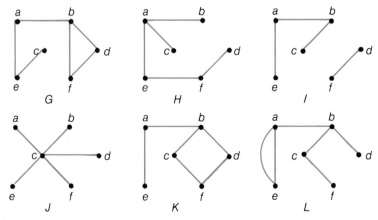

Figure 10.54

2.   Consider the graph G in Figure 10.55. Which of the graphs H, I, J, K, and L in Figure 10.55 are (a) partial graphs of G, (b) spanning trees of G?

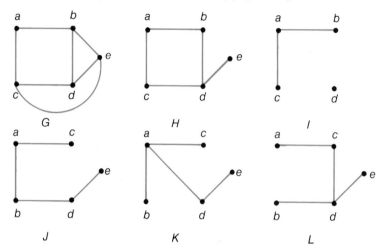

Figure 10.55

3.   Find a minimal spanning tree for each of the graphs in Figure 10.56.

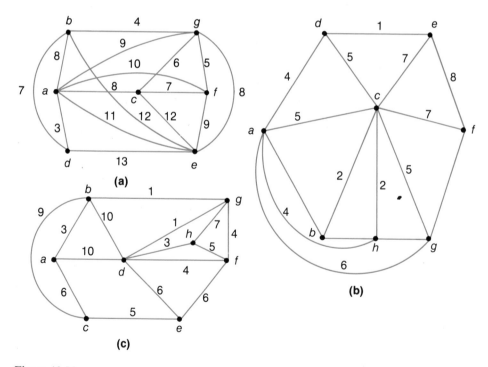

Figure 10.56

4.   A company wishes to build an intercommunication system connecting its eight branches. The distances between the branches are given in Table 10.8. For

	a	b	c	d	e	f	g	h
a	0	23	42	31	27	38	29	33
b		0	29	36	25	34	44	31
c			0	41	33	28	31	38
d				0	33	36	45	37
e					0	31	36	40
f						0	41	32
g							0	25

Table 10.8

example, the distance between branches a and f is 38, which is the number in the intersection of the row labeled a and the column labeled f.

Assume that the cost of directly connecting two branches is some constant k times the distance between them. (In other words, the further apart two branches are, the more it will cost to connect them.) Also assume that messages from one branch can be relayed to another without any loss of efficiency.

(a) Find the cheapest way to build the system.

(b) What is the total cost?

Key Word Review

Each key word is followed by the number of the section in which it is defined.

graph   10.1
vertices   10.1
arcs   10.1
edges   10.1
directed graph   10.1
undirected graph   10.1
initial vertex   10.1, 10.2
terminal vertex   10.1, 10.2
adjacent vertices   10.1
isomorphic graphs   10.1
path   10.2
circuit   10.2
simple path   10.2
elementary sircuit   10.2

Euler path   10.4
Euler circuit   10.4
degree (of a vertex)   10.4
connected graph   10.4
incoming degree   10.4
outgoing degree   10.4
length of a path   10.5
critical path   10.5
critical activities   10.5
partial graph   10.6
tree   10.6
spanning tree   10.6
minimal spanning tree   10.6

CHAPTER 10   **REVIEW EXERCISES**

1.   Define each word in the Key Word Review.

2.   What do the following symbols mean?

  (a) $a \bullet \longrightarrow \bullet b$

  (b) $a \bullet \longrightarrow \bullet b$

  (c) (a,b)

  (d) $p(x)$

  (e) $p_i(x)$

  (f) $p_o(x)$

3.   Draw a graph with five vertices, a, b, c, d, and e; three arcs, (a,d), (b,c), and (a,e); and two edges, (d,b) and (c,e).

4.   Draw a graph with six vertices labeled 2, 3, 4, 5, 6, and 10. Let the arc (a,b) represent the fact that b is divisible by a and $a \neq b$. Draw all such arcs to complete the graph.

5.   Which of the graphs in Figure 10.57 (a) are connected, (b) contain an Euler path, (c) contain an Euler circuit, (d) are trees, (e) contain spanning trees, (f) are undirected?

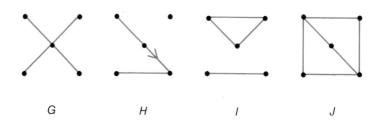

  G H I J

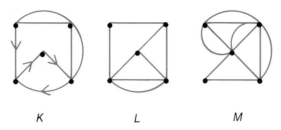

K   L   M

Figure 10.57

6.   For the graph in Figure 10.58, find (a) $p_i(a)$, (b) $p_o(b)$, and (c) a path from a to e.

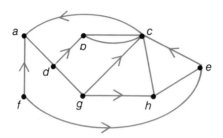

Figure 10.58

7.   Let G be the graph in Figure 10.58. If possible, find—
     (a) a path from f to h that is not simple
     (b) an elementary circuit beginning at a
     (c) a simple path from f to g
     (d) an Euler circuit
     (e) an Euler path that is not a circuit.

8.   Which of the following pairs of graphs are isomorphic? Explain your answers.

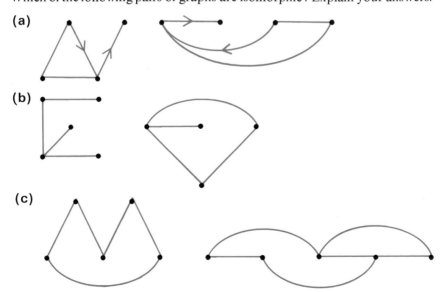

9. Find all simple paths from *a* to *f* in the graph in Figure 10.59.

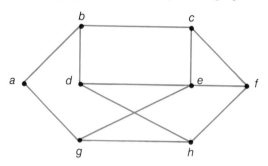

Figure 10.59

10. If possible, draw the picture in Figure 10.60 without lifting your pencil from the paper or retracing any lines.

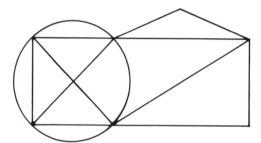

Figure 10.60

11. A highway inspector wishes to plan a route that will allow him to start from his office, inspect each of the highways on the map in Figure 10.61 exactly once, and return to his office.

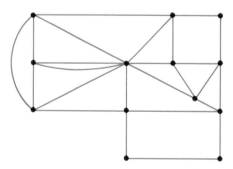

Figure 10.61

(a) Is such a route possible? Explain.

(b) Where should his office be located?

(c) Find the route.

12. A road inspector must inspect each of the roads on the map in Figure 10.62. If possible, plan a route that will allow him to inspect each one-way street (indicated by an arc) exactly once, and each two-way street (indicated by an edge) exactly once in each direction. (He does not have to return to his starting point.)

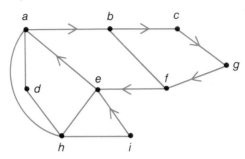

Figure 10.62

13. The following operations are required to start a taxi service:

Activity	Time (months)	Immediate Predecessors
Start (a)	0	None
Rent office and garage (b)	6	a
Buy cars (c)	2	a
Paint cars and install two-way radios (d)	1	c, e
Hire drivers (e)	6	b, c
Advertise (f)	5	d, e
Train drivers and office personnel (g)	4	e
Finish (h)	0	f, g

(a) Draw a graph having eight vertices, one for each activity. Let the arc (x,y) show that activity x is an immediate predecessor of activity y.

(b) What is the shortest possible time in which a taxi service can be started?

(c) Find the critical path and the critical activities.

(d) Schedule the activities so that the work can be completed in the shortest possible time. Show the earliest and latest possible starting dates for each activity.

14. A motorist wishes to travel from city a to city k. The distances between the cities are tabulated in Table 10.9. (A dash in the table means that there is no direct route between the corresponding cities.)

(a) Find the shortest possible route.

(b) What is the length of this route? (Hint: Draw a graph with eleven vertices, one for each city. Draw the edge (i,j) if and only if $i \neq j$ and there is a direct route between i and j. Label the edge x if the distance between *i* and *j* is *x*. List all simple paths from *a* to *k* and their corresponding mileages.)

Table 10.9

	a	b	c	d	e	f	g	h	i	j	k
a	0	6	3	5	—	—	—	—	—	—	—
b		0	—	—	8	1	—	—	—	—	—
c			0	2	4	1	2	—	—	—	—
d				0	—	5	2	—	—	—	—
e					0	4	—	—	6	2	—
f						0	2	2	4	—	—
g							0	8	—	—	—
h								0	—	—	1
i									0	2	2
j										0	5
k											0

15.   **(a)** Find two spanning trees for the graph in Figure 10.63.

       (b) What is the length of the minimal spanning tree?

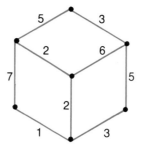

Figure 10.63

16.   A network of customers must be supplied with a certain utility that requires pipes. The graph in Figure 10.64 shows the distances between the different customers and the utility company, *a*. Suppose that the cost of the pipe is some constant *k* times the pipe length.

       (a) Find the cheapest way to build the system.

       (b) What is the total cost of the system?

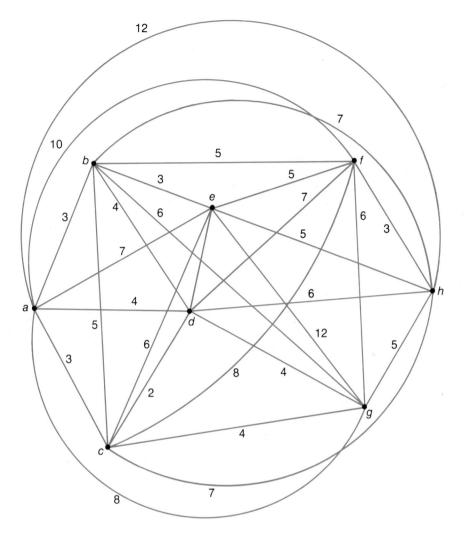

Figure 10.64

Suggested Readings

1. Bellman, Richard E.; Cooke, Kenneth L.; and Lockett, J. *Algorithms, Graphs and Computers.* New York: Academic Press, 1970.
2. Busacker, Robert G., and Saaty, T. L. *Finite Graphs and Networks: An Introduction with Applications.* New York: McGraw-Hill Book Co., 1965.
3. David, H. A. "Tournaments and Paired Comparison," *Biometrika* 46 (1959): 139–49.
4. Euler, Leonhard, "The Koenigsberg Bridges," In *Mathematics in the Modern World.* Readings from *Scientific American.* San Francisco: W. H. Freeman & Co., 1968.
5. Harary, Frank; Norman, R. Z.; and Cartwright, C. C. *Structural Models: An Introduction to the Theory of Directed Graphs.* New York: John Wiley & Sons, 1965.
6. Moon, John W. *Topics on Tournaments.* New York: Holt, Rinehart & Winston, 1968.
7. Ore, Oyenstein. *Graphs and Their Uses.* New York: Random House, 1963.
8. Tucker, A. "Perfect Graphs and an Application to Optimizing Municipal Services." *SIAM Review* 15 (July 1973).
9. Wiest, J., and Levy, F. *A Management Guide to Pert-CPM.* Englewood Cliffs, N.J.: Prentice-Hall, 1969.
10. Wilson, Robert J. *Introduction to Graph Theory.* New York: Academic Press, 1972.

Port of Los Angeles

CHAPTER 11 Networks and Flows

INTRODUCTION

If two cities need oil and there is oil available at several seaports, then the obvious solution to the oil shortage is to load the oil onto tankers like the one pictured at left and send the oil to the cities that need it. However, before this can be done, several practical questions must be answered: Do the seaports have enough oil to supply the cities? Are there enough tankers available to ship all of the needed oil? Is there a direct sea route from each seaport to each city that needs oil? If there are several possible sea routes, which one is best? Which seaports should ship oil to which cities?

In this chapter we will extend our knowledge of graph theory to answer questions like these. We will develop the concepts of *transportation network*, *flow*, and *matching* and use these concepts to analyze and solve transportation and assignment problems. We will develop techniques that will enable us, for example, to plan the cheapest shipping route between a mine and a steel mill and to make the best match between jobs that need doing and people who can do them.

11.1 TRANSPORT NETWORKS AND FLOWS

In this section we will learn how to apply graph theory to problems dealing with transporting goods from suppliers to customers. In our first example we will be concerned with shipping oil from three seaports to two cities. Our object will be to see that enough oil is shipped each week to meet the two cities' needs. We will assume that, as in real life, there is a limit on the amount of oil that can be shipped from one place to another each week.

Example 1   Oil is being exported to cities x and y from seaports a, b, and c. Each week, 20,000, 40,000, and 60,000 gallons of oil are available at a, b, and c, respectively, while 30,000 gallons are needed by x and 20,000 are needed by y.

The sea routes connecting seaports a, b, c and cities x, y are denoted by arcs on the graph in Figure 11.1(a). Each arc on this graph is labeled with a number

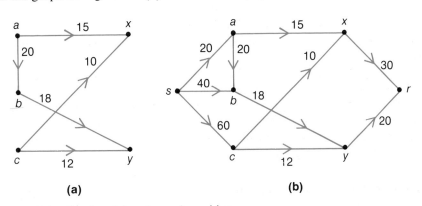

Figure 11.1 Graphs of the oil supply problem

that represents the maximum amount of oil (in thousands of gallons) that can be shipped via that route in a week.

In Figure 11.1(b) two extra vertices s and r have been added to the graph. These vertices represent the supply and the demand. Each arc (s,z) is labeled with a number that represents the amount of oil that seaport z can supply each week. Each arc (z,r) is labeled with a number that represents the amount of oil city z requires each week. The arc (s,a), for example, is labeled 20 because seaport a can supply 20,000 gallons of oil each week; the arc (x,r) is labeled 30 because x requires 30,000 gallons of oil each week.

The graph of Figure 11.1(b) is called a *transport network*. Notice that Figure 11.1(b) contains no loops (arcs that begin and end at the same vertex).

Definition 11.1   A directed graph that is connected and contains no loops is called a **transport network** if and only if—

(a)   There is exactly one vertex, called the **source**, that has no incoming arcs.

(b)   There is exactly one vertex, called the **sink**, that has no outgoing arcs.

(c)   There is a nonnegative number associated with each arc called the **capacity**. The capacity of the arc (i,j) is denoted $C(i,j)$.

It is clear that the graph of Figure 11.1(b) satisfies Definition 11.1. For example, the capacity $C(a,x)$ of the arc (a,x) is 15; $C(c,y) = 12$. The source and the sink are the vertices s and r, respectively. A vertex z of the graph is *adjacent* to the source s if it is connected to s by an arc (s,z). In Figure 11.1(b), a, b, and c are adjacent to the source.

A practical question that one might ask about a situation such as the one described in Example 1 is the following: Is it possible to organize the shipment of oil so that the demands of seaports x and y are met? If not, how should the shipments be organized so that the maximum possible amounts of oil will reach x and y?

Although there are only three suppliers and two customers, any attempt to solve this problem logically shows how complicated it is. We might, for example, begin by reasoning as follows: The total supply of oil available at seaport a is 20,000 gallons. (This is indicated on the graph of Figure 11.1(b) by the fact that a has only one incoming arc (s,a) and the capacity of this arc is 20.) Clearly, a cannot distribute more oil than it has. Now a has *two* outgoing arcs, (a,b) and (a,x). The endpoints of these arcs, b and x, are the potential recipients of a's oil. Thus, a can send all 20,000 (gallons of oil) to b and 0 to x; or a can send 5000 to b and 15,000 to x; or a can send 12,500 to b and 7500 to x; and so on. Seaport a cannot, however, send 0 to b and 20 to x because the capacity of route (a,x) is only 15.

Although we have not even begun to consider how seaports b and c should distribute their oil, we already have a large number of possible plans. The completion of each of these plans by logic alone is clearly impractical. In the next example we will see how certain graphs called *flows* can help us organize the data for possible oil shipments. In Section 11.2 we will see how these graphs can also be used to find the best possible shipment plan.

Example 2   Find an arrangement for the shipment of oil that fits the requirements of Example 1.

Solution   In order to organize our thoughts about this problem, let us first redraw Figure 11.1(b) omitting all of the numbers. Our decisions about the way that the oil should be distributed will be indicated on this graph as follows:

(a) All arcs (s,z) from the source to vertex z will be labeled with the amount of oil that seaport z supplies.

(b) All arcs (z,r) from vertex z to the sink r will be labeled with the amount of oil that city z receives.

(c) All arcs (z,w) from vertex z to vertex w where z is not s and w is not r will be labeled with the amount that seaport z sends to city or seaport w.

In the discussion preceding this example, we saw that there are many ways that seaport a can distribute its oil. Of these possible ways, suppose we arbitrarily decide to have seaport a send 15,000 gallons of oil to x and the remainder 5000 gallons to b. This decision is indicated in Figure 11.2(a) by the number 15 on the arc (a,x) and the number 5 on the arc (a,b). The fact that the entire $15 + 5 = 20,000$ gallons originally came from a's own supply is indicated by the number 20 on the arc (s,a).

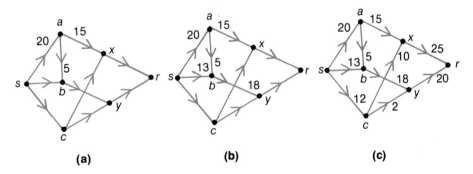

Figure 11.2   Possible solution for the oil supply problem

We must now decide how seaport b is to distribute its $40 + 5 = 45,000$ gallons of oil (40,000 from b's original supply and 5000 from seaport a). Notice that vertex b has only one outgoing arc. This arc, (b,y), has capacity 18. Hence, although y requires 20,000 gallons, b can send y at most 18,000. Since b has 45,000 gallons, we can arbitrarily decide to let b send 18,000 gallons to y. This decision is shown by placing the number 18 on the arc (b,y). (See Figure 11.2(b).) Of these 18,000 gallons, 5000 originally came from a to b and the remaining 13,000 originally came from b's own supply. This information is shown in Figure 11.2(b) by the numbers 5 on (a,b) and 13 on (s,b).

In a similar manner, we can now decide to have seaport c send 2000 gallons to y and 10,000 to x. This decision is shown in Figure 11.2(c) by the numbers 2 on the arc (c,y) and 10 on the arc (c,x). The fact that the entire $10 + 2 = 12,000$ gallons originally came from c's own supply is indicated by the number 12 on the arc (s,c). Figure 11.2(c) also indicates that x receives $15 + 10 = 25,000$ gallons (indicated on the arc (x,r)) and y receives $18 + 2 = 20,000$ gallons (indicated on the arc (y,r)).

The following is a summary of the schedule we have developed:

a sends 20,000	x receives 25,000
b sends 13,000	y receives 20,000
c sends 12,000	Total 45,000
Total 45,000	

These figures are all indicated in Figure 11.2(c). Notice that the total amount received by x and y is 45,000 gallons, 5000 less than required. We will have to wait until Section 11.2 to see if this is the best possible schedule.

In Example 2 we tried to arrange the shipments so that a maximum amount of oil will reach seaports x and y. The data for the problem are given in the network of Figure 11.1(b). A possible solution (perhaps not the best one) is shown in Figure 11.2(c). This possible-solution graph was found by using the network of Figure 11.1(b) and the following rules:

(a) The amount of material to be shipped along a route cannot be greater than the capacity of the route.
(b) Except at the source and the sink, the amount of material flowing into a vertex must equal the amount of material flowing out of the vertex.

Graphs that are formed from a network in this manner are called *flows*.

Definition 11.2   Let N be a transport network. A graph F having the same vertices and arcs as N is called a **flow** for N if and only if there is a nonnegative number associated with each arc of F called its **flow**. The flow of the arc (i,j) is denoted $F(i,j)$ and satisfies the following requirements:

(a)   The flow of each arc is less than or equal to its capacity.
(b)   If x is any vertex except the source or the sink, then the sum of all incoming flows at x equals the sum of all outgoing flows at x.

Part (a) of Definition 11.2 implies that

$$F(i,j) \leq C(i,j) \tag{1}$$

for every arc (i,j) in F.

The reader can easily verify that Figure 11.2(c) is a flow for the network of Figure 11.1(b). For example, the flow of the arc (s,a) is 20. This is less than or equal to the capacity of this arc, which is also 20. Note also that $F(c,y) = 2 \leq C(c,y) = 12$.

The following definitions are needed before we can use the concept of flow to solve transportation problems such as the one in Example 1.

Definition 11.3   Let F be a flow. An arc (i,j) is a **saturated arc** if and only if $F(i,j) = C(i,j)$; it is an **unsaturated arc** if $F(i,j) < C(i,j)$.

Definition 11.4   Let F be a flow and let $a_1, a_2, \ldots, a_k$ be all the vertices that are adjacent to the source s. The quantity

$$F_v = F(s,a_1) + F(s,a_2) + \cdots + F(s,a_k)$$

is called the **value of the flow**.

In other words, the value of the flow is the total outgoing flow at the source.

Definition 11.5   A **maximal flow** for a transport network N is a flow that has the largest possible value.

Example 3 illustrates these concepts.

Example 3   Consider the flow of Figure 11.2(c).

(a) Comparing the capacities of Figure 11.1(b) with the flow of Figure 11.2(c), we see that

$$F(s,a) = C(s,a), \qquad F(a,x) = C(a,x), \qquad F(b,y) = C(b,y)$$
$$F(c,x) = C(c,x), \qquad \text{and} \qquad F(y,r) = C(y,r)$$

It follows that the arcs (s,a), (a,x), (b,y), (c,x), and (y,r) are saturated. The remaining arcs are unsaturated because

$$F(s,c) < C(s,c), \qquad F(s,b) < C(s,b), \quad \text{etc.}$$

(b) There are three arcs, (s,a), (s,b), and (s,c), that are adjacent to the source s. Using Figure 11.2(c) and Definition 11.4, the value of the flow is

$$F_v = F(s,a) + F(s,b) + F(s,c) = 20 + 13 + 12 = 45$$

Notice that in Figure 11.2(c) the total outgoing flow at the source s ($F_v = 45$) is equal to the total incoming flow at the sink r ($F(y,r) + F(x,r) = 20 + 25 = 45$). Since everything that is sent must be received by someone, it is intuitively clear that this will always be true. Hence, *the total incoming flow at the sink is always equal to F_v*.

Returning to the oil-shipment problem, it is also clear that the amount of oil received by x and y will be maximized if the total incoming flow at the sink is made as large as possible. Since we just saw that the total incoming flow at the sink is equal to the total outgoing flow at the source, F_v, the problem of Example 2 can be solved by finding a maximal flow for the network of Figure 11.1(b).

In general, many practical problems can be solved by finding a maximal flow for a network. The next example shows how this flow can sometimes be found by reasoning from the given information and using the concepts that we have just discussed.

Example 4   A steel company wishes to transport iron ore from its mine, a, to its plant, b. The railway connections between the mine, the plant, and the intermediate cities (c, d, e, and f) are shown in Figure 11.3. The number appearing on each arc represents the available space, in tons of ore, on the train that runs

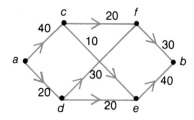

Figure 11.3   Graph for the iron-ore transport problem

between the two cities. Find the maximum amount of ore that can be transported from a to b.

Solution   The graph in Figure 11.3 is a network with source a and sink b. To solve the transport problem represented by this graph, we need to find a maximal flow for this network. We know that the flow will be maximal if the total outgoing flow at the source a is as large as possible; that is, if

$$F_v = F(a,c) + F(a,d)$$

is maximized.

    Let us now try to make $F(a,c)$ as large as possible. Notice that there is exactly one arc, (a,c), that enters vertex c, and exactly two arcs, (c,e) and (c,f), that leave c. Definition 11.2(b) therefore implies that

$$F(a,c) = F(c,e) + F(c,f)$$

Hence, it follows from equation (1) that

$$F(a,c) = F(c,e) + F(c,f) \leq C(c,e) + C(c,f) = 10 + 20 = 30$$

Since $C(a,c) > 30$, Definition 11.2 will be satisfied if we let $F(a,c) = 30$. Notice that this is the maximum value that $F(a,c)$ can have. If $F(a,c)$ is 30, then we must let $F(c,e) = 10$ and $F(c,f) = 20$. These figures have been placed on the graph in Figure 11.4(a).

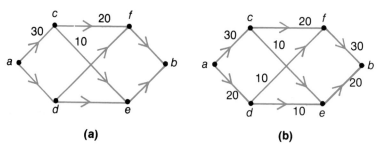

(a)   **(b)**

Figure 11.4   Solution of the iron-ore transport problem

    Now let us try to maximize the value of $F(a,d)$. Since

$$F(a,d) \leq C(a,d) = 20$$

$F(a,d)$ can be at most 20. Notice also that $F(a,d)$ is the only incoming flow at vertex d, and remember that the sum of the incoming flows at d must equal the sum of the outgoing flows at d. We must therefore find the sum of the outgoing flows before assigning a value to $F(a,d)$. From Figure 11.3 and Definition 11.2(a), the sum of the outgoing flows is

$$F(d,f) + F(d,e) \leq C(d,f) + C(d,e) = 30 + 20 = 50$$

It follows that $F(a,d)$ must also be less than or equal to 50. Since $20 \leq 50$, we may let $F(a,d) = 20$.

We now have many choices for $F(d,f)$ and $F(d, e)$. We may choose any numbers satisfying

$$F(d,f) + F(d,e) = 20, \qquad F(d,f) \leq 30, \qquad \text{and} \qquad F(d,e) \leq 20$$

Suppose we arbitrarily let $F(d,f) = F(d,e) = 10$. Then, since the sum of all incoming flows at e and f must equal the sum of all outgoing flows,

$$F(f,b) = 10 + 20 = 30 \qquad \text{and} \qquad F(e,b) = 10 + 10 = 20$$

These figures have all been placed on the graph of Figure 11.4(b).

The graph of Figure 11.4(b) satisfies the requirements of Definition 11.2 and hence is a flow for the network of Figure 11.3. This flow is maximal because it maximizes F_v, the total outgoing flow of the source. It follows that the maximum amount of ore that can be transported is

$$F_v = F(a,c) + F(a,d) = 30 + 20 = 50 \text{ tons}$$

This amount of ore will be transported if the transportation schedule indicated by Figure 11.4(b) is used.

In the next section we will learn a different technique for finding a maximal flow for a network.

11.1 EXERCISES

1. Which of the graphs in Figure 11.5 are networks? Explain your answer.

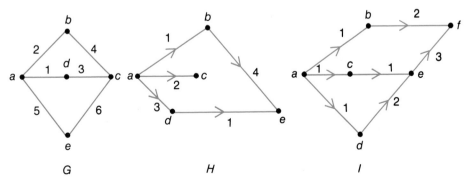

Figure 11.5

2.   Consider the network in Figure 11.6.

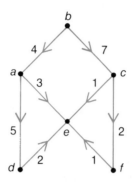

Figure 11.6

(a) What is the source of this graph? What is the sink?
(b) Find $C(a,d)$ and $C(f,e)$.

3.   Figure 11.7 is a flow for the network in Figure 11.6. Find—
(a) $F(a,d)$ (b) $F(c,e)$ (c) F_v

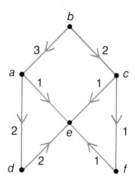

Figure 11.7

4.   Which of the graphs in Figure 11.8 are flows for the network in Figure 11.6? Explain.

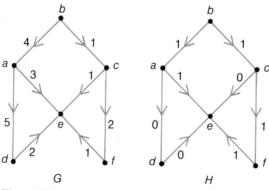

Figure 11.8

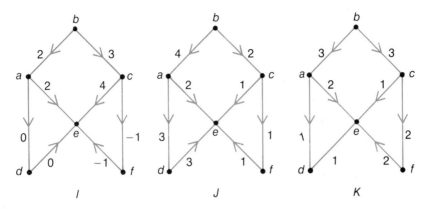

Figure 11.8

5. Consider the network of Figure 11.6.
   (a) Find two maximal flows G and H.
   (b) Find F_v for G and for H.
   (c) Consider the flow G. What is the total outgoing flow of the source? What is the total ingoing flow of the sink?

6. Seven kinds of tractors, $t_1, t_2, \ldots$, and t_7, are to be flown to a destination by five cargo planes. There are 4, 4, 3, 5, 3, 5, 4 tractors of types $t_1, t_2, \ldots, t_7$, respectively. The capacities of the planes are 7, 7, 6, 4, and 4 tractors, respectively.
   (a) Draw a transport network that represents the following problem: Can the tractors be loaded so that no two tractors of the same type are on one plane? (Hint: Draw a graph with fourteen vertices: five vertices p_i representing the planes, seven vertices t_i representing the kinds of tractors, and two vertices, a and b, representing the sink and the source.)
   (b) Solve the problem stated in part **(a)**.

7. The vertices of the graph in Figure 11.9 represent tutors t_i and students s_j at a university learning center. The arc (t_i, s_j) indicates that student s_j would like to be tutored by tutor t_i.

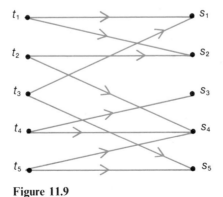

Figure 11.9

(a) Draw a transport network T that represents the following problem: Can the five tutors be assigned to the five students in a way that will allow each student to be tutored by a different tutor that he or she wants?

(b) Let s be the source and r be the sink of T. Verbally state what $C(s,t_1) = 1$, $C(s_2,r) = 1$, and $C(t_1,s_2) = 1$ represent.

(c) Find a maximal flow for T.

(d) Use the maximal flow F to arrange the required tutor-student assignments.

11.2 MAXIMAL FLOWS

We saw in Section 11.1 that the solution of many practical problems requires finding a maximal flow for a network. Sometimes, as in Example 4, this can be done by careful inspection. Other problems, such as Example 1, are more complicated.

Consider the network of Figure 11.1(b), page **481**. Finding a flow for this network is relatively simple. (See Figure 11.2(c).) However, deciding whether this flow is maximal is quite difficult, given only the techniques we have developed so far.

Suppose we let T_1 be the total capacity of the source and T_2 be the total capacity of the sink in Figure 11.1(b). Then

$$T_1 = C(s,a) + C(s,b) + C(s,c)$$
$$= 20 + 40 + 60 = 120$$

and

$$T_2 = C(x,r) + C(y,r)$$
$$= 30 + 20 = 50$$

It is clear that the value of any flow cannot be bigger than the smaller of these two numbers; that is, it cannot be bigger than 50. Hence, if we find a flow F with value 50, we know that F is maximal. However, this does *not* mean that the flow of Figure 11.2(c) is not maximal, just because its value is less than 50. (Its value is 45.) It is possible that no flow of higher value exists. We therefore need a method for determining whether a flow is maximal when its value is less than the smaller of T_1 and T_2.

To develop such a method, let us consider the network N of Figure 11.10(a) and the flow F_1 of Figure 11.10(b). (Note that a is the source of N and c is the sink.) If this

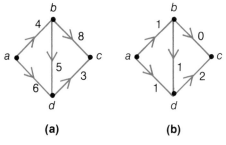

Figure 11.10   (a) Network N (b) First flow F_1 for N

flow is not maximal, we should be able to find another flow with a larger value. Since the value of any flow for N is

$$F_v = F(a,b) + F(a,d)$$

We can increase the value of F_1 by increasing $F(a,b)$ or $F(a,d)$.

Of course, these numbers cannot be increased arbitrarily. We must be certain that the requirements of Definition 11.2 are satisfied. For example, we know that the total incoming flow of any vertex must equal its total outgoing flow. Applying this rule to vertex d of N yields

$$F(a,d) + F(b,d) = F(d,c)$$

Hence, if we increase the flow from a to d by some number D, we must either

(a) increase the flow from d to c by D

or

(b) decrease the flow from b to d by D.

Suppose that we choose the first of these two alternatives. This means that we increase the flow from a to d by D and also increase the flow from d to c by D. In other words, we are increasing the flow of each arc on the path a,d,c by D.

We must now decide how large this number D can be. Since the flow of (a,d) must be less than its capacity, the flow from a to d cannot be increased by more than

$$C(a,d) - F(a,d) = 6 - 1 = 5$$

Similarly, the flow from d to c cannot be increased by more than

$$C(d,c) - F(d,c) = 3 - 2 = 1$$

Since we wish to increase the flows of both (a,d) and (d,c) by the same number D, the largest possible value of D is the smaller of these two numbers; that is,

$$D = \min(5,1) = 1$$

where "$\min(a,b)$" denotes the minimum (smaller) of the two numbers a and b. Increasing the flow of each arc on the path a,d,c by $D = 1$ yields the flow F_2 of Figure 11.11.

Notice that the value of the second flow, F_2, is $1 + 2 = 3$, while the value of the first flow is $1 + 1 = 2$. It follows that the first flow, F_1, is not maximal.

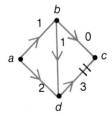

Figure 11.11   Second flow F_2 for N

In general, let F be any flow, let (i,j) be any arc, and let

$$D(i,j) = C(i,j) - F(i,j)$$

be the difference between the capacity of (i,j) and its flow. The preceding discussion suggests the following method for increasing the value of the flow:

(1) Choose any simple path $P = s, a_1, a_2, a_3, \ldots, a_n, r$ from the source s to the sink r.

(2) Calculate

$$D(s,a_1), D(a_1,a_2), \ldots, D(a_n,r)$$

and let D be the smallest of these numbers.

(3) Increase the value of the flow of each arc on the path P by D.

This method can now be applied to the second flow for N, F_2. (See Figure 11.11.) We must first choose a simple path P from a to c. Suppose that (d,c) is on this path. Since (d,c) is saturated, its capacity and flow are equal. Hence,

$$D(d,c) = C(d,c) - F(d,c) = 0$$

It follows that D is also 0, and so adding D to the flow of each arc on P will not change the value of these flows. Therefore, if we wish to increase the flows, our path must not include any saturated arcs. In Figures 11.11–11.13 a saturated arc is indicated by double lines ($=$) placed on the arc.

Comparing Figure 11.11 with Figure 11.10(a), we see that we can choose the path a,b,c because this path does not contain any saturated arcs. Since

$$D(a,b) = C(a,b) - F(a,b) = 4 - 1 = 3$$

and

$$D(b,c) = C(b,c) - F(b,c) = 8 - 0 = 8$$

it follows that

$$D = \min(3,8) = 3$$

Increasing the flow of each arc on the path a,b,c by 3 yields the flow F_3 shown in Figure 11.12. Notice that arc (a,b) is now saturated. Notice also that the value of this flow is $4 + 2 = 6$. It follows that the second flow F_2, whose value is 3, is not maximal.

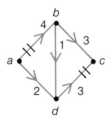

Figure 11.12 Third flow F_3 for N

The method we have used to derive F_2 and F_3 cannot be applied to the third flow for N because every simple path from a to c now traverses a saturated arc. We are, however, not finished. We have the second alternative on page 471 to consider. This alternative indicates that the value of a flow can sometimes be increased by *decreasing* the flows of certain arcs.

We should not, of course, decrease $F(a,b)$, $F(a,d)$, $F(b,c)$, or $F(d,c)$, because decreasing any of these will decrease the value of the flow. (Remember that $F_v = F(a,b) + F(a,d) = F(b,c) + F(d,c)$.) Our only choice, therefore, is to decrease $F(b,d)$ by some number D.

Since

$$F(a,d) + F(b,d) = F(d,c)$$

and

$$F(a,b) = F(b,d) + F(b,c)$$

if we decrease $F(b,d)$ by D, we must also increase $F(a,d)$ by D and increase $F(b,c)$ by D to avoid decreasing $F(d,c)$ and $F(a,b)$. Hence, the value of the flow can be increased by

> increasing $F(a,d)$ by D
>
> decreasing $F(b,d)$ by D

and

> increasing $F(b,c)$ by D

Referring to Figures 11.12 and 11.10(a), we see that

$$D(a,d) = C(a,d) - F(a,d) = 6 - 2 = 4$$

and

$$D(b,c) = C(b,c) - F(b,c) = 8 - 3 = 5$$

The number D cannot be more than the smaller of these two numbers; that is, $D \leq 4$. However, since $F(b,d) = 1$ and a flow cannot be negative, $F(b,d)$ cannot be decreased by more than 1. It follows that $D = 1$.

Increasing $F(a,d)$ and $F(b,c)$ by 1 and decreasing $F(b,d)$ by 1 yields the flow F_4 pictured in Figure 11.13. Since the value of this flow is $4 + 3 = 7$, the third flow for

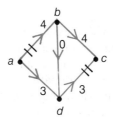

Figure 11.13   Fourth flow F_4 for N

N, F_3, is not maximal. However, F_4 is a maximal flow, as we will see after we generalize the process for increasing the flow of a network.

In order to generalize the process we have used to increase the flow for N, let us reconsider the arcs whose flows we increased. Notice that these arcs, (a,d), (b,d), and (b,c), do not form a path. However, if we reverse the direction of the arc whose flow we decreased, we obtain the arc (d,b), and arcs (a,d), (d,b), and (b,c) do form a path from a to c.

An arc like (d,b) that is formed by reversing the direction of an arc of a graph G is called a **pseudoarc** of G. The *flow of the pseudoarc* (x,y) is defined to be the flow of the arc (y,x); that is,

$$F(x,y) = F(y,x)$$

The *capacity of the pseudoarc* (x,y) is defined to be the capacity of the arc (y,x); that is,

$$C(x,y) = C(y,x)$$

A sequence of arcs and pseudoarcs (a_1,a_2), (a_2,a_3), ..., (a_{n-1},a_n) is called a **pseudopath** from a_1 to a_n. Using the same notation as that used for paths, the pseudopath (a_1,a_2), (a_2,a_3), ..., (a_{n-1},a_n) is also denoted $a_1,a_2, ..., a_n$. If $a_1,a_2,...,$ and a_n are all different, the pseudopath $a_1,a_2,...,a_n$ is a **simple pseudopath**. An example of a simple pseudopath is a,d,b,c in Figure 11.10(a).

Using these concepts, the process we used on the network N suggests the following method for increasing the value of the flow of a network:

(1) Choose any simple pseudopath P from the source to the sink.
(2) Suppose that the arcs of P are (b_1,b_2), $(b_3,b_4),...,(b_{k-1},b_k)$ and the pseudoarcs are (c_1,c_2), $(c_3,c_4),...,(c_{t-1},c_t)$. Determine

$$D(b_1,b_2), \ D(b_3,b_4), \ ..., \ D(b_{k-1},b_k)$$

and

$$F(c_1,c_2), \ F(c_3,c_4), \ ..., \ F(c_{t-1},c_t)$$

and let D be the smallest of these numbers.
(3) *Increase* the flow of each arc of P by D and *decrease* the flow of each pseudoarc of P by D.

The reasoning that we used with regard to paths also shows that we should not choose a pseudopath that contains either a saturated arc or a pseudoarc whose flow is zero. A careful examination of Figure 11.13 shows that every simple path contains a saturated arc and every simple pseudopath contains a pseudoarc whose flow is zero. It follows that this flow is maximal, because there is no simple pseudopath or path we can use to increase the value of the flow.

The following method summarizes our discussion of flows. It is illustrated by finding the maximal flow for the network in Figure 11.14.

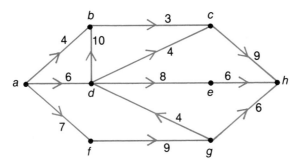

Figure 11.14

Method for Finding a Maximal Flow

(1)   Draw the flow having $F(i,j) = 0$ for all arcs (i,j).
(See Figure 11.15.)

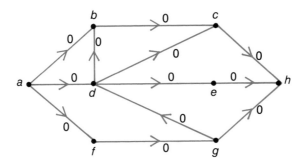

Figure 11.15   Zero flow for Figure 11.14

(2)   (a) Choose a simple path from the source to the sink that does not contain any
saturated arcs. Calculate $D(i,j) = C(i,j) - F(i,j)$ for each arc on the chosen
path, and let D be the smallest of the calculated values.

For example, we might choose a,d,b,c,h. (A different maximal flow may be
generated if we choose a different simple path, e.g., a,f,g,d,e,h.) Since

$$D(a,d) = 6 - 0 = 6, \qquad D(d,b) = 10 - 0 = 10$$
$$D(b,c) = 3 \quad \text{and} \quad D(c,h) = 9$$

D must be 3.

(b) Increase the flow of each arc on the chosen path by D. Place a double line
$(=)$ on each saturated arc.

Increasing the flow of each arc on the path a,d,b,c,h by $D = 3$ yields the
flow in Figure 11.16.

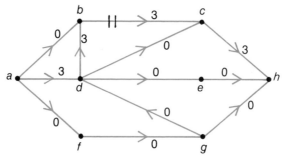

Figure 11.16 Each arc on path a,d,b,c,h has had its flow increased by $D = 3$

(3)   Repeat step 2 until all simple paths from the source to the sink contain a saturated arc.

Applying step 2 to the paths a,f,g,h, a,d,e,h, and a,f,g,d,e,h of Figure 11.16 yields the flows in Figure 11.17(a), (b), and (c).

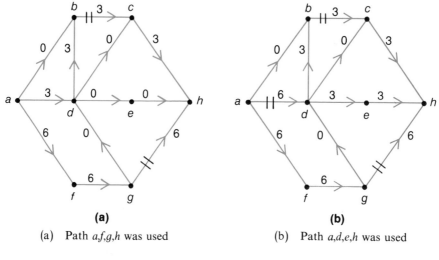

(a)   Path a,f,g,h was used   (b)   Path a,d,e,h was used

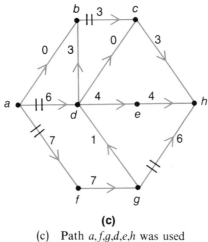

(c)   Path a,f,g,d,e,h was used

Figure 11.17   Step (2) has been applied to increase the flows on the arcs of three simple paths

(4)   (a) Choose a simple pseudopath P from the source to the sink that does not contain any saturated arc and that does not contain any pseudoarc whose flow is zero. Calculate $D(i,j)$ for each arc on P and determine $F(i,j)$ for each pseudoarc on P. Let D be the smallest of these numbers.

The pseudopath a,b,d,c,h of Figure 11.17(c) may be selected. Since

$$D(a,b) = 4 - 0 = 4, \qquad F(b,d) = 3$$

$$D(d,c) = 4 - 0 = 4 \qquad \text{and} \qquad D(c,h) = 9 - 3 = 6$$

D must be 3.

(b) Increase the flow of each arc on P by D, and place a double line $(=)$ on each saturated arc. Decrease the flow of each pseudoarc on P by D.

Applying this step with $D = 3$ to the pseudopath a,b,d,c,h of Figure 11.17 yields the flow of Figure 11.18.

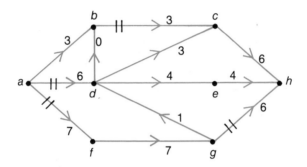

Figure 11.18   Step (4) has been applied to the pseudopath a,b,d,c,h

(5)   Repeat step 4 until all simple pseudopaths from the source to the sink contain a saturated arc or contain a pseudoarc whose flow is zero. The last flow drawn is maximal.

A careful inspection of the flow drawn in step 4 (Figure 11.18) shows that every simple pseudopath contains a saturated arc or a pseudoarc whose flow is zero. It follows that this flow is maximal.

Steps 3 and 5 of this method give us a test for determining whether a flow is maximal. This test is stated as the next theorem.

THEOREM   11.1   Let N be a network with source s and sink r. A flow F for N is maximal if and only if—

(a) every simple path from s to r contains a saturated arc

and

(b) every simple pseudopath from s to r either contains a saturated arc or contains a pseudoarc whose flow is zero.

In general, let F be any flow for a network N. Theorem 11.1 can be used to determine whether F is maximal. If F is not maximal, steps 2–5 of the method can be used to find a maximal flow.

The question raised in the beginning of this section may now be answered by applying Theorem 11.1: Is the flow of Figure 11.2(c) maximal for the network of Figure 11.1(b)? In other words, will the schedule developed in the solution of Example 2 (Section 11.1) result in the maximum shipments of oil to cities x and y? The following example answers these questions.

Example 5   For convenience, the network of Figure 11.1(b) and the flow of Figure 11.2(c) have been redrawn in Figures 11.19(a) and (b), respectively. All saturated arcs have been indicated on the second diagram. Is the flow of Figure 11.19(b) maximal?

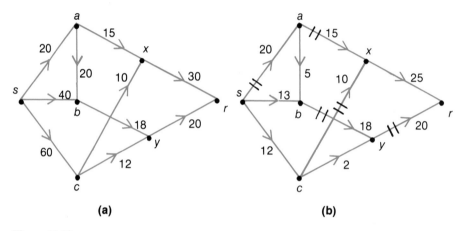

Figure 11.19

Solution   A careful inspection of Figure 11.19(b) indicates that every simple path and every simple pseudopath from source to sink contains a saturated arc. It follows from Theorem 11.1 that this flow is maximal. This means that the schedule of Example 2 will ship the maximum amount of oil to cities x and y.

11.2 EXERCISES

1.   Consider the network N and flow F in Figure 11.20.

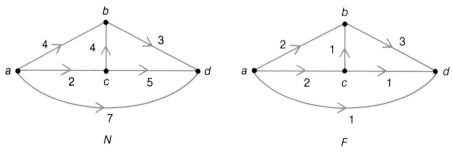

Figure 11.20

(a) Find $F(a,d)$, $C(a,d)$, and $D(a,d)$.

(b) Which arcs are saturated?

(c) Find two simple paths from the source to the sink.

(d) Find a simple pseudopath from the source to the sink.

(e) Find a flow of larger value by applying step 4 of the method developed in this section.

2.   Determine whether or not each of the flows in Figure 11.21 is maximal.

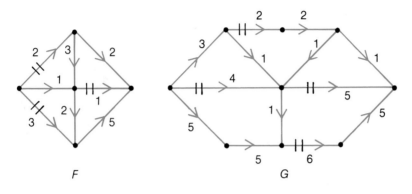

F   G

Figure 11.21

3.   (a) Find a second maximal flow for the network of Figure 11.19(a). (Hint: Apply the method of this section to the paths s,b,y,r, s,a,x,r, and s,c,y,r.)

(b) Using this flow, determine a second schedule for shipping the maximum amounts of oil to x and y.

   Answer the following questions for the transport networks of Exercises 4–6. (The flow on undirected edges can be in either direction.)

(a) Find a maximal flow.

(b) Find the value of the maximal flow.

4.

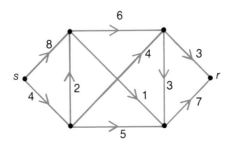

5.

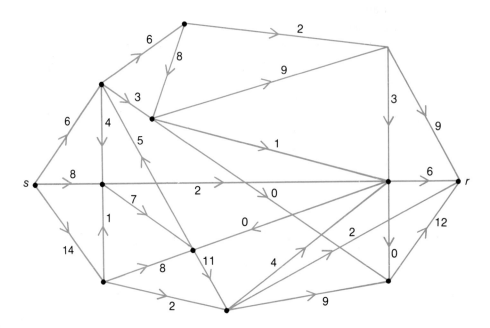

6.

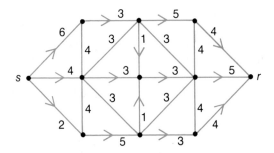

7. Find two maximal flows for the network shown in Figure 11.22.

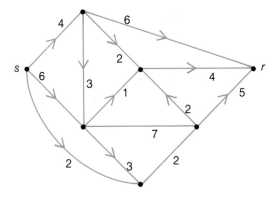

Figure 11.22

8. (a) Use the method developed in this section to find a maximal flow for the network of Example 4, Section 11.1.

(b) What is the value of this maximal flow?

9. Merchandise warehoused in three cities, v, t, and u, must be shipped to three other cities, c, d, and e. Cities v, t, and u can provide a maximum of 15, 25, and 45 units of merchandise, respectively; cities c, d, and e require 20, 25, and 15 units, respectively. Rail connections between the two cities are shown in Figure 11.23.

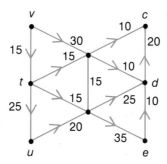

Figure 11.23

(a) Draw a transport network N that reflects the information given in this exercise.

(b) Find a maximal flow for N.

(c) How many units should each city send and each city receive so that as much merchandise as possible is transported to cities c, d, and e?

11.3 ASSIGNMENT PROBLEMS AND MATCHINGS

In Section 11.2 we saw that networks and flows can be used to find the best way to transport merchandise. In this section we will see that networks and flows can also be used to make job assignments.

Suppose that there are k positions that need to be filled and w people, each of whom is qualified for at least one of these jobs. Assignment problems, which we will now study, are concerned with the following questions: Is it possible to assign every one of these people to a job for which he or she is qualified? If not, what is the largest number of people that can be assigned to jobs? How should these assignments be made?

In the next example we will see how a simple assignment problem can be solved with the help of graphs.

Example 6 Suppose that there are three people, w_1, w_2, and w_3, willing to work, and four jobs, j_1, j_2, j_3, and j_4, that need to be done. Suppose further that worker w_1 is qualified for jobs j_2 and j_3; worker w_2 is qualified for job j_1; and worker w_3 is qualified for jobs j_2 and j_4. If possible, assign each worker to one job for which he or she is qualified.

Solution   The given information for this problem is conveniently represented by the graph in Figure 11.24(a). The people and jobs are represented by the vertices. Vertices w_i and j_k are joined by an edge if and only if worker w_i is qualified for job j_k.

A careful inspection of this graph reveals that the job assignment indicated by Figure 11.24(b) will solve the problem. This assignment matches w_1 to j_2, w_2 to j_1, and w_3 to j_4. Notice that although everyone is assigned to a job, no one is assigned to job j_3.

A second solution to the problem is indicated by the graph in Figure 11.24(c).

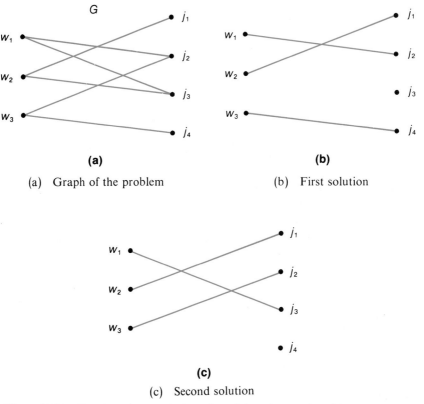

(a)   Graph of the problem   (b)   First solution

(c)   Second solution

Figure 11.24   Graph and two solutions for a job-assignment problem

Although the assignment problem of Example 6 was solved by inspection, most assignment problems are not so simple. We will see that more complicated assignment problems can be solved by finding a maximal flow for a network. In order to see how this might be done, let us reconsider Example 6.

In Example 6 we saw that the relationships between people and the jobs for which they are qualified can be represented by an undirected graph G. The vertices of this graph can be partitioned into two disjoint subsets, $W = \{w_1, w_2, w_3\}$ and $J = \{j_1, j_2, j_3, j_4\}$. Notice that no two vertices of W are connected by an edge and, similarly, no two vertices of J are connected by an edge. Graphs like G that can be partitioned in this manner are called *bipartite graphs*.

Definition 11.6 A **bipartite graph** is an undirected graph whose vertices can be divided into two nonempty, disjoint subsets X and Y which have the following properties:

(a) No two vertices of X are adjacent.

(b) No two vertices of Y are adjacent.

Returning to Example 6, we now see that the relationships between people and the jobs for which they are qualified can be represented by a bipartite graph G whose vertices are divided into the disjoint subsets W and J. The problem of assigning people to jobs for which they are qualified is now equivalent to the problem of selecting a subset of the edges of G that connects each vertex w in W to exactly one vertex j in J. This selection is called a *matching* of W into J and can be represented by graphs such as those in Figures 11.24(b) and (c).

Definition 11.7 Let X and Y be the two disjoint subsets of vertices in a bipartite graph G. Let M be an undirected graph whose vertices are the same as those of G and whose edges are all edges of G. (There may be some edges of G that are not edges of M.) Then M is a **matching** from X into Y if there is a subset X_o of X and a subset Y_o of Y such that—

(a) for every vertex x in X_o there is exactly one vertex y in Y_o for which (x,y) is an edge of M

and

(b) for every vertex y in Y_o there is exactly one vertex x in X_o for which (y,x) is an edge of M.

If vertices x and y are adjacent vertices of M, then x is said to be *matched* to y, and y is said to be matched to x.

Definition 11.8 Let G be a bipartite graph with disjoint subsets of vertices X and Y and let M be a matching between a subset X_o of X and a subset Y_o of Y. Then—

(a) M is a **complete matching** if and only if $X_o = X$

(b) M is a **maximal matching** if a maximum number of vertices in X are matched with vertices of Y.

In particular, every complete matching is maximal. The following examples illustrate these new concepts.

Example 7

(a) The matching of Figure 11.24(b) is a complete matching of W into J because $W = W_o = \{w_1, w_2, w_3\}$. Since it is complete, this matching is also maximal.

(b) The matching of Figure 11.24(b) is not a complete matching of J into W because $J_o = \{j_1, j_2, j_4\}$ does not equal $J = \{j_1, j_2, j_3, j_4\}$. However, this matching is maximal because no matching of J into W can match more than three vertices of J to the three vertices of W.

(c) Similar remarks hold for the matching of Figure 11.24(c).

Example 8   The graph G of Figure 11.25 is bipartite with disjoint subsets $X = \{a,b,c\}$ and $Y = \{d,e,f\}$. Determine which of the graphs in Figure 11.26 are matchings of G. Also determine which of the matchings are complete.

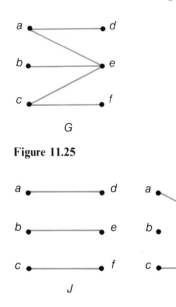

Figure 11.25

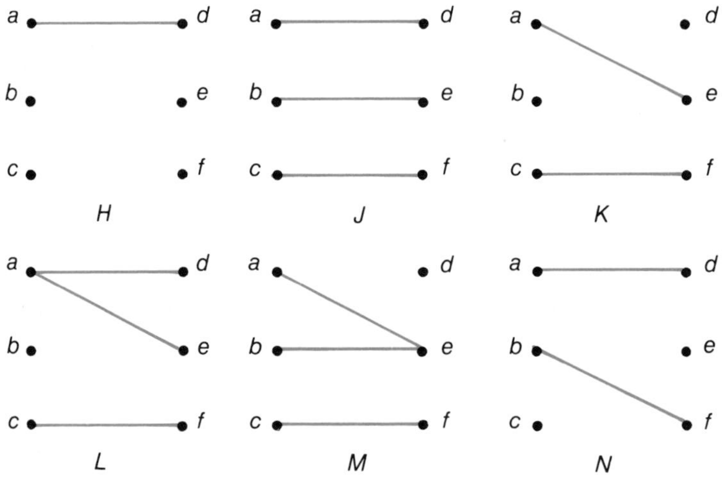

Figure 11.26

Solution   The graphs H, J, and K are matchings of G. For example, the subsets X_o and Y_o of H are $X_o = \{a\}$ and $Y_o = \{d\}$. The only complete matching is J, because only in J is $X = X_o$.

The graph L is not a matching because vertex a is adjacent to two vertices (d and e). Similarly, M is not a matching because vertex e is adjacent to two vertices. The graph N is not a matching because the edge (b,f) is not an edge of the bipartite graph G.

Example 9   By inspection, we see that a complete matching from X into Y is not possible for any of the bipartite graphs in Figure 11.27. However, complete matchings from Y into X can be found for graphs G, H, and I.

Let G be a bipartite graph with disjoint subsets of vertices X and Y. We have seen that, in many simple cases, a careful inspection of G will determine whether a complete matching of X into Y exists. Suppose, however, that G is a complicated graph, so that this is not easy to do. We will now see that, if a complete matching exists, it can be found by constructing a certain network and then finding a maximal

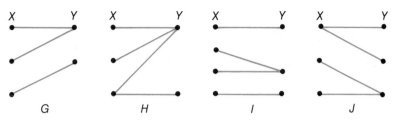

Figure 11.27

flow for that network. If a complete matching does not exist, this method will produce a matching in which a maximum number of vertices of X are matched to the vertices of Y.

We begin by constructing the required network.

Method for Constructing a Transport Network from a Bipartite Graph G

Let X and Y be the disjoint subsets of vertices of G.

(1) Add a source s and a sink r to G.

(2) Join each vertex in X to the source s by an arc (s,x).

(3) If y is any vertex in Y that is not adjacent to some vertex x in X, join y to the source s by an arc (s,y).

(4) Join each vertex in Y to the sink r by an arc (y,r).

(5) If x is any vertex in X that is not adjacent to some vertex y in Y, join x to the sink r by an arc (x,r).

(6) Give each edge (x,y) of G a direction so that it becomes arc (x,y).

(7) Let the capacity of each arc be 1.

The next example shows how to apply this method.

Example 10 Applying the method for constructing a network to the bipartite graph of Figure 11.28(a) yields the transport network of Figure 11.28(b). The

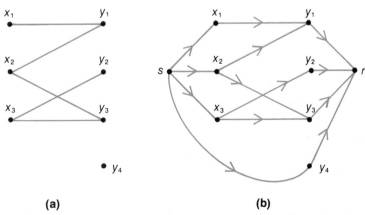

(a)

(a) Bipartite graph G

(b)

(b) Corresponding network N with $C(i,j) = 1$ for all arcs (i,j)

Figure 11.28

capacities of the arcs have been omitted in the network for brevity; no confusion is possible, because the capacity of each arc is 1.

In general, let G be any bipartite graph with disjoint subsets of vertices X and Y. Using the method we have discussed, we can construct a corresponding transport network N. By applying the method developed in Section 11.2, we can then find a maximal flow F for N. This flow F can then be used to obtain a complete matching for G using the following steps:

To Obtain a Complete Matching for G—

(1) draw a graph having the same vertices as F except for the source s and the sink r

(2) draw an edge (x,y) if and only if $F(x,y) = 1$.

Example 11 shows how to apply this method.

Example 11   Use the network N of Figure 11.28(b) to find a complete matching for the graph G of Figure 11.28(a).

Solution   Applying the method developed in Section 11.2 to N yields the maximal flow F shown in Figure 11.29(a). The paths that were used in finding the maximal flow were s,x_1,y_1,r, s,y_4,r, s,x_3,y_2,r, and s,x_2,y_3,r. Applying the steps for finding a complete matching C from F yields Figure 11.29(b). For

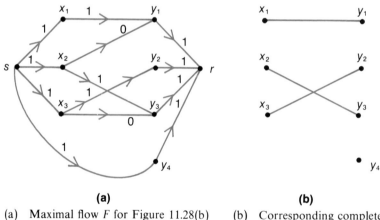

(a)   **(b)**

(a)   Maximal flow F for Figure 11.28(b)   (b)   Corresponding complete
                                                      matching C for Figure
Figure 11.29   11.28(a)

example, the edge (x_1,y_1) appears in Figure 11.29(b) because $F(x_1,y_1) = 1$ in Figure 11.29(a). However, the edge (x_3,y_3) does not appear in the matching because $F(x_3,y_3) = 0$ in Figure 11.29(a). The reader can easily verify that C is a complete matching of X into Y for G.

In general, let G be any bipartite graph with disjoint subsets of vertices X and Y. Summarizing the results of this section, a complete matching of X into Y (if it exists) can be found as follows:

(1) Construct the corresponding network N. (See page 504.)

(2) Use the method of Section 11.2 to find a maximal flow F for N.

(3) Construct the corresponding matching. (See page 506.)

If a complete matching does not exist, this method will produce a matching in which a maximum number of vertices of X are matched to the vertices of Y; that is, this approach will always produce a maximal matching.

11.3 EXERCISES

1. Which of the graphs in Figure 11.30 are bipartite?

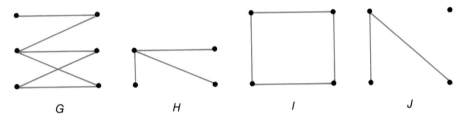

G H I J

Figure 11.30

2. Let G be the bipartite graph in Figure 11.31. The two disjoint subsets of vertices for G are $X = \{1,2,3,4\}$ and $Y = \{5,6,7\}$. Which of the graphs in Figure 11.32 are (a) matchings of X into Y, (b) complete matchings of X into Y, (c) maximal matchings of X into Y, (d) complete matchings of Y into X?

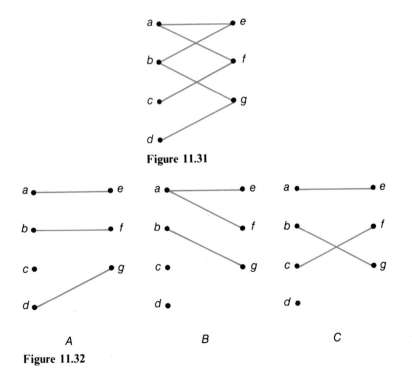

Figure 11.31

A B C

Figure 11.32

3. Let G be the bipartite graph in Figure 11.33. Use the network-construction method to find a maximal matching of X into Y. Show (a) the corresponding network, (b) a maximal flow, and (c) the corresponding maximal matching.

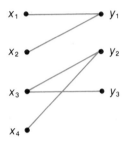

Figure 11.33

4. A set of "words," $\{bcd, aef, abef, abdf, abc\}$, is to be transmitted. Is it possible to represent each word by one of the letters in the word so that the words will be uniquely represented? If so, how?

5. Five students have been hired to fix up a house. The house must be painted, wallpapered, and cleaned. In addition, heavy furniture must be moved and new curtains made. Suppose that Peter can paint and move heavy furniture; Robert can make new curtains; Henrietta can paint, wallpaper, and move heavy furniture; Ruth can clean and wallpaper; and Lewis can paint and wallpaper. Give an assignment of tasks that will match as many people as possible with jobs that they can do, assuming that each person does a different job.

11.4 EXISTENCE OF COMPLETE MATCHINGS

In the preceding sections we saw that many assignment problems can be solved by a careful inspection of their graphs. This method would be greatly simplified if we knew the answers to the following questions before beginning work on the problem:

(a) Does a complete matching exist?

(b) If a complete matching does not exist, what is the maximum number of vertices of the first set X that can be matched to the vertices of the second set Y?

In this section we will introduce three theorems which answer these questions.

Let G be a bipartite graph with disjoint subsets of vertices X and Y. The following notation will enable us to state the first theorem concisely:

If S is a subset of X, then—

(a) $V(S)$ is the set of vertices Y that are adjacent to vertices in S

(b) $|V(S)|$ is the number of elements in $V(S)$

(c) $|S|$ is the number of elements in S.

The notation $|S|$ means the same as $n(S)$, but is simpler. The following example illustrates its use.

Example 12   Let G be the graph of Figure 11.34. Then G is a bipartite graph with $X = \{x_1, x_2, x_3, x_4\}$ and $Y = \{y_1, y_2, y_3\}$. Suppose $S = \{x_1, x_2\}$. Then $V(S) = \{y_1, y_2\}$ because (x_1, y_1) and (x_2, y_2) are edges of G and x_1 and x_2 are not adjacent to any other vertices of Y. For this set, $|S| = |V(S)| = 2$.

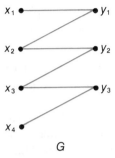

G

Figure 11.34

Several other subsets S of X are given in Table 11.1, along with $|S|$, $V(S)$, and $|V(S)|$.

Table 11.1

S	$|S|$	$V(S)$	$|V(S)|$
$\varnothing$	0	$\varnothing$	0
$\{x_1\}$	1	$\{y_1\}$	1
$\{x_1, x_3\}$	2	$\{y_1, y_2, y_3\}$	3
$\{x_2, x_3, x_4\}$	3	$\{y_1, y_2, v_3\}$	3
$X = \{x_1, x_2, x_3, x_4\}$	4	$Y = \{y_1, y_2, y_3\}$	3

gives us our first rule for determining whether a matching is complete. A second rule will be given later.

THEOREM 11.2   Let G be a bipartite graph with disjoint subsets of vertices X and Y. A complete matching of X into Y exists if and only if $|S| \le |V(S)|$ for every subset S of X.

The following example shows how to use Theorem 11.2.

Example 13   Let G be the graph of Figure 11.34. Determine whether a complete matching of X into Y exists.

Solution   Table 11.1 lists several subsets S of X, together with $|S|$, $V(S)$, and $|V(S)|$. Let us imagine that this table has been completed by listing every subset of X. (The second counting principle can be used to show that X has sixteen subsets.) According to Theorem 11.2, a complete matching of X into Y exists if and only if *every* entry in the second column of the completed table is less than or equal to the corresponding entry in the fourth column. Since the fourth entry in the second column of Table 11.1 is 4, while the corresponding entry in the fourth column is 3, a complete matching of X into Y does not exist.

Applying Theorem 11.2 to an actual problem can be a tedious process because a table like Table 11.1 must be made. This usually entails calculating and listing a large number of entries. In certain cases, the following theorem, which requires less calculation, can be used instead.

THEOREM 11.3 Let G be a bipartite graph with disjoint subsets of vertices X and Y. A complete matching of X into Y exists if there is a positive integer n such that—

(a) every vertex in X is adjacent to *n or more* vertices in Y

(b) every vertex in Y is adjacent to *n or less* vertices in X.

Notice that Theorem 11.3 states that, if there is a number n that satisfies requirements (a) and (b), then G has a complete matching. It does *not* state what happens if there is no such number. In this case, Theorem 11.2 must be applied. The following examples illustrate these points.

Example 14 Six people, a, b, c, d, e, and f, have been organized into five committees: $\{a,b,c\}$, $\{b,e\}$, $\{d,a,c\}$, $\{d,f\}$, and $\{e,f\}$. Can a chairperson be selected for each committee if no person can chair more than one committee?

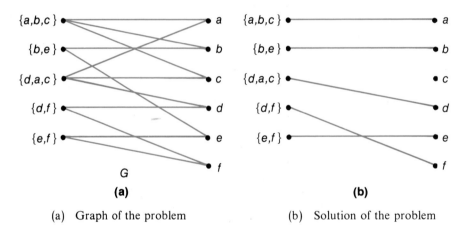

(a) Graph of the problem   (b) Solution of the problem

Figure 11.35 Graph and solution for the committee problem

Solution The bipartite graph G of Figure 11.35(a) conveniently represents the conditions of this problem. In this graph X is the set of committees and Y is the set of people. An edge joins a committee in X to a person in Y if and only if the person is on the committee. The problem will be solved if we can determine whether a complete matching of X into Y exists.

The vertices of X and Y and the number of adjacent vertices of each are listed in Table 11.2.

From this table it is clear that each vertex in X is adjacent to 2 or more vertices of Y, and each vertex in Y is adjacent to 2 or less vertices in X. It follows from Theorem 11.3 with $n = 2$ that a complete matching of X into Y

Table 11.2

Vertex of X	Number of Adjacent Vertices	Vertex of Y	Number of Adjacent Vertices
$\{a, b, c\}$	3	a	2
$\{b, e\}$	2	b	2
$\{d, a, c\}$	3	c	2
$\{d, f\}$	2	d	2
$\{e, f\}$	2	e	2
		f	2

exists. One such matching is given in Figure 11.35(b). The answer to our original problem is therefore yes; Figure 11.35(b) indicates one way that the chairpersons can be selected.

Example 15   Determine whether a complete mapping of X into Y exists for the bipartite graph of Figure 11.36.

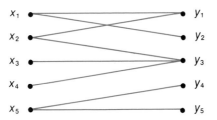

Figure 11.36

Solution   The vertices of X and Y and the number of adjacent vertices of each are listed in Table 11.3.

Table 11.3

Vertex of X	Number of Adjacent Vertices	Vertex of Y	Number of Adjacent Vertices
x_1	2	y_1	2
x_2	2	y_2	1
x_3	1	y_3	3
x_4	1	y_4	1
x_5	2	y_5	1

From the table we see that every vertex of X is adjacent to 1 or more vertices of Y, but every vertex of Y is adjacent to 3 or less vertices of X. It is clear that an integer n satisfying the requirements of Theorem 11.3 cannot be found, so Theorem 11.3 cannot be applied. We must therefore use Theorem 11.2.

Let $S = \{x_3, x_4\}$. Then $V(S) = \{y_3\}$, $|S| = 2$, and $|V(S)| = 1$. Hence, $|V(S)| < |S|$, so Theorem 11.2 implies that a complete matching of X into Y does not exist.

Let G be a bipartite graph with disjoint subsets of vertices X and Y. If a complete matching of X into Y does not exist, one usually wishes to match as many of the vertices in X with vertices in Y as possible. Theorem 11.4 gives a formula for determining this maximum number of vertices. To simplify the statement of this theorem, we need the following definition:

Definition 11.9   Let G be a bipartite graph with disjoint subsets of vertices X and Y, and let S be a subset of X.

(a)   The *deficiency* of S, denoted $\delta(S)$, is defined as

$$\delta(S) = |S| - |V(S)|$$

(b)   The **deficiency of the graph** G, denoted $\delta(G)$, is the maximum value of the deficiencies of the subsets of X.

The symbol δ is the Greek letter *delta*.
Notice that the deficiency of a set can be positive, negative, or zero. However, the deficiency of G, $\delta(G)$ must be nonnegative, because $\delta(G) \geq \delta(\varnothing)$, and $\delta(\varnothing) = 0$.

THEOREM   11.4   Let G be a bipartite graph. The maximum number N of vertices in X that can be matched to vertices in Y is

$$N = |X| - \delta(G)$$

Since a complete matching of X into Y exists if and only if $N = |X|$, Theorem 11.4 implies that a complete matching exists if and only if $\delta(G) = 0$.
The remaining examples in this section illustrate the use of Theorem 11.4.

Example 16   Consider the bipartite graph in Figure 11.37. Determine the maximum number of vertices in X that can be matched to vertices in Y.

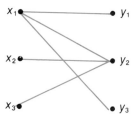

Figure 11.37

Solution   From the figure, $X = \{x_1, x_2, x_3\}$. Table 11.4 lists the deficiencies of the subsets S of X.

By definition, $\delta(G)$ is the largest number appearing in the last column of Table 11.4, that is, $\delta(G) = 1$. Hence, applying Theorem 11.4, the maximum number of vertices in X that can be matched to vertices in Y is $|X| - \delta(G) = 3 - 1 = 2$. Three examples of maximal matchings are shown in Figure 11.38.

Table 11.4	S	$\lvert S\rvert$	$V(S)$	$\lvert V(S)\rvert$	$\delta(S)=\lvert S\rvert-\lvert V(S)\rvert$
	$\varnothing$	0	$\varnothing$	0	0
	$\{x_1\}$	1	$\{y_1, y_2, y_3\}$	3	-2
	$\{x_2\}$	1	$\{y_2\}$	1	0
	$\{x_3\}$	1	$\{y_2\}$	1	.0
	$\{x_1, x_2\}$	2	$\{y_1, y_2, y_3\}$	3	-1
	$\{x_1, x_3\}$	2	$\{y_1, y_2, y_3\}$	3	-1
	$\{x_2, x_3\}$	2	$\{y_2\}$	1	1
	$\{x_1, x_2, x_3\}$	3	$\{y_1, y_2, y_3\}$	3	0

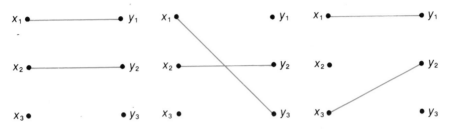

Figure 11.38 Three maximal matchings of X into Y for Figure 11.37

Example 17 Thirty manufacturers use a certain type of miniature circuit in their minicomputers. There are twenty-four electronics companies that make this type of circuit. Because transportation costs for the circuits are high, the computer manufacturers would like to buy their circuits from nearby electronics companies.

The computer manufacturers can be divided into three transportation categories with ten members each: The manufacturers in category 1 are within 50 miles of four electronics companies; those in category 2 are within 50 miles of two electronics companies; and those in category 3 are within 50 miles of one electronics company. No electronics company is within 50 miles of more than three computer manufacturers, and no electronics company is able to supply more than one manufacturer.

Given this information, what is the smallest number of supply contracts that can be signed between computer manufacturers and electronics companies that are within 50 miles of each other?

Solution Let G be a bipartite graph in which X is the set of computer manufacturers and Y is the set of electronics companies. Each edge of G indicates that the companies it connects are within 50 miles of each other. Since we have not been told exactly which companies are within 50 miles of one another, we do not know exactly what the graph looks like. However, Figure 11.39 shows a portion of one possibility. Notice that each vertex representing a computer manufacturer is connected with 1, 2, or 4 vertices representing electronics companies, and that each vertex representing an electronics company is connected to at most 3 vertices representing computer manufacturers.

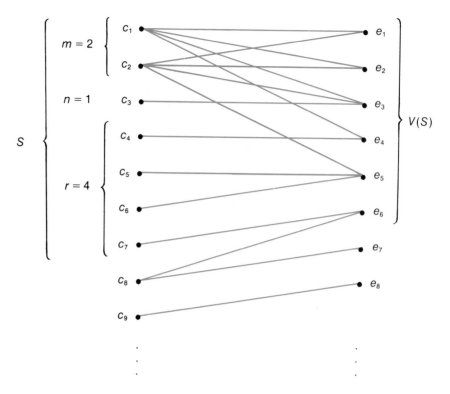

Figure 11.39 Partial graph of computer manufacturers (c_i) and electronics company (e_i) that are within fifty miles of each other

According to Theorem 11.4, the maximum number of vertices in X that can be matched to vertices in Y is

$$N = |X| - \delta(G)$$

We do not have enough information to determine N because we do not know exactly which computer manufacturers are within 50 miles of which electronics companies. However, we can show that N is at least as big as some number N_1. This number N_1 will be the solution to our problem.

Since N depends on $|X|$ and $\delta(G)$, let us first calculate how large $\delta(G)$ might be. Recall that $\delta(G)$ is the maximum value of the deficiencies $\delta(S)$ of subsets S of X.

Suppose that S is any subset of X. Since X is the set of computer manufacturers, S contains m category 1 manufacturers, n category 2 manufacturers, and r category 3 manufacturers, where m, n, and r must be greater than or equal to 0 and less than or equal to 10. Therefore, by Definition 11.9(a),

$$\delta(S) = |S| - |V(S)| = m + n + r - |V(S)| \tag{2}$$

Recall that $|V(S)|$ is the number of vertices in Y that are adjacent to vertices in S. This means that $|V(S)|$ is the number of electronics companies that are within 50 miles of the computer manufacturers in S. Since each category 1 manufacturer is within 50 miles of four electronics companies, each category 2 manufacturer is within 50 miles of two electronics companies, and each category 3 manufacturer is within 50 miles of one electronics company, there are

$$4m + 2n + r$$

edges of G that begin at vertices of S. (See Figure 11.39 for an example, with $S = \{c_1, c_2, \ldots, c_7\}$.)

This does not mean that $|V(S)|$ is equal to $4m + 2n + r$, because two or more manufacturers may be within 50 miles of the same electronics firm. (For example, in Figure 11.39, c_1 and c_2 are both adjacent to e_1.) However, since each electronics company is within 50 miles of at most three computer manufacturers, the number of edges that end at vertices in $V(S)$ is at most $3|V(S)|$. Since the number of edges beginning at vertices in S is the same as the number of edges ending at vertices in $V(S)$,

$$4m + 2n + r \leq 3|V(S)|$$

Therefore,

$$\frac{4m + 2n + r}{3} \leq |V(S)|$$

If we multiply by -1,

$$-\left(\frac{4m + 2n + r}{3}\right) \geq -|V(S)| \tag{3}$$

Using equations (2) and (3), we have

$$\delta(S) = m + n + r - |V(S)| \leq m + n + r - \left(\frac{4m + 2n + r}{3}\right)$$

That is,

$$\delta(S) \leq -\frac{m}{3} + \frac{n}{3} + \frac{2r}{3} \tag{4}$$

It follows from inequality (4) that $\delta(S)$ will be the largest when m is smallest and n and r are largest. Since $0 \leq m \leq 10$, $0 \leq n \leq 10$, and $0 \leq r \leq 10$, $\delta(S)$ will be largest when $m = 0$, $n = 10$, and $r = 10$. Substituting these values into inequality (4), we have

$$\delta(S) \leq -\frac{0}{3} + \frac{10}{3} + 2\left(\frac{10}{3}\right) = \frac{30}{3} = 10 \tag{5}$$

for all subsets S of X.

By Definition 11.9(b) and inequality (5),

$$\delta(G) \leq 10$$

Therefore, Theorem 11.4 implies that N_1 is at least as big as

$$|X| - 10 = 30 - 10 = 20$$

Thus, at least 20 supply contracts can be signed between firms that are within 50 miles of each other, assuming that no electronics company can supply more than one computer manufacturer.

11.4 EXERCISES

1.   Let G be the bipartite graph in Figure 11.40.

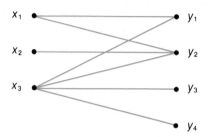

Figure 11.40

(a) Calculate $V(S)$, $|S|$, $|V(S)|$, and $\delta(S)$ for each subset of $X = \{x_1, x_2, x_3\}$.

(b) Calculate $\delta(G)$.

(c) Does a complete matching of X into Y exist? If it does, give such a matching.

(d) Does a complete matching of Y into X exist? If it does, give such a matching.

(e) Find a maximal matching of X into Y.

2.   Find the deficiency of and a maximal matching for each of the graphs in Figure 11.41.

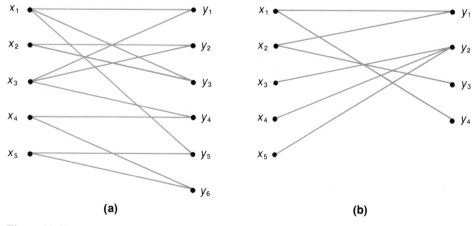

(a)   (b)

Figure 11.41

3. **(a)** Determine whether Theorem 11.3 can be applied to each of the graphs in Figure 11.42.

(b) Use either Theorem 11.3 or Theorem 11.2 to determine whether a complete matching exists for each of these graphs.

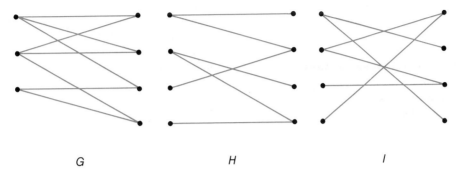

   G   H   I

Figure 11.42

4. A law firm has five partners and five important clients. Each partner has a good relationship with two of the important clients, and each important client has a good relationship with two of the partners. Is it possible to assign each partner to a different important client with whom he or she has a good relationship? Explain.

5. Rework Exercise 4 replacing five by p and two by q.

6. Andrew, Barbara, Carole, Daniel, and Ellen wish to form a car pool to drive their children to school. Andrew can drive Mondays or Tuesdays, Barbara can drive Mondays or Wednesdays, Carole can drive Tuesdays or Thursdays, Daniel can drive Wednesdays or Fridays, and Ellen can drive Thursdays or Fridays.

(a) Draw a graph that represents this situation.

(b) Use Theorem 11.3 to show that a complete matching of people with days exists.

(c) Find one such complete matching.

Key Word Review

Each key word is followed by the number of the section in which it is defined.

transport network   11.1
source   11.1
sink   11.1
capacity   11.1
flow (for a transport network)   11.1
flow (of an arc)   11.1
saturated arc   11.1
unsaturated arc   11.1
value of the flow   11.1

maximal flow   11.1
pseudoarc   11.2
pseudopath   11.2
simple pseudopath   11.2
bipartite graph   11.3
matching   11.3
complete matching   11.3
maximal matching   11.3
deficiency of a graph   11.4

CHAPTER 11   REVIEW EXERCISES

1. Define each word in the Key Word Review.

2. What do the following symbols mean?
    (a) $C(i,j)$ (b) $F(i,j)$ (c) F_v (d) $D(i,j)$
    (e) $V(S)$ (f) $|S|$ (g) $\delta(G)$ (h) $\delta(S)$

3. A company has two warehouses, w and x, and three distributors, d, e, and f.
    Warehouses w and x have on hand 50,000 and 60,000 tons of steel, respec-
    tively, and distributors d, e, and f require 40,000, 20,000, and 70,000 tons of
    steel, respectively. The train routes connecting the warehouses and distribu-
    tors are represented by edges in the graph in Figure 11.43. (The trains can
    travel in either direction.) Each edge is labeled with the maximum amount of
    steel (in thousands of tons) that can be transported on that route.

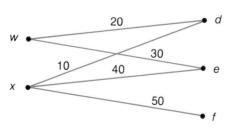

Figure 11.43

(a) Draw a transport network reflecting this situation.
(b) How many tons of steel should each warehouse ship and each distribu-

tor receive in order that as much steel as possible will be transported to the distributors?

(c) Draw a flow that shows how the shipments should be routed to achieve the maximal schedule of part **(b)**.

4.   Which of the graphs in Figure 11.44 are networks? Explain.

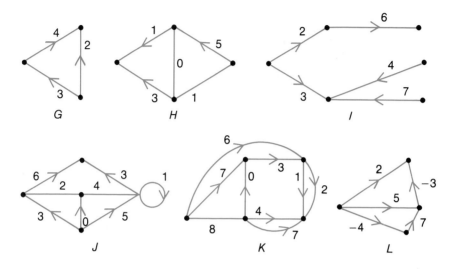

Figure 11.44

5.   Consider the transport network pictured in Figure 11.45.
   (a) What are the source and the sink of this network?
   (b) Find $C(d,e)$ and $C(b,f)$.
   (c) Find a maximal flow for this network.
   (d) What is the maximum amount of material that can be transported from the source to the sink?

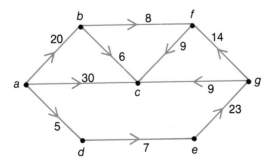

Figure 11.45

6. Figure 11.46 shows a flow for the network of Exercise 5 (Figure 11.45).
  (a) Find—
    (*i*) $F(b,c)$ (*ii*) $F(d,e)$ (*iii*) F_v (*iv*) $D(f,c)$ (*v*) $D(e,g)$
  (b) Which arcs are saturated?
  (c) Give a simple path from the source to the sink.
  (d) Give a simple pseudopath from the source to the sink.

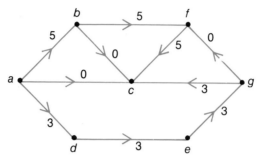

Figure 11.46

7. Which of the graphs in Figure 11.47 are flows for the network in Figure 11.45? Explain your answer.

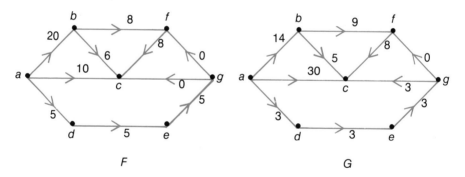

Figure 11.47

8. Which of the graphs in Figure 11.48 are bipartite?

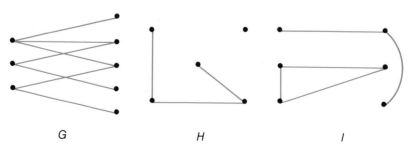

Figure 11.48

9. Consider the bipartite graph B in Figure 11.49. Which of the graphs in Figure 11.50 are—

(a) matchings from X into Y

(b) matchings from Y into X

(c) complete matchings from X into Y

(d) complete matchings from Y into X?

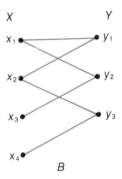

Figure 11.49

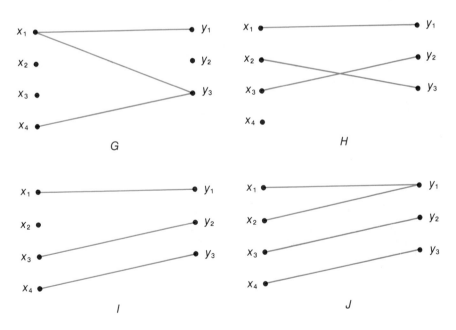

Figure 11.50

10. Let B be the bipartite graph of Exercise 9 (Figure 11.49) and let $S = \{x_1, x_2\}$.

(a) Calculate—

(i) $|S|$ (ii) $V(S)$ (iii) $|V(S)|$ (iv) $\delta(S)$ (v) $\delta(B)$

(b) What is the maximum number of vertices in X that can be matched to vertices in Y?

11. Find the deficiency of and a maximal mapping for each of the graphs in Figure 11.51.

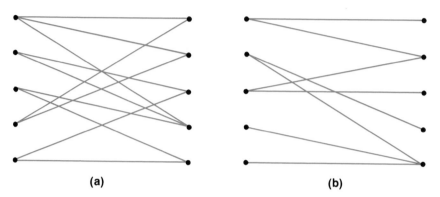

(a)   **(b)**

Figure 11.51

12. At a girls' summer camp the following chores must be assigned: sweeping (s), cooking (c), setting the table (t), washing the dishes (w), and feeding the animals (f). Anne (a), Barbara (b), and Carole (k) are allergic to dust and cannot sweep; Barbara, Debby (d), and Elaine (e) have colds and cannot cook; Anne and Carole are clumsy and cannot set the table or wash dishes; Anne and Elaine are allergic to animals and so cannot feed them.

(a) Draw a bipartite graph that displays this information.

(b) Design a schedule that will distribute a maximum number of jobs to girls capable of performing them. (No girl can be assigned more than one task.)

Suggested Readings

1. Berge, Claude. *The Theory of Graphs and Its Applications.* New York: John Wiley & Sons, 1962.
2. Busacker, Robert G., and Saaty, T. L. *Finite Graphs and Networks: An Introduction with Applications.* New York: McGraw-Hill Book Co., 1965.
3. Ford, Lester R., Jr., and Fulkerson, D. R. *Flows in Networks.* Princeton, N.J.: Princeton University Press, 1962.

Marshall Licht

12 Dynamic
Programming

INTRODUCTION

The truck in the picture on the facing page is part of a complex system designed to deliver produce from the wholesaler to a number of retail outlets. An efficient delivery system is one that meets the needs of the retailers while keeping costs at a minimum. When several warehouses and several retail outlets are involved, the problem of planning an efficient, money-saving delivery system can be a difficult one. Fortunately, this type of problem is one of many that can be formulated as a sequence of decisions and that can be solved using a method called *dynamic programming.*

In this chapter we will develop the basic techniques of dynamic programming. These techniques can be used to solve problems in transportation similar to those we solved using networks and flows in Chapter 11. However, dynamic programming can also be used to plan the allocation of limited dollar resources—for example, in an investment plan or city budget—and to plan manufacturing schedules that minimize production and storage costs.

12.1 DECISIONS AND POLICIES

We will begin our study of dynamic programming by presenting a problem that can be easily solved by methods we have already discussed. However, it will be clear that these earlier methods are unsatisfactory for more complicated problems of the same type. We will then see that the problem in Example 1 involves a sequence of decisions and that it can be solved by dynamic programming.

Example 1   Figure 12.1 shows all the major highways connecting cities $a, b, c, \ldots,$ and g and the mileages between them. (For example, the mileage between a and b is 40 miles.) Plan the shortest possible route from a to g.

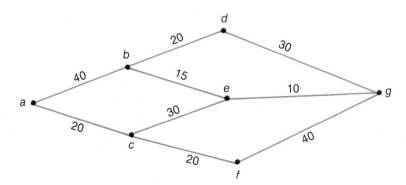

Figure 12.1   Highway map for Example 1

Solution   This problem can be easily solved by listing all simple paths from a to g and their associated mileages:

Simple Path	Mileage
a,b,d,g	$40 + 20 + 30 = 90$
a,b,e,g	$40 + 15 + 10 = 65$
a,c,e,g	$20 + 30 + 10 = 60$
a,c,f,g	$20 + 20 + 40 = 80$

It is clear from this list that the shortest possible route is a,c,e,g, and that the smallest possible total mileage is 60 miles.

Although Example 1 was easily solved by listing all possible routes and their mileages, this method is clearly impractical if there are a large number of cities (say 25) and a large number of highways (say 100). However, the method of dynamic programming works for both simple problems like Example 1 and for more difficult ones.

We first note that Example 1 involves a sequence of three decisions, as illustrated in Figure 12.2.

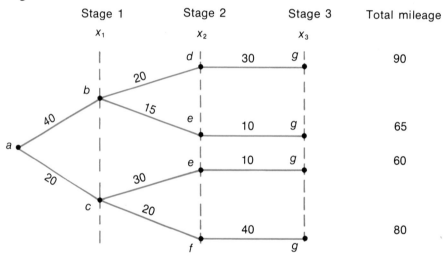

Figure 12.2 Diagram of the decision stages in Example 1

(1) Initially we are at a. In stage 1 we must decide whether to go to city b or to city c. The possible outcomes, or *states*, of stage 1 are therefore b and c. Notice that these states are listed under stage 1 in Figure 12.2.

(2) We are now either at city b or at city c. The decision that we make in stage 2 therefore depends upon the decision that we made in stage 1. If b was chosen in stage 1, we must decide to go to d or to e. If c was chosen in stage 1, we must decide whether to go to e or to f. The states of stage 2 are therefore d, e, and f. These states have been listed under stage 2 in Figure 12.2.

Notice that, although one might be tempted to go from c to f because the mileage from c to f is smaller than the mileage from c to e, the total mileage is better if we go from c to e.

(3) In stage 3, we must decide to go to g. Stage 3 therefore has only one state, g.

In general, we have the following definition of a *policy* for this type of problem:

Definition 12.1   Let m be a positive integer. Suppose that a problem can be solved in m steps, or **stages**, so that at each stage k, a choice must be made from a finite number of possibilities, and this choice depends only upon the choices made in the previous stages. Then—

(a)   The choice made at stage k is called **decision** d_k, and the possible choices are called the **states** of stage k

(b)   A sequence of decisions $d_1, d_2, \ldots, d_m$ is called a **policy** P

(c)   If k is 1, 2, …, or m, a sequence of decisions $d_1, d_2, \ldots, d_k$ that forms the beginning part of a policy P is called a **subpolicy** of P.

Notice that a policy is a sequence of m decisions: the first decision is made in stage 1, the second decision is made in stage 2 and depends upon the first decision, the third decision is made in stage 3 and depends upon the first two decisions, and so on. The last decision is made in stage m.

An example of a policy for Example 1 is:

d_1:   Go from a to b

d_2:   Go from b to e

d_3:   Go from e to g

A subpolicy of this policy is d_1, d_2.

In general, consider any problem that satisfies Definition 12.1. Suppose some standard is given for determining which policy is best. In Example 1, for example, the best policy is the one that gives the shortest possible route. In other examples we may wish to minimize costs or maximize profits. The best policy for a particular problem is called an **optimal policy**. The method we will develop for finding this optimal policy is called **dynamic programming**. We will shortly state a theorem that can be used to find an optimal policy. The concise statement of this theorem requires the following definition:

Definition 12.2   Let

$$Q = d_1, d_2, \ldots, d_k$$

be a subpolicy of

$$P = d_1, d_2, \ldots, d_k, d_{k+1}, \ldots, d_m$$

Moreover, assume that the outcome of decision d_k is state E. Then Q is an **optimal subpolicy** of P if no better policy exists for arriving at state E after k decisions have been made.

The following example illustrates this definition.

Example 2   Using the results of Example 1, we see that the policy

d_1:   Go from a to c

d_2:   Go from c to e

d_3:   Go from e to g

is optimal. Show that the subpolicy $Q = d_1, d_2$ is also optimal.

Solution   Decision d_2 results in state e. Using Figure 12.1, we see that there are exactly two ways of reaching state e after two decisions. These policies and their associated mileages are as follows:

Policy	Mileage
(1) $Q = d_1, d_2$	50
(2) Go from a to b	
Go from b to e	55

This list clearly indicates that there is no better policy for arriving at state e after two decision are made. It follows from Definition 12.1 that Q is an optimal subpolicy of P.

We are now able to state the basic principle of dynamic programming.

THEOREM   12.1   (*The Optimality Theorem*)   An optimal policy contains only optimal subpolicies.

Proof   Suppose that $P = d_1, d_2, \ldots, d_m$ is an optimal policy, but that subpolicy $Q = d_1, d_2, \ldots, d_k$ $(k \leq m)$ is not optimal. Moreover, assume that state E is reached after decision d_k. Since Q is not optimal, there must be a better policy $R = e_1, e_2, \ldots, e_k$ for arriving at state E after k decisions have been made. It then follows that

$$e_1, e_2, \ldots, e_k, d_{k+1}, d_{k+2}, \ldots, d_m$$

is a better policy than P. Since this contradicts the fact that P is optimal, Q must be optimal also.

Before we can apply Theorem 12.1, we must define two important functions that will be used throughout the remainder of this chapter. These functions are

$$\min[\ \ ] \quad \text{and} \quad \max[\ \ ]$$

The function $\min[\ \ ]$ is the smallest (i.e., *minimum*) of the numbers that appear within the brackets. For example,

$$\min[2,3,4] = 2 \quad \text{and} \quad \min[-1,5,6,-7] = -7$$

The function $\max[\ \ ]$ is the largest (i.e., *maximum*) of the numbers that appear within the brackets. For example,

$$\max[2,3,4] = 4 \quad \text{and} \quad \max[2,3,-5] = 3$$

The item within the brackets of $\min[\ \ ]$ and $\max[\ \ ]$ may sometimes be a function. For example,

$$\min_{x=1,\,2,\,3} [x^2 + 2x]$$

is the smallest value of

$$x^2 + 2x \tag{1}$$

when x is 1, 2, or 3. (The allowable values of x are written under "min.") Substituting $x = 1$, $x = 2$, and $x = 3$ into equation (1) yields

$$1^2 + 2(1) = 3 \qquad \text{when } x = 1$$

$$2^2 + 2(2) = 8 \qquad \text{when } x = 2$$

and

$$3^2 + 2(3) = 15 \qquad \text{when } x = 3$$

Since the smallest of these numbers is 3, it follows that

$$\min_{x=1,\,2,\,3} \; [x^2 + 2x] = \min[3,8,15] = 3$$

Notice that the value of

$$\min_{x=1,\,2,\,3} \; [x^2 + 2x]$$

is given by the function within the brackets, $x^2 + 2x$, when it is evaluated at one of the allowable values of x, $x = 1$. In general, the particular value of x that yields the minimum or maximum of a function is denoted x^*.

Since the concepts of min[ ] and max[ ] are of such great importance, we will illustrate them by another example.

Example 3   Find—

(a) $\max\limits_{x=-1,\,0,\,1} \; [x^3 - 2x]$ **(b)** $\min\limits_{x=-1,\,0,\,1} \; [1 - x^2]$

Also find the corresponding values of x^*.

Solution

(a) $\max\limits_{x=-1,\,0,\,1} \; [x^3 - 2x]$ is the largest value of

$$x^3 - 2x$$

when this function is evaluated at $x = -1$, $x = 0$, and $x = 1$. Substituting these values into $x^3 - 2x$ yields

$$(-1)^3 - 2(-1) = 1 \qquad \text{when } x = -1$$

$$0^3 - 2(0) = 0 \qquad \text{when } x = 0$$

$$1^3 - 2(1) = -1 \qquad \text{when } x = 1$$

It follows that

$$\max_{x=-1,\,0,\,1} \; [x^3 - 2x] = \max[1,0,-1] = 1$$

Since the value of the function is 1 when $x = -1$, the corresponding value of x^* is -1.

(b) $\min\limits_{x=-1,\,0,\,1} \; [1 - x^2] = \min[0,1,0] = 0$. Since $1 - x^2$ is 0 when $x = -1$ and when $x = 1$, there are two values of x^*, $x^* = -1$, and $x^* = 1$.

In Example 4 we will see how Theorem 12.1 and the minimum function can be applied to the problem of Example 1. First, however, we will introduce some new notation.

Let x and y be two cities in Example 1 that are directly connected by a highway, and let $h(x,y)$ be the mileage between x and y. For example, using Figure 12.1,

$$h(a,b) = 40, \qquad h(b,e) = 15, \qquad \text{and} \qquad h(e,g) = 10$$

Notice that $h(e,f)$ is undefined because there is no highway directly connecting cities e and f.

The situation of Example 1 has three stages. We will now define three functions, $F_1, F_2,$ and F_3, that correspond to these stages. Let x_1 be a state of stage 1, let x_2 be a state of stage 2, and let x_3 be a state of stage 3. Specifically, x_1 can be b or c, x_2 can be d, e, or f, and x_3 must be g. Then

$\qquad F_1(x_1)$ is the minimum mileage between a and x_1

$\qquad F_2(x_2)$ is the minimum mileage between a and x_2

and

$\qquad F_3(x_3)$ is the minimum mileage between a and x_3

Figure 12.2 clearly indicates that $F_1(c) = 20$, $F_1(b) = 40$, and $F_1(d)$ is undefined. The problem outlined in Example 1 will be solved when we find $F_3(g)$.

We will now solve Example 1 using the optimality theorem.

Example 4 Use the optimality theorem to solve Example 1.

Solution We must find $F_3(g)$, the minimum mileage between a and g. We will see that $F_3(g)$ can be found by (a) first finding the values of $F_1(x_1)$, (b) using these values to find the values of $F_2(x_2)$, and then (c) using these values to find $F_3(g)$. This will be done by finding formulas for $F_2(x_2)$ and $F_3(x_3)$. The values of $F_1(x_1)$ can be found from Figure 12.1.

We will first use the optimality theorem to find a formula for $F_3(x_3)$. Suppose that

$$R = a, x_1, x_2, g$$

is any route from a to g. Then R is the result of the following policy P:

$\qquad d_1:$ Go from a to x_1

$\qquad d_2:$ Go from x_1 to x_2

$\qquad d_3:$ Go from x_2 to g

If R is a route of minimum mileage, then P is optimal and its associated mileage is $F_3(g)$. Moreover, Theorem 12.1 implies that the subpolicy $Q = d_1, d_2$, is also optimal; that is, the corresponding route

$$T = a, x_1, x_2$$

is a route of minimum mileage between a and x_2. By definition, the length of route T is $F_2(x_2)$.

Clearly,

the length of route $R =$ the length of route T

$+$ the distance from x_2 to g $\qquad(2)$

(See Figure 12.3.)

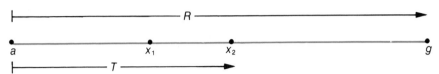

Figure 12.3   The length of R is equal to the length of T plus the distance from x_2 to g

We have seen that the length of route R is $F_3(g)$, and the length of route T is $F_2(x_2)$. By definition, the distance from x_2 to g is $h(x_2,g)$. Substituting these values into equation (2) yields

$$F_3(g) = F_2(x_2) + h(x_2,g)$$

for some particular state x_2.

Of course, $F_3(g)$ cannot be calculated unless we know the value of x_2. Let us suppose that x_2 is the particular state $x_2{}^*$. Since x_2 is a state of stage 2, $x_2{}^*$ must be d, e, or f. Moreover, $x_2{}^*$ is the state that yields the smallest value of

$$F_2(x_2) + h(x_2,g)$$

To find $x_2{}^*$, we must therefore calculate

$$F_2(d) + h(d,g)$$

$$F_2(e) + h(e,g)$$

and

$$F_2(f) + h(f,g)$$

The smallest of these numbers is $F_3(g)$; the corresponding state d, e, or f is $x_2{}^*$. In other words,

$$F_3(g) = \min_{x_2 = d, e, f} [F_2(x_2) + h(x_2,g)] \qquad(3)$$

and $x_2{}^*$ is the corresponding value of x. Notice that formula (3) cannot be applied until we find $F_2(d)$, $F_2(e)$, and $F_2(f)$.

A similar discussion shows that

$$F_2(x_2) = \min_{x_1 = b, c} [F_1(x_1) + h(x_1,x_2)] \qquad \text{for } x_2 = d, e, f \qquad(4)$$

(Remember that $h(x_1,x_2)$ does not exist for certain values of x_1 and x_2; e.g., $x_1 = b$ and $x_2 = f$. These pairs (x_1,x_2) are not considered when calculating $F_2(x_2)$.)

Formulas (3) and (4) and Figure 12.1 can now be used to calculate $F_3(g)$.

We first find the values of $F_1(x_1)$ from Figure 12.1. Remember that $F_1(x_1)$ is the minimum distance between a and x_1. Since there is only one route between a and x_1, $F_1(x_1)$ is just the distance between a and x_1. The values of $F_1(x_1)$ are listed in Table 12.1 for convenience.

Table 12.1

x_1	$F_1(x_1)$	x_2	$F_2(x_2)$	$x_1{}^*$	x_3	$F_3(x_3)$	$x_2{}^*$
b	40	d	60	b	g	60	e
c	20	e	50	c			
		f	40	c			

Table 12.1, formula (4), and Figure 12.1 can now be used to calculate $F_2(x_2)$. The values of $h(x_1,x_2)$ can be found from Figure 12.1 because $h(x_1,x_2)$ is the distance between x_1 and x_2.

For example, in order to find $F_2(d)$, we can apply formula (4) with $x_2 = d$. This yields

$$F_2(d) = \min_{x_1 = b, c} [F_1(x_1) + h(x_1,d)]$$

Substituting $x_1 = b$ and $x_1 = c$ into $F_1(x_1) + h(x_1,d)$ in turn, we have

$$F_1(b) + h(b,d) = 40 + 20 = 60$$

and

$$F_1(c) + h(c,d) \text{ is undefined}$$

It follows that $F_2(d) = 60$ with $x_1{}^* = b$.

Similarly, applying formula (4) with $x_2 = e$ yields

$$F_2(e) = \min_{x_1 = b, c} [F_1(x_1) + h(x_1,e)]$$

Substituting $x_1 = b$ and then $x_2 = c$ into $F_1(x_1) + h(x_1,e)$, we have

$$F_1(b) + h(b,e) = 40 + 15 = 55$$

and

$$F_1(c) + h(c,e) = 20 + 30 = 50$$

Since $F_2(e)$ is the smaller of these two numbers, $F_2(e) = 50$ with $x_1{}^* = c$.

The reader can similarly verify that $F_2(f) = 20 + 20 = 40$ with $x_1{}^* = c$. These values of $F_2(x_2)$ and $x_1{}^*$ are also listed in Table 12.1.

We can now use Table 12.1, formula (3), and Figure 12.1 to calculate $F_3(g)$ and $x_2{}^*$. Substituting $x_2 = d$, $x_2 = e$, and $x_2 = f$ into $F_2(x_2) + h(x_2,g)$ in turn, we have

$$F_2(d) + h(d,g) = 60 + 30 = 90$$

$$F_2(e) + h(e,g) = 50 + 10 = 60$$

and

$$F_2(f) + h(f,g) = 40 + 40 = 80$$

It follows that $F_3(g) = 60$ with $x_2^* = e$. The minimum mileage is therefore 60 miles.

We can now quickly find the optimal route by noting that the value $F_3(g)$ is the value of $F_2(x_2) + h(x_2,g)$ when $x_2 = x_2^* = e$. We therefore must visit city e directly before visiting city g. Similarly, since $F_2(e) = F_1(x_1) + h(x_1,e)$ when $x_1 = x_1^* = c$, we must visit city c directly before city e. It follows that the required route is a,c,e,g.

In general, consider any problem of the type described in Definition 12.1. Moreover, assume that there is some standard for determining which policy is optimal. Then an optimal policy can be found by using the following method.

Steps for Applying the Dynamic-Programming Method

(1) Let x_1 represent a state of stage 1, let x_2 represent a state of stage 2, and so on.

(2) Define a function h or several functions $h_1, h_2, \ldots, h_m$ similar to the function h of Example 4. The exact definition of these functions will depend upon the requirements of the problem.

(3) Define m functions $F_1(x_1), F_2(x_2), \ldots, F_m(x_m)$ corresponding to stages 1, 2, ..., m. These functions will be similar to those used in Example 4; their exact definitions will depend upon the requirements of the problem.

(4) Use the optimality theorem to find a formula for $F_2(x_2)$ in terms of $F_1(x_1)$ and h. Then the theorem to find a formula for $F_3(x_3)$ in terms of $F_2(x_2)$ and h, and so on.

(5) Use the information given in the problem to find the values of $F_1(x_1)$ and h. Use the values of $F_1(x_1)$ and h and the formula for $F_2(x_2)$ to find the values of $F_2(x_2)$. Then use the values of $F_2(x_2)$ and h and the formula for $F_3(x_3)$ to find the values of $F_3(x_3)$, and so on.

(6) Use $F_m(x_m)$ to find the optimum amount. Use $F_m(x_m)$ to find x_{m-1}^*; use $F_{m-1}(x_{m-1}^*)$ to find x_{m-2}^*; and so on. Then $x_{m-1}^*, x_{m-2}^*, \ldots, x_2^*, x_1^*$ is the optimal policy.

The reader may object that the dynamic-programming method is tedious and that equations like formulas (3) and (4) are difficult to obtain. We will see, however, that with a little experience these equations are relatively easy to find. In the remaining sections of this chapter we will study certain classes of problems requiring a sequence of decisions. We will see that, once a formula has been found for one problem in a certain class, it can be applied to all problems in that class.

The actual calculations that the dynamic-programming method requires can easily be performed by a computer. If the problem is complex, considerable valuable computer time can be saved by using the dynamic-programming method instead of listing all possible simple paths. In addition, dynamic programming can even be applied to problems having an infinite number of decisions at some stage. The method of listing cannot, of course, be applied to such situations.

12.1 EXERCISES

1. List all subpolicies of the policy d_1, d_2, d_3, d_4, d_5.

2. Find—

(a) min[2,4,0.5]   (b) max[2,4,0.5]   (c) max[4,4,4,4]   (d) min[4,4,4,4]

3. Find—

(a) $\min_{x=-1,0,1} [2x^2 - 1]$ (b) $\max_{x=-1,0,1} [2x^2 - 1]$

Also find x^* for **(a)** and **(b)**.

4. An independent trucker is hired to transport a perishable cargo across country. Although his starting point a and final destination k are fixed, he is given the choice of any possible route from a to k, as shown in Figure 12.4. The

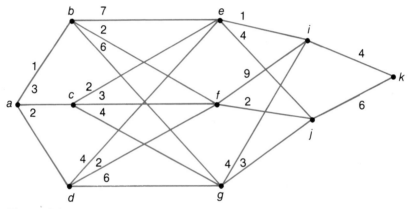

Figure 12.4

cost in hundreds of dollars of insuring the cargo between cities x and y is shown on the edge of (x,y). The route between a and k with the cheapest total insurance policy is the route on which there is the least chance that the cargo will spoil due to delays or bad weather. The trucker wishes to find the route which minimizes the total insurance cost.

(a) Show that this problem involves a sequence of four decisions.

(b) How many stages are there?

(c) Find a policy P and a subpolicy Q.

(d) Let x_1 be a state of stage 1, x_2 be a state of stage 2, and so on. What values can x_1, x_2, and x_3 assume?

(e) Define a function $h(x,y)$ similar to that used in Example 4.

(f) Calculate $h(b,f)$, $h(g,i)$, and $h(d,i)$.

(g) Define functions $F_1(x_1)$, $F_2(x_2)$, ..., $F_m(x_m)$ similar to those used in Example 4. Calculate $F_1(x_1)$.

(h) Use the optimality theorem to find a formula for $F_2(x_2)$ in terms of $F_1(x_1)$ and $h(x_1,x_2)$. Then use this theorem to find a formula for $F_3(x_3)$ in terms of $F_2(x_2)$ and $h(x_2,x_3)$, and so on.

(i) Construct a table for this problem similar to Table 12.1.

(j) What is the minimal cost of insuring the cargo from a to k?

5.   Which route should the trucker of Exercise 4 use?

12.2 ALLOCATION PROCESSES

In many practical situations a fixed amount of some resource must be allocated divided among several different projects. A city government, for example, must allocate portions of its tax revenues to police protection, fire protection, education, welfare, health, and so on. A method for allocating a fixed amount of some resource is called on **allocation process**. In this section we will see that dynamic programming can be used to find the best allocation process for a given situation. The following example shows how this is done.

Example 5   A bank has 5 million dollars available for investment in four enterprises, a, b, c, and d. The money can be invested in multiples of 1 million dollars and the return varies with the amount spent. The profit (in millions of dollars) for each investment is given in Table 12.2. For example, if the bank invests 1 million dollars in enterprise a, the profit will be 0.25 million dollars because 0.25 is the number that appears in the intersection of the row labeled 1 and the column labeled enterprise a. Similarly, if the bank invests 2 million dollars in enterprise d, the profit will be 0.30 million dollars.

How should the bank allocate its money if it wishes to obtain the largest possible profit?

Table 12.2   **Profit (in millions of dollars)**

Investment (in millions of dollars)	Enterprise a $h_1(x) = F_1(x)$	Enterprise b $h_2(x)$	Enterprise c $h_3(x)$	Enterprise d $h_4(x)$
0	0	0	0	0
1	0.25	0.20	0.15	0.23
2	0.40	0.35	0.25	0.30
3	0.65	0.55	0.60	0.50
4	0.75	0.65	0.75	0.60
5	0.90	0.75	0.85	0.70

Solution   This situation involves a sequence of four decisions. In stage 1 we must decide how much money to invest in enterprise a. In stage 2, we must decide how much to invest in enterprise b, and so on. Since this problem satisfies Definition 12.1, we will use the dynamic-programming method outlined on page 513.

(1) Let x_1 represent a state of stage 1, let x_2 represent a state of stage 2, and so on. Since 0, 1, 2, 3, 4, or 5 million dollars may be invested, x_1 may be any one of 0, 1, 2, 3, 4, 5; x_2 may be any one of 0, 1, 2, 3, 4, 5; and so on.

(2) Let $h_1(x)$ be the profit that will be made if x million dollars are invested in enterprise a, and let $h_2(x)$ be the profit that will be made if x million dollars are invested in enterprise b. Define $h_3(x)$ and $h_4(x)$ similarly.

(3) Let $F_1(x)$ be the maximum profit that can be made if x million dollars are invested in enterprise a. (Notice that $F_1(x) = h_1(x)$.) Let $F_2(x)$ be the maximum profit that can be made if x million dollars are invested in enterprises a and b. Let $F_3(x)$ be the maximum profit that can be made if x million dollars are invested in enterprises a, b, and c. Let $F_4(x)$ be the maximum profit that can be made if x million dollars are invested in enterprises a, b, c, and d.

The problem will be solved when we have calculated $F_4(5)$.

(4) We must find a formula for calculating $F_4(x)$. This is the maximum profit that can be made if x million dollars are invested in a, b, c, and d. In general, this investment can be made by investing x_1 in a, x_2 in b, x_3 in c, and x_4 in d, where

$$x_1 + x_2 + x_3 + x_4 = x$$

and each x_i is one of the numbers $0, 1, \ldots, 5$. The associated policy P is

$d_1:$ Invest x_1 in a

$d_2:$ Invest x_2 in b $\Big\}$ $x - x_4$ is invested in a, b, and c

$d_3:$ Invest x_3 in c

$d_4:$ Invest x_4 in d

Suppose P is optimal. The optimality theorem then implies that the subpolicy $Q = d_1, d_2, d_3$ is also optimal. This means that investing x_1 in a, x_2 in b, and x_3 in c yields the maximum profit that can be made when $x - x_4$ is invested in a, b, and c. In other words, the profit associated with subpolicy d_1, d_2, d_3 is $F_3(x - x_4)$. This value is shown in Table 12.3.

Table 12.3	Enterprise	Amount	Profit
	a, b, c	$x - x_4$	$F_3(x - x_4)$
	d	x_4	$h_4(x_4)$
	Total	x	$F_3(x - x_4) + h_4(x_4)$

By definition, the profit that is made when x_4 is invested in d is $h_4(x_4)$. This value has also been listed in Table 12.3. Table 12.3 clearly indicates that the total profit associated with the optimal policy P is

$$F_3(x - x_4) + h_4(x_4) \tag{5}$$

where x_4 is $0, 1, 2, \ldots$, or x.

Since P is optimal, the total profit associated with P is $F_4(x)$. This means that

$$F_4(x) = F_3(x - x_4) + h_4(x_4)$$

for some particular value of x_4. This particular value of x_4 is denoted $x_4{}^*$. It follows that $F_4(x)$ and $x_4{}^*$ can be found by substituting $x_4 = 0, 1, \ldots, x$ into expression (5) and deciding which value of x_4 results in the largest value; that is,

$$F_4(x) = \max_{x_4 = 0, 1, \ldots, x} [F_3(x - x_4) + h_4(x_4)] \tag{6}$$

Similarly,

$$F_3(x) = \max_{x_3 = 0, 1, \ldots, x} [F_2(x - x_3) + h_3(x_3)] \tag{7}$$

and

$$F_2(x) = \max_{x_2 = 0, 1, \ldots, x} [F_1(x - x_2) + h_2(x_2)] \tag{8}$$

(5) The values of $F_1(x)$ and $h_2(x)$ are given in Table 12.2. Using these values and formula (8), we can now find the values of $F_2(x)$.

Substituting $x = 0$ into equation (8) yields

$$F_2(0) = \max_{x_2 = 0} [F_1(-x_2) + h_2(0)]$$
$$= F_1(0) + h_2(0) = 0 \qquad \text{with } x_2{}^* = 0$$

Substituting $x = 1$ into equation (8) yields

$$F_2(1) = \max_{x_2 = 0, 1} [F_1(1 - x_2) + h_2(x_2)]$$

Substituting $x_2 = 0$ and then $x_2 = 1$ into $F_1(1 - x_2) + h_2(x_2)$, we have

$$F_1(1 - 0) + h_2(0) = 0.25 + 0 = 0.25$$

and

$$F_1(1 - 1) + h_2(1) = 0 + 0.20 = 0.20$$

The largest of these values is 0.25, so

$$F_2(1) = 0.25 \qquad \text{with } x_2{}^* = 0$$

Substituting $x = 2$ into equation (8) yields

$$F_2(2) = \max_{x_2 = 0, 1, 2} [F_1(2 - x_2) + h_2(x_2)]$$

Substituting $x_2 = 0$, $x_2 = 1$, and $x_2 = 2$ into $F_1(2 - x_2) + h_2(x_2)$ in turn, we have

$$F_1(2 - 0) + h_2(0) = 0.40 + 0 = 0.40$$
$$F_1(2 - 1) + h_2(1) = 0.25 + 0.20 = 0.45$$
$$F_1(2 - 2) + h_2(2) = 0 + 0.35 = 0.35$$

It follows that $F_2(2) = \max(0.40,0.45,0.35) = 0.45$, with $x_2{}^* = 1$.
Similarly, the reader can check that

$$F_2(3) = \max[0.65,0.60,0.60,0.55] = 0.65 \qquad \text{with } x_2{}^* = 0$$

$$F_2(4) = \max[0.75,0.85,0.75,0.80,0.65] = 0.85 \qquad \text{with } x_2{}^* = 1$$

and

$$F_2(5) = 1.00 \qquad\qquad\qquad\qquad \text{with } x_2{}^* = 2$$

The values of $F_2(x)$ for are listed in Table 12.4.

Table 12.4	x	$F_2(x)$	$x_2{}^*$	$F_3(x)$	$x_3{}^*$
	0	0.00	0	0.00	0
	1	0.25	0	0.25	0
	2	0.45	1	0.45	0
	3	0.65	0	0.65	0
	4	0.85	1	0.85	0 and 3
	5	1.00	2	1.05	3

Table 12.4, Table 12.2, and equation (7) can now be used to find the values of $F_3(x)$. Substituting $x = 0$ into equation (7) yields

$$F_3(0) = \max_{x_3=0}[F_2(0 - x_3) + h_3(x_3)]$$

$$= F_2(0 - 0) + h_3(0) = 0 + 0 = 0 \qquad \text{with } x_3{}^* = 0$$

Moreover, since

$$F_2(1 - 0) + h_3(0) = 0.25 + 0 = 0.25$$

and

$$F_2(1 - 1) + h_3(1) = 0 + 0.15 = 0.15$$

it follows that $F_3(1) = \max(0.25,0.15) = 0.25$, with $x_3{}^* = 0$. These and the remaining values of $F_3(x)$ have been placed in Table 12.4.

Finally, applying equation (6), Table 12.4, and Table 12.2, we get

$$F_4(5) = \max[1.05 + 0,0.85 + 0.23,0.65 + 0.30,0.45$$

$$+ 0.50,0.25 + 0.60,0 + 0.70]$$

$$= 1.08 \qquad \text{with } x_4{}^* = 1 \tag{9}$$

(6) Using equation (9), the maximum profit that can be made is 1.08 million dollars.

Example 6   Consider the problem of Example 5. Find a distribution of investments that yields the maximum profit.

Solution   According to equation (9), the maximum profit will be obtained if $x_4{}^* = 1$; that is, if 1 million dollars is invested in enterprise d. This leaves $x = 4$ million dollars to be invested in enterprises a, b, and c. We must now see how $F_3(4)$ was obtained. According to Table 12.4, $F_3(4) = 0.85$ when $x_3{}^* = 0$ or 3. It follows that the maximum profit will be obtained if 0 or 3 million dollars is invested in c.

If 0 is invested in c, there will be 4 million dollars to invest in a and b. Since $F_2(4) = 0.85$ when $x_2{}^* = 1$, 1 million dollars must be invested in b. (See Table 12.4.) This leaves 3 million dollars to be invested in a. This distribution of investments, along with the alternative optimal distribution that is obtained when 3 million dollars is invested in enterprise c, is displayed in Table 12.5. (The reader should check the second optimal policy.)

Table 12.5   Optimal Investments (in millions of dollars)

Enterprise	a	b	c	d
Policy 1	3	1	0	1
Policy 2	1	0	3	1

The reader can check that each of these distributions yields the maximum profit of 1.08 million dollars.

Formulas similar to equations (6)–(8) can be used to solve many other allocation problems. Instead of money, the resource to be allocated may be men, machines, hydroelectric power, spaceship fuel, and so on.

Example 5 can also be solved by listing all possible investments. However, this procedure requires that 224 additions be carried out, whereas the dynamic-programming method only requires 33 additions. Moreover, if 100 million dollars were available for investment, the listing method would require 646,800 additions. The dynamic-programming approach, however, would require only 1377 additions.

12.2 EXERCISES

1. Consider Example 6.
    (a) Show that policy 2 of Table 12.5 results if 3 million dollars is invested in enterprise c.
    (b) Verify that each of the distributions listed in Table 12.5 yields the maximum profit of 1.08 million dollars.

2. Nineteen yards of material are available for making three kinds of dresses: style a, style b, and style c. The fabric requirements and profit for each type of dress are given in the following table:

Dress Style	Fabric (in yards)	Profit (in dollars)
a	2	1.5
b	3	2.5
c	5	4.0

(a) The problem of determining how much material should be used to make dresses of each type in order to maximize the total profit is a problem requiring a sequence of three decisions. What decision must be made at stage 1? At stage 2? At stage 3?

(b) Let x_1, x_2, and x_3 be the amounts of material allocated to dress styles a, b, and c, respectively. What values can x_1, x_2, and x_3 assume?

(c) Let $F_1(x)$ be the maximum profit that can be made if at most x yards are used to make style a dresses. Let $F_2(x)$ be the maximum profit that can be made if at most x yards are used to make styles a and b. Let $F_3(x)$ be the maximum profit that can be made if at most x yards are used to make styles a, b, and c. Find $F_1(4)$, $F_1(6)$, and $F_1(18)$.

(d) Let $h_1(x)$, $h_2(x)$, and $h_3(x)$ be the profits that will be made if x yards are used to make styles a, b, and c, respectively. Calculate $h_2(3)$, $h_3(5)$, and $h_1(10)$.

3.   (a) Using the functions defined in Exercise 2, find a formula for calculating $F_3(x)$ in terms of F_2 and h_3. Find a similar formula for finding $F_2(x)$.

(b) What values of $F_2(x)$ are needed to calculate $F_3(19)$?

(c) What is the maximum profit that can be made?

(d) How many yards should be used to make dresses of each style in order to maximize profits?

4.   A company has three market areas (A, B, and C) and six salespeople. Table 12.6 gives the profits that will be obtained according to the number of salespeople assigned to each area.

Table 12.6 Market Area

Market Area	Number of Salespeople						
	0	1	2	3	4	5	6
A	10	12	15	18	24	30	33
B	12	18	20	22	24	25	26
C	6	12	20	22	24	26	28

(a) Find the maximum possible profit.

(b) Determine the number of salespeople who should be assigned to each market area to maximize the total profit.

5.   A student must take examinations in mathematics, physics, and history. Each course has the same number of credits. He decides to study a total of 12 hours in three blocks of 4 hours each. Each block of time will be devoted to one course. His estimate of his grades ($A = 4$, $B = 3$, $C = 2$, $D = 1$, $F = 0$) according to the various numbers of hours devoted to each course are given in the following table:

	Hours of Study	0	4	8	12
Estimated Grades	Mathematics	0	1	1	3
	Physics	1	1	3	4
	Chemistry	0	1	3	3

For example, if he studies mathematics for 8 hours, he will receive a 1 (D); if he studies chemistry for 12 hours, he will receive a 3 (B).

(a) What is the maximum number of grade points that the student can obtain? (An A is 4 points, a B is 3 points, etc.)

(b) How many hours should he devote to each subject in order to maximize his total grade points?

6.   The weight (in pounds) and value (in dollars) of three items, A, B, and C, that can be put into a knapsack are as follows:

Item	A	B	C
Weight/Unit	1	3	5
Value/Unit	1	5	8

Assume that several units of each item can be carried, but that the knapsack cannot hold more than 19 pounds.

(a) Determine the maximum value of the items that can be put into the knapsack.

(b) How should the knapsack be filled if the total value of its contents is to be maximized?

12.3 TRANSPORTATION PROBLEMS

One way that businesses can cut costs is by minimizing shipping costs. Typically, a business will have several factories or warehouses that supply the needs of several distributors or retail outlets. The problem of fulfilling the distributors' needs while keeping shipping costs at a minimum is called the **transportation problem**. In this section dynamic programming will be used to solve a transportation problem. The following example illustrates how this is done.

Example 7   Two warehouses W_1 and W_2, hold 8 and 14 tons of flour, respectively, while six distributors, D_1, D_2, ..., and D_6, require 5, 4, 3, 4, 2, and 4 tons of flour, respectively. Let $h_i(x)$ be the cost of

(a)   shipping x tons from warehouse W_1 to D_i

and

(b)   shipping the remaining tons required by D_i from warehouse W_2.

The values of $h_i(x)$ are given in Table 12.7. For example, distributor D_2 requires 4 tons of flour. It follows that $h_2(3)$ is the cost of shipping 3 of these tons from W_1 and shipping the remaining $4 - 3 = 1$ ton from W_2. The value of $h_2(3)$ is found in the intersection of the row labeled 3 and the column labeled $h_2(x)$; therefore, $h_2(3) = 33$. Notice that $h_2(5)$, $h_2(6)$, $h_2(7)$, and $h_2(8)$ are undefined because D_2 requires only 4 tons of flour.

Find the minimum cost of shipping the entire 22 tons of flour from the warehouses to the distributors.

Table 12.7	x	$h_1(x) = F_1(x)$	$h_2(x)$	$h_3(x)$	$h_4(x)$	$h_5(x)$	$h_6(x)$
	0	44	38	17	17	13	24
	1	49	41	23	25	11	31
	2	48	37	27	32	5	30
	3	47	33	26	39	—	29
	4	46	27	—	41	—	24
	5	41	—	—	—	—	—

Solution A sequence of six decisions must be made. In stage 1 we must decide how many tons to ship from W_1 to D_1; in stage 2 we must decide how many tons to ship from W_1 to D_2; and so on. It follows that the dynamic-programming method outlined in Section 12.1 can be used.

(1) Let x_1 be the amount shipped from W_1 to D_1; let x_2 be the amount shipped from W_1 to D_2; and so on. From the information given, we see that x_1 can be 0, 1, 2, 3, 4 or 5; x_2 can be 0, 1, 2, 3, or 4; x_3 can be 0, 1, 2, or 3; x_4 can be 0, 1, 2, 3, or 4; and so on.

(2) The functions $h_i(x)$ are defined in Table 12.7.

(3) Let $F_1(x)$ be the minimum cost of shipping x tons from W_1 to D_1 and shipping the remaining flour required by D_1 from W_2. Notice that $F_1(x) = h_1(x)$. Let $F_2(x)$ be the minimum cost of shipping x tons from W_1 to D_1 and D_2 and shipping the remaining flour required by D_1 and D_2 from W_2. Let $F_3(x)$ be the minimum cost of shipping x tons from W_1 to D_1, D_2, and D_3, and shipping the remaining flour required by D_1, D_2, and D_3 from W_2.

Similar definitions can be made for $F_4(x)$, $F_5(x)$, and $F_6(x)$. To solve the problem, we must calculate $F_6(8)$. This is the minimum cost of shipping 8 tons of flour to $D_1, D_2, \ldots, D_6$ from W_1 and shipping the remaining $22 - 8 = 14$ tons of flour from W_2.

(4) Applying the optimality theorem as we did in Example 5 yields

$$F_6(x) = \min_{x_6 = 0, 1, \ldots, x} [F_5(x - x_6) + h_6(x_6)] \tag{10}$$

$$F_5(x) = \min_{x_5 = 0, 1, \ldots, x} [F_4(x - x_5) + h_5(x_5)] \tag{11}$$

$$F_4(x) = \min_{x_4 = 0, 1, \ldots, x} [F_3(x - x_4) + h_4(x_4)] \tag{12}$$

$$F_3(x) = \min_{x_3 = 0, 1, \ldots, x} [F_2(x - x_3) + h_3(x_3)] \tag{13}$$

$$F_2(x) = \min_{x_2 = 0, 1, \ldots, x} [F_1(x - x_2) + h_2(x_2)] \tag{14}$$

(Compare these equations with equations (6)–(8).)

(5) The values of $F_1(x)$ are given in Table 12.7. The values of $F_2(x)$ can be calculated from these values, equation (14), and the values of $h_2(x)$ given in Table 12.7. The values of $F_3(x)$ can be calculated from the values of $F_2(x)$, equation (13), and the values of $h_3(x)$ in Table 12.7. Continuing in this manner also yields the values of $F_4(x)$, $F_5(x)$, and $F_6(x)$.

Since the computations involved in finding $F_2(x)$, $F_3(x)$, ..., $F_6(x)$ are similar to those of Example 5, we will omit the details. However, the results are summarized in Table 12.8. As before, $x_k{}^*$ is the value of x_k that gives the optimal (in this case, minimum) value of $F_{k-1}(x - x_k) + h_k(x_k)$.

Table 12.8

x	$F_2(x)$	$x_2{}^*$	$F_3(x)$	$x_3{}^*$	$F_4(x)$	$x_4{}^*$	$F_5(x)$	$x_5{}^*$
0	82	0	99	0	116	0	129	0
1	85	1	102	0	119	0	127	1
2	81	2	98	0	115	0	121	2
3	77	3	94	0	111	0	124	0 or 2
4	71	4	88	0	105	0	118	0
5	76	4	93	0	110	0	116	1 or 2
6	75	4	92	0	109	0	110	2
7	74	4	91	0	108	0	115	2
8	73	4	90	0	107	0	114	2

(6) Using Tables 12.7 and 12.8 and equation (10), the minimum transportation cost is

$$F_6(8) = \min[114 + 24, 115 + 31, 110 + 30, 116 + 29, 118 + 24]$$
$$= 138 \tag{15}$$

with $x_6{}^* = 0$. Thus, the minimum cost is 138 dollars.

Example 8 Find how much flour should be shipped from each warehouse to each distributor in Example 7 in order to minimize the transportation costs.

Solution The minimum cost $F_6(8) = 138$ dollars is achieved when $x_6{}^* = 0$. (See equation (15).) It follows that 0 tons should be shipped from W_1 to D_6. This value has been placed in Table 12.9.

Table 12.9

	D_1	D_2	D_3	D_4	D_5	D_6
W_1	2	4	0	0	2	0
W_2	3	0	3	4	0	4

Remember that warehouse W_1 originally has 8 tons of flour. Therefore, shipping 0 units from W_1 to D_6 leaves $8 - 0 = 8$ tons that must be shipped from W_1 to $D_1, D_2, \ldots, D_5$. In Table 12.8 we see that $F_5(8)$ is achieved when $x_5{}^* = 2$. This means that 2 tons should be shipped from W_1 to D_5. (This value has also been placed in Table 12.9.) There are now $8 - 2 = 6$ tons that must be shipped from W_1 to D_1, D_2, D_3, D_4. Continuing our calculations in this manner completes the first row of Table 12.9.

The second row of Table 12.9 can now be completed by remembering that D_i must receive all the flour that it requires from W_1 and W_2. For example, since D_1 requires 5 tons, and it receives 2 tons from W_1, it must receive the remaining $5 - 2 = 3$ tons from W_2. This number has been placed in Table 12.9. The remaining values in the second row are calculated similarly.

Table 12.9 gives an optimal distribution of the flour. Using this table, we see that the minimum cost will be achieved if 2, 4, 0, 0, 2, 0 tons of flour are shipped from W_1 to $D_1, D_2, \ldots, D_6$, respectively. The remaining flour, 3, 0, 3, 4, 0, 4 tons, must be shipped from W_2 to $D_1, D_2, \ldots, D_6$, respectively.

The transportation costs of Example 7 are displayed in a single table, Table 12.7. These costs can also be displayed in two matrices, **C** and **K**:

$$
\mathbf{C} = \begin{array}{c} \\ W_1 \\ W_2 \end{array} \begin{array}{ccc} D_1 & D_2 & \cdots & D_6 \\ \begin{pmatrix} c_{11} & c_{12} & \cdots & c_{16} \\ c_{21} & c_{22} & \cdots & c_{26} \end{pmatrix} \end{array},
$$

$$
\mathbf{K} = \begin{array}{c} \\ W_1 \\ W_2 \end{array} \begin{array}{ccc} D_1 & D_2 & \cdots & D_6 \\ \begin{pmatrix} k_{11} & k_{12} & \cdots & k_{16} \\ k_{21} & k_{22} & \cdots & k_{26} \end{pmatrix} \end{array}
$$

The first matrix, **C**, gives the cost c_{ij} of transporting *each* ton of flour from W_i to D_j. The second matrix, **K**, gives the fixed cost k_{ij} that must be paid if anything at all is shipped from W_i to D_j. For example, k_{ij} might be the cost of the driver's salary, gas, and tolls; this cost remains the same whether 1 or 1000 tons of flour are shipped.

Suppose, for example, that $c_{21} = 4$ and $k_{21} = 3$. This means that the cost of transporting each unit from W_2 to D_1 is 4 and the fixed transportation cost is 3. The total cost of transporting x tons is therefore x times the cost of transporting 1 ton, $4x$, plus the fixed transportation cost, 3. In other words, the total transportation cost is

$$4x + 3$$

In general, the total transportation cost of shipping x tons from W_i to D_j is

$$c_{ij}x + k_{ij} \tag{16}$$

In the next example we will see how a table like Table 12.7 can be calculated from the matrices **C** and **K**. Once this table is determined, the method used in Example 7 can be applied to find the minimum transportation cost.

Example 9   Let W_1, W_2, and $D_1, \ldots, D_6$ be the warehouses and distributors of Example 7. Furthermore, assume that the capacities of the warehouses and the requirements of the distributors are those given in that problem. However, assume that the costs c_{ij} of shipping 1 ton of flour are the entries in the matrix

$$
\mathbf{C} = (c_{ij}) = \begin{array}{c} \\ W_1 \\ W_2 \end{array} \begin{array}{cccccc} D_1 & D_2 & D_3 & D_4 & D_5 & D_6 \\ \begin{pmatrix} 7 & 5 & 8 & 10 & 1 & 4 \\ 8 & 9 & 4 & 3 & 6 & 5 \end{pmatrix} \end{array}
$$

while the fixed transportation charges k_{ij} are the entries in the matrix

$$\mathbf{K} = (k_{ij}) = \begin{array}{c} \\ W_1 \\ W_2 \end{array} \begin{array}{cccccc} D_1 & D_2 & D_3 & D_4 & D_5 & D_6 \\ \begin{pmatrix} 6 & 7 & 2 & 1 & 3 & 8 \\ 4 & 2 & 5 & 5 & 1 & 4 \end{pmatrix} \end{array}$$

Let $h_1(x)$ be the cost of shipping x tons of flour from W_1 to D_1 and shipping the remainder from W_2. Determine the values of $h_1(x)$.

Solution   If $x = 0$ tons are shipped from W_1 to D_1, the remaining $5 - 0 = 5$ tons must be shipped from W_2. The cost of shipping 5 tons from W_2 to D_1 is 5 times the cost of shipping 1 ton, $5c_{21}$, plus the fixed transportation charge k_{21}; that is,

$$h_1(0) = 5c_{21} + k_{21} \tag{17}$$

The values c_{21} and k_{21} are found in the intersection of row 2, column 1 of **C** and **K**, respectively. Substituting $c_{21} = 8$ and $k_{21} = 4$ into equation (17) yields

$$h_1(0) = 5(8) + 4 = 44$$

If $x = 1$ ton is shipped from W_1 to D_1, the remaining $5 - 1 = 4$ tons must be shipped from W_2. Using expression (16), the cost of shipping 1 ton from W_1 to D_1 is $1c_{11} + k_{11}$, while the cost of shipping 4 tons from W_2 to D_1 is $4c_{21} + k_{21}$. It follows that

$$h_1(1) = 1c_{11} + k_{11} + 4c_{21} + k_{21}$$
$$= 1(7) + 6 + 4(8) + 4 = 49$$

Similarly,

$$h_1(2) = 2c_{11} + k_{11} + 3c_{21} + k_{21} = 2(7) + 6 + 3(8) + 4 = 48$$
$$h_1(3) = 3(7) + 6 + 2(8) + 4 = 47$$
$$h_1(4) = 4(7) + 6 + 1(8) + 4 = 46$$

and

$$h_1(5) = 5(7) + 6 = 41$$

In Example 9 the matrices **C** and **K** were used to calculate the values of $h_1(x)$. In a similar manner, the reader can calculate the values of $h_2(x)$, $h_3(x)$, ..., $h_6(x)$. These values can all be placed in a table like Table 12.7. Formulas (10)–(14) can then be used to calculate the minimum transportation cost.

The reader has probably noticed that the values of $h_1(x)$ that we calculated in Example 9 are the same as the values of $h_1(x)$ listed in Table 12.7. In fact, all the values $h_2(x)$, $h_3(x)$, ..., $h_6(x)$ for Example 9 are the same as those given in Table 12.7. This means that the minimum transportation cost for Example 9 is also 138 dollars and that the distribution of flour given in Table 12.9 is optimal.

12.3 EXERCISES

1. Show that the distribution of Table 12.9 results in the optimal transportation cost of 138 dollars.

2. Calculate $h_2(x)$ for Example 9.

3. A test-tube factory has two branches, A and B. Branch A has five cases of test tubes on hand, and branch B has twelve cases. The factory has received orders from three customers, C_1, C_2, and C_3. Customer C_1 requires two cases of test tubes, customer C_2 requires seven cases, and customer C_3 requires eight cases. Table 12.10 gives the cost (in dollars) of shipping x cases from branch A to C_i and shipping the remaining cases required by C_i from branch B. This cost has been labeled $h_i(x)$.

Table 12.10	x	$h_1(x)$	$h_2(x)$	$h_3(x)$
	0	40	38	40
	1	42	45	38
	2	44	50	36
	3	—	48	38
	4	—	46	40
	5	—	40	41

(a) Define three functions $F_1(x)$, $F_2(x)$, and $F_3(x)$ similar to those of Example 7.

(b) Find formulas for calculating $F_2(x)$ and $F_3(x)$ similar to equations (10)–(14).

(c) Find the minimum cost of shipping the seventeen cases to the three customers.

(d) How should the cases be shipped if the factory wishes to minimize its shipping costs?

4. Let $A = (a_1, a_2)$ represent the number of units of a certain product held by warehouses W_1 and W_2 (i.e., W_1 holds a_1 units, and W_2 holds a_2 units). Let $B = (b_1, b_2, \ldots, b_n)$ represent the number of units required by distributors D_1, D_2, ..., D_n. Find the number of units held by W_2 and the number of units required by D_2 if—

(a) $A = (1,2)$, $B = (2,1)$ (b) $A = (1,4)$, $B = (2,5,4,3,7)$

5. Let matrix $C = (c_{ij})$ where c_{ij} is the cost of transporting one unit from W_i to D_j and let matrix $K = (k_{ij})$ where k_{ij} is the fixed cost that must be paid if anything at all is shipped from W_i to D_j. Moreover, let A and B be matrices like those defined in Exercise 4. Suppose that

$$A = (8,7), \qquad B = (6,3,4,2)$$

$$C = (c_{ij}) = \begin{pmatrix} 1 & 3 & 5 & 4 \\ 3 & 0 & 4 & 4 \end{pmatrix}, \qquad K = (k_{ij}) = \begin{pmatrix} 1 & 4 & 1 & 5 \\ 0 & 1 & 3 & 2 \end{pmatrix}$$

(a) How many distributors are there?

(b) Determine the cost of shipping—
   (i) 2 units from W_1 to D_2
   (ii) 2 units from W_2 to D_1
   (iii) 0 units from W_1 to D_3
   (iv) 3 units from W_1 to D_3 and the remainder required by D_3 from W_2.

(c) How many units are required by D_1?

(d) How many units are held in W_1?

(e) What is the total number of units available from the two warehouses?

(f) Determine the total number of units required by the distributors.

6. Using the information given in Exercise 5—
   (a) calculate $h_i(x)$, the cost of shipping x units from W_1 to D_i and shipping the remainder required by D_i from W_2, for all values of i and x
   (b) calculate $F_k(x)$, $k = 1, 2, 3$, the minimal cost of shipping x units to $D_1, D_2,$ ..., D_k from W_1 and shipping the remainder from W_2
   (c) calculate $x_k{}^*$, the value of x_k that yields $F_k(x)$
   (d) Find the minimum cost of transporting the units to the distributors.

7. (a) Find all optimal policies for Exercise 5.
   (b) Consider an optimal policy found in part (a). Find the cost of shipping the required units to $D_1, D_2, D_3,$ and D_4 if this policy is used. What is the sum of these costs?

8. (a) Using the notation of Exercises 4 and 5, find the minimum transportation cost if

$$A = (7,6), \qquad B = (3,4,3,2,1)$$

$$C = \begin{pmatrix} 2 & 3 & 3 & 1 & 4 \\ 2 & 5 & 2 & 2 & 5 \end{pmatrix}, \qquad K = \begin{pmatrix} 3 & 4 & 2 & 3 & 2 \\ 2 & 4 & 1 & 2 & 3 \end{pmatrix}$$

   (b) Find all optimal policies and the total cost associated with each.

12.4 INVENTORY PROBLEMS

Suppose that a manufacturing company has orders for its product (say, a hand calculator) that must be delivered within the next few months. For example, suppose that some of the calculators must be delivered in January; some must be delivered in February; and so on. The company must decide whether to produce all of the calculators in January, or to produce some in January, some in February, and so on. If the company produces all of the calculators in January, it will incur storage costs. These might include rent for warehouse space, salaries for warehouse workers, and the expense of keeping records on the stored items. On the other hand, if the company produces some calculators each month, it will incur set-up costs. For example, if the company produces more than one model of calculator, it may have to reset its machinery to the specifications of each model.

The problem of balancing storage and set-up costs so that the total of these costs is minimized is called the **inventory problem**. In this section we will see how inventory problems can be solved using dynamic programming. To see how this is done, we will use the following example.[1]

Example 10   A company has orders for calculators that must be delivered within the next five months. The total number of calculators that must be delivered at the end of each month are shown in Table 12.11. (Notice that January is month 5 instead of month 1. This will make the calculations simpler.) Suppose that the company has the following costs:

Table 12.11

i	Month	Calculators Required r_i
5	January	400
4	February	300
3	March	400
2	April	300
1	May	200

(a) Each calculator costs 3 dollars to produce.

(b) There is a set-up cost of 30 dollars per run whenever calculators are produced.

(c) It costs 10 cents to store a calculator for 1 month.

In addition, suppose that the initial and final inventory is 0 calculators, that the company's facilities permit only one production run a month, and that each run produces one or more groups of 100 calculators each. Find the minimum total storage, set-up, and production cost.

Solution   The total cost of producing the 1600 calculators is 4800 dollars, regardless of when they are produced. Therefore, to minimize total cost, we must minimize the storage and set-up costs.

Keeping the supply equal to the demand month by month requires a set-up cost of $30(5) = 150$ dollars. On the other hand, producing more than the current demand involves storage costs. We must find an optimal policy that minimizes the total cost by balancing these two factors.

This problem can be considered to require a sequence of five decisions. In stage 1 we must decide how many calculators to produce in month 1 (May); in stage 2 we must decide how many calculators to produce in month 2 (April); and so on. Therefore, we can use the dynamic-programming method outlined in Section 12.1.

(1) Let x_i be the number of calculators produced during month i. Since 400 calculators are needed at the end of January and the initial inventory is 0, x_5 must be at least 400. Since 1600 calculators are needed altogether and

---

1. The techniques of this section, while of great practical value, are more complex than those of preceding sections.

the final inventory is 0, x_5 cannot be more than 1600. Therefore, the fact that calculators are produced in groups of 100 implies that

x_5 must be one of   400, 500, ..., 1600

Similarly,

x_4 must be one of   0, 100, ..., 1200

x_3 must be one of   0, 100, ..., 900

x_2 must be one of   0, 100, ..., 500

x_1 must be one of   0, 100, 200

(2) We need to define functions h_i that will represent costs incurred during month i. These costs depend on the number x_i of calculators produced during the month and also on the number of calculators stored during the month. Therefore, we must define a new variable, y_i, that will represent the number of calculators in the warehouse at the beginning of month i. Then we can define the functions h_i by letting $h_i(x_i,y_i)$ be the storage and set-up cost (in dollars) for month i. The reader can check that

y_5 must be 0

y_4 must be one of   0, 100, ..., 1200

y_3 must be one of   0, 100, ..., 900

and so on. The reader can also check that

$$h_i(x_i,y_i) = (0.1)y_i + 30 \qquad \text{if } x_i > 0$$
$$h_i(x_i,y_i) = (0.1)y_i \qquad\qquad \text{if } x_i = 0 \tag{18}$$

(3) Let $F_i(x)$ be the minimum storage and set-up cost for months 1 through i if x calculators are in storage at the beginning of month i. For example, $F_3(x)$ is the minimum storage and set-up cost for months 1 through 3 (i.e., May, April, and March) if x calculators are in storage at the beginning of month 3 (March). The last of these functions, $F_5(x)$, is the minimum storage and set-up cost for all five months if there are x calculators in storage at the beginning of month 5 (January). Since we know that there will be 0 calculators in storage at the beginning of January, x must be 0. Therefore, to solve the inventory problem—that is, to find the minimum total storage and set-up cost—we must calculate $F_5(0)$.

(4) Since the value of $F_5(0)$ will depend on what decisions are made for months 1, 2, 3, and 4, we must find formulas for calculating F_1, F_2, F_3, and F_4. To find a formula for F_1, let us review the definitions:

$F_1(x)$ is the minimum storage and set-up cost for month 1 (May) if x calculators are in storage at the beginning of May

x_1 is the number of calculators produced during May

y_1 is the number of calculators in storage at the beginning of May

$h_1(x_1,y_1)$ is the storage and set-up cost for May if x_1 calculators are produced in May and y_1 calculators are stored during May

If x calculators are in storage at the beginning of May, then by definition

$$y_1 = x \tag{19}$$

Furthermore, since 200 calculators are needed by the distributors in May (see Table 12.11), and no calculators are to be left in storage at the end of May, the company must manufacture $200 - x$ calculators in May; that is,

$$x_1 = 200 - x \tag{20}$$

Using the relationships between y_1, x_1, and x given in equations (19) and (20) and the definitions of F_1 and h_1, we have that

$$F_1(x) = h_1(x_1, y_1) = h_1(200 - x, x) \tag{21}$$

The definitions and the optimality theorem can now be used to find a formula for $F_2(x)$. The function $F_2(x)$ is the minimum storage and set-up cost for months 1 and 2 (April and May) if x calculators are in storage at the beginning of month 2 (April). April's inventory is the result of the following policy P:

d_1: Produce x_1 calculators in May

d_2: Produce x_2 calculators in April

Suppose that P is an optimal policy. Then $x_1{}^* = x_1$, $x_2{}^* = x_2$, and the optimality theorem implies that

$$Q = d_1$$

is an optimal subpolicy. Since x calculators are in storage at the beginning of April, $x_2{}^*$ calculators are produced in April, and 300 calculators are distributed in April (see Table 12.11), it follows that

$$x + x_2{}^* - 300$$

calculators are in storage at the beginning of May. It follows that the optimal cost associated with policy Q is $F_1(x + x_2{}^* - 300)$. This cost is recorded in Table 12.12.

Table 12.12

Month	Amount Produced	Inventory	Cost for the Month
April	$x_2{}^*$	x	$h_2(x_2{}^*, x)$
May	$x_1{}^*$	$x + x_2{}^* - 300$	$F_1(x + x_2{}^* - 300)$
Total			$F_1(x + x_2{}^* - 300)$ $+ h_2(x_2{}^*, x)$

Since $x_2{}^*$ calculators are produced and x are stored during April, the set-up and storage cost for April is $h_2(x_2{}^*, x)$. This value has also been listed in Table 12.12. This table clearly indicates that the total cost associated with the optimal policy P is

$$F_1(x + x_2{}^* - 300) + h_2(x_2{}^*,x)$$

Since P is an optimal policy, its total cost is by definition equal to $F_2(x)$. It follows that

$$F_2(x) = F_1(x + x_2{}^* - 300) + h_2(x_2{}^*,x) \qquad (22)$$

To find $x_2{}^*$ and $F_2(x)$, we must therefore calculate

$$F_1(x + x_2 - 300) + h_2(x_2,x)$$

for all allowable values of x_2 and then take the smallest value; that is,

$$F_2(x) = \min[F_1(x + x_2 - 300) + h_2(x_2,x)] \qquad (23)$$

Similarly,

$$F_3(x) = \min[F_2(x + x_3 - 400) + h_3(x_3,x)] \qquad (24)$$
$$F_4(x) = \min[F_3(x + x_4 - 300) + h_4(x_4,x)] \qquad (25)$$

and

$$F_5(x) = \min[F_4(x + x_5 - 400) + h_5(x_5,x)] \qquad (26)$$

We have omitted the values of x_i under "min" for convenience. It is understood that all allowable values of x_i will be considered.

(5) The values of $F_1(x)$ can now be computed from equations (18) and (21). Clearly, the amount in storage, x, at the beginning of May cannot exceed 200 because only 200 units are required in May and the inventory at the end of May must be 0. This can also be seen by the fact that the term $200 - x$ in equation (21) cannot be negative. Since units of 100 are stored, x can only be 0, 100, or 200. We must therefore calculate $F_1(0)$, $F_1(100)$, and $F_1(200)$. Using equations (18) and (21),

$$F_1(0) = h_1(200,0) = (0.1)(0) + 30 = 30 \qquad \text{with } x_1{}^* = 200$$
$$F_1(100) = h_1(100,100) = (0.1)(100) + 30 = 40 \qquad \text{with } x_1{}^* = 100$$
$$F_1(200) = h_1(0,200) = (0.1)(200) = 20 \qquad \text{with } x_1{}^* = 0$$

These figures have been placed in Table 12.13 for easy reference.

The values of $F_2(x)$ can now be calculated from the values of $F_1(x)$ and equations (18) and (23). Setting $x = 0$ in equation (23) and listing the permissible values of x_2 yields

$$F_2(0) = \min_{x_2 = 300, 400, 500} [F_1(x_2 - 300) + h_2(x_2,0)]$$

Table 12.13

x	$F_1(x)$	$x_1{}^*$	$F_2(x)$	$x_2{}^*$	$F_3(x)$	$x_3{}^*$	$F_4(x)$	$x_4{}^*$
0	30	200	50	500	80	400	110	300
100	40	100	60	400	90	300	120	200
200	20	0	70	300	100	200	130	100
300	—	—	60	0	110	100	110	0
400	—	—	80	0	90	0	130	0
500	—	—	70	0	110	0	150	0
600	—	—	—	—	130	0	170	0
700	—	—	—	—	130	0	160	0
800	—	—	—	—	160	0	190	0
900	—	—	—	—	160	0	220	0
1000	—	—	—	—	—	—	230	0
1100	—	—	—	—	—	—	270	0
1200	—	—	—	—	—	—	280	0

Substituting $x_2 = 300$, $x_2 = 400$, and $x_2 = 500$ into $F_1(x_2 - 300) + h_2(x_2,0)$ in turn, we have

$$F_1(300 - 300) + h_2(300,0) = 30 + (0.1)(0) + 30 = 60 \quad (x_2 = 300)$$

$$F_1(400 - 300) + h_2(400,0) = 40 + (0.1)(0) + 30 = 70 \quad (x_2 = 400)$$

and

$$F_1(500 - 300) + h_2(500,0) = 20 + (0.1)(0) + 30 = 50 \quad (x_2 = 500)$$

The smallest value of $F_1(x_2 - 300) + h_2(x_2,0)$ is 50 and occurs when $x_2 = 500$. In other words, $F_2(0) = 50$, with $x_2{}^* = 500$.

Similarly, the reader can show that

$$F_2(100) = \min_{x_2 = 200, 300, 400} [F_1(x_2 - 200) + h_2(x_2,100)]$$

$$= \min[70,80,60] = 60 \quad \text{with } x_2{}^* = 400$$

Notice that, although x_2 can be any one of the numbers 0, 100, ..., 500, we need only consider $x_2 = 200, 300, 400$ in calculating $F_2(100)$. This is because $F_1(x_2 - 200)$ is only defined for $x_2 - 200 = 0, 100, 200$ (see Table 12.13 under $F_1(x)$). Similar comments apply to the calculations of $F_2(200)$, ..., $F_2(500)$:

$$F_2(200) = \min_{x_2 = 100, 200, 300} [F_1(x_2 - 100) + h_2(x_2,200)]$$

$$= \min[80,90,70] = 70 \quad \text{with } x_2{}^* = 300$$

$$F_2(300) = \min[60,100,80] = 60 \quad \text{with } x_2{}^* = 0$$

$$F_2(400) = \min[80,90] = 80 \quad \text{with } x_2{}^* = 0$$

$$F_2(500) = \min[70] = 70 \quad \text{with } x_2{}^* = 0$$

The values of $F_2(x)$ have been placed in Table 12.13. The values of $F_3(x)$ and $F_4(x)$ can be calculated similarly and are also listed in Table 12.13.

(6) Applying equation (26) with $x = 0$, and using the values in Table 12.13, the minimum total storage and set-up cost is

$$F_5(0) = \min[F_4(x_5 - 400) + h_5(x_5,0)] \qquad (27)$$

$$= 140 \qquad\qquad \text{with } x_5{}^* = 400 \text{ or } 700$$

Therefore, the minimum total storage, set-up, and production cost is

$$140 + 4800 = 4940 \text{ dollars}$$

Example 11   Find an optimal policy for Example 10.

Solution   Looking at equation (27), we see that the optimal cost occurs when x_5 is either 400 or 700. Suppose that we let $x_5 = 400$ and enter this value in Table 12.14. (Another optimal policy will be formed if we let $x_5 = 700$.)

Table 12.14

	January	February	March	April	May	Total
Month i	5	4	3	2	1	
Amount produced x_i	400	300	400	500	0	1600
Amount stored y_i	0	0	0	0	200	200

Since the costs depend on the amount stored as well as the amount produced, Table 12.14 also lists the optimal amounts that must be stored each month. The amount $y_5 = 0$ is placed in Table 12.14 because the initial inventory is 0.

We can now use x_5 and y_5 to calculate y_4:

January 1	Stored	y_5
	Produced	x_5
	Distributed	$-r_5$
February 1	Stored	$\overline{y_4 = y_5 + x_5 - r_5}$

$$(28)$$

The amount in storage on February 1 is the result of the following three factors (see Table 12.14): (1) the amount in storage on January 1, (2) the amount produced during January, and (3) the amount distributed during January. Substituting $y_5 = 0$, $x_5 = 400$, and $r_5 = 400$ (see Tables 12.11 and 12.14) in equation (28) yields

$$y_4 = 0 + 400 - 400 = 0$$

This value is listed in Table 12.14.

The amount in storage at the beginning of February is therefore 0. Since Table 12.13 indicates that $F_4(0)$ occurs when $x_4{}^* = 300$, it follows that x_4 must be 300 if our policy is to be optimal. This value for x_4 is also shown in Table 12.14.

Using the same reasoning,

$$y_3 = y_4 + x_4 - r_4$$
$$y_2 = y_3 + x_3 - r_3 \tag{29}$$

and

$$y_1 = y_2 + x_2 - r_2$$

Tables 12.14 and 12.11, and formula (29) can now be used to find the amount y_3 in storage at the beginning of March. Table 12.13 can then be used to find the corresponding amount x_3 that must be produced. This process can be continued to find y_2, x_2, y_1, and x_1, in that order.

The reader can check that the policy listed in Table 12.14 yields an inventory cost of 20 dollars and a 120-dollar set-up cost. This results in the required 140-dollar total storage and set-up cost calculated in equation (27).

12.4 EXERCISES

1.   Check the values of $F_3(x)$ listed in Table 12.13.

2.   Show that the policy listed in Table 12.14 results in a total storage and set-up cost of 140 dollars.

3.   **(a)** Find a second optimal policy for Example 10.
   (b) Give a verbal statement of this policy.
   (c) What are the storage, set-up, and total charges for this policy?

4.   A company must deliver 1500 radios within the next six months. The total number to be delivered at the end of each month is shown in the following table:

Month	Requirement
January	500
February	400
March	300
April	100
May	200

Suppose that each radio costs 6 dollars to produce. In addition, there is a set-up cost of 10 dollars per run if any radios are produced and a storage charge of 10 cents per radio per month. Suppose that the initial and final inventory is 0, and production is run only once a month, in units of 100 radios.
   (a) Make a table similar to Table 12.13.
   (b) Find the minimum total cost.
   (c) Find all optimal policies and their associated production, set-up, and storage costs.

Key Word Review

Each key word is followed by the number of the section in which it is defined.

stages   12.1

decision   12.1

states   12.1

policy   12.1

subpolicy   12.1

optimal policy   12.1

dynamic programming   12.1

optimal subpolicy   12.1

allocation process   12.2

transportation problem   12.3

inventory problem   12.4

CHAPTER 12   **REVIEW EXERCISES**

1.   Define each word in the Key Word Review.

2.   Calculate—
   (a) $\min[1, 0.1, -1]$ (b) $\max[1, 0.1, -1]$
   (c) $\min[2, 2, 2, 2]$ (d) $\max[2, 2, 2, 2]$

3.   Calculate—
   (a) $\min\limits_{x = -2, 0, 2}[x^2 + 2x - 1]$ (b) $\max\limits_{x = -2, 0, 2}[x^2 + 2x - 1]$
   Also find x^* for parts (a) and (b).

4.   A company has eight employees whom it must assign to three projects. The expected profits (in hundreds of dollars) that can be attributed to assigning various numbers of employees to each project are given in the following table.

Number of Employees	0	1	2	3	4	5	6	7	8
Profit from Project 1	0	5	7	9	10	11	12	13	14
Profit from Project 2	0	7	10	14	16	18	20	21	22
Profit from Project 3	0	2	4	8	12	16	18	20	21

   (a) Let $F_1(x)$ be the maximum profit if x employees are assigned to project 1. Calculate $F_1(x)$.
   (b) Let $h_i(x)$ be the profit that will be made if x employees are assigned to project i. Calculate $h_1(3)$, $h_2(4)$, and $h_3(1)$.
   (c) Let $F_2(x)$ be the maximum profit that can be made if a total of x employees are assigned to projects 1 and 2. Let $F_3(x)$ be the maximum profit that can be made if a total of x employees are assigned to projects 1, 2, and 3. Find a formula for finding $F_2(x)$ and $F_3(x)$.

5.   (a) Make a table listing x, $F_1(x)$, x_1^*, $F_2(x)$, and x_2^* for Exercise 4.
   (b) Use this table to find the maximum possible profit.

6.   Find the number of employees that should be assigned to each project in Exercise 4 if profits are to be maximized. Show that this assignment results in the profit calculated in Exercise 5(b).

7.   Two warehouses, W_1 and W_2 hold 6 and 8 tons of goose down, respectively, while five distributors, D_1, D_2, ..., D_5, require 2, 3, 2, 4, and 3 tons, respectively. Table 12.15 gives the cost $h_i(x)$ (in dollars) of shipping x tons of down from W_1 to D_i and shipping the remaining tons required by D_i from W_2.

Table 12.15

x	$h_1(x)$	$h_2(x)$	$h_3(x)$	$h_4(x)$	$h_5(x)$
0	30	21	15	10	12
1	40	32	13	18	10
2	34	30	10	17	9
3	—	20	—	18	10
4	—	—	—	9	13

(a) Let $F_1(x)$ be the minimum cost of shipping x tons of down from W_1 to D_1 and shipping the remaining down required by D_1 from W_2. Find $F_1(0)$, $F_1(2)$, and $F_1(4)$.

(b) Let $F_2(x)$ be the minimum cost of shipping a total of x tons of down from W_1 to D_1 and D_2 and shipping the remaining down required by D_1 and D_2 from W_2. Define $F_3(x)$, $F_4(x)$, and $F_5(x)$ similarly. Find formulas for calculating $F_2(x)$, $F_3(x)$, $F_4(x)$, and $F_5(x)$.

(c) Make a table listing the values of x, $F_1(x)$, $x_1{}^*$, $F_2(x)$, $x_2{}^*$, $F_3(x)$, $x_3{}^*$, $F_4(x)$, and $x_4{}^*$.

8.   Find the number of tons of down that must be shipped from each warehouse to each distributor in order to minimize the transportation costs. Show that this distribution results in the transportation cost calculated in Exercise 7.

9.   Using the notation developed in Exercises 4 and 5 of Section 12.3, let

$$\mathbf{A} = (6,9), \qquad \mathbf{B} = (4,3,2,4,2)$$

$$\mathbf{C} = (c_{ij}) = \begin{pmatrix} 1 & 2 & 4 & 3 & 1 \\ 3 & 2 & 1 & 4 & 2 \end{pmatrix},$$

$$\mathbf{K} = (k_{ij}) = \begin{pmatrix} 1 & 3 & 2 & 4 & 1 \\ 2 & 2 & 1 & 1 & 5 \end{pmatrix}$$

(a) Find the cost of shipping—
    (i) 2 units from W_1 to D_3
    (ii) 3 units from W_2 to D_1
    (iii) 3 units from W_1 to D_4 and the remainder required by D_4 from W_2.

(b) Let $h_i(x)$ be the cost of shipping x units from warehouse W_1 to D_i and shipping the remainder required by D_i from W_2. Make a table listing x and $h_1(x)$, $h_2(x)$, ..., $h_5(x)$.

(c) Describe how you would go about finding the minimum transportation cost.

Suggested Readings

1. Bellman, Richard E. *Dynamic Programming*. Princeton, N.J.: Princeton University Press, 1957.
2. Bellman, Richard E., and Dreyfus, Stuart E. *Applied Dynamic Programming*. Princeton, N.J.: Princeton University Press, 1962.
3. Hadley, George F. *Nonlinear and Dynamic Programming*. Reading, Mass.: Addison-Wesley Publishing Co., 1964.
4. Nemhauser, George L. *Introduction to Dynamic Programming*. New York: John Wiley & Sons, 1966.

Appendix

Table A.1 The Binomial Density Function

$$b(x; n, p) = {}_nC_x\, p^x q^{n-x} = P(S_n = x)$$

n	x	0.01	0.02	0.04	0.06	0.08	0.10	0.20	0.30	0.40	0.50	x	n
5	0	0.95099	0.90392	0.81537	0.73390	0.65908	0.59049	0.32768	0.16807	0.07776	0.03125	5	5
5	1	0.04803	0.09224	0.16987	0.23422	0.28656	0.32805	0.40960	0.36015	0.25920	0.15625	4	5
5	2	0.00097	0.00376	0.01416	0.02990	0.04984	0.07290	0.20480	0.30870	0.34560	0.31250	3	5
5	3	0.00001	0.00008	0.00059	0.00191	0.00433	0.00810	0.05120	0.13230	0.23040	0.31250	2	5
5	4	0.00000	0.00000	0.00001	0.00006	0.00019	0.00045	0.00640	0.02835	0.07680	0.15625	1	5
5	5	0.00000	0.00000	0.00000	0.00000	0.00000	0.00001	0.00032	0.00243	0.01024	0.03125	0	5
6	0	0.94148	0.88584	0.78276	0.68987	0.60636	0.53144	0.26214	0.11765	0.04666	0.01563	6	6
6	1	0.05706	0.10847	0.19569	0.26421	0.31636	0.35429	0.39322	0.30253	0.18662	0.09375	5	6
6	2	0.00144	0.00553	0.02038	0.04216	0.06877	0.09841	0.24576	0.32413	0.31104	0.23438	4	6
6	3	0.00002	0.00015	0.00113	0.00359	0.00797	0.01458	0.08192	0.18522	0.27648	0.31250	3	6
6	4	0.00000	0.00000	0.00004	0.00017	0.00052	0.00121	0.01536	0.05953	0.13824	0.23438	2	6
6	5	0.00000	0.00000	0.00000	0.00000	0.00002	0.00005	0.00154	0.01021	0.03686	0.09375	1	6
6	6	0.00000	0.00000	0.00000	0.00000	0.00000	0.00000	0.00006	0.00073	0.00410	0.01563	0	6
7	0	0.93207	0.86813	0.75145	0.64848	0.55785	0.47830	0.20972	0.08235	0.02799	0.00781	7	7
7	1	0.06590	0.12402	0.21917	0.28975	0.33956	0.37201	0.36700	0.24706	0.13064	0.05469	6	7
7	2	0.00200	0.00759	0.02740	0.05548	0.08858	0.12400	0.27525	0.31765	0.26127	0.16406	5	7
7	3	0.00003	0.00026	0.00190	0.00590	0.01284	0.02296	0.11469	0.22689	0.29030	0.27344	4	7
7	4	0.00000	0.00001	0.00008	0.00038	0.00112	0.00255	0.02867	0.09724	0.19354	0.27344	3	7
7	5	0.00000	0.00000	0.00000	0.00001	0.00006	0.00017	0.00430	0.02500	0.07741	0.16406	2	7
7	6	0.00000	0.00000	0.00000	0.00000	0.00000	0.00001	0.00036	0.00357	0.01720	0.05469	1	7
7	7	0.00000	0.00000	0.00000	0.00000	0.00000	0.00000	0.00001	0.00022	0.00164	0.00781	0	7
8	0	0.92274	0.85076	0.72139	0.60957	0.51322	0.43047	0.16777	0.05765	0.01680	0.00391	8	8
8	1	0.07457	0.13890	0.24046	0.31127	0.35702	0.38264	0.33554	0.19765	0.08958	0.03125	7	8
8	2	0.00264	0.00992	0.03507	0.06954	0.10866	0.14880	0.29360	0.29647	0.20902	0.10938	6	8
8	3	0.00005	0.00040	0.00292	0.00888	0.01890	0.03307	0.14680	0.25412	0.27869	0.21875	5	8
8	4	0.00000	0.00001	0.00015	0.00071	0.00205	0.00459	0.04588	0.13614	0.23224	0.27344	4	8
8	5	0.00000	0.00000	0.00001	0.00004	0.00014	0.00041	0.00918	0.04668	0.12386	0.21875	3	8
8	6	0.00000	0.00000	0.00000	0.00000	0.00001	0.00002	0.00115	0.01000	0.04129	0.10938	2	8
8	7	0.00000	0.00000	0.00000	0.00000	0.00000	0.00000	0.00008	0.00122	0.00786	0.03125	1	8
8	8	0.00000	0.00000	0.00000	0.00000	0.00000	0.00000	0.00000	0.00007	0.00066	0.00391	0	8

n	x	0.50	0.60	0.70	0.80	0.90	0.92	0.94	0.96	0.98	0.99	x	n
9	9	0.00195	0.01008	0.04035	0.13422	0.38742	0.47216	0.57299	0.69253	0.83375	0.91352	0	9
9	8	0.01758	0.06047	0.15565	0.30199	0.38742	0.36952	0.32917	0.25970	0.15314	0.08305	1	9
9	7	0.07031	0.16124	0.26683	0.30199	0.17219	0.12853	0.08404	0.04328	0.01250	0.00336	2	9
9	6	0.16406	0.25082	0.26683	0.17616	0.04464	0.02608	0.01252	0.00421	0.00060	0.00008	3	9
9	5	0.24609	0.25082	0.17153	0.06606	0.00744	0.00340	0.00120	0.00026	0.00002	0.00000	4	9
9	4	0.24609	0.16721	0.07351	0.01652	0.00083	0.00030	0.00008	0.00001	0.00000	0.00000	5	9
9	3	0.16406	0.07432	0.02100	0.00275	0.00006	0.00002	0.00000	0.00000	0.00000	0.00000	6	9
9	2	0.07031	0.02123	0.00386	0.00029	0.00000	0.00000	0.00000	0.00000	0.00000	0.00000	7	9
9	1	0.01758	0.00354	0.00041	0.00002	0.00000	0.00000	0.00000	0.00000	0.00000	0.00000	8	9
9	0	0.00195	0.00026	0.00002	0.00000	0.00000	0.00000	0.00000	0.00000	0.00000	0.00000	9	9
10	10	0.00098	0.00605	0.02825	0.10737	0.34868	0.43439	0.53862	0.66483	0.81707	0.90438	0	10
10	9	0.00977	0.04031	0.12106	0.26844	0.38742	0.37773	0.34380	0.27701	0.16675	0.09135	1	10
10	8	0.04395	0.12093	0.23347	0.30199	0.19371	0.14781	0.09875	0.05194	0.01531	0.00415	2	10
10	7	0.11719	0.21499	0.26683	0.20133	0.05740	0.03427	0.01681	0.00577	0.00083	0.00011	3	10
10	6	0.20508	0.25082	0.20012	0.08808	0.01116	0.00522	0.00188	0.00042	0.00003	0.00000	4	10
10	5	0.24609	0.20066	0.10292	0.02642	0.00149	0.00054	0.00014	0.00002	0.00000	0.00000	5	10
10	4	0.20508	0.11148	0.03676	0.00551	0.00014	0.00004	0.00001	0.00000	0.00000	0.00000	6	10
10	3	0.11719	0.04247	0.00900	0.00079	0.00001	0.00000	0.00000	0.00000	0.00000	0.00000	7	10
10	2	0.04395	0.01062	0.00145	0.00007	0.00000	0.00000	0.00000	0.00000	0.00000	0.00000	8	10
10	1	0.00977	0.00157	0.00014	0.00000	0.00000	0.00000	0.00000	0.00000	0.00000	0.00000	9	10
10	0	0.00098	0.00010	0.00001	0.00000	0.00000	0.00000	0.00000	0.00000	0.00000	0.00000	10	10
11	11	0.00049	0.00363	0.01977	0.08590	0.31381	0.39964	0.50630	0.63824	0.80073	0.89534	0	11
11	10	0.00537	0.02661	0.09322	0.23622	0.38355	0.38226	0.35549	0.29253	0.17976	0.09948	1	11
11	9	0.02686	0.08868	0.19975	0.29528	0.21308	0.16620	0.11345	0.06094	0.01834	0.00502	2	11
11	8	0.08057	0.17737	0.25682	0.22146	0.07103	0.04336	0.02173	0.00762	0.00112	0.00015	3	11
11	7	0.16113	0.23649	0.22013	0.11073	0.01578	0.00754	0.00277	0.00063	0.00005	0.00000	4	11
11	6	0.22559	0.22072	0.13208	0.03876	0.00246	0.00092	0.00025	0.00004	0.00000	0.00000	5	11
11	5	0.22559	0.14715	0.05661	0.00969	0.00027	0.00008	0.00002	0.00000	0.00000	0.00000	6	11
11	4	0.16113	0.07007	0.01733	0.00173	0.00002	0.00000	0.00000	0.00000	0.00000	0.00000	7	11
11	3	0.08057	0.02336	0.00371	0.00022	0.00000	0.00000	0.00000	0.00000	0.00000	0.00000	8	11
11	2	0.02686	0.00519	0.00053	0.00002	0.00000	0.00000	0.00000	0.00000	0.00000	0.00000	9	11
11	1	0.00537	0.00069	0.00005	0.00000	0.00000	0.00000	0.00000	0.00000	0.00000	0.00000	10	11
11	0	0.00049	0.00004	0.00000	0.00000	0.00000	0.00000	0.00000	0.00000	0.00000	0.00000	11	11

p

Table A.1 (*cont.*)

n	x	0.01	0.02	0.04	0.06	0.08	p 0.10	0.20	0.30	0.40	0.50
12	0	0.88638	0.78472	0.61271	0.47592	0.36767	0.28243	0.06872	0.01384	0.00218	0.00024
12	1	0.10744	0.19218	0.30635	0.36453	0.38365	0.37657	0.20616	0.07118	0.01741	0.00293
12	2	0.00597	0.02157	0.07021	0.12797	0.18349	0.23013	0.28347	0.16779	0.06385	0.01611
12	3	0.00020	0.00147	0.00975	0.02723	0.05318	0.08523	0.23622	0.23970	0.14189	0.05371
12	4	0.00000	0.00007	0.00091	0.00391	0.01041	0.02131	0.13288	0.23114	0.21284	0.12085
12	5	0.00000	0.00000	0.00006	0.00040	0.00145	0.00379	0.05315	0.15850	0.22703	0.19336
12	6	0.00000	0.00000	0.00000	0.00003	0.00015	0.00049	0.01550	0.07925	0.17658	0.22559
12	7	0.00000	0.00000	0.00000	0.00000	0.00001	0.00005	0.00332	0.02911	0.10090	0.19336
12	8	0.00000	0.00000	0.00000	0.00000	0.00000	0.00000	0.00052	0.00780	0.04204	0.12085
12	9	0.00000	0.00000	0.00000	0.00000	0.00000	0.00000	0.00006	0.00149	0.01246	0.05371
12	10	0.00000	0.00000	0.00000	0.00000	0.00000	0.00000	0.00000	0.00019	0.00249	0.01611
12	11	0.00000	0.00000	0.00000	0.00000	0.00000	0.00000	0.00000	0.00001	0.00030	0.00293
12	12	0.00000	0.00000	0.00000	0.00000	0.00000	0.00000	0.00000	0.00000	0.00002	0.00024
13	0	0.87752	0.76902	0.58820	0.44737	0.33825	0.25419	0.05498	0.00969	0.00131	0.00012
13	1	0.11523	0.20403	0.31861	0.37122	0.38237	0.36716	0.17867	0.05398	0.01132	0.00159
13	2	0.00698	0.02498	0.07965	0.14217	0.19950	0.24477	0.26801	0.13881	0.04528	0.00952
13	3	0.00026	0.00187	0.01217	0.03327	0.06361	0.09972	0.24567	0.21813	0.11068	0.03491
13	4	0.00001	0.00010	0.00127	0.00531	0.01383	0.02770	0.15354	0.23371	0.18446	0.08728
13	5	0.00000	0.00000	0.00010	0.00061	0.00216	0.00554	0.06910	0.18029	0.22135	0.15710
13	6	0.00000	0.00000	0.00001	0.00005	0.00025	0.00082	0.02303	0.10302	0.19676	0.20947
13	7	0.00000	0.00000	0.00000	0.00000	0.00002	0.00009	0.00576	0.04415	0.13117	0.20947
13	8	0.00000	0.00000	0.00000	0.00000	0.00000	0.00001	0.00108	0.01419	0.06559	0.15710
13	9	0.00000	0.00000	0.00000	0.00000	0.00000	0.00000	0.00015	0.00338	0.02429	0.08728
13	10	0.00000	0.00000	0.00000	0.00000	0.00000	0.00000	0.00001	0.00058	0.00648	0.03491
13	11	0.00000	0.00000	0.00000	0.00000	0.00000	0.00000	0.00000	0.00007	0.00118	0.00952
13	12	0.00000	0.00000	0.00000	0.00000	0.00000	0.00000	0.00000	0.00000	0.00013	0.00159
13	13	0.00000	0.00000	0.00000	0.00000	0.00000	0.00000	0.00000	0.00000	0.00001	0.00012

n	x	0.50	0.60	0.70	0.80	0.90	0.92	0.94	0.96	0.98	0.99
14	14	0.00006	0.00078	0.00678	0.04398	0.22877	0.31119	0.42052	0.56467	0.75364	0.86875
14	13	0.00085	0.00731	0.04069	0.15393	0.35586	0.37884	0.37579	0.32939	0.21533	0.12285
14	12	0.00555	0.03169	0.11336	0.25014	0.25701	0.21413	0.15591	0.08921	0.02856	0.00807
14	11	0.02222	0.08452	0.19433	0.25014	0.11423	0.07448	0.03981	0.01487	0.00233	0.00033
14	10	0.06110	0.15495	0.22903	0.17197	0.03490	0.01781	0.00699	0.00170	0.00013	0.00001
14	9	0.12219	0.20660	0.19631	0.08599	0.00776	0.00310	0.00089	0.00014	0.00001	0.00000
14	8	0.18329	0.20660	0.12620	0.03224	0.00129	0.00040	0.00009	0.00001	0.00000	0.00000
14	7	0.20947	0.15741	0.06181	0.00921	0.00016	0.00004	0.00001	0.00000	0.00000	0.00000
14	6	0.18329	0.09182	0.02318	0.00202	0.00002	0.00000	0.00000	0.00000	0.00000	0.00000
14	5	0.12219	0.04081	0.00662	0.00034	0.00000	0.00000	0.00000	0.00000	0.00000	0.00000
14	4	0.06110	0.01360	0.00142	0.00004	0.00000	0.00000	0.00000	0.00000	0.00000	0.00000
14	3	0.02222	0.00330	0.00022	0.00000	0.00000	0.00000	0.00000	0.00000	0.00000	0.00000
14	2	0.00555	0.00055	0.00002	0.00000	0.00000	0.00000	0.00000	0.00000	0.00000	0.00000
14	1	0.00085	0.00006	0.00000	0.00000	0.00000	0.00000	0.00000	0.00000	0.00000	0.00000
14	0	0.00006	0.00000	0.00000	0.00000	0.00000	0.00000	0.00000	0.00000	0.00000	0.00000
15	15	0.00003	0.00047	0.00475	0.03518	0.20589	0.28630	0.39529	0.54209	0.73857	0.86006
15	14	0.00046	0.00470	0.03052	0.13194	0.34315	0.37343	0.37847	0.33880	0.22609	0.13031
15	13	0.00320	0.02194	0.09156	0.23090	0.26690	0.22731	0.16910	0.09882	0.03230	0.00921
15	12	0.01389	0.06339	0.17004	0.25014	0.12851	0.08565	0.04677	0.01784	0.00286	0.00040
15	11	0.04166	0.12678	0.21862	0.18760	0.04283	0.02234	0.00896	0.00223	0.00017	0.00001
15	10	0.09164	0.18594	0.20613	0.10318	0.01047	0.00427	0.00126	0.00020	0.00001	0.00000
15	9	0.15274	0.20660	0.14724	0.04299	0.00194	0.00062	0.00013	0.00001	0.00000	0.00000
15	8	0.19638	0.17708	0.08113	0.01382	0.00028	0.00007	0.00001	0.00000	0.00000	0.00000
15	7	0.19638	0.11806	0.03477	0.00345	0.00003	0.00001	0.00000	0.00000	0.00000	0.00000
15	6	0.15274	0.06121	0.01159	0.00067	0.00000	0.00000	0.00000	0.00000	0.00000	0.00000
15	5	0.09164	0.02449	0.00298	0.00010	0.00000	0.00000	0.00000	0.00000	0.00000	0.00000
15	4	0.04166	0.00742	0.00058	0.00001	0.00000	0.00000	0.00000	0.00000	0.00000	0.00000
15	3	0.01389	0.00165	0.00008	0.00000	0.00000	0.00000	0.00000	0.00000	0.00000	0.00000
15	2	0.00320	0.00025	0.00001	0.00000	0.00000	0.00000	0.00000	0.00000	0.00000	0.00000
15	1	0.00046	0.00002	0.00000	0.00000	0.00000	0.00000	0.00000	0.00000	0.00000	0.00000
15	0	0.00003	0.00000	0.00000	0.00000	0.00000	0.00000	0.00000	0.00000	0.00000	0.00000

p

Table A.2   The Binomial Distribution Function

$$B(x;n, p) = P(S_n \le x) = \sum_{k=0}^{x} {}_nC_k\, p^k\, q^{n-k}$$

n	x	$p = 0.20$	$p = 0.25$	$p = 0.30$	$p = 0.35$	$p = 0.40$	$p = 0.45$
5	0	0.32768	0.23730	0.16807	0.11603	0.07776	0.05033
5	1	0.73728	0.63281	0.52822	0.42841	0.33696	0.25622
5	2	0.94208	0.89648	0.83692	0.76483	0.68256	0.59313
5	3	0.99328	0.98438	0.96922	0.94598	0.91296	0.86878
5	4	0.99968	0.99902	0.99757	0.99475	0.98976	0.98155
5	5	1.00000	1.00000	1.00000	1.00000	1.00000	1.00000
10	0	0.10737	0.05631	0.02825	0.01346	0.00605	0.00253
10	1	0.37581	0.24403	0.14931	0.08595	0.04636	0.02326
10	2	0.67780	0.52559	0.38278	0.26161	0.16729	0.09956
10	3	0.87913	0.77588	0.64961	0.51383	0.38228	0.26604
10	4	0.96721	0.92187	0.84973	0.75150	0.63310	0.50440
10	5	0.99363	0.98027	0.95265	0.90507	0.83376	0.73844
10	6	0.99914	0.99649	0.98941	0.97398	0.94524	0.89801
10	7	0.99992	0.99958	0.99841	0.99518	0.98771	0.97261
10	8	1.00000	0.99997	0.99986	0.99946	0.99832	0.99550
10	9	1.00000	1.00000	0.99999	0.99997	0.99990	0.99966
10	10	1.00000	1.00000	1.00000	1.00000	1.00000	1.00000
15	0	0.03518	0.01336	0.00475	0.00156	0.00047	0.00013
15	1	0.16713	0.08018	0.03527	0.01418	0.00517	0.00169
15	2	0.39802	0.23609	0.12683	0.06173	0.02711	0.01065
15	3	0.64816	0.46129	0.29687	0.17270	0.09050	0.04242
15	4	0.83577	0.68649	0.51549	0.35194	0.21728	0.12040
15	5	0.93895	0.85163	0.72162	0.56428	0.40322	0.26076
15	6	0.98194	0.94338	0.86886	0.75484	0.60981	0.45216
15	7	0.99576	0.98270	0.94999	0.88677	0.78690	0.65350
15	8	0.99922	0.99581	0.98476	0.95781	0.90495	0.81824
15	9	0.99989	0.99921	0.99635	0.98756	0.96617	0.92307
15	10	0.99999	0.99988	0.99933	0.99717	0.99065	0.97453
15	11	1.00000	0.99999	0.99991	0.99952	0.99807	0.99367
15	12	1.00000	1.00000	0.99999	0.99994	0.99972	0.99889
15	13	1.00000	1.00000	1.00000	1.00000	0.99997	0.99988
15	14	1.00000	1.00000	1.00000	1.00000	1.00000	0.99999
15	15	1.00000	1.00000	1.00000	1.00000	1.00000	1.00000

Table A.3 The Poisson Density Function

$$p(x;\mu) = \frac{\mu^x e^{-\mu}}{x!}$$

	μ									
	0.1	0.2	0.3	0.4	0.5	0.6	0.7	0.8	0.9	1.0
0	0.9048	0.8187	0.7408	0.6703	0.6065	0.5488	0.4966	0.4493	0.4066	0.3679
1	0.0905	0.1637	0.2222	0.2681	0.3033	0.3293	0.3476	0.3595	0.3659	0.3679
2	0.0045	0.0164	0.0333	0.0536	0.0758	0.0988	0.1217	0.1438	0.1647	0.1839
3	0.0002	0.0011	0.0033	0.0072	0.0126	0.0198	0.0284	0.0383	0.0494	0.0613
4	0.0000	0.0001	0.0003	0.0007	0.0016	0.0030	0.0050	0.0077	0.0111	0.0153
5	0.0000	0.0000	0.0000	0.0001	0.0002	0.0004	0.0007	0.0012	0.0020	0.0031
6	0.0000	0.0000	0.0000	0.0000	0.0000	0.0000	0.0001	0.0002	0.0003	0.0005
7	0.0000	0.0000	0.0000	0.0000	0.0000	0.0000	0.0000	0.0000	0.0000	0.0001

	1.1	1.2	1.3	1.4	1.5	1.6	1.7	1.8	1.9	2.0
0	0.3329	0.3012	0.2725	0.2466	0.2231	0.2019	0.1827	0.1653	0.1496	0.1353
1	0.3662	0.3614	0.3543	0.3452	0.3347	0.3230	0.3106	0.2975	0.2842	0.2707
2	0.2014	0.2169	0.2303	0.2417	0.2510	0.2584	0.2640	0.2678	0.2700	0.2707
3	0.0738	0.0867	0.0998	0.1128	0.1255	0.1378	0.1496	0.1607	0.1710	0.1804
4	0.0203	0.0260	0.0324	0.0395	0.0471	0.0551	0.0636	0.0723	0.0812	0.0902
5	0.0045	0.0062	0.0084	0.0111	0.0141	0.0176	0.0216	0.0260	0.0309	0.0361
6	0.0008	0.0012	0.0018	0.0026	0.0035	0.0047	0.0061	0.0078	0.0098	0.0120
7	0.0001	0.0002	0.0003	0.0005	0.0008	0.0011	0.0015	0.0020	0.0027	0.0034
8	0.0000	0.0000	0.0001	0.0001	0.0001	0.0002	0.0003	0.0005	0.0006	0.0009
9	0.0000	0.0000	0.0000	0.0000	0.0000	0.0000	0.0001	0.0001	0.0001	0.0002

	2.1	2.2	2.3	2.4	2.5	2.6	2.7	2.8	2.9	3.0
0	0.1225	0.1108	0.1003	0.0907	0.0821	0.0743	0.0672	0.0608	0.0550	0.0498
1	0.2572	0.2438	0.2306	0.2177	0.2052	0.1931	0.1815	0.1703	0.1596	0.1494
2	0.2700	0.2681	0.2652	0.2613	0.2565	0.2510	0.2450	0.2384	0.2314	0.2240
3	0.1890	0.1966	0.2033	0.2090	0.2138	0.2176	0.2205	0.2225	0.2237	0.2240
4	0.0992	0.1082	0.1169	0.1254	0.1336	0.1414	0.1488	0.1557	0.1622	0.1680
5	0.0417	0.0476	0.0538	0.0602	0.0668	0.0735	0.0804	0.0872	0.0940	0.1008
6	0.0146	0.0174	0.0206	0.0241	0.0278	0.0319	0.0362	0.0407	0.0455	0.0504
7	0.0044	0.0055	0.0068	0.0083	0.0099	0.0118	0.0139	0.0163	0.0188	0.0216
8	0.0011	0.0015	0.0019	0.0025	0.0031	0.0038	0.0047	0.0057	0.0068	0.0081
9	0.0003	0.0004	0.0005	0.0007	0.0009	0.0011	0.0014	0.0018	0.0022	0.0027
10	0.0001	0.0001	0.0001	0.0002	0.0002	0.0003	0.0004	0.0005	0.0006	0.0008
11	0.0000	0.0000	0.0000	0.0000	0.0000	0.0001	0.0001	0.0001	0.0002	0.0002
12	0.0000	0.0000	0.0000	0.0000	0.0000	0.0000	0.0000	0.0000	0.0000	0.0001

	3.1	3.2	3.3	3.4	3.5	3.6	3.7	3.8	3.9	4.0
0	0.0450	0.0408	0.0369	0.0334	0.0302	0.0273	0.0247	0.0224	0.0202	0.0183
1	0.1397	0.1304	0.1217	0.1135	0.1057	0.0984	0.0915	0.0850	0.0789	0.0733
2	0.2165	0.2087	0.2008	0.1929	0.1850	0.1771	0.1692	0.1615	0.1539	0.1465
3	0.2237	0.2226	0.2209	0.2186	0.2158	0.2125	0.2087	0.2046	0.2001	0.1954
4	0.1734	0.1781	0.1823	0.1858	0.1888	0.1912	0.1931	0.1944	0.1951	0.1954
5	0.1075	0.1140	0.1203	0.1264	0.1322	0.1377	0.1429	0.1477	0.1522	0.1563
6	0.0555	0.0608	0.0662	0.0716	0.0771	0.0826	0.0881	0.0936	0.0989	0.1042
7	0.0246	0.0278	0.0312	0.0348	0.0385	0.0425	0.0466	0.0508	0.0551	0.0595
8	0.0095	0.0111	0.0129	0.0148	0.0169	0.0191	0.0215	0.0241	0.0269	0.0298
9	0.0033	0.0040	0.0047	0.0056	0.0066	0.0076	0.0089	0.0102	0.0116	0.0132

Table A.3 (*cont.*)

x	3.1	3.2	3.3	3.4	3.5	3.6	3.7	3.8	3.9	4.0
10	0.0010	0.0013	0.0016	0.0019	0.0023	0.0028	0.0033	0.0039	0.0045	0.0053
11	0.0003	0.0004	0.0005	0.0006	0.0007	0.0009	0.0011	0.0013	0.0016	0.0019
12	0.0001	0.0001	0.0001	0.0002	0.0002	0.0003	0.0003	0.0004	0.0005	0.0006
13	0.0000	0.0000	0.0000	0.0000	0.0001	0.0001	0.0001	0.0001	0.0002	0.0002
14	0.0000	0.0000	0.0000	0.0000	0.0000	0.0000	0.0000	0.0000	0.0000	0.0001

x	4.1	4.2	4.3	4.4	4.5	4.6	4.7	4.8	4.9	5.0
0	0.0166	0.0150	0.0136	0.0123	0.0111	0.0101	0.0091	0.0082	0.0074	0.0067
1	0.0679	0.0630	0.0583	0.0540	0.0500	0.0462	0.0427	0.0395	0.0365	0.0337
2	0.1393	0.1323	0.1254	0.1188	0.1125	0.1063	0.1005	0.0948	0.0894	0.0842
3	0.1904	0.1852	0.1798	0.1743	0.1687	0.1631	0.1574	0.1517	0.1460	0.1404
4	0.1951	0.1944	0.1933	0.1917	0.1898	0.1875	0.1849	0.1820	0.1789	0.1755
5	0.1600	0.1633	0.1662	0.1687	0.1708	0.1725	0.1738	0.1747	0.1753	0.1755
6	0.1093	0.1143	0.1191	0.1237	0.1281	0.1323	0.1362	0.1398	0.1432	0.1462
7	0.0640	0.0686	0.0732	0.0778	0.0824	0.0869	0.0914	0.0959	0.1002	0.1044
8	0.0328	0.0360	0.0393	0.0428	0.0463	0.0500	0.0537	0.0575	0.0614	0.0653
9	0.0150	0.0168	0.0188	0.0209	0.0232	0.0255	0.0280	0.0307	0.0334	0.0363
10	0.0061	0.0071	0.0081	0.0092	0.0104	0.0118	0.0132	0.0147	0.0164	0.0181
11	0.0023	0.0027	0.0032	0.0037	0.0043	0.0049	0.0056	0.0064	0.0073	0.0082
12	0.0008	0.0009	0.0011	0.0014	0.0016	0.0019	0.0022	0.0026	0.0030	0.0034
13	0.0002	0.0003	0.0004	0.0005	0.0006	0.0007	0.0008	0.0009	0.0011	0.0013
14	0.0001	0.0001	0.0001	0.0001	0.0002	0.0002	0.0003	0.0003	0.0004	0.0005
15	0.0000	0.0000	0.0000	0.0000	0.0001	0.0001	0.0001	0.0001	0.0001	0.0002

x	5.1	5.2	5.3	5.4	5.5	5.6	5.7	5.8	5.9	6.0
0	0.0061	0.0055	0.0050	0.0045	0.0041	0.0037	0.0033	0.0030	0.0027	0.0025
1	0.0311	0.0287	0.0265	0.0244	0.0255	0.0207	0.0191	0.0176	0.0162	0.0149
2	0.0793	0.0746	0.0701	0.0659	0.0618	0.0580	0.0544	0.0509	0.0477	0.0446
3	0.1348	0.1293	0.1239	0.1185	0.1133	0.1082	0.1033	0.0985	0.0938	0.0892
4	0.1719	0.1681	0.1641	0.1600	0.1558	0.1515	0.1472	0.1428	0.1383	0.1339
5	0.1753	0.1748	0.1740	0.1728	0.1714	0.1697	0.1678	0.1656	0.1632	0.1606
6	0.1490	0.1515	0.1537	0.1555	0.1571	0.1584	0.1594	0.1601	0.1605	0.1606
7	0.1086	0.1125	0.1163	0.1200	0.1234	0.1267	0.1298	0.1326	0.1353	0.1377
8	0.0692	0.0731	0.0771	0.0810	0.0849	0.0887	0.0925	0.0962	0.0998	0.1033
9	0.0392	0.0423	0.0454	0.0486	0.0519	0.0552	0.0586	0.0620	0.0654	0.0688
10	0.0200	0.0220	0.0241	0.0262	0.0285	0.0309	0.0334	0.0359	0.0386	0.0413
11	0.0093	0.0104	0.0116	0.0129	0.0143	0.0157	0.0173	0.0190	0.0207	0.0225
12	0.0039	0.0045	0.0051	0.0058	0.0065	0.0073	0.0082	0.0092	0.0102	0.0113
13	0.0015	0.0018	0.0021	0.0024	0.0028	0.0032	0.0036	0.0041	0.0046	0.0052
14	0.0006	0.0007	0.0008	0.0009	0.0011	0.0013	0.0015	0.0017	0.0019	0.0022
15	0.0002	0.0002	0.0003	0.0003	0.0004	0.0005	0.0006	0.0007	0.0008	0.0009
16	0.0001	0.0001	0.0001	0.0001	0.0001	0.0002	0.0002	0.0002	0.0003	0.0003
17	0.0000	0.0000	0.0000	0.0000	0.0000	0.0001	0.0001	0.0001	0.0001	0.0001

Table A.3 (*cont.*)

x						μ				
	6.1	6.2	6.3	6.4	6.5	6.6	6.7	6.8	6.9	7.0
0	0.0022	0.0020	0.0018	0.0017	0.0015	0.0014	0.0012	0.0011	0.0010	0.0009
1	0.0137	0.0126	0.0116	0.0106	0.0098	0.0090	0.0082	0.0076	0.0070	0.0064
2	0.0417	0.0390	0.0364	0.0340	0.0318	0.0296	0.0276	0.0258	0.0240	0.0223
3	0.0848	0.0806	0.0765	0.0726	0.0688	0.0652	0.0617	0.0584	0.0552	0.0521
4	0.1294	0.1249	0.1205	0.1162	0.1118	0.1076	0.1034	0.0992	0.0952	0.0912
5	0.1579	0.1549	0.1519	0.1487	0.1454	0.1420	0.1385	0.1349	0.1314	0.1277
6	0.1605	0.1601	0.1595	0.1586	0.1575	0.1562	0.1546	0.1529	0.1511	0.1490
7	0.1399	0.1418	0.1435	0.1450	0.1462	0.1472	0.1480	0.1486	0.1489	0.1490
8	0.1066	0.1099	0.1130	0.1160	0.1188	0.1215	0.1240	0.1263	0.1284	0.1304
9	0.0723	0.0757	0.0791	0.0825	0.0858	0.0891	0.0923	0.0954	0.0985	0.1014
10	0.0441	0.0469	0.0498	0.0528	0.0558	0.0588	0.0618	0.0649	0.0679	0.0710
11	0.0245	0.0265	0.0285	0.0307	0.0330	0.0353	0.0377	0.0401	0.0426	0.0452
12	0.0124	0.0137	0.0150	0.0164	0.0179	0.0194	0.0210	0.0227	0.0245	0.0264
13	0.0058	0.0065	0.0073	0.0081	0.0089	0.0098	0.0108	0.0119	0.0130	0.0142
14	0.0025	0.0029	0.0033	0.0037	0.0041	0.0046	0.0052	0.0058	0.0064	0.0071
15	0.0010	0.0012	0.0014	0.0016	0.0018	0.0020	0.0023	0.0026	0.0029	0.0033
16	0.0004	0.0005	0.0005	0.0006	0.0007	0.0008	0.0010	0.0011	0.0013	0.0014
17	0.0001	0.0002	0.0002	0.0002	0.0003	0.0003	0.0004	0.0004	0.0005	0.0006
18	0.0000	0.0001	0.0001	0.0001	0.0001	0.0001	0.0001	0.0002	0.0002	0.0002
19	0.0000	0.0000	0.0000	0.0000	0.0000	0.0000	0.0000	0.0001	0.0001	0.0001

x	7.1	7.2	7.3	7.4	7.5	7.6	7.7	7.8	7.9	8.0
0	0.0008	0.0007	0.0007	0.0006	0.0006	0.0005	0.0005	0.0004	0.0004	0.0003
1	0.0059	0.0054	0.0049	0.0045	0.0041	0.0038	0.0035	0.0032	0.0029	0.0027
2	0.0208	0.0194	0.0180	0.0167	0.0156	0.0145	0.0134	0.0125	0.0116	0.0107
3	0.0492	0.0464	0.0438	0.0413	0.0389	0.0366	0.0345	0.0324	0.0305	0.0286
4	0.0874	0.0836	0.0799	0.0764	0.0729	0.0696	0.0663	0.0632	0.0602	0.0573
5	0.1241	0.1204	0.1167	0.1130	0.1094	0.1057	0.1021	0.0986	0.0951	0.0916
6	0.1468	0.1445	0.1420	0.1394	0.1367	0.1339	0.1311	0.1282	0.1252	0.1221
7	0.1489	0.1486	0.1481	0.1474	0.1465	0.1454	0.1442	0.1428	0.1413	0.1396
8	0.1321	0.1337	0.1351	0.1363	0.1373	0.1382	0.1388	0.1392	0.1395	0.1396
9	0.1042	0.1070	0.1096	0.1121	0.1144	0.1167	0.1187	0.1207	0.1224	0.1241
10	0.0740	0.0770	0.0800	0.0829	0.0858	0.0887	0.0914	0.0941	0.0967	0.0993
11	0.0478	0.0504	0.0531	0.0558	0.0585	0.0613	0.0640	0.0667	0.0695	0.0722
12	0.0283	0.0303	0.0323	0.0344	0.0366	0.0388	0.0411	0.0434	0.0457	0.0481
13	0.0154	0.0168	0.0181	0.0196	0.0211	0.0227	0.0243	0.0260	0.0278	0.0296
14	0.0078	0.0086	0.0095	0.0104	0.0113	0.0123	0.0134	0.0145	0.0157	0.0169
15	0.0037	0.0041	0.0046	0.0051	0.0057	0.0062	0.0069	0.0075	0.0083	0.0090
16	0.0016	0.0019	0.0021	0.0024	0.0026	0.0030	0.0033	0.0037	0.0041	0.0045
17	0.0007	0.0008	0.0009	0.0010	0.0012	0.0013	0.0015	0.0017	0.0019	0.0021
18	0.0003	0.0003	0.0004	0.0004	0.0005	0.0006	0.0006	0.0007	0.0008	0.0009
19	0.0001	0.0001	0.0001	0.0002	0.0002	0.0002	0.0003	0.0003	0.0003	0.0004
20	0.0000	0.0000	0.0001	0.0001	0.0001	0.0001	0.0001	0.0001	0.0001	0.0002
21	0.0000	0.0000	0.0000	0.0000	0.0000	0.0000	0.0000	0.0000	0.0001	0.0001

Table A.3 (*cont.*)

x	8.1	8.2	8.3	8.4	8.5	8.6	8.7	8.8	8.9	9.0
0	0.0003	0.0003	0.0002	0.0002	0.0002	0.0002	0.0002	0.0002	0.0001	0.0001
1	0.0025	0.0023	0.0021	0.0019	0.0017	0.0016	0.0014	0.0013	0.0012	0.0011
2	0.0100	0.0092	0.0086	0.0079	0.0074	0.0068	0.0063	0.0058	0.0054	0.0050
3	0.0269	0.0252	0.0237	0.0222	0.0208	0.0195	0.0183	0.0171	0.0160	0.0150
4	0.0544	0.0517	0.0491	0.0466	0.0443	0.0420	0.0398	0.0377	0.0357	0.0337
5	0.0882	0.0849	0.0816	0.0784	0.0752	0.0722	0.0692	0.0663	0.0635	0.0607
6	0.1191	0.1160	0.1128	0.1097	0.1066	0.1034	0.1003	0.0972	0.0941	0.0911
7	0.1378	0.1358	0.1338	0.1317	0.1294	0.1271	0.1247	0.1222	0.1197	0.1171
8	0.1395	0.1392	0.1388	0.1382	0.1375	0.1366	0.1356	0.1344	0.1332	0.1318
9	0.1256	0.1269	0.1280	0.1290	0.1299	0.1306	0.1311	0.1315	0.1317	0.1318
10	0.1017	0.1040	0.1063	0.1084	0.1104	0.1123	0.1140	0.1157	0.1172	0.1186
11	0.0749	0.0776	0.0802	0.0828	0.0853	0.0878	0.0902	0.0925	0.0948	0.0970
12	0.0505	0.0530	0.0555	0.0579	0.0604	0.0629	0.0654	0.0679	0.0703	0.0728
13	0.0315	0.0334	0.0354	0.0374	0.0395	0.0416	0.0438	0.0459	0.0481	0.0504
14	0.0182	0.0196	0.0210	0.0225	0.0240	0.0256	0.0272	0.0289	0.0306	0.0324
15	0.0098	0.0107	0.0116	0.0126	0.0136	0.0147	0.0158	0.0169	0.0182	0.0194
16	0.0050	0.0055	0.0060	0.0066	0.0072	0.0079	0.0086	0.0093	0.0101	0.0109
17	0.0024	0.0026	0.0029	0.0033	0.0036	0.0040	0.0044	0.0048	0.0053	0.0058
18	0.0011	0.0012	0.0014	0.0015	0.0017	0.0019	0.0021	0.0024	0.0026	0.0029
19	0.0005	0.0005	0.0006	0.0007	0.0008	0.0009	0.0010	0.0011	0.0012	0.0014
20	0.0002	0.0002	0.0002	0.0003	0.0003	0.0004	0.0004	0.0005	0.0005	0.0006
21	0.0001	0.0001	0.0001	0.0001	0.0001	0.0002	0.0002	0.0002	0.0002	0.0003
22	0.0000	0.0000	0.0000	0.0000	0.0001	0.0001	0.0001	0.0001	0.0001	0.0001

x	9.1	9.2	9.3	9.4	9.5	9.6	9.7	9.8	9.9	10
0	0.0001	0.0001	0.0001	0.0001	0.0001	0.0001	0.0001	0.0001	0.0001	0.0000
1	0.0010	0.0009	0.0009	0.0008	0.0007	0.0007	0.0006	0.0005	0.0005	0.0005
2	0.0046	0.0043	0.0040	0.0037	0.0034	0.0031	0.0029	0.0027	0.0025	0.0023
3	0.0140	0.0131	0.0123	0.0115	0.0107	0.0100	0.0093	0.0087	0.0081	0.0076
4	0.0319	0.0302	0.0285	0.0269	0.0254	0.0240	0.0226	0.0213	0.0201	0.0189
5	0.0581	0.0555	0.0530	0.0506	0.0483	0.0460	0.0439	0.0418	0.0398	0.0378
6	0.0881	0.0851	0.0822	0.0793	0.0764	0.0736	0.0709	0.0682	0.0656	0.0631
7	0.1145	0.1118	0.1091	0.1064	0.1037	0.1010	0.0982	0.0955	0.0928	0.0901
8	0.1302	0.1286	0.1269	0.1251	0.1232	0.1212	0.1191	0.1170	0.1148	0.1126
9	0.1317	0.1315	0.1311	0.1306	0.1300	0.1293	0.1284	0.1274	0.1263	0.1251
10	0.1198	0.1210	0.1219	0.1228	0.1235	0.1241	0.1245	0.1249	0.1250	0.1251
11	0.0991	0.1012	0.1031	0.1049	0.1067	0.1083	0.1098	0.1112	0.1125	0.1137
12	0.0752	0.0776	0.0799	0.0822	0.0844	0.0866	0.0888	0.0908	0.0928	0.0948
13	0.0526	0.0549	0.0572	0.0594	0.0617	0.0640	0.0662	0.0685	0.0707	0.0729
14	0.0342	0.0361	0.0380	0.0399	0.0419	0.0439	0.0459	0.0479	0.0500	0.0521
15	0.0208	0.0221	0.0235	0.0250	0.0265	0.0281	0.0297	0.0313	0.0330	0.0347
16	0.0118	0.0127	0.0137	0.0147	0.0157	0.0168	0.0180	0.0192	0.0204	0.0217
17	0.0063	0.0069	0.0075	0.0081	0.0088	0.0095	0.0103	0.0111	0.0119	0.0128
18	0.0032	0.0035	0.0039	0.0042	0.0046	0.0051	0.0055	0.0060	0.0065	0.0071
19	0.0015	0.0017	0.0019	0.0021	0.0023	0.0026	0.0028	0.0031	0.0034	0.0037

Table A.3 (*cont.*)

x	μ									
	9.1	9.2	9.3	9.4	9.5	9.6	9.7	9.8	9.9	10
20	0.0007	0.0008	0.0009	0.0010	0.0011	0.0012	0.0014	0.0015	0.0017	0.0019
21	0.0003	0.0003	0.0004	0.0004	0.0005	0.0006	0.0006	0.0007	0.0008	0.0009
22	0.0001	0.0001	0.0002	0.0002	0.0002	0.0002	0.0003	0.0003	0.0004	0.0004
23	0.0000	0.0001	0.0001	0.0001	0.0001	0.0001	0.0001	0.0001	0.0002	0.0002
24	0.0000	0.0000	0.0000	0.0000	0.0000	0.0000	0.0000	0.0001	0.0001	0.0001

x	11	12	13	14	15	16	17	18	19	20
0	0.0000	0.0000	0.0000	0.0000	0.0000	0.0000	0.0000	0.0000	0.0000	0.0000
1	0.0002	0.0001	0.0000	0.0000	0.0000	0.0000	0.0000	0.0000	0.0000	0.0000
2	0.0010	0.0004	0.0002	0.0001	0.0000	0.0000	0.0000	0.0000	0.0000	0.0000
3	0.0037	0.0018	0.0008	0.0004	0.0002	0.0001	0.0000	0.0000	0.0000	0.0000
4	0.0102	0.0053	0.0027	0.0013	0.0006	0.0003	0.0001	0.0001	0.0000	0.0000
5	0.0224	0.0127	0.0070	0.0037	0.0019	0.0010	0.0005	0.0002	0.0001	0.0001
6	0.0411	0.0255	0.0152	0.0087	0.0048	0.0026	0.0014	0.0007	0.0004	0.0002
7	0.0646	0.0437	0.0281	0.0174	0.0104	0.0060	0.0034	0.0018	0.0010	0.0005
8	0.0888	0.0655	0.0457	0.0304	0.0194	0.0120	0.0072	0.0042	0.0024	0.0013
9	0.1085	0.0874	0.0661	0.0473	0.0324	0.0213	0.0135	0.0083	0.0050	0.0029
10	0.1194	0.1048	0.0859	0.0663	0.0486	0.0341	0.0230	0.0150	0.0095	0.0058
11	0.1194	0.1144	0.1015	0.0844	0.0663	0.0496	0.0355	0.0245	0.0164	0.0106
12	0.1094	0.1144	0.1099	0.0984	0.0829	0.0661	0.0504	0.0368	0.0259	0.0176
13	0.0926	0.1056	0.1099	0.1060	0.0956	0.0814	0.0658	0.0509	0.0378	0.0271
14	0.0728	0.0905	0.1021	0.1060	0.1024	0.0930	0.0800	0.0655	0.0514	0.0387
15	0.0534	0.0724	0.0885	0.0989	0.1024	0.0992	0.0906	0.0786	0.0650	0.0516
16	0.0367	0.0543	0.0719	0.0866	0.0960	0.0992	0.0963	0.0884	0.0772	0.0646
17	0.0237	0.0383	0.0550	0.0713	0.0847	0.0934	0.0963	0.0936	0.0863	0.0760
18	0.0145	0.0256	0.0397	0.0554	0.0706	0.0830	0.0909	0.0936	0.0911	0.0844
19	0.0084	0.0161	0.0272	0.0409	0.0557	0.0699	0.0814	0.0887	0.0911	0.0888
20	0.0046	0.0097	0.0177	0.0286	0.0418	0.0559	0.0692	0.0798	0.0866	0.0888
21	0.0024	0.0055	0.0109	0.0191	0.0299	0.0426	0.0560	0.0684	0.0783	0.0846
22	0.0012	0.0030	0.0065	0.0121	0.0204	0.0310	0.0433	0.0560	0.0676	0.0769
23	0.0006	0.0016	0.0037	0.0074	0.0133	0.0216	0.0320	0.0438	0.0559	0.0669
24	0.0003	0.0008	0.0020	0.0043	0.0083	0.0144	0.0226	0.0328	0.0442	0.0557
25	0.0001	0.0004	0.0010	0.0024	0.0050	0.0092	0.0154	0.0237	0.0336	0.0446
26	0.0000	0.0002	0.0005	0.0013	0.0029	0.0057	0.0101	0.0164	0.0246	0.0343
27	0.0000	0.0001	0.0002	0.0007	0.0016	0.0034	0.0063	0.0109	0.0173	0.0254
28	0.0000	0.0000	0.0001	0.0003	0.0009	0.0019	0.0038	0.0070	0.0117	0.0181
29	0.0000	0.0000	0.0001	0.0002	0.0004	0.0011	0.0023	0.0044	0.0077	0.0125
30	0.0000	0.0000	0.0000	0.0001	0.0002	0.0006	0.0013	0.0026	0.0049	0.0083
31	0.0000	0.0000	0.0000	0.0000	0.0001	0.0003	0.0007	0.0015	0.0030	0.0054
32	0.0000	0.0000	0.0000	0.0000	0.0001	0.0001	0.0004	0.0009	0.0018	0.0034
33	0.0000	0.0000	0.0000	0.0000	0.0000	0.0001	0.0002	0.0005	0.0010	0.0020
34	0.0000	0.0000	0.0000	0.0000	0.0000	0.0000	0.0001	0.0002	0.0006	0.0012
35	0.0000	0.0000	0.0000	0.0000	0.0000	0.0000	0.0000	0.0001	0.0003	0.0007
36	0.0000	0.0000	0.0000	0.0000	0.0000	0.0000	0.0000	0.0001	0.0002	0.0004
37	0.0000	0.0000	0.0000	0.0000	0.0000	0.0000	0.0000	0.0000	0.0001	0.0002
38	0.0000	0.0000	0.0000	0.0000	0.0000	0.0000	0.0000	0.0000	0.0000	0.0001
39	0.0000	0.0000	0.0000	0.0000	0.0000	0.0000	0.0000	0.0000	0.0000	0.0001

Table A.4 The Mean and Variance of Some Common Density Functions

Random Variable	Symbol	Density Function	$\{x \mid f(x) \neq 0\}$	Mean	Variance
Binomial	S_n	$b(x;n,p) = {}_nC_x\, p^x q^{n-x}$	$\{0,1,...,n\}$	np	npq
Geometric	G	$q^{x-1}p$	$\{1,2,3,...\}$	$\dfrac{1}{p}$	$\dfrac{q}{p^2}$
Hypergeometric	H	$h(x;n,b,w) = \dfrac{{}_bC_x \cdot {}_wC_{n-x}}{{}_{b+w}C_n}$ where $n \le b+w$	where $M =$ larger of 0 and $n-w$ $m =$ smaller of n and b $\{M,M+1,...,m\}$	np where $p = \dfrac{b}{b+w}$	$npq\left(\dfrac{b+w-n}{b+w-1}\right)$ where $q = \dfrac{w}{b+w}$
Pascal	N_k	${}_{x-1}C_{k-1}\, p^k q^{x-k}$	$\{k,k+1,k+2,...\}$	$\dfrac{k}{p}$	$\dfrac{kq}{p^2}$
Poisson		$p(x;\mu) = \dfrac{e^{-\mu}\mu^x}{x!}$	$\{0,1,2,...\}$	μ	μ
Uniform	U	$\dfrac{1}{n}$	$\{1,2,3,...,n\}$	$\dfrac{n+1}{2}$	$\dfrac{n^2-1}{12}$

Answers to Odd-numbered Exercises

1.1 EXERCISES

1. Only A_2, A_3, A_5, A_7, and A_8 are sets.

3. **(a)** $(i)\ 0$ $(ii)\ 1$
    (b) No

5. **(a)** $V = W$ and $X = Z$
    (b) Only (i), (iii), and (iv) are true.
    (c) $(i)\ 3$ $(ii)\ 2$

7. There are many answers. Some examples are: $\{5,6,\ldots,11\}$, $\{5,7,9,11,10,8,6\}$, $\{x\,|\,4 < x < 12, x \text{ is an integer}\}$, $\{x\,|\,5 \le x \le 11, x \text{ is an integer}\}$

1.2 EXERCISES

1. **(a)** $\varnothing$, $\{a\}$, $\{b\}$, $\{c\}$, $\{a,b\}$, $\{a,c\}$, $\{b,c\}$, $\{a,b,c\}$
    (b) All except $\{a,b,c\}$

3. Only **(c)**, **(d)**, **(f)**, and **(g)** are true.

5. **(a)** $\{d,e,f,\ldots,z\}$ **(b)** $\{d,e,f,\ldots,z\}$ **(c)** $\varnothing$
    (d) $\{d\}$ **(e)** $\{a,b,e,f,g,\ldots,z\}$ **(f)** $\{a,b\}$
    (g) $\{c\}$ **(h)** A **(i)** $\varnothing$
    (j) $\{d\}$ **(k)** $\varnothing$

7. **(a)** A' **(b)** $\varnothing$ **(c)** A **(d)** A **(e)** $\varnothing$ **(f)** A' **(g)** A

1.3 EXERCISES

1. **(a)** $\{1,2,4,5,6\}$ **(b)** $\{1,2\}$ **(c)** $\{4\}$
    (d) $\{3,7,8,9,10\}$ **(e)** $\{1,2,4\}$ **(f)** $\{1,2,4,5,6,7\}$
    (g) 3 **(h)** 6 **(i)** 4
    (j) $\{1,2,4,5,6,7\}$ **(k)** $\varnothing$

3. **(a)** $\{7,8,9\}$ **(b)** $\{1,2,3,4\}$ **(c)** $\{5\}$
    (d) $\{1,2,3,\ldots,12\}$ **(e)** $\{3,4,\ldots,9\}$ **(f)** $\varnothing$

5. **(a)**   **(b)**   **(c)**

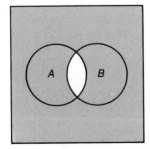
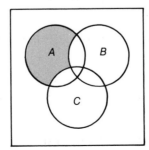
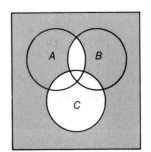

(d)   **(e)** Same as **(d)**   **(f)**

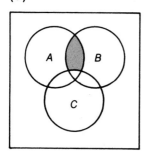   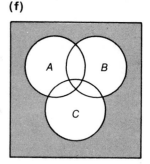

7.   **(a)** The set of all people who are either intelligent or who like math
   (b) The set of all intelligent people who like math
   (c) The set of all people who are not intelligent
   (d) The set of all intelligent people who do not like math
   (e) The set of all intelligent college professors who like math
   (f) The set of all people who are either intelligent or who do not like math or who are not college professors
   (g) The set of all college professors who are not intelligent and who do not like math

9. Ω = The set of all students; A = The set of all students who do no homework; B = The set of all students who do at least 5 hours of homework per week. Two other relevant disjoint subsets of Ω are:
   A = The set of all students who watch at least 25 hours of television per week
   B = The set of all students who watch less than 5 hours of television per week
   Obviously, other relevant disjoint subsets can be defined.

11.   **(a)** Ω **(b)** $\varnothing$ **(c)** A **(d)** A' **(e)** A' **(f)** A **(g)** A **(h)** $A' \cup B'$

13.   **(a)** The set of all people   (There are other correct answers.)

   (b) A = All people who have high blood pressure; B = All people who do not have high blood pressure   (There are other correct answers.)

   (c) C = All people who are obese; D = all people who have had at least one heart attack   (There are other correct answers.)

1.4 EXERCISES

1.   **(a)** $\{a,b,c,e,f\}$ **(b)** $\{a,b,c,e,f\}$ **(c)** $\{b\}$
   (d) $\{b\}$ **(e)** $\{a,b,c,e,f\}$ **(f)** $\{a,b,c,e,f\}$
   (g) $\{a,b,c,e,f\}$

3.   **(a)** Ω **(b)** A' **(c)** A **(d)** Ω **(e)** $A' \cap B'$ **(f)** $A \cap (B \cup C)$

1.5 EXERCISES

1.   **(a)** $\{(1,2,4),(1,3,4),(2,2,4),(2,3,4),(3,2,4),(3,3,4)\}$
   (b) $\{(1,4,2),(1,4,3),(2,4,2),(2,4,3),(3,4,2),(3,4,3)\}$

(c) $|\{(1,4,4),(2,4,4),(3,4,4)\}$

(d) $\{(4,4)\}$

(e) $\varnothing$

(f) $\{(4,4,4,4,4,4,4,4)\}$

3. (a) If $A = B$, $A = \varnothing$, or $B = \varnothing$

 (b) If $A = \varnothing$ or $B = \varnothing$

5. (a) A^3 (b) A^5 (c) A^{100}

CHAPTER 1   REVIEW EXERCISES

3. (a) $\{2,3,4\}$ (b) $\{1,2,3,4,5\}$ (c) $\{1\}$

 (d) $\{5,6,7,\ldots,20\}$ (e) $\{(1,4),(2,4),(3,4),(4,4)\}$ (f) $\{4\}$

 (g) $\{1,2,3,4,5\}$ (h) $\{(4,4)\}$ (i) $\{(4,4,4)\}$

 (j) 17   (k) 5   (l) $\varnothing$

5. (a) $\{7,14,21\}$ (b) $\varnothing$

 (c) $\{1,2,3,5,6,9,10,15\}$ (d) $\{1,2,\ldots,9,10,12,14,15,16,18,21,24,27\}$

 (e) $\varnothing$

7.

 (a)   (b)   (c)   (d)

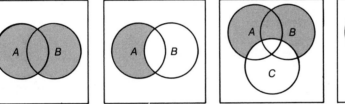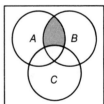

9. (a) $C - (A \cup B) = A' \cap B' \cap C$ (b) A (c) $\varnothing$

11. (a) $\Omega = $ All employees   (There are other correct answers.)

 (b) $A = $ All male employees; $B = $ All female employees

 (c) $C = $ All employees earning over \$10,000 per year; $D = $ All employees who are department heads

 (d) (i) All employees who are department heads or who earn over \$10,000 per year.

 (ii) All female employees

 (iii) All female employees

 Other subsets A,B,C,D of Ω can be used to answer this exercise.

13. (a) The set of all male employees

 (b) The set of all employees who are either female or who were absent at least ten days last year

 (c) The set of all female employees who were absent at least ten days last year

 (d) The set of all female employees who were not absent at least ten days last year

 (e) The set of all male employees who were not absent at least ten days last year

CHAPTER 2

2.1 EXERCISES

1. $\{0,1,2,3,4\}$

3. $\{\infty,2,3,4,5,\ldots\}$, where the integers represent the number of tosses needed to achieve the first repetition, and $\{\infty\}$ is the event "the same result never appears twice in succession."

5. $\{0,1,2,\ldots,500\}$

7. $\{0,1,2\}$

There are other correct answers for Exercises 1–7.

9. $\{5,6,7,\ldots,40\}$, $\{5.0,5.5,6.0,6.5,7.0,\ldots,40.0\}$, $\{5.0,5.25,5.50,5.75,\ldots,40.0\}$

There are other correct answers.

2.2 EXERCISES

1.   **(a)** $\{2,3,4,5\}$ **(b)** $\{1\}$

3.   **(a)** $\{d,l,e\}$
(b) (*i*) A teacher is chosen. (There are other correct answers.)   (*ii*) A doctor, lawyer, or executive is chosen.

5.   **(a)** Exactly one tail occurs; exactly two tails occur. (There are other correct answers.)
(b) Exactly one tail occurs; at least one head occurs. (There are other correct answers.)
(c)   (*i*) $\{(t,t,t),(t,t,h),(t,h,t),(t,h,h),(h,t,t),(h,t,h),(h,h,t)\}$
       (*ii*) $\{(t,t,t),(t,t,h),(h,t,t),(h,t,h)\}$
       (*iii*) $\{(h,t,t),(h,t,h)\}$

7.   **(a)** No
(b) $E_2' \subset E_1'$
(c) E_1' is the event "the circuit is not operative."
     E_2' is the event "neither C_1 nor C_2 is operative."

2.3 EXERCISES

1.   **(a)** 0.05   **(b)** 0.70

3.   **(a)** (*i*) 0.54   (*ii*) 0.3   (*iii*) 0.74   (*iv*) 0   (*v*) 0.7
(b) (*i*) 180   (*ii*) 120   (*iii*) 20

2.4 EXERCISES

3.   No

5.   (a) $\frac{13}{52}$ (b) $\frac{16}{52}$ (c) $\frac{26}{52}$

9.   (a) $\Omega = \{(h,h,h),(h,t,h),(h,h,t),(h,t,t),(t,t,h),(t,t,t),(t,h,t),(t,h,h)\}$
        (b) (i) $\frac{4}{8}$ (ii) $\frac{4}{8}$ (iii) $\frac{7}{8}$ (iv) $\frac{3}{8}$

11. (a) 0.08   (b) 0.60   (c) 0.18   (d) 0.72   (e) 0.62   (f) 0.36

2.5 EXERCISES

1.   (a) $\frac{3}{4}$ (b) $\frac{1}{2}$

3.   (a) $\frac{1}{4}$ (b) $\frac{2}{4}$

5.   (a) $Na \times Na$ (b) $\frac{1}{4}$

7.   (a) $\frac{1}{4}$ (b) $\frac{1}{4}$ (c) $\frac{1}{2}$

9. $\frac{1}{2}$

11. (a) $\frac{1}{2}$ (b) $\frac{1}{2}$

13. (a) $\frac{1}{4}$ (b) $\frac{1}{4}$ (c) $\frac{1}{2}$ (d) $\frac{1}{4}$

2.6 EXERCISES

1.   0.55

3.   (a) 0.50   (b) 0.91   (c) 0.955   (d) 0.455   (e) 0.09   (f) 0

5.   (a) 0.35   (b) 0.75   (c) 0.6   (d) 0

2.7 EXERCISES

1.   (a) $\frac{3}{8}$ (b) $5:3$

3.   (a) $\frac{1}{4}$ (b) $\frac{1}{5}$ (c) $\frac{1}{4}$

5. $1:9$ for A, $9:1$ for A'

CHAPTER 2   **REVIEW EXERCISES**

3.   (a) $\Omega = \{(x,y) \mid x,y = 1,2,\dots,6\}$
        (b) $A = \{(3,1),(3,2),(3,3),(3,4),(3,5),(3,6)\}$
            $B = \{(1,1),(1,3),(1,5),(2,2),(2,4),(2,6),\dots,(6,2),(6,4),(6,6)\}$
        (c) (i) The total is even or the first toss is a 3.
            (ii) The first toss is not a 3.

    (d) The first toss is a 3; the first toss is not a 3. (There are other correct answers.)

    (e) 36

    (f) The first toss is a 2 or 4; the first toss is a 2 or 6. (There are other correct answers.)

5. $11:19$

7. **(a)** $\frac{3}{4}$ **(b)** $\frac{1}{2}$

9. **(a)** 0.3   **(b)** 0.05   **(c)** 0.45

11. **(a)** 0.70   **(b)** 0.25   **(c)** 0.35   **(d)** 0.26   **(e)** 0.95   **(f)** 0.20

CHAPTER 3

3.1 EXERCISES

1. $(26)^2(10)^3 = 676{,}000$

3. $\frac{1}{24}$

5.   320

7. $2^8 - 2 = 254$

9. $\frac{1}{24}$

11.   **(a)** 21,000   **(b)** 5000   **(c)** 45,000

13.   **(a)** 350   **(b)** 100   **(c)** 150   **(d)** Insufficient data

15.   **(a)** 0.1   **(b)** 0.5   **(c)** 0.3   **(d)** 0.8

3.2 EXERCISES

1.   **(a)** 40,320   **(b)** 840   **(c)** 336   **(d)** $7! = 5040$ **(e)** 28   **(f)** 999,000

5.   **(a)** 9!   **(b)** $\frac{1}{9}$

7.   **(a)** 1176   **(b)** 420

9.   **(a)** 5!3!   **(b)** 4!3!

11. $\dfrac{2!\,3!\,4!\,5!}{2}$

3.3 EXERCISES

1.   **(a)** $\dfrac{6!}{6^6}$ **(b)** $\dfrac{28}{6^6}$ **(c)** $\dfrac{30}{6^6}$

3.   **(a)** $\dfrac{6!}{6^5}$ **(b)** $\dfrac{5!}{2!3!6^5}$ **(c)** $\dfrac{5^5}{6^5}$

3.4 EXERCISES

1.   **(a)** Only B and D **(b)** Only C and E

3.   126

5.

n	$P(X=n)$	$P(X\le n)$
45	0.00002	0.00002
46	0.00002	0.00004
47	0.00004	0.00008
48	0.00006	0.00014
49	0.00010	0.00025
50	0.00014	0.00039

7. $\dfrac{82}{48{,}620}$

9.   **(a)** 0,   **(b)** $\dfrac{2}{10{,}002}$

*11.   **(a)** (i) $\frac{1}{70}$ (ii) $\frac{17}{70}$

        (b) (i) $\dfrac{6!6!}{12!}=\dfrac{1}{924}$ (ii) $\frac{37}{924}$

3.5 EXERCISES

1.   **(a)** 35   **(b)** 5   **(c)** 1999   **(d)** 1   **(e)** 1   **(f)** 1

3. $_{40}C_{10}$

5. $_{14}C_5$

7.   **(a)** $_{100}C_{10}$ **(b)** $\dfrac{_{97}C_7+{_{97}C_8}\cdot 3}{_{100}C_{10}}$

9.   **(a)** $\dfrac{4}{_{52}C_{13}}$ **(b)** $\dfrac{_{39}C_{13}}{_{52}C_{13}}$ **(c)** $1-\dfrac{_{32}C_{13}}{_{39}C_{13}}$

3.6 EXERCISES

1.   **(a)** 15   **(b)** 21   **(c)** 35   **(d)** 28   **(e)** 28   **(f)** 56

3.   **(a)** (i) $\frac{1}{512}$ (ii) $\frac{130}{512}$ (iii) $\frac{84}{512}$ (iv) $\frac{9}{512}$

        (b) Using a 5-percent significance level and a 1-percent high significance level, (i) is highly significant, (iv) is significant, and (ii) and (iii) are not significant.

3.7 EXERCISES

1.   **(a)** $_{69}C_{30}$ **(b)** $\dfrac{40}{_{69}C_{30}}$

3. $_{26}C_3 \cdot {_7}C_2 \cdot 4$

3.8 EXERCISES

1.   **(a)** $a^7 + 7a^6b + 21a^5b^2 + 35a^4b^3 + 35a^3b^4 + 21a^2b^5 + 7ab^6 + b^7$
        (b) $x^7 - 7x^6y + 21x^5y^2 - 35x^4y^3 + 35x^3y^4 - 21x^2y^5 + 7xy^6 - y^7$
        (c) $(2^7)x^7 - (21 \cdot 2^6)x^6y + (21 \cdot 2^5 \cdot 9)x^5y^2 - (35 \cdot 2^4 \cdot 3^3)x^4y^3 + (35 \cdot 8 \cdot 81)x^3y^4$
            $- (21 \cdot 4 \cdot 243)x^2y^5 + (7 \cdot 2 \cdot 3^6)xy^6 - (3^7)y^7$
        (d) $x^{14} - 7x^{12}y + 21x^{10}y^2 - 35x^8y^3 + 35x^6y^4 - 21x^4y^5 + 7x^2y^6 - y^7$

3. 1.0510100501

5.   **(a)** $\displaystyle\sum_{k=1}^{50} k$ **(b)** $\displaystyle\sum_{k=1}^{99} (-1)^{k+1}\frac{1}{k}$ **(c)** $\displaystyle\sum_{n=2}^{10} n^2$ **(d)** $\displaystyle\sum_{k=1}^{28} (2k-1)$

7.   **(a)** 41 **(b)** 4 **(c)** $\frac{9}{10}$ **(d)** 62

3.9 EXERCISES

1.   Only **(b)** is arithmetic, and only **(c)** is geometric.

3.   **(a)** $3 + 5 + 7 + 9 + 11 + 13$ **(b)** $3 + 2 + 1 + 0 - 1 - 2$

5.   **(a)** 1860 **(b)** 1760 **(c)** $4[1 - (\frac{1}{2})^{15}]$ **(d)** 1769

7. (a) 4**(a)**; 5**(a)**,**(b)**,**(d)**; 6**(a)**,**(b)**
        (b) 4**(b)**,**(c)**; 5**(c)**

*9.   **(a)** $1 - (\frac{1}{2})^{20}$ **(b)** $\frac{2}{3}[1 - (\frac{1}{4})^{10}]$ **(c)** $\dfrac{1}{2^{12}}[1 - (\frac{1}{2})^8]$

CHAPTER 3   **REVIEW EXERCISES**

3.   **(a)** $20 \cdot 19 \cdot 18 = 6840$ **(b)** 91 **(c)** $5! = 120$
        (d) 1 **(e)** $999{,}000$ **(f)** $499{,}500$
        (g) n

5.   **(a)** $\dfrac{11!}{3!3!2!2!}$ **(b)** $\dfrac{\dfrac{9!}{3!2!2!}}{\dfrac{11!}{3!3!2!2!}} = \dfrac{3}{55}$

7.

Hand	Probability
(1) five of a kind	$\dfrac{6}{6^5} = \dfrac{6}{7776}$
(2) four of a kind	$\dfrac{6 \cdot 5 \cdot 5}{6^5} = \dfrac{150}{7776}$
(3) straight	$\dfrac{2 \cdot 5!}{6^5} = \dfrac{240}{7776}$
(4) full house	$\dfrac{6 \cdot 5\left(\dfrac{5!}{3!2!}\right)}{6^5} = \dfrac{300}{7776}$
(5) nothing	$\dfrac{6! - 240}{6^5} = \dfrac{480}{7776}$
(6) three of a kind	$\dfrac{6 \cdot {}_5C_2\left(\dfrac{5!}{3!}\right)}{6^5} = \dfrac{1200}{7776}$
(7) two pairs	$\dfrac{6 \cdot {}_5C_2\left(\dfrac{5!}{2!2!}\right)}{6^5} = \dfrac{1800}{7776}$
(8) one pair	$\dfrac{6 \cdot {}_5C_3\left(\dfrac{5!}{2!}\right)}{6^5} = \dfrac{3600}{7776}$

9. (a) (i) ${}_5C_2$ (ii) ${}_6C_2$ (b) (i) $5 \cdot 4$ (ii) 5^2

11. (a) $7!$ (b) $7! - 2 \cdot 6!$

13. ${}_{20}C_2 \cdot {}_{10}C_2$

15. (a) $\frac{70}{256}$ (b) $\frac{93}{256}$ (c) $\frac{1}{256}$

17. $\dfrac{19!}{2}$

19. ${}_{20}P_{10}$

21. (a) $b_1 + b_2 + \cdots + b_5$ (b) $a_2 + a_3 + a_4$ (c) $2 + 2^2 + 2^3 + \cdots + 2^{100}$

23. (a) 5050 (b) $2[2^{100} - 1]$ (c) $10,400$

25. 1.040604

CHAPTER 4

4.1 EXERCISES

1. (a) $\frac{3}{10}$ (b) $\frac{1}{10}$ (c) $\frac{20}{47}$ (d) $\frac{2}{3}$ (e) $\frac{26}{100}$ (f) $\frac{2}{10}$ (g) 0

3. (a) $\frac{1}{2}$ (b) 0 (c) $\frac{1}{13}$

5. (a) $\frac{3}{14}$ (b) $\frac{12}{44}$ (c) $\frac{6}{14}$ (d) $\frac{10}{26}$

7. (a) $\dfrac{1}{2^{10}}$ (b) $\dfrac{1}{2^4}$ (c) $\dfrac{1}{_{10}C_5} = \dfrac{1}{252}$

9. (a) $\dfrac{_{39}C_{26}}{_{39}C_{13}}$ (b) $1 - \dfrac{_{39}C_{26}}{_{39}C_{13}}$

11. False

13. (a) Undefined   (b) 0   (c) 0.1   (d) 1   (e) 0

4.2 EXERCISES

1. Only the events of part (c) are independent. The events of part (a) are not independent because one spoiled apple in a barrel will tend to spoil the other apples in the barrel. Similarly, the events of part (d) are not independent because tuberculosis is a contagious disease, so that the second child can catch it from the first. The events of part (b) are also not independent. If the first card is a king, then three of the remaining fifty-one cards are kings; but if the first card is *not* a king, then four of the remaining fifty-one cards are kings.

3. (a) 0.2   (b) 0.02   (c) 0.1   (d) 0.1   (e) 0.72

5. $(\frac{1}{2})^{10}$

7. $(\frac{1}{45})^{27,000}$

9. (a) 0.9996   (b) 0.0392

11. (a) $\frac{3}{8000}$ (b) $\frac{126}{8000}$ (c) $\frac{272}{400}$ (d) $\frac{15}{8000}$

13. (a) $(\frac{1}{4})^2 = \frac{1}{16}$ (b) $1 - (\frac{1}{4})^4 = \frac{255}{256}$

15. (a) $\frac{1}{6}$ (b) $\frac{1}{6}$ (c) At most 2   (d) $\frac{4}{5}$

19. $(0.99647)(0.99854) = 0.9950$

4.3 EXERCISES

1. (*a*) $\frac{241}{546}$ (*b*) $\frac{305}{546}$

3. (*a*) $\frac{2}{3}$ (*b*) $\frac{1}{3}$

5. No; the probability of success is 0.75

7. (a) (*i*) $\frac{81}{105}$ (*ii*) $\frac{17}{105}$ (*iii*) $\frac{1}{15}$ (*iv*) 0
(b) 1

11. 0.0212

13. (a) $\frac{2}{47}$ (b) $\frac{2}{47}$ (c) $\frac{2}{47}$

4.4 EXERCISES

1. (a) $\frac{99}{189}$ (b) $\frac{10}{9811}$

3. $\frac{27}{69}$

5. **(a)** $\frac{1}{2}$ **(b)** $\frac{1}{2}$ **(c)** 0

7. **(a)** $\frac{15}{37}$ **(b)** $\frac{18}{37}$ **(c)** $\frac{4}{37}$

CHAPTER 4   **REVIEW EXERCISES**

3. **(a)** $\frac{1}{2}$ **(b)** $\frac{1}{3}$ **(c)** $\frac{1}{2}$ **(d)** $\frac{1}{4}$ **(e)** 1
      (f) 0 **(g)** Undefined **(h)** 0.3 **(i)** 1

5. **(a)** $\frac{5}{36}$ **(b)** $\frac{1}{6}$ **(c)** $\frac{1}{6}$ **(d)** $\frac{5}{18}$ **(e)** 0

7. **(a)** (i) $\frac{422}{846} = 0.4989$ (ii) $\frac{118}{846} = 0.1394$ (iii) $\frac{70}{846} = 0.0827$
      (iv) $\frac{70}{422} = 0.1659$ (v) $\frac{70}{118} = 0.5932$
      (b) No; $P(CF) \neq P(C)P(F)$

11. **(a)** (i) $\frac{1}{900}$ (ii) $\frac{1}{3600}$
       (b) (i) $\frac{1}{4}$ (ii) $\frac{1}{16}$ (iii) $1 - (\frac{3}{4})^3 = 0.58$

13. **(a)** 0.1025 **(b)** $\frac{175}{1025}$

CHAPTER 5

5.1 EXERCISES

1. **(a)** {1,2,3,4} **(b)** {1,3,5} **(c)** 5

3. **(a)** 1 **(b)** 81 **(c)** 625 **(d)** k^4

5. **(a)** 4 **(b)** 1 **(c)** 0

7. **(a)** 6 **(b)** 720 **(c)** 1

5.2 EXERCISES

1. **(a)** {1,2,3,4} **(b)** (i) 0 (ii) 1 (iii) 2

3. **(a)** {1,2,3,...,100}
      (b) {∞,1,2,3,...} where {∞} denotes the event "a person who enjoys Health-Crisp Cereal is never found."

5.3 EXERCISES

1. (a) 0.1 (b) 0 (c) 0.6 (d) 0.4

3.

x	0	1	2	3	4	5
$F(x)$	0.1	0.3	0.6	0.8	0.9	1.0

5.

x	0	1	2	3	4	5
$f(x)$	0.30	0.20	0.15	0.15	0.10	0.10

5.4 EXERCISES

1. (a) $-\frac{1}{8}$ (b) $\frac{71}{64}$ (c) -1 (d) $\frac{\sqrt{71}}{8}$

3. 0 dollars

5. (a) $P(N = 1) = (0.9)^{10}$, $P(N = 11) = 1 - (0.9)^{10}$ (b) $11 - 10(0.9)^{10}$

7. (a) $\frac{36}{13}$ (b) 2 (c) $\frac{108}{169}$ (d) $\frac{\sqrt{108}}{13}$

9. Most people prefer game 4 under these conditions. After a single toss, one risks losing $20,000. However, since one expects to gain $10,000 each time, the risk of losing any money after 200 tosses is negligible (zero to at least four decimal places).

5.5 EXERCISES

1. 3.5

3. $\frac{23}{30}$

5. (a) $\frac{51}{2}$ (b) $\frac{2499}{12}$ (c) $\frac{1}{50}$ (d) $\frac{15}{50}$

5.6 EXERCISES

1. (a) 0.05120 (b) 0.99328 (c) 0.01311
(d) 1.0000 (e) 0.23347 (f) 0.26272

3. (a) 0.22520 (b) 0.94338 (c) 0.68649 (d) 0.31351 (e) 0.98652

5. (a) $(0.98)^3 = 0.94$ (b) (i) 0.18293 (ii) 0.00003

7. (a) $_{10}C_9(\frac{1}{3})^9(\frac{2}{3}) + (\frac{1}{3})^{10} = 0.0004$ (b) $3\frac{1}{3}$

9. The probability $(0.5)^{202}$ is zero to sixty decimal places. This low value indicates that the dolphin probably used sonar.

11. (a) $b(300;1000,0.3)$ (b) $b(500;1000,0.3)$ (c) $1 - B(299;1000,0.3)$

13. (a) $1 - (\frac{18}{19})^5$ (b) $_5C_3(\frac{1}{19})^3(\frac{18}{19})^2$ (c) $\sum_{k=3}^{5} {}_5C_k(\frac{1}{19})^k(\frac{18}{19})^{5-k}$

5.7 EXERCISES

1. $1 - \dfrac{_{95}C_{10}}{_{100}C_{10}}$

3. $\dfrac{_6C_3 \cdot _{14}C_2 + _6C_4 \cdot _{14}C_1 + _6C_5}{_{20}C_5}$

5. **(a)** Hypergeometric   **(b)** 0,1,2,3,4

    (c) $f(x) = \dfrac{_4C_x \cdot _{20}C_{5-x}}{_{24}C_5}$, $x = 0,1,2,3,4$ **(d)** $\dfrac{_4C_3 \cdot _{20}C_2}{_{24}C_5}$, $\frac{5}{6}$, $\frac{25}{36}$

7. **(a)** (i) 0.07290   (ii) 0.00856   (iii) 0.08145

    (b) Because $\dfrac{n}{b+w} = \dfrac{5}{1000} < 0.1$ **(c)** 0

5.8 EXERCISES

1. **(a)** 0.1008   **(b)** 0.1404   **(c)** 0.0378   **(d)** 0.8571   **(e)** 0.3987

3. **(a)** 0.0948   **(b)** 0.7852

5. **(a)** 0.1839   **(b)** 0.6321   **(c)** 0.3679

7. **(a)** approximation: 0.6321; exact: 0.6340
    (b) approximation: 0.0613; exact: 0.0610

5.9 EXERCISES

1. **(a)** $(0.4)^4(0.6)$ **(b)** $1 - (0.4)^4$ **(c)** $6(0.4)^2(0.6)^3$

3. **(a)** $3(0.9)^2(0.1)^2$ **(b)** 2

5. **(a)** (i) $2(0.2)(0.8)^2$ (ii) $6(0.2)^5(0.8)^2$ (iii) $(x-1)(0.2)^{x-2}(0.8)^2$, $x = 2,3,4,\ldots$
    (b) 2   **(c)** 0.99328

7. **(a)** 2   **(b)** 1

CHAPTER 5   REVIEW EXERCISES

3. **(a)** 9   **(b)** $x^2 + 2x$ **(c)** k^2 **(d)** $\dfrac{x}{2}$

5. **(a)** 0.03427   **(b)** 0.99942   **(c)** 0.1873   **(d)** 0.04007

7. **(a)** 0.2   **(b)** 0   **(c)** 2.25   **(d)** 1.5875   **(e)** $\sqrt{1.5875}$ **(f)** 2 and 3

9. **(a)** 8.3   **(b)** 8   **(c)** 0.0112

11. **(a)** 0.19371   **(b)** 0.99985   **(c)** 1   **(d)** 1

13. **(a)** $(0.4)^3(0.6) = 0.0384$ **(b)** $(0.4)^6 = 0.0041$ **(c)** $1 - (0.4)^5 = 0.98976$

15. $_8C_3(0.6)^4(0.4)^5$

17. **(a)** 0.00243, 0.30870, 0.99757   **(b)** Since $\dfrac{n}{b+w} = \dfrac{5}{100} = 0.05 < 0.1$

CHAPTER 6

6.1 EXERCISES

1. A and E are equal; B and F are equal.

3. (a) 1   (b) -1 (c) 4   (d) 5   (e) Impossible   (f) 0

5. (a) $\begin{pmatrix} 1 & 1 & 1 \\ 1 & 1 & 1 \end{pmatrix}$ (b) $\begin{pmatrix} 2 & 3 & 4 \\ 3 & 4 & 5 \end{pmatrix}$ (c) $\begin{pmatrix} 1 & 1 & 1 \\ 2 & 2 & 2 \end{pmatrix}$

7. (a) $x = 1, y = 6$ (b) $x = 1, y = 4$

9. (a) 5   (b) 25   (c) 35   (d) 15

6.2 EXERCISES

1. (a) $\begin{pmatrix} 5 & 14 & 5 \\ 3 & 5 & 9 \end{pmatrix}$ (b) $\begin{pmatrix} 5 & 14 & 5 \\ 3 & 5 & 9 \end{pmatrix}$

(c) Impossible; A and B do not have the same number of rows.

(d) $\begin{pmatrix} 0 & 5 \\ -2 & -3 \end{pmatrix}$ (e) $\begin{pmatrix} -3 & 0 & 5 \\ 1 & -5 & -3 \end{pmatrix}$

(f) Impossible; C and A do not have the same number of columns.

(g) $\begin{pmatrix} 6 & 13 & 5 \\ 1 & 9 & 11 \end{pmatrix}$ (h) Same as (g)

(i) $\begin{pmatrix} 6 & 21 & 10 \\ 5 & 5 & 12 \end{pmatrix}$ (j) Same as (i)

3. (a) $\begin{pmatrix} 0 & 0 & 0 & 0 & 0 \\ 0 & 0 & 0 & 0 & 0 \end{pmatrix}$

5. (a) $(9,6,23)$ (b) $(2,-10,-10)$ (c) $\begin{pmatrix} 1 & 0 & 1 \\ 2 & 4 & 8 \end{pmatrix}$

(d) Impossible   (e) Φ

7. (a) $(x,y) = (-1,-1)$ (b) $(x,y) = (2,-1)$ (c) $(x,y) = (2,4)$

9. (a) $\begin{pmatrix} 300 & 250 & 130 \\ 230 & 250 & 65 \end{pmatrix}$ (b) $\begin{pmatrix} 1200 & 800 & 410 \\ 910 & 800 & 205 \end{pmatrix}$ (c) \$350

6.3 EXERCISES

1. Only AB, BA, AI_2, and $A\Phi$

3. 11

7. Φ

9. **(a)** Cost

$$\begin{matrix} I \\ C \\ A \end{matrix} \begin{pmatrix} 59.5 \\ 77.4 \\ 130.0 \end{pmatrix}$$

(b) The production cost of 10 shirts

(c) (i) \$59.50   (ii) \$390.00

11. \$2111.50

6.4 EXERCISES

1. Only **(a)** and **(d)**

3. Only **(a)** and **(d)**

5. Only **(b)**

7. **(a)** One solution, $(32.5, 5)$
   (b) An infinite number of solutions, including $(0,5)$, $(1,4)$, and $(3,2)$
   (c) No solution
   (d) One solution, $(4,2)$

6.5 EXERCISES

1. Only **(a)**, **(b)**, and **(d)**

3. 100 gallons of regular and 80 gallons of extra-chocolaty

6.6 EXERCISES

1. **(a)** $\begin{pmatrix} 2 & -0.3 & 8 \\ 4 & -90 & 7 \end{pmatrix}$ **(b)** $\begin{pmatrix} 14 & 3 & 12 \\ 6 & -3 & 15 \end{pmatrix}$

   (c) $\begin{pmatrix} 4 & 3 & 6 \\ 2 & 7 & 4 \end{pmatrix}$

3. **(a)** $\begin{pmatrix} 4 & 12 & 20 \\ 4 & 7 & 2 \end{pmatrix}$ **(b)** $\begin{pmatrix} 4 & 7 & 2 \\ 1 & 3 & 5 \end{pmatrix}$ **(c)** $\begin{pmatrix} 1 & 3 & 5 \\ 0 & -5 & -18 \end{pmatrix}$

5. $F_1 = 6, F_2 = 4$

7. **(a)** $\begin{pmatrix} 3 \\ 1 \end{pmatrix}$ **(b)** $\begin{pmatrix} 2 \\ 4 \end{pmatrix}$

9. Two sandwiches and one piece of fruit

6.7 EXERCISES

1. The system has an infinite number of solutions because **K** has a zero row.

3.   In each case, the first matrix was transformed into the second by the following elementary row operations:

(a) Rows 1 and 2 were interchanged.

(b) Row 2 was replaced by the sum of row 2 and -2 times row 1, and row 3 was replaced by the sum of row 3 and -2 times row 1.

(c) Row 2 was divided by 9.

(d) Row 1 was replaced by the sum of row 1 and 3 times row 2, and row 3 was replaced by the sum of row 3 and -7 times row 2.

(e) All of the above operations were performed in the indicated order.

5.   (a) $\begin{pmatrix} 1 & 1 & 2 & 3 \\ 2 & -1 & 3 & 3 \\ 1 & 3 & -4 & 5 \end{pmatrix}$ (b) $\begin{pmatrix} 1 & 1 & 2 & 3 \\ 1 & -1 & 4 & 4 \\ 3 & -1 & 10 & 11 \end{pmatrix}$

(c) $\begin{pmatrix} 1 & 2 & 3 & 0 \\ 1 & -1 & 1 & 1 \\ 4 & 5 & 10 & 2 \end{pmatrix}$

7. $\begin{pmatrix} 1 \\ 0 \\ 1 \end{pmatrix}$

9. $(1,2,1,0)$

11.   All of them

13.   30 pounds of peanuts, 20 pounds of raisins, 50 pounds of chocolate chips

15.   (a) 60 cc^3 of 10%-solution and 40 cc^3 of 15%-solution   (b) Impossible

CHAPTER 6   REVIEW EXERCISES

3.   (a) $\begin{pmatrix} 2 & 7 \\ 5 & 4 \end{pmatrix}$ (b) $\begin{pmatrix} 2 & 7 \\ 5 & 4 \end{pmatrix}$ (c) $\begin{pmatrix} 5 & 13 \\ 3 & 15 \end{pmatrix}$

(d) $\begin{pmatrix} 16 & 2 \\ 14 & 4 \end{pmatrix}$ (e) Impossible   (f) $(7,2)$

(g) $\begin{pmatrix} 2 & 4 \\ 6 & 0 \end{pmatrix}$ (h) $\begin{pmatrix} 7 & 2 \\ 3 & 6 \end{pmatrix}$ (i) Impossible

5.   (a) $(3,1)$ (b) $(2,3)$ (c) $(3,2)$

7. $\begin{matrix} S \\ W \\ P \\ H \end{matrix} \begin{pmatrix} 4.00 \\ 5.00 \\ 3.00 \\ 4.50 \end{pmatrix}$

9.   (a) $(0.48,0.83,1.15,0.60)$ (b) $6.44

11.   Only (b) and (d)

13.   (a) $\begin{pmatrix} 2 & 5 & 12 \\ 3 & -4 & -5 \end{pmatrix}$ (b) $\begin{pmatrix} 1 & 1 & -1 & 3 \\ 2 & -2 & 1 & -2 \\ 1 & 2 & 3 & 5 \end{pmatrix}$

(c) $\begin{pmatrix} 1 & 2 & 3 \\ 2 & 4 & 5 \end{pmatrix}$ (d) $\begin{pmatrix} 1 & 1 & 2 & 1 \\ 2 & 3 & -4 & 2 \\ 5 & 6 & 2 & 5 \end{pmatrix}$

15.   1 pound of brand A, 2 pounds of brand B, 1 pound of brand C

CHAPTER 7

7.1 EXERCISES

1. Only **(a)** and **(c)** are homogeneous Markov chains. Process **(b)** is a nonhomogeneous Markov chain because the probability that a red ball is picked depends upon the number of the trial. Process **(d)** is not a Markov chain because the probability that the nth ball is red depends upon the outcome of all the previous $n - 1$ trials.

3. **(a)** (i) The customer will be using brand A in 3 weeks.
    (ii) The customer will be using brand A in 2 weeks and brand B in 3 weeks.
    (iii) The customer will be using brand A in 3 weeks if he is using brand A in 2 weeks.

    (b) (i) 1   (ii) 0.3   (iii) 0.3   (iv) 0.3,   (v) 0.2

5. **(a)** (i) 0.2   (ii) 0.3   (iii) 0.5   (iv) 0.3

    (b)

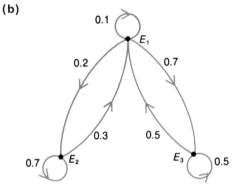

    (c) (i) 0.7   (ii) 0   (iii) 0.3

7. $\frac{1}{6}$

9. **(a)**

$$\begin{array}{c c} & \begin{array}{c c} T & M \end{array} \\ \begin{array}{c} T \\ M \end{array} & \begin{pmatrix} 0.7 & 0.3 \\ 0.2 & 0.8 \end{pmatrix} \end{array}$$

    where T is the state "the mouse took two minutes or less to finish the maze," and M is the state "the mouse took more than two minutes to finish the maze."

    (b) 0.55

7.2 EXERCISES

1. **(a)** $\frac{1}{4}$ **(b)** $\frac{3}{4}$ **(c)** $\frac{11}{16}$ **(d)** $\frac{1}{4}$
    (e) $(\frac{177}{576}, \frac{399}{576})$ **(f)** $\frac{177}{576}$ **(g)** $\frac{11}{16}$ **(h)** $\frac{177}{576}$

    (i) $\frac{11}{16}$ **(j)** $\frac{11}{16}$ **(k)** $\frac{3}{4}$ **(l)** $\begin{pmatrix} \frac{5}{16} & \frac{11}{16} \\ \frac{11}{36} & \frac{25}{36} \end{pmatrix}$

3. **(a)**

$$\begin{array}{c c c c} & A & B & C \\ \begin{array}{c} A \\ B \\ C \end{array} & \begin{pmatrix} 0.5 & 0.3 & 0.2 \\ 0 & 0.6 & 0.4 \\ 0.2 & 0.4 & 0.4 \end{pmatrix} \end{array}$$

(b) The probability p_{12} is the probability that a customer who is now using brand A will be using brand B next week. The probability $p_{12}(2)$ is the probability that a customer who is now using brand A will be using brand B in 2 weeks.

5.   (a)   C   I

$$\begin{array}{c} C \\ I \end{array} \begin{pmatrix} 0.9 & 0.1 \\ 0.3 & 0.7 \end{pmatrix}$$

(b) $\begin{pmatrix} 0.84 & 0.16 \\ 0.48 & 0.52 \end{pmatrix}$

7. $\frac{3}{8}$

*9.   (a) 0.438   (b) 0.4479   (c) 0.428

7.3 EXERCISES

1.   Only (a), (b), (d), and (e)

3.   (a) Only (i), (iii), and (iv)
     (b) Only (ii) and (iv)

5.   (a) 0.49   (b) 0.60

7.   (a) 50.4%, 44.0%, 0%, 5.6%, 0%, 0%
     (b) 0%, 0%, 0%, 6.4%, 93.6%

7.4 EXERCISES

1.   Only (b), (d), and (e)

3.   (a) After 3.5 moves   (b) 14.3%

5.   (a) $\frac{14}{41}$, $\frac{16}{41}$, and $\frac{11}{41}$ (b) $\frac{1}{3}$, $\frac{1}{3}$, and $\frac{1}{3}$ (c) Branch D; $\frac{2}{3}$

7.   (a) (i) 100% (ii) 1 (if the original parents are called generation 0)
     (b) (i) $\frac{1}{2}$, $\frac{1}{4}$, $\frac{1}{8}$, $\frac{1}{2^n}$ (ii) 0%, 100%, 0%
     (c) (i) 1, $\frac{1}{2}$, $\frac{1}{4}$, $\frac{1}{2^{n-1}}$ (ii) 0%, 100%, 0%

CHAPTER 7   REVIEW EXERCISES

3.   (a) Stochastic, probability vector   (b) Stochastic, transition
     (c) Stochastic, transition, regular   (d) None of these
     (e) Stochastic   (f) Stochastic, transition, regular
     (g) None of these

5.

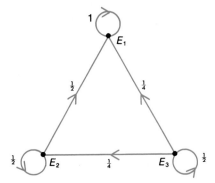

7.   **(a)** 0.1   **(b)** 0.02   **(c)** 0.2   **(d)** 0.1

9.   **(a)** Yes   **(b)** Yes   **(c)**
$$L = \begin{pmatrix} \frac{1}{5} & \frac{2}{5} & \frac{2}{5} \\ \frac{1}{5} & \frac{2}{5} & \frac{2}{5} \\ \frac{1}{5} & \frac{2}{5} & \frac{2}{5} \end{pmatrix}, \quad v = \left(\tfrac{1}{5},\tfrac{2}{5},\tfrac{2}{5}\right)$$

(d) $\frac{3}{8}, \frac{9}{32}$ **(e)** $\frac{1}{5}, \frac{2}{5}$ **(f)** After 5 moves

CHAPTER 8

8.1 EXERCISES

1.

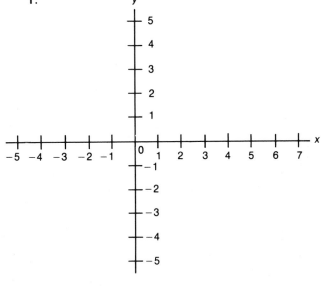

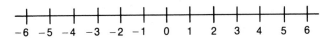

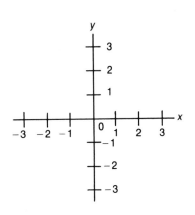

There are other correct answers.

3.

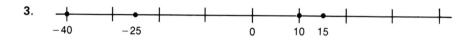

5.

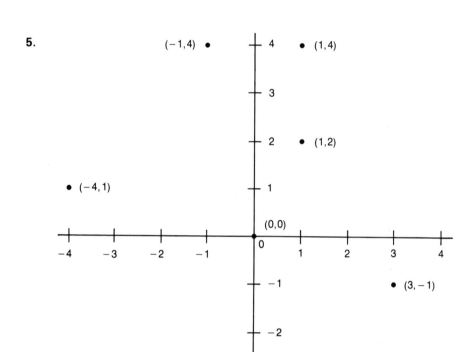

7.   (**a**) 0 cm, 2.54 cm   (**b**) 0.787 in.
(**c**) Note that $c \geq 0$ and $i \geq 0$ for the graph.

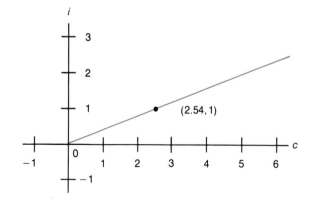

9.   The lines are identical.

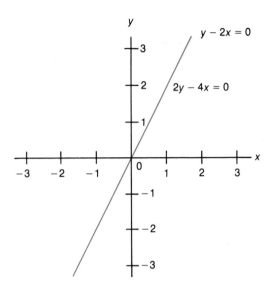

11.

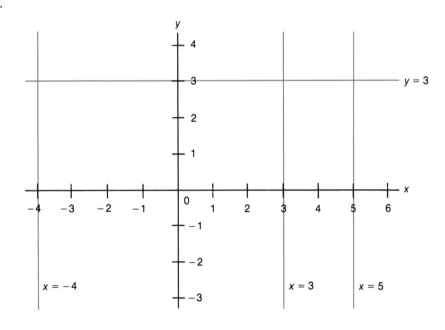

8.2 EXERCISES

1.   Only **(a)**, **(c)**, **(e)**, **(g)**, and **(h)**

3.   **(a)**, **(c)**, and **(d)**.

5. **(a)**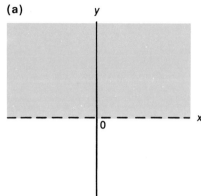

(b)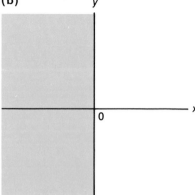

(c)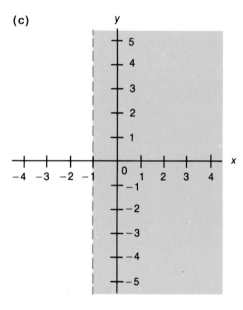

7. **(a)**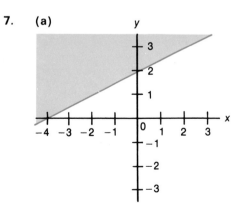

(b)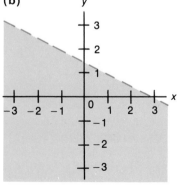

(c)

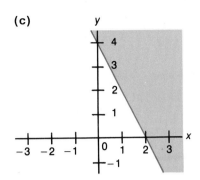

9.   **(a)**

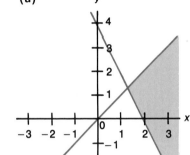

(b)

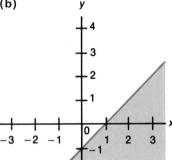

(c)

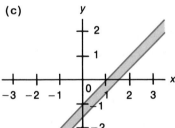

(d)

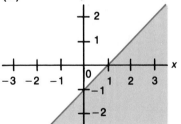

8.3 EXERCISES

1.   **(a)**

	Half	Full	Number Available
Trucks	1	1	20
Employees	2	3	45
Garbage	0.5	1	

(b) *Find the maximum value of*

$$F(x,y) = (0.5)x + y$$

Subject to the constraints

$$x + y \leq 20$$
$$2x + 3y \leq 45$$
$$x \geq 0,\ y \geq 0$$

(c) (*i*) 5.5 tons   (*ii*) 6.5 tons

3.   *Find the maximum value of*

$$F(x,y,z) = 2x + (1.5)y + z$$

Subject to the constraints

$$x \geq 400,\ y \geq 600,\ z \geq 500,$$
$$x + y + z \leq 2000$$
$$z \geq 2x$$

8.4 EXERCISES

1.   **(a)** $F(x,y) = 2x + y$
(b) $x + y \leq 3,\ x - y \leq 2,\ x \geq 0,\ y \geq 0$
(c) Only $(1,2)$ and $(0,3)$
(d) 4, 6, 3, 1
(e) Two feasible solutions are $(1,2)$ and $(0,3)$. There are other correct answers.

3. $0;\ (0,0)$

5.   **(a)** 0, 8   **(b)** 2, 5   **(c)** 5, no maximum   **(d)** 4, no maximum

7.   0 half teams and 15 full teams

8.5 EXERCISES

1.   **(b)** $\left(\frac{13}{4}, -\frac{3}{4}\right)$, $\left(0, \frac{6}{5}\right)$, $(2,0)$, $(0, -4)$, $(4,0)$, $(0,0)$
(c) Only $\left(0, \frac{6}{5}\right)$, $(2,0)$, $(4,0)$, and $(0,0)$

3.   **(a)** $\left(0, \frac{6}{5}, \frac{18}{5}\right)$, $(6,0,0)$, $(0,0,3)$, $(0,0,0)$, $(0,12,0)$, $(0, -6,0)$, $(0,0,4)$
(b) Only $\left(0, \frac{6}{5}, \frac{18}{5}\right)$, $(6,0,0)$, $(0,0,3)$, $(0,0,0)$, and $(0,12,0)$

(c) Maximum $= 30$; minimum $= 0$

(d) Maximum occurs at $(0, \frac{6}{5}, \frac{18}{5})$ and $(6,0,0)$; minimum occurs at $(0,0,0)$

5. **(a)** 12 units of A and 0 units of B; 96 g

(b) 0 units of A and 5 units of B, or $\frac{20}{3}$ units of A and 0 units of B; 20 g

(c) $\frac{12}{7}$ units of A and $\frac{26}{7}$ units of B, or 0 units of A and 6 units of B; $1.80

7. 0 pounds of A, 0 pounds of B, and 16,000 pounds of C; $3200

8.6 EXERCISES

1. **(a)** Because x, y, and x are all nonnegative,

$$|x| = x \leq x + y + 2z \leq 1$$
$$|y| = y \leq x + y + 2z \leq 1$$
$$|z| = z \leq 2z \leq x + y + 2z \leq 1$$

(b) $x + y + 2z + r = 1$

$x - y + 3z + s = 2$

$2x + 3y - z + t = 3$

(c) Corner points (*i*), (*iii*), and (*iv*) are feasible; corner point (*ii*) is not.

3. **(a)** $p = 1$ or 3, in column 2 of **I**

(b) Using $p = 1$ as the pivot yields

$$
\mathbf{J} = \begin{array}{c} \\ y \\ s \\ t \\ F \end{array}
\begin{array}{ccc} x & r & z \\ \end{array}
\left(
\begin{array}{ccc|c}
1 & 1 & 2 & 1 \\
2 & 1 & 5 & 3 \\
-1 & -3 & -7 & 0 \\
1 & 3 & 5 & 3
\end{array}
\right)
$$

(There is another correct answer when 3 is used as the pivot.)

(c) $x = 0$, $r = 0$, $z = 0$, or $(0,1,0)$ **(d)** 3

(e) Yes, because none of the entries in the bottom row of **J** are negative.

5. See the answer to Exercise 5(b), Section 8.5.

*7. **(a)** She should invest $5000 in industrial stocks and $500 in airline stocks.

(b) 6.5%

CHAPTER 8   **REVIEW EXERCISES**

3.   **(a)** 1   **(b)** 7

5.   **(a)**

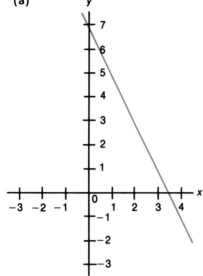

(b)

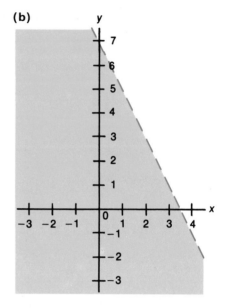

(c)

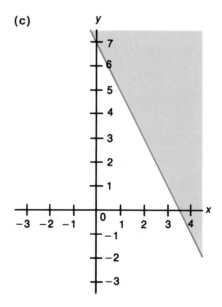

7.   **(a)**   **(b)**

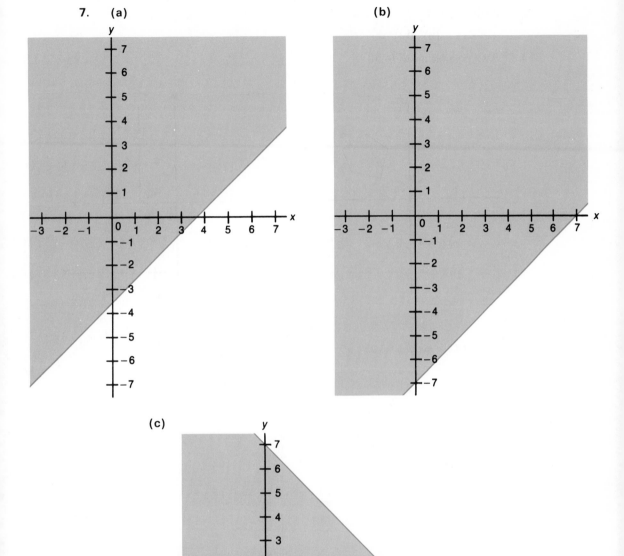

(c)

9. Only **(b)** and **(c)**

11. **(a)** 21   **(b)** (0,4)

13. Two possible answers are—

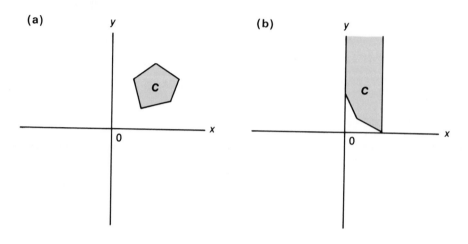

(a)

(b)

15. **(a)** *Find the maximum value of*

$F(x,y) = 3x + 4y$

Subject to the constraints

$2x + 2y \leq 80$

$\frac{5}{4}x + 3y \leq 120$

$x \geq 0, y \geq 0$

(b)

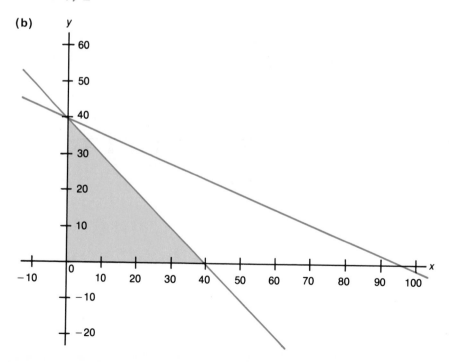

(c) 0 standard and 40 deluxe bowls   **(d)** (i) 80   (ii) 120   **(e)** \$160.00

17.   **(a)** $p = 3$, in row 3, column 1 of **I**

(b)

$$
\mathbf{J} = \begin{array}{c} \\ z \\ s \\ t \\ F \end{array}
\begin{array}{cccccc}
x & y & r & w & \\
\left(\begin{array}{cccc|c}
\frac{1}{3} & \frac{1}{3} & 1 & \frac{1}{3} & \frac{1}{3} \\
0 & 3 & 1 & 0 & 1 \\
0 & -2 & -1 & 1 & 3 \\
-\frac{2}{3} & -\frac{5}{3} & \frac{4}{3} & \frac{1}{3} & \frac{13}{3}
\end{array}\right)
\end{array}
$$

(c) $x = 0$, $y = 0$, $r = 0$, $w = 0$, or $(0,0,\frac{1}{3},0)$ **(d)** $\frac{13}{3}$

(e) The optimal solution is $(\frac{2}{3},\frac{1}{3},0,0)$ and the maximum value of F is $\frac{16}{3}$.

CHAPTER 9

9.1 EXERCISES

1.   **(a)** Move 1; Red wins   **(b)** Move 1 or 3; draw

3.   **(a)** Mini

   (b) At least 70%; if his competitor makes a mistake, the manufacturer may get more than 70%.

5.   No; the players do not have perfect information.

9.2 EXERCISES

1.   **(a)** No; the players do not have perfect information.

   (b) Yes; the four criteria of a matrix game are satisfied.

   (c)

$$
\begin{array}{c} 1 \\ 2 \\ 3 \end{array}
\begin{array}{ccc}
1 & 2 & 3 \\
\left(\begin{array}{ccc}
1 & -1 & 1 \\
-1 & 1 & -1 \\
1 & -1 & 1
\end{array}\right)
\end{array}
$$

3.   **(a)** (i) 1   (ii) 3

   (b) (i) No one pays anything.   (ii) Black pays Red \$2.   (iii) Black pays Red \$1.   (iv) No one pays anything.

5.

Defenders

High Low

$$
\text{Attackers} \begin{array}{c} \text{High} \\ \text{Low} \end{array}
\left(\begin{array}{cc}
1 & 4 \\
7 & 3
\end{array}\right)
$$

9.3 EXERCISES

1.

$$
\mathbf{A} = \left(\begin{array}{ccc}
\boxed{1} & 2 & 3 \\
-4 & 1 & 5 \\
-2 & -1 & -3
\end{array}\right), \quad
\mathbf{B} = \left(\begin{array}{ccc}
2 & 1 & -4 \\
0 & -2 & \boxed{-3} \\
1 & 3 & \boxed{-3}
\end{array}\right)
$$

$$C = \begin{pmatrix} 4 & ② & 3 \\ -3 & -1 & 5 \end{pmatrix}, \quad D = \begin{pmatrix} 2 & 3 & 6 & 6 \\ 4 & 1 & -3 & 4 \\ 2 & 5 & -7 & 1 \end{pmatrix}$$

3 (a) $b_{21} = b_{23} = b_{31} = b_{33} = 0$ (b) 0 (c) Yes; $v = 0$

(d) Red should always play r_2 or r_3; Black should always play b_1 or b_3.

(e) At least \$0 (f) (r_2,b_1), (r_2,b_3), (r_3,b_1), (r_3,b_3)

5. (a)

Nature

Goes bad Does not go bad

Manufacturer Inexpensive component $\begin{pmatrix} -10 & -4 \\ -8 & -6 \end{pmatrix}$
Expensive component

(b) The expensive component

(c) At most \$8

7. (a) Both stations should reduce their prices 5¢ per gallon.

(b) She should buy the certificate.

9.4 EXERCISES

1. (a) $(0.1, 0.9)$ (b) $(1,0)$

(c) $\begin{pmatrix} 0.1 \\ 0.4 \\ 0.5 \end{pmatrix}$

3. Only **A**

5. (a) 40% (b) 30% (c) 0%

7. He should plant tomatoes every year.

9.5 EXERCISES

1. (a) $R^* = (\frac{8}{14}, \frac{6}{14})$, $B^* = \begin{pmatrix} \frac{13}{14} \\ \frac{1}{14} \end{pmatrix}$

(b) $\frac{8}{14}$

(c) Red should play r_1 eight-fourteenths of the time and r_2 six-fourteenths of the time.

(d) Red will win about $\frac{8}{14}$ dollars per game.

3. (a) $R^* = (\frac{3}{4}, \frac{1}{4})$, $B^* = \begin{pmatrix} \frac{1}{12} \\ \frac{11}{12} \end{pmatrix}$

(b) $\frac{33}{12}$

(c) Red should play r_1 75% of the time and r_2 25% of the time.

(d) Red will win about \$2.75 per game.

5. $R^* = (\frac{4}{7}, \frac{3}{7})$, $B^* = \begin{pmatrix} \frac{1}{7} \\ \frac{6}{7} \end{pmatrix}$; $\frac{25}{7}$

9.6 EXERCISES

1. **(a)** $\mathbf{R^*} = (\frac{8}{9}, 0, \frac{1}{9})$, $\mathbf{B^*} = \begin{pmatrix} \frac{1}{9} \\ 0 \\ \frac{8}{9} \end{pmatrix}$

   (b) $\frac{44}{9} = 4.89$

   (c) Red should play r_1 eight-ninths of the time, r_3 one-ninth of the time, and r_2 never.

   (d) Red will win about $4.89 per game.

3. **(a)** $\mathbf{R^*} = (0, \frac{2}{15}, \frac{13}{15})$, $\mathbf{B} = \begin{pmatrix} \frac{7}{15} \\ 0 \\ \frac{8}{15} \\ 0 \end{pmatrix}$

   (b) $\frac{46}{15} = 3.07$

   (c) Red should play r_2 two-fifteenths of the time, r_3 thirteen-fifteenths of the time, and r_1 never.

   (d) Red will win about $3.07 per game.

5. **(a)** Buy an inexpensive gift.   **(b)** Same as **(a)**

9.7 EXERCISES

1. To find an optimal strategy for Red:

   Find the maximum value of w
   Subject to the constraints

   $$w \le x_1 + 2x_2 - x_3$$
   $$w \le 2x_1 - x_2 - 4x_3$$
   $$w \le 4x_1 - 3x_2 + 10x_3$$
   $$x_1 + x_2 + x_3 = 1$$
   $$x_1 \ge 0, \; x_2 \ge 0, \; x_3 \ge 0$$

   To find an optimal strategy for Black:

   Find the minimum value of w
   Subject to the constraints

   $$w \ge x_1 + 2x_2 + 4x_3$$
   $$w \ge 2x_1 - x_2 - 3x_3$$
   $$w \ge -x_1 - 4x_2 + 10x_3$$
   $$x_1 + x_2 + x_3 = 1$$
   $$x_1 \ge 0, \; x_2 \ge 0, \; x_3 \ge 0$$

3. $\mathbf{R^*} = (\frac{2}{7}, \frac{5}{7})$, $\mathbf{B^*} = \begin{pmatrix} 0 \\ \frac{6}{7} \\ \frac{1}{7} \end{pmatrix}$, $v = \frac{23}{7}$

5. To find an optimal strategy for Red:

Find the maximum value of w
Subject to the constraints

$$w \leq 2x_1 - x_2 + 13x_4$$

$$w \leq x_1 + 5x_2 - 4x_4$$

$$x_1 + x_2 + x_4 = 1$$

$$x_1 \geq 0,\ x_2 \geq 0,\ x_4 \geq 0$$

To find an optimal strategy for Black:

Find the minimum value of w
Subject to the constraints

$$w \geq 2x_1 + x_2$$

$$w \geq -x_1 + 5x_2$$

$$w \geq 13x_1 - 4x_2$$

$$x_1 + x_2 = 1$$

$$x_1 \geq 0,\ x_2 \geq 0$$

CHAPTER 9   **REVIEW EXERCISES**

3. **(a)** $-\frac{13}{8}$

(b) $\mathbf{R}^* = (\frac{1}{13}, \frac{12}{13})$, $\mathbf{B}^* = \begin{pmatrix} 0 \\ \frac{8}{13} \\ \frac{5}{13} \end{pmatrix}$, $v = -\frac{34}{13}$

(c) (*i*) Only **B** and **F**   (*ii*) Only **D** and **G**

5. $\mathbf{R}^* = (\frac{1}{6}, \frac{5}{6}, 0)$, $\mathbf{B}^* = \begin{pmatrix} 0 \\ \frac{2}{3} \\ \frac{1}{3} \end{pmatrix}$, $v = \frac{7}{3}$

7. $\mathbf{R}^* = (0, \frac{3}{8}, 0, \frac{5}{8})$, $\mathbf{B}^* = \begin{pmatrix} 0 \\ \frac{1}{8} \\ \frac{7}{8} \\ 0 \end{pmatrix}$ or $\begin{pmatrix} \frac{7}{8} \\ \frac{1}{8} \\ 0 \\ 0 \end{pmatrix}$, $v = \frac{11}{8}$

9. **(a)**

$$\begin{array}{cc} & \begin{array}{cc} A & C \end{array} \\ \begin{array}{c} A \\ C \end{array} & \begin{pmatrix} 50 & 70 \\ 30 & 50 \end{pmatrix} \end{array}$$

(b) Both Red and Black should build in town A.
(c) 50%

11. (a) There is no saddle point, there are no recessive rows or columns, and the game is not 2×2.

(b) To find an optimal strategy for Red:

Find the maximum value of w
Subject to the constraints

$w \le x_1 + 3x_2$

$w \le 2x_1 - 4x_2$

$w \le -x_1 + 5x_2$

$x_1 + x_2 = 1$

$x_1 \ge 0, x_2 \ge 0$

To find an optimal strategy for Black:

Find the minimum value of w
Subject to the constraints

$w \ge x_1 + 2x_2 - x_3$

$w \ge 3x_1 - 4x_2 + 5x_3$

$x_1 + x_2 + x_3 = 1$

$x_1 \ge 0, x_2 \ge 0, x_3 \ge 0$

(c) To find an optimal strategy for Red:

Find the maximum value of w
Subject to the constraints

$w \le 3 - 2x_1$

$w \le -4 + 6x_1$

$w \le 5 - 6x_1$

$x_1 + x_2 = 1$

$x_1 \ge 0, x_2 \ge 0$

To find an optimal strategy for Black:

Find the minimum value of w
Subject to the constraints

$w \ge -1 + 2x_1 + 3x_2$

$w \ge 5 - 2x_1 - 9x_2$

$x_1 + x_2 + x_3 = 1$

$x_1 \ge 0, x_2 \ge 0, x_3 \ge 0$

(d) $\mathbf{R}^* = (\frac{3}{4}, \frac{1}{4})$, $v = \frac{1}{2}$

CHAPTER 10

Answer to the maze on p. 426:

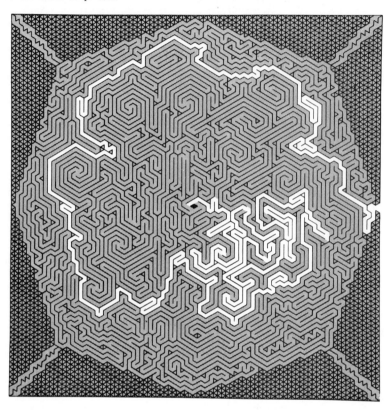

10.1 EXERCISES

1.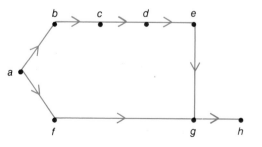

3.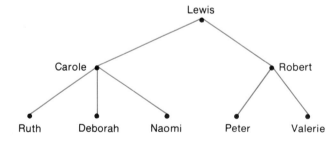

5. **(a)** (*i*) No; *G* has 2 vertices while *H* has 3.

(*ii*) Yes; the required correspondence is $r \leftrightarrow x$, $s \leftrightarrow v$, $t \leftrightarrow w$, $u \leftrightarrow y$. There are other correct correspondences.

(*iii*) No; vertex *s* of graph *G* is the initial vertex of an arc that is not an edge and the terminal vertex of a second arc that is not an edge; graph *H* has no vertex with similar properties.

(b) Only the graphs of part (*iii*) are directed.

10.2 EXERCISES

1. **(a)** *a* and *e*

(b) No; the path passes through *e* twice.

(c) No; the path does not begin and end with the same vertex.

3. The farmer should cross to the other side with the goat. He should then row himself back to the original side. He should then row the dog to the other side. Then he should row the goat to the original side. He should then row the cabbage to the other side, row himself to the first side, and finally row the goat to the other side.

5. 33, 31, 32, 30, 31, 11, 22, 02, 03, 01, 11, 00. There are other correct answers.

10.3 EXERCISES

1. Fill jug B from jug A; fill jug C from jug B. Pour jug C into jug A; pour jug B into jug C. Fill jug B from jug A; fill jug C from jug B. Pour jug C into jug A.

3. (12,0,0), (5,7,0), (5,3,4), (9,3,0), (9,0,3), (2,7,3), (2,6,4), (6,6,0), where (*a,b,c*) denotes the fact that jugs A, B, C have *a*, *b*, *c* quarts of wine.

5. **(a)** *a,e*; *a,g,e*; *a,b,g,e*

(b) *a,b*; *b,f*; *f,c*; *c,d*; *b,d*; *b,g*; *g,e*; *a,e*; *a,g*

(c) *b,f,c,d*; *b,d,c,f*; *b,g,e,a*; *b,g,a,e*; *b,a,g,e*; *b,a,e,g*

10.4 EXERCISES

1. **(a)**

x	*a*	*b*	*c*	*d*	*e*
$p(x)$	4	2	1	3	4

(b) No; the vertex *c* has odd degree.

(c) Yes; one Euler path is *c,a,b,e,e,d,a,d*.

3. **(a)**

x	*a*	*b*	*c*	*d*	*e*
$p_i(x)$	1	5	2	3	2
$p_o(x)$	3	4	1	2	3

   (b) (*i*) 13   (*ii*) 13
   (c) No; $p_i(a) \neq p_o(a)$
   (d) No; $p_i(x) \neq p_o(x)$ for more than two vertices x.

5.   Theorem 10.6 applies because we must traverse each edge in two directions. One solution is $a,b,a,d,a,g,a,e,d,e,c,b,c,f,g,b,d,b,g,f,c,e,a$.

9.

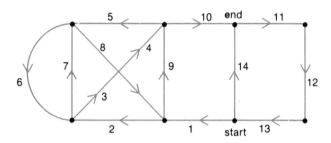

10.5 EXERCISES

1.   **(a)** 12   **(b)** a,c,e,f,i,j **(c)** a, c, e, f, i and j **(d)** 19 hours
   (e)

Activity	Starting Time
a	0
b	0–2
c	0
d	$(B + 2)$–6, where B is the starting time of b
e	4
f	11
g	14–15
h	14–17
i	14
j	19

3.   **(a)** The graph drawn should be isomorphic to this one:

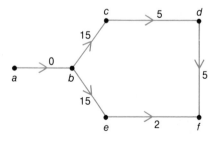

   (b) a,b,c,d,f **(c)** 25 minutes

(d)

Activity	Starting Time
a	0
b	0
c	15
d	20
e	15–23
f	25

5.   **(a)** The graph drawn should be isomorphic to this one:

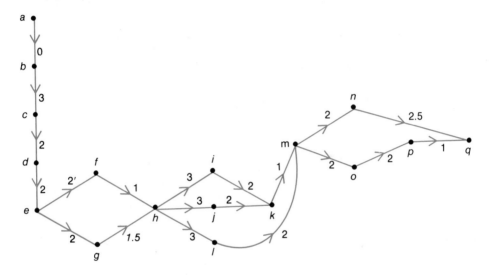

(b) There are two critical paths, *a,b,c,d,e,g,h,j,k,m,o,p,q* and *a,b,c,d,e,g,h,i,k,m,o,p,q*.
(c) 21.5 weeks
(d)

Activity	Starting Time
a	0
b	0
c	3
d	5
e	7
f	9–9.5
g	9
h	10.5
i	13.5
j	13.5
k	15.5
l	13.5–14.5
m	16.5
n	18.5–19
o	18.5
p	20.5
q	21.5

10.6 EXERCISES

1. Only H and J are trees; these graphs satisfy the criteria of Definition 10.11. Graphs G, K, and L are not trees because they contain the circuits b,d,f,b,b,c,f,d,b and a,e,a, respectively. Graph I is not a tree because it is not connected.

3. **(a)**   **(b)**

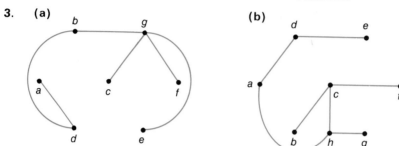

  (c)

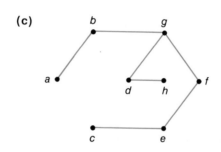

CHAPTER 10   **REVIEW EXERCISES**

3. The graph drawn should be isomorphic to this one:

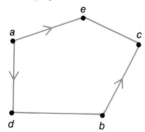

5. (a) Only G, J, K, L, and M (b) Only J, K, L, and M (c) Only M
  (d) Only G (e) Only G, J, L, and M (f) Only G, I, J, L, and M

7. **(a)** f,e,h,e,h **(b)** a,d,b,c,a **(c)** f,a,d,g
  (d) None exists   **(e)** None exists

9. a,b,c,f; a,b,d,e,f; a,b,d,h,f; a,b,d,h,g,e,f; a,g,h,f; a,b,d,e,c,f; a,g,h,d,e,f; a,g,h,d,e,c,f; a,g,h,d,b,c,f; a,g,e,c,f; a,g,e,f; a,g,e,d,h,f; a,g,e,d,b,c,f; a,g,h,d,b,c,e,f; a,b,c,e,f; a,b,c,e,d,h,f; a,b,c,e,g,h,f; a,b,d,h,g,e,c,f; a,b,d,e,g,h,f; a,g,e,c,b,d,h,f

11.   **(a)** Yes   **(b)** At any highway intersection
         (c)

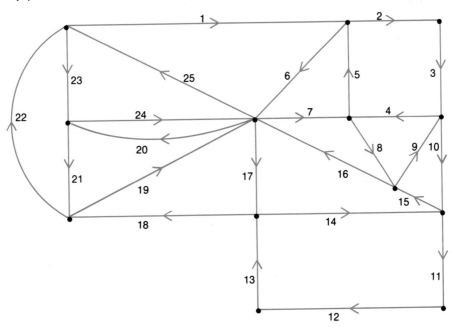

There are other correct answers.

13.   **(a)** The graph drawn should be isomorphic to this one:

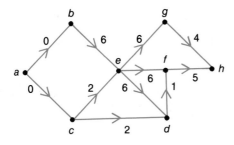

(b) 18 months
(c) a,b,e,d,f,h
(d)

Activity	Starting time
a	0
b	0
c	0–4
d	12
e	6
f	13
g	12–14
h	18

15.   **(a)**

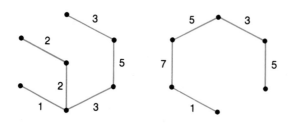

(b) 16

CHAPTER 11

11.1 EXERCISES

1.   Only I is a network; it satisfies the criteria of Definition 11.1. Graph G is not a network because it is undirected; H is not a network because there are two vertices, c and e, that have no outgoing arcs.

3.   **(a)** 2   **(b)** 1   **(c)** 5

5.   **(a)**

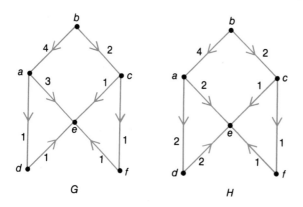

There are other correct answers.
(b) $F_v = 6$ for both G and H.   **(c)** Both flows are 6.

7. **(a)** The network is

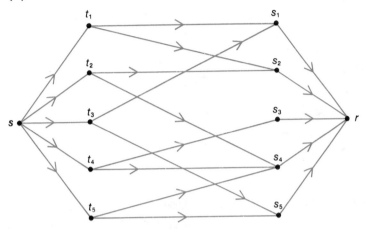

The capacity of each arc is 1.

(b) Tutor 1 wants to tutor 1 student; student 2 wants to be tutored; student 2 and tutor 1 can have 0 or 1 tutoring assignment together.

(c) One maximal flow (there are others) is

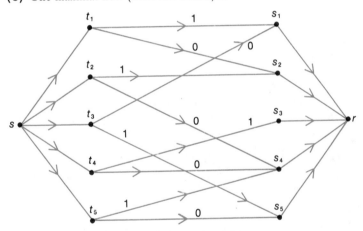

(d) Using the maximal flow from part **(c)**, the assignments are: tutor 1 tutors student 1, tutor 2 tutors student 2, tutor 3 tutors student 5, tutor 4 tutors student 3, and tutor 5 tutors student 4.

11.2 EXERCISES

1. **(a)** 1, 7, 6 **(b)** Only (b,d) and (a,c) **(c)** a,d; a,c,d **(d)** a,b,c,d

(e)

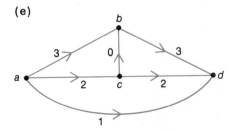

3.   **(a)**

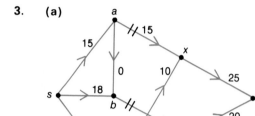

(b) a sends   15,000 x receives   25,000

b sends   18,000 y receives   20,000

c sends   12,000   Total   45,000

Total   45,000

5.   **(a)**

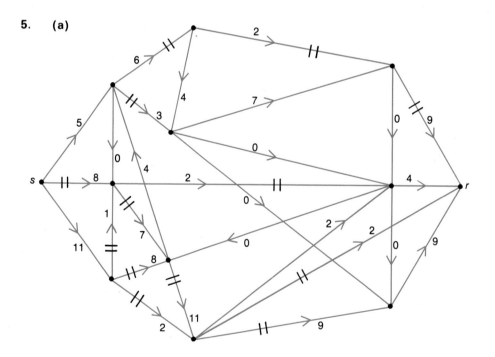

(b) 24

7.

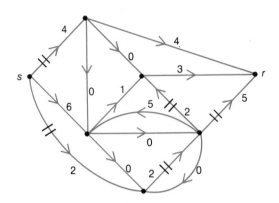

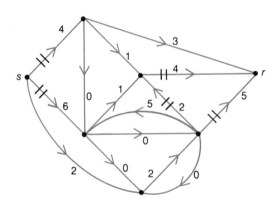

9. (a)

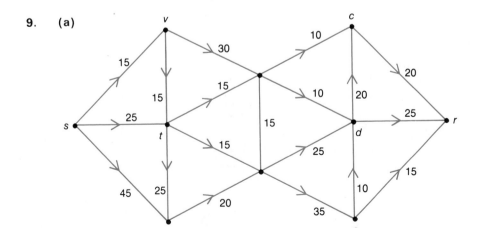

(b)

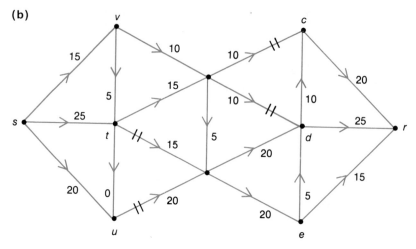

(c)

city s sends	15		city c receives	20
city t sends	25		city d receives	20
city u sends	20		city e receives	20
Total	60		Total	60

11.3 EXERCISES

1. Only G, H, and J are bipartite.

3. **(a)**

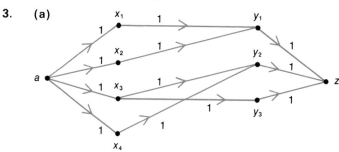

(b)

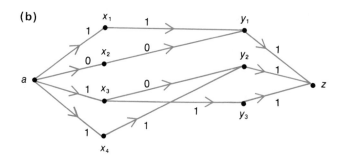

There are other correct answers

(c)

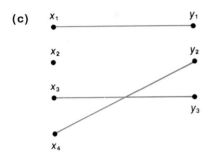

5. Yes. The required matching is:

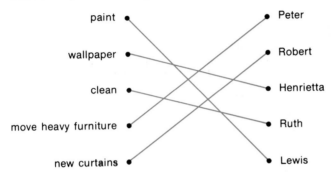

paint • ———— • Peter

wallpaper • ———— • Robert

clean • ———— • Henrietta

move heavy furniture • ———— • Ruth

new curtains • ———— • Lewis

11.4 EXERCISES

1. **(a)**

S	$V(S)$	$|S|$	$|V(S)|$	$\delta(S)$
$\varnothing$	$\varnothing$	0	0	0
$\{x_1\}$	$\{y_1, y_2\}$	1	2	-1
$\{x_2\}$	$\{y_2\}$	1	1	0
$\{x_3\}$	Y	1	4	-3
$\{x_1, x_2\}$	$\{y_1, y_2\}$	2	2	0
$\{x_1, x_3\}$	Y	2	4	-2
$\{x_2, x_3\}$	Y	2	4	-2
X	Y	3	4	-1

(b) 0   **(c)** Yes

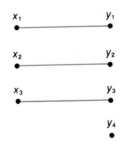

(d) No   **(e)** The complete matching of part **(c)** is maximal.

3. (a) Theorem 11.3 may only be applied to G.

(b) Only G and H have complete matchings.

5. Yes; this result follows from Theorem 11.3 with $n = q$.

CHAPTER 11 REVIEW EXERCISES

3. (a)

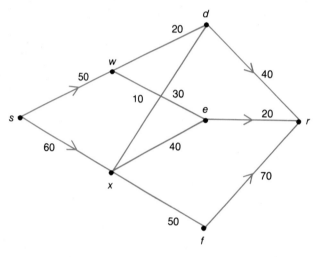

(b)

w sends	50,000		d receives	30,000
x sends	50,000		e receives	20,000
Total	100,000		f receives	50,000
			Total	100,000

There may be other correct answers. However, the total must be 100,000.

(c)

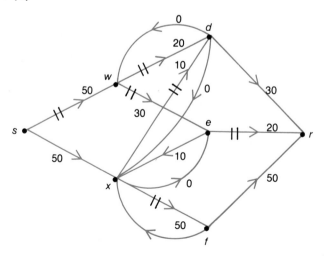

5.   **(a)** The source is vertex a, and the sink is vertex c.

(b) $C(d,e) = 7$; $C(b,f) = 8$

(c)

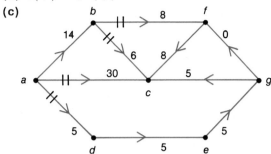

(d) 49

7.   Neither one. In F, the flow into b is not equal to the flow out of b; in G, $C(b,f) < F(b,f)$.

9.   **(a)** Only H and I
        (b) Only H and I
        (c) There are none.
        (d) Only H and I

11.   **(a)** $\delta(G) = 0$

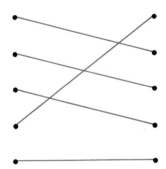

(b) $\delta(H) = 1$

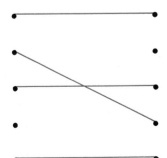

CHAPTER 12

12.1 EXERCISES

1. d_1, d_2, d_3, d_4, d_5; d_1, d_2, d_3, d_4; d_1, d_2, d_3; d_1, d_2; d_1

3. **(a)** -1; $x^* = 0$ **(b)** 1; $x^* = -1$ or 1

5. a, c, e, i, k

12.2 EXERCISES

3. **(a)** $F_3(x) = \max_{x_3 = 0.5, 10, 15} [F_2(x - x_3) + h_3(x_3)]$

$F_2(x) = \max_{x_2 = 0, 3, 6, \ldots, 18} [F_1(x - x_2) + h_2(x_2)]$

(b) $F_2(19)$, $F_2(14)$, $F_2(9)$, and $F_2(4)$

(c) $15.50

(d)

Style	a	b	c
Policy 1	4	15	0
Policy 2	0	9	10

5. **(a)** 5

(b) He should devote 4 hours to math, 0 hours to physics, and 8 hours to chemistry.

12.3 EXERCISES

3. **(a)** $F_1(x)$ is the minimal cost of shipping x units from branch A to C_1 and shipping the remaining units required by customer C_1 from branch B. $F_2(x)$ is the minimal cost of shipping x units from branch A to C_1 *and* C_2 and shipping the remaining units required by C_1 and C_2 from branch B. $F_3(x)$ is the minimal cost of shipping x units from branch A to C_1, C_2, and C_3 and shipping the remaining units required by C_1, C_2, and C_3 from branch B.

(b) $F_2(x) = \min_{x_2 = 0, 1, \ldots, x} [F_1(x - x_2) + h_2(x_2)]$

$F_3(x) = \min_{x_3 = 0, 1, \ldots, x} [F_2(x - x_3) + h_3(x_3)]$

(c) $119

(d) Branch A should ship 5 cases to customer C_3; branch B should ship 2 cases to customer C_1 and 7 cases to customer C_2.

5. **(a)** 4

(b) (*i*) 10 (*ii*) 6 (*iii*) 0 (*iv*) 23

(c) 6 **(d)** 8 **(e)** 15 **(f)** 15

7. **(a)–(b)**

	D_1	D_2	D_3	D_4	Total
Ship from W_1	6	0	0	2	8
Ship from W_2	0	3	4	0	7
Cost	7	1	19	13	40

12.4 EXERCISES

3. **(a)**

Month	January	February	March	April	May	Total
i	5	4	3	2	1	
Amount produced x_i	700	0	400	500	0	1600
Amount stored y_i	0	300	0	0	200	500

(b) Produce 700 units in January, 400 units in March, and 500 units in April. This means that 300 units will be stored in February, and 200 units will be stored in May.

(c) $50, $90, and $4940

CHAPTER 12 REVIEW EXERCISES

3. **(a)** $-1; x^* = -2$ or 0 **(b)** $7; x^* = 2$

5. **(a)**

x	$F_1(x)$	$x_1{}^*$	$F_2(x)$	$x_2{}^*$
0	0	0	0	0
1	5	1	7	1
2	7	2	12	1
3	9	3	15	2
4	10	4	19	3
5	11	5	21	3, 4
6	12	6	23	3, 4, 5
7	13	7	25	4, 5, 6
8	14	8	27	5, 6

(b) $3100

7. **(a)** 30, 34, undefined

(b) $F_5(x) = \min_{x_5 = 0, 1, \ldots, x} [F_4(x - x_5) + h_5(x_5)]$

$F_4(x) = \min_{x_4 = 0, 1, \ldots, x} [F_3(x - x_4) + h_4(x_4)]$

$F_3(x)$ and $F_2(x)$ are similar.

(c)

x	$F_1(x)$	x_1^*	$F_2(x)$	x_2^*	$F_3(x)$	x_3^*	$F_4(x)$	x_4^*
0	30	0	51	0	66	0	76	0
1	40	1	61	0	64	1	74	0
2	34	2	55	0	61	2	71	0
3	—	—	50	3	65	0	75	0
4	—	—	60	3	63	1	73	0
5	—	—	54	3	60	2	70	0
6	—	—	—	—	67	1	70	4

9. **(a)** (i) 10 (ii) 11 (iii) 18

(b)

x	$h_1(x)$	$h_2(x)$	$h_3(x)$	$h_4(x)$	$h_5(x)$
0	14	8	3	17	9
1	13	11	8	20	9
2	11	11	10	19	3
3	9	9	—	18	—
4	5	—	—	16	—

(c) (1) Define $F_1(x), F_2(x), \ldots, F_5(x)$ as in Exercise 7(a).

(2) Use the table given in part **(d)** and the formulas given in Exercise 7(b) to consecutively find $F_1(x), F_2(x), \ldots, F_4(x)$, and $F_5(6)$. The answer is $F_5(6)$.

INDEX

LIST OF SYMBOLS

$\in$	element of
Ω	universal set
$\varnothing$	empty set
$\{x \mid \dots\}$	set of all x such that $\dots$
$\subseteq$	subset
$\subset$	proper subset
A'	complement of set A
$A - B$	relative complement of set B in set A
$A \cup B$	union of sets A and B
$A \cap B$	intersection of sets A and B
$n(A)$	number of elements in set A
(a,b)	ordered pair a and b, or arc from a to b
$A \times B$	Cartesian product of sets A and B
∞	infinity
$f_n(A)$	relative frequency of event A
$\approx$	approximately equal
$P(A)$	probability of event A
$_nP_r$	number of permutations of n things taken r at a time
$n!$	n factorial
$_nC_r$	number of combinations of n things taken r at a time
Σ	summation symbol
$P(B \mid A)$	probability of event B given event A
$f(a)$	value of function f at a
$\mathbf{R}$	set of real numbers
$\mathbf{R}^*$	set of extended real numbers
$(X = x)$	event "X is x occurs"